6
Springer Series on Fluorescence

Methods and Applications

Series Editor: O. S. Wolfbeis

Springer Series on Fluorescence
Series Editor: O. S. Wolfbeis

Recently Published and Forthcoming Volumes

Standardization and Quality Assurance in Fluorescence Measurements II
Bioanalytical and Biomedical Applications
Volume Editor: Resch-Genger, U.
Vol. 6, 2008

Standardization and Quality Assurance in Fluorescence Measurements I
Techniques
Volume Editor: Resch-Genger, U.
Vol. 5, 2008

Fluorescence of Supermolecules, Polymeres, and Nanosystems
Volume Editor: Berberan-Santos, M. N.
Vol. 4, 2007

Fluorescence Spectroscopy in Biology
Volume Editor: Hof, M.
Vol. 3, 2004

Fluorescence Spectroscopy, Imaging and Probes
Volume Editor: Kraayenhof, R.
Vol. 2, 2002

New Trends in Fluorescence Spectroscopy
Volume Editor: Valeur, B.
Vol. 1, 2001

Standardization and Quality Assurance in Fluorescence Measurements II

Bioanalytical and Biomedical Applications

Volume Editor: Ute Resch-Genger

With contributions by

P. E. Barker · F. Brakenhoff · V. Buschmann · A. Dixon · A. Esposito
F. W. Frueh · A. K. Gaigalas · K. Garsha · H. C. Gerritsen
F. M. Goodsaid · H.-J. He · T. Heinlein · I. Hemmilä · F. Hillger
K. Hoffmann · R. A. Hoffman · M. J. Holden · D. Hyde · F. Koberling
C. Maercker · J. N. Miller · D. Nettels · W. Nietfeld · R. Nitschke
V. Ntziachristos · E. P. Petrov · U. Resch-Genger · H. Schneckenburger
B. Schuler · P. Schwille · M. Seydack · L. Shi · W. Tong · L. Wang
R. Wolleschensky · F. S. Wouters · Y. Xiao · Y. Zong · S. Zou
R. M. Zucker · J. Zwier

Springer

Fluorescence spectroscopy, fluorescence imaging and fluorescent probes are indispensible tools in numerous fields of modern medicine and science, including molecular biology, biophysics, biochemistry, clinical diagnosis and analytical and environmental chemistry. Applications stretch from spectroscopy and sensor technology to microscopy and imaging, to single molecule detection, to the development of novel fluorescent probes, and to proteomics and genomics. The Springer Series on Fluorescence aims at publishing state-of-the-art articles that can serve as invaluable tools for both practitioners and researchers being active in this highly interdisciplinary field. The carefully edited collection of papers in each volume will give continuous inspiration for new research and will point to exciting new trends.

ISBN 978-3-540-70570-3 e-ISBN 978-3-540-70571-0
DOI 10.1007/978-3-540-70571-0

Springer Series on Fluorescence ISSN 1617-1306

Library of Congress Control Number: 2008934168

Cover design: WMXDesign GmbH, Heidelberg
Typesetting and Production: le-tex publishing services oHG, Leipzig

Printed on acid-free paper

9 8 7 6 5 4 3 2 1 0

springer.com

Series Editor

Prof. Dr. Otto S. Wolfbeis

Institute of Analytical Chemistry,
Chemo- and Biosensors
University of Regensburg
93040 Regensburg, Germany
otto.wolfbeis@chemie.uni-regensburg.de

Volume Editor

Dr. Ute Resch-Genger

Bundesanstalt für Materialforschung und -prüfung (BAM)
Arbeitskreis "Optical Spectroscopy"
Richard-Willstaetter-Str. 11
12489 Berlin
Germany
ute.resch@bam.de

Preface

In the booming fields of the life and material sciences, advances are taking place on all fronts and often involve the use of luminescence techniques as analytical tools and detection methods due to their high sensitivity, intrinsic selectivity, noninvasive (or at least minimally invasive) character, comparative ease of use, potential for multiplexing applications, and remote accessibility of signals. Despite the fact that the measurement of fluorescence—with its birth marked by the study of Sir Stokes on quinine sulfate in 1852—is not a new technique and many fluorescence techniques have matured to a state where quantification is desired, standardization of the broad variety of fluorescence methods and applications is still in its infancy as compared to other prominent (bio)analytical methods.

It is still often overlooked that all types of fluorescence measurements yield signals containing both analyte-specific and instrument-specific contributions. Furthermore, the absorption and fluorescence of most fluorophores is sensitive to their microenvironment, and this can hamper quantification based on measurements of relative fluorescence intensities as well as accurate measurements of absolute fluorescence intensities. Hence, the realization of a truly quantitative measurement is inherently challenging. This situation renders quality assurance in fluorometry very important, especially with respect to the increasing complexity of instrumentation, and the blackbox-type of present-day instruments and software. This may compromise future applications of fluorescence techniques in strongly regulated areas like medical diagnostics and clinical chemistry that are within reach.

As a result, there is an ever increasing need for (a) recommendations and guidelines for the characterization and performance validation of fluorescence instrumentation and the performance of typical fluorescence measurements, and (b) for an improved understanding of fluorescence-inherent sources of error. This is closely linked to the availability of suitable and easily handled standards that can be operated under routine analytical conditions, are adequately characterized, and meet overall accepted quality criteria.

Within this context, the aim of this book is to provide a unique overview on the current state of instrumentation and application of a very broad variety of fluorescence techniques employed in the material and especially in the life sciences thereby highlighting the present state of quality assurance and the need for future standards. Methods included span microfluorometric techniques used for immunoassays, fluorescence microscopic and imaging techniques including single molecule spectroscopy, flow cytometry and fluorescence in situ hybridization to the microarray technology and technologies used in biomedical diagnostics like in vivo fluorescence imaging. Method-inherent advantages, limitations, and sources of uncertainties are addressed, often within the context of typical and upcoming applications. The ultimate goal is to make users of fluorescence techniques more aware of necessary steps to improve the overall reliability and comparability of fluorescence data to encourage the further broadening of fluorescence applications.

I wish to express my appreciation and special thanks to the individuals who insisted and encouraged me in the preparation of this book. These include Dr. K. Hoffmann, Dr. R. Nitschke, Dr. L. Wang, Dr. R. Zucker, and especially Prof. Dr. O. Wolfbeis for help with the choice of authors and reviewers. And finally, Jürgen and Claudia, for their continuous support and encouragement.

Berlin, July 2008 Dr. Ute Resch-Genger

Contents

Part III
Fluorescence-Based Microarray Technology: Applications, Future Trends, and Need for Standardization

Part IV
Flow Cytometry

Part V
Fluorescence Immunoassays

Contributors

Barker, Peter E.
Biochemical Sciences Division
National Institute of Standards
and Technology
100 Bureau Drive
Gaithersburg, MD 20899-8311, USA
SAIC
4301 N. Fairfax Drive, Suite 200
Arlington, VA 22203, USA

Brakenhoff, Fred
Swammerdam Institute for Life Sciences
University of Amsterdam
Kruislaan 316
1098 SM Amsterdam, The Netherlands

Buschmann, Volker
PicoQuant GmbH
Rudower Chaussee 29
12489 Berlin, Germany

Dixon, Andrew
Carl Zeiss MicroImaging GmbH
Carl Zeiss Promenade 10
07745 Jena, Germany

Esposito, Alessandro
Laser Analytics Group
Department of Chemical Engineering
University of Cambridge
New Museums Site, Pembroke
Cambridge CB2 3RA, UK
Physiological Laboratory
Department of Physiology
Development and Neuroscience
University of Cambridge
Downing Street
Cambridge CB2 3EG, UK

Frueh, Felix W.
Center for Drug Evaluation and Research
U.S. Food and Drug Administration
10903 New Hampshire Avenue
Silver Spring, Maryland 20903, USA

Gaigalas, A. K.
National Institute of Standards
and Technology (NIST)
100 Bureau Drive MS 8312
Gaithersburg, MD 20899-8312, USA

Garsha, Karl
Roper Bioscience
Advanced Microimaging Group
3440 E. Britannia Drive
Tucson, AZ 85706, USA

Gerritsen, Hans C.
Debye Institute
Utrecht University
PO Box 80 000
NL 3508 TA Utrecht, The Netherlands

Goodsaid, Federico M.
Center for Drug Evaluation and Research
U.S. Food and Drug Administration
10903 New Hampshire Avenue
Silver Spring, Maryland 20903, USA

He, Hua-Jun
Biochemical Science Division
National Institute of Standards
and Technology (NIST)
100 Bureau Drive MS 8312
Gaithersburg, Maryland 20899-8312, USA

Heinlein, Thomas
Carl Zeiss MicroImaging GmbH
Carl Zeiss Promenade 10
07745 Jena, Germany

Hemmilä, Ilkka
PerkinElmer Life and Analytical Sciences
Wallac Oy
PO Box 10
FIN 20101 Turku, Finland

Hillger, Frank
Universität Zürich
Biochemisches Institut
Winterthurerstrasse 190
8057 Zürich, Switzerland

Hoffmann, Katrin
Federal Institute for Materials Research
and Testing (BAM)
Richard-Willstaetter-Str. 11
12489 Berlin, Germany

Hoffmann, Robert A.
BD Biosciences
2350 Qume Drive
San Jose, CA 95131, USA

Holden, Marcia J.
National Institute of Standards
and Technology
Biochemical Science Division
100 Bureau Drive, Mail Stop 8311
Gaithersburg, MD 20899, USA

Hyde, Damon
Center for Molecular Imaging Research
Massachusetts General Hospital
and Harvard Medical School
Building 149
13th Street 5406
Charlestown, MA 02129-2060, USA

Koberling, Felix
PicoQuant GmbH
Rudower Chaussee 29
12489 Berlin, Germany

Maercker, Christian
University of Applied Sciences
Department of Biotechnology
Paul Wittsack Strasse 10
68163 Mannheim, Germany
German Cancer Research Center
Genomics and Proteomics Core Facilities
Im Neuenheimer Feld 515
69120 Heidelberg, Germany

Miller, James N.
Department of Chemistry
Loughborough University
Loughborough LE11 3TU, UK

Nettels, Daniel
Universität Zürich
Biochemisches Institut
Winterthurerstrasse 190
8057 Zürich, Switzerland

Nietfeld, Wilfried
Department Vertebrate Genomics
Max-Planck-Institute
for Molecular Genetics
Ihnestr. 63–73
14195 Berlin, Germany

Nitschke, Roland
Life Imaging Center,
Center for Systems Biology
Albert-Ludwigs-University Freiburg
Hauptstr. 1
79104 Freiburg im Breisgau, Germany

NTZIACHRISTOS, VASILIS
Center for Molecular Imaging Research
Massachusetts General Hospital
and Harvard Medical School
Building 149
13th Street 5406
Charlestown, MA 02129-2060, USA

PETROV, E. P.
Biophysics, BIOTEC
Technische Universität Dresden
Tatzberg 47–51
01307 Dresden, Germany

RESCH-GENGER, UTE
Federal Institute for Materials Research
and Testing (BAM)
Richard-Willstaetter-Str. 11
12489 Berlin, Germany

SCHNECKENBURGER, HERBERT
Hochschule Aalen
Institut für Angewandte Forschung
Anton-Huber-Str. 21
73430 Aalen, Germany

SCHULER, BENJAMIN
Universität Zürich
Biochemisches Institut
Winterthurerstrasse 190
8057 Zürich, Switzerland

SCHWILLE, P.
Biophysics, BIOTEC
Technische Universität Dresden
Tatzberg 47–51
01307 Dresden, Germany

SEYDACK, MATTHIAS
8sens.biognostic GmbH
Robert-Roessle-Str. 10
13125 Berlin, Germany

SHI, LEMING
National Center for Toxicological Research
US Food and Drug Administration (FDA)
3900 NCTR Road
Jefferson, AR 72079, USA

TONG, WEIDA
National Center for Toxicological Research
US Food and Drug Administration (FDA)
3900 NCTR Road
Jefferson, AR 72079, USA

WANG, LILI
Laboratory of Experimental Gerontology
National Institute on Aging
National Institutes of Health (NIH)
5600 Nathan Shock Drive
Baltimore, MD 21224, USA
Biochemical Science Division
National Institute of Standards
and Technology (NIST)
100 Bureau Drive MS 8312
Gaithersburg, Maryland 20899-8312, USA

WOLLESCHENSKY, RALF
Carl Zeiss MicroImaging GmbH
Carl Zeiss Promenade 10
07745 Jena, Germany

WOUTERS, FRED S.
European Neuroscience Institute Göttingen
and DFG Center for Molecular Physiology
of the Brain (CMPB)
Waldweg 33
37073 Göttingen, Germany
Laboratory for Molecular
and Cellular Systems
Department of Neuro-
and Sensory Physiology
Institute for Physiology and
Pathophysiology
University Medicine Göttingen
Humboldtallee 23
37073 Göttingen, Germany

XIAO, YAN
SAIC
4301 N. Fairfax Drive, Suite 200
Arlington, VA 22203, USA

ZONG, YAPING
Full Moon Biosystems
754 N. Pastoria Avenue
Sunnyvale, CA 94085, USA

Zou, Sige
Laboratory of Experimental Gerontology
National Institute on Aging
National Institutes of Health (NIH)
5600 Nathan Shock Drive
Baltimore, MD 21224, USA

Zucker, Robert M.
U.S. Environmental Protection Agency
Office of research and Development
National Health and Environmental
Effects Research Laboratory
Reproductive Toxicology Division (MD-67)
Research Triangle Park
NC 27711, USA

Zwier, Jurriaan
Swammerdam Institute for Life Sciences
University of Amsterdam
Kruislaan 316
1098 SM Amsterdam, The Netherlands

Part I
Fluorescence Microscopy

Springer Ser Fluoresc (2008) 6: 3–24
DOI 10.1007/4243_2008_026

Published online: 6 March 2008

Need for Standardization of Fluorescence Measurements from the Instrument Manufacturer's View

Andrew Dixon · Thomas Heinlein · Ralf Wolleschensky (✉)

Carl Zeiss MicroImaging GmbH, Carl Zeiss Promenade 10, 07745 Jena, Germany
Wolleschensky@zeiss.de

Abstract Characterization of fluorescence imaging systems from the manufacturer's view creates several challenges. What are the key parameters for which characterization is appropriate? How can the standardization procedures developed for use during manufacture be applied during installation and application? With so many instrument variables, how can procedures be developed that give precise diagnostic information? These are not simply questions of "standardized tests". There are also issues of finding shared confidence in the tests amongst the different users of the systems. Ideally such tests should also allow objective comparison of the performance of systems of different design or from different manufacturers.

This chapter first discusses the factors that affect performance of fluorescence imaging systems and for which standardization tests are required. In many cases the performance in one respect is inter-dependent on the performance in another. The need to develop tests that uncouple these dependencies is discussed.

The chapter then discusses in more detail the particular issue of signal detection sensitivity and the development of standardized tests that are usable and acceptable both during manufacture and for demonstration of performance during installation and ongoing use of the instrument. It is shown that featureless test samples have significant advantages. They enable a range of performance tests to be made with a single sample in a way that is equally accessible to the manufacturer and end user.

Keywords Confocal · Fluorescence · Instrumentation · Laser Scanning · Microscopy · Multiphoton · Standards

1
Introduction

Having just sold his thousandth microscope in 1866, Carl Zeiss (1816–1888) quoted that even the best technical knowledge is insufficient when trying to reach perfect optical systems by manually trying out and not using calculations ("Pröbeln") [5]. This motivated Zeiss to contact the mathematician and physicist Ernst Abbe (1840–1905), who during the next years laid the theoretical basis for optics design, upon them the wave theory and the Abbe sine condition [6].

From a manufacturer's point of view Abbe's achievements enabled one, for the first time, to measure and control the properties of single optical elements as well as whole instruments. This can be seen as the first standardization tools for optical instrumentation.

During the last decade, fluorescence has become the most rapidly expanding analytical technique available, used both in the medical and biological sciences [1]. A fluorescence microscope image is an enormously rich source of information. It will commonly reveal the spatial organisation of structural elements of a cell or organism, such as the nucleus, the cytoskeleton, or the cell membrane. These structural landmarks may be correlated with other information such as the distribution and co-localisation of particular proteins as visualised, using fluorescent probes with discrete spectral signatures. In fluorescence microscopy, modern techniques, such as colocalization using linear unmixing [10–12], Förster resonance energy transfer (FRET) [13–15], fluorescence correlation spectroscopy (FCS) [16–18], fluorescence recovery after photobleaching (FRAP) [19–22], or fluorescence loss in photobleaching (FLIP) [20, 23] have been established. In case of the study of live material fluorescence probes may be used to provide physiological information regarding ionic concentration, pH, or membrane potential. A sequence of images may provide time-course information both on the occurrence of physiological changes and on changes of cell or organism structures as occurring during various developmental stages. The observer of these images cannot help but be enchanted, even bewitched, by such visual delight. Probably no other technique in biological studies has such a strong aesthetic element.

The richness of fluorescence images as a source of information is both their strength and weakness. It is unavoidable that the interpretation of such images will, at first, be qualitative. In order to obtain quantitative information an intense reduction of the information must occur. One can then ask such questions as: a) What is the size of a particular structure? b) Are two proteins acting independently or is their activity correlated? c) What is the time-course of a particular physiological response?

This complex relationship between qualitative and quantitative content of fluorescence images also expresses itself in the way that commercial instruments are assessed. On the one hand it is unavoidable that users initially are strongly influenced by the visual quality of the images presented by the instrument. On the other hand the longer term scientific value of the instrument depends crucially both on the quality of the visual information and the effectiveness with which it can be reduced for quantitative analysis. This is a very different "dynamic" from how, for example, a flow cytometer system is assessed. In this case the information produced by the instrument is intrinsically quantitative. The very first view of the data is a quantitative presentation of counts per second correlated with intensity in different detection channels. The strength of flow cytometry is the consistency of measurements. Without this the technique could not have gained its very wide acceptance as a tool for medical diagnostics. Indeed the scattergrams so widely used to present flow cytometry data act almost as real time diagnostics of instrument performance. Any change in performance from day to day will almost certainly be immediately apparent and, in any case, a calibrated bead sample can be run at any time to check that the system is performing within specification [26–31]. The reason for this simplicity is that the data in a flow cytometer are already reduced (in the meaning described above); thus they are immediately amenable to the normal protocols for ensuring the system is performing to specification.

If only it were this easy with fluorescence images! The sample preparation required for fluorescence imaging is itself subject to considerable variability between different groups and even from experiment to experiment when a "standard" protocol is used [7–9]. There are many factors that affect the apparent sensitivity of a fluorescence imaging system that the manufacturer must understand and control. These are briefly described in the appendix. The problem, to put it simply, is that it is not readily possible to establish that an image delivers the full available information content from the raw data. This really is a case, at least in comparison with flow cytometry, of more being less. The increased complexity of a fluorescent image (in terms of the number of dimensions that can be measured), and the variability of sample preparation, makes it less easy to assess quantitatively the quality of the image. This may seem a surprising, or at least somewhat bleak, assessment. But consider this: In the 20 years or more since introduction of laser scanning confocal microscopes it has not been feasible, from published data, to assess how these systems compare (in terms of absolute units associated with measurements). No one can assert with confidence that instrument A in use in 1990 has better or worse sensitivity than instrument B operating in 2006. There is, therefore, a paramount need for standardised test samples and procedures for their use.

2 Standardization – but Which Parameters?

At the heart of any discussion about fluorescence images is the question of which variables influence the information content of an image. How can these performance parameters be optimised and how can the performance of a manufactured instrument be characterised to ensure that it meets its required specifications?

In laser scanning microscopy a widely accepted representation of the kinds of information available is the “eternal” triangle as illustrated in Fig. 1 [36, 37]. The diagram shows the three main kinds of information that are available, i.e., information relating to signal intensity (photometric sensitivity), information relating to structural detail (spatial resolution), and information relating to dynamic changes (temporal resolution). More importantly the diagram shows that these sources of information are interrelated. It is the interaction of these three factors that influence the *contrast* of an image. If resolution of structural detail is of most importance then the speed of acquisition may need to be relatively slow thereby sacrificing temporal resolution. If, on the other hand, what is important is to follow dynamic changes such as a physiological response then some loss of resolution may be required in order to achieve sufficient signal. These are essential tradeoffs in fluorescence imaging.

A good illustration of this interrelationship is provided by the example of Stelzer [34] showing how the cut-off frequency, which determines the limit of spatial resolution of the microscope is influenced by the signal-to-noise (S/N) ratio. Figure 2 shows the optical transfer functions for a wide-field and a confocal fluorescence microscope. Since the signal is lower in confocal microscopy the transfer function is also lower. Hence, any noise (indicated by the line in Fig. 2) reduces the contrast of a confocal microscope more dramatically in comparison to a wide-field microscope. This is directly relevant to practical fluorescence microscopy where signal levels are generally very low.

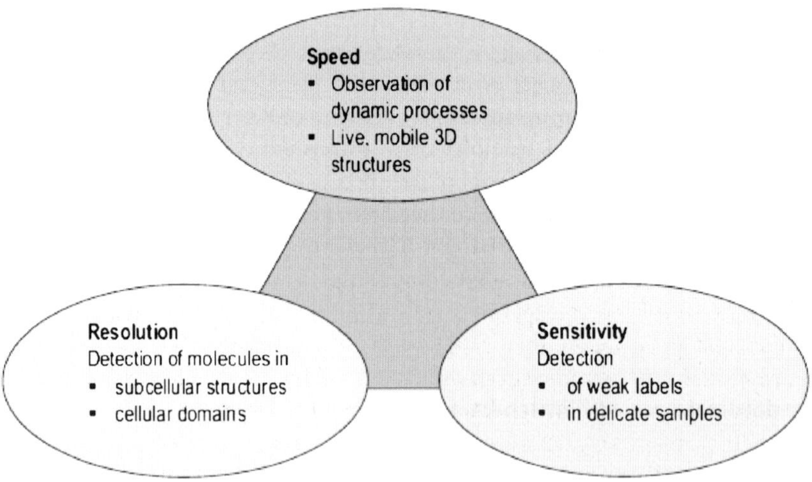

Fig. 1 Eternal triangle showing the interrelationship between important parameters that influence the image quality

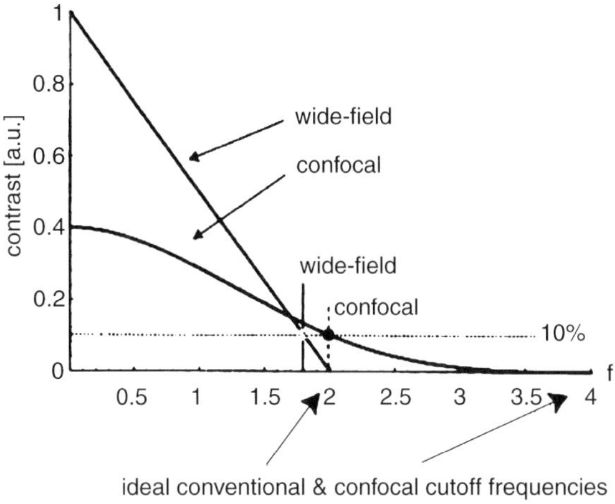

Fig. 2 Optical transfer functions of fluorescent microscopes: The fluorescent signal noise level determines the resolution and contrast of the measured image (adapted from Stelzer et al. [34])

In a point scanning confocal microscope the number of detected photons per pixel of the image commonly ranges from a few 10s to a few 1000s of photons. The higher figure might apply for a brightly labelled (usually fixed) sample. The lower figure would apply for a weakly labelled live sample, as might be prepared using genetic expression of fluorescent proteins. At these levels statistical variations in the detected signal are relatively large, since for a signal of average intensity N photons, the standard deviation around this number is $\sqrt{N}$. The effect of this on the available information content can be severe as illustrated by this example of Stelzer.

Thus even if the system has very high detection efficiency the full resolution may not be achieved in practice due to inability to collect sufficient signal, i.e., to achieve a good enough signal-to-noise ratio. For this reason the sensitivity of the system may be evaluated by determining the signal to noise performance under a given set of imaging conditions.

The interrelationship shown here raises an important question as how best to characterise system performance. How can one measure performance in one aspect without conflicting effects from another aspect? The "eternal" triangle gives some encouragement how this can be approached, which is to develop test procedures that characterize the performance in each respect separately. In this way measurements can be devised which provide more objective assessment of that aspect of instrument performance.

At this point it should be emphasised that there are several levels of increasing sophistication in methods of characterising system performance. At the most straightforward level is day to day characterisation of the performance of an individual system. At the second level is comparison of instruments build to the same or a directly comparable specification. At the third and most difficult level is the comparison of instruments of different manufactures built to different specifications. In some cases the instruments being compared may have very different technological approaches e.g., comparison of a Nipkow spinning disk confocal system with a fast-raster point scanning confocal.

3
Overview on Calibration Methods for Confocal Microscopy

Routine characterisation of an individual system is of most relevance to an individual user or core facility. Once the system is installed and commissioned, the user needs to maintain confidence that the instrument continues to deliver consistent performance. Ensuring conformance to specifications for imaging workstations of the same or very similar design is of most relevance to the manufacturer who must maintain consistent product performance.

The goal of meaningful comparisons of performance between the performance of instruments built to different specifications or from different manufacturers is the most difficult to achieve. However, this would certainly be a requirement if imaging systems were to be used in the clinical arena for quantitative diagnostic applications. It is also probably true that objective comparison of systems from different manufacturers would stimulate innovation, even if the comparisons were not always welcomed. In all that follows these three levels of system characterisation should be kept in mind.

We have already seen that an attempt to characterize resolution when there is insufficient signal is unsatisfactory. What other approach might be adopted where one can be assured of a very bright signal? In confocal microscopy there is an elegant way of addressing this issue which is to use the axial resolution performance as the key characterisation of resolution. Lateral and axial resolution are interrelated; therefore, axial resolution can be used as a metric of lateral resolution. Axial resolution also has the advantage that is more sensitive than lateral resolution to optical (especially spherical) aberration [38]. In this way it is possible to specify a simple test sample such as a mirror or fluorescent "sea".

For characterisation of sensitivity one similarly should develop a sample that gives results that are not influenced by resolution. Thus a block of fluorescent material would not fully meet this requirement since the fluorescence signal is sensitive to the axial resolution, which itself depends on the confocality setting in the system, e.g., the size of the confocal aperture. What is preferable is a sub-resolution sample that can be set up so that the collected signal is insensitive to the confocal aperture setting. This sample could in principle be made of fluorescent beads. However there are difficulties in using such beads for imaging, and in acquiring sufficient signal, or more precisely signal from sufficient samples, for accurate analysis. It is also difficult with such samples to easily monitor the photo-bleaching effect, which can lead to understatement of the sensitivity. With sub-resolution beads there is the additional risk that they drift out of focus during measurement. Larger beads overcome these problems to a degree but may replace them with uncertainties introduced by spherical aberration and lensing artifacts unless great care is taken to match the refractive index of the bead material to the lens design.

Probably the popularity of beads as a test sample follows their widespread use as calibration standards in flow cytometry. There is no doubt that they are useful for characterising instrument performance as exemplified by the work of Zucker [3, 4, 25, 35]. However, a flow cytometer has very different optical and signal collection properties in comparison to a microscope. In reality it analyses cells essentially as structureless particles not dissimilar from beads and collects signal from the entire volume of the cell (or test) bead, and thus is not troubled by resolution issues. It is our experience that for characterisation of sensitivity of a fluorescence imaging system there are other samples that might have advantages over the use of beads.

A different sample which shows promise for characterisation of sensitivity is an ultra-thin film of fluorescent material which has been used by Wolf [32] and Brakenhoff [2, 33]. The sample comprises a thin fluorescent film spin-coated onto a glass coverslip. The film thickness is smaller than the optical section resolution of the system. Thus in making measurements the confocal sectioning can be enlarged so that the collected signal is insensitive to the precise focus position of the sample. The requirement of developing a sample for characterising sensitivity free from influence of resolution effects is, therefore, met.

4 Use of a Thin Fluorescent Film Sample to Determine the Signal-to-Noise Ratio in a Confocal Microscope

In what follows we present characterisation of sensitivity in a confocal LSM using a thin fluorescent film sample as described above. We do not claim that it meets all the requirements of a calibration sample for sensitivity as itemised in Table 1 but believe it is a step in the right direction. In particular, we believe it can meet the need for comparisons between systems of the same design and, with further development, open the way to comparison of systems from different manufacturers of different design.

Consider a simple question. How can one check that the fluorescence signal in a CLSM increases linearly with the intensity setting, and determine over what range of intensity this applies? In the factory this will be done with the help of a calibrated power meter and access to a software mapping function. Upon installation the service engineer may want to confirm this. And

Table 1 Key requirements of a test sample for characterisation of sensitivity in a LSM

- Simple to use
- Well defined and agreed protocols for appropriate tests
- Free of interfering effects due to uncertainty of confocal setting (axial resolution)
- Used at instrument settings (laser intensity, spectral filter, photomultiplier gain etc.) similar to those used normally for biological samples
- Long term stability
- Suitable both for instrument manufacturer and users of the instrument
- Available from independent supplier who can guarantee standardisation
- Bleach resistant or well characterised bleach properties
- Fluorescence saturation only at illumination intensities higher than for normal imaging applications

from time to time, the user may wish to check the linearity. The thin film sample enables this test to be made very simply, with operating conditions similar to those used in practice, and additionally reveals other valuable information about the system.

The basic measurement is to mount the sample in the focal plane and set the imaging to achieve similar conditions of laser intensity, filter selection, photomultiplier gain etc. to those used in normal imaging. These conditions can be predetermined by the manufacturer so that they can be quickly recalled by the software. For a range of illumination intensities a set of images is then obtained. Note that the sample has been designed so that the image should be of uniform brightness, apart from statistical fluctuations due to the low photon signal. In practice there will be some variation of intensity over the image due to field uniformity effects. However, with a region of interest of a few thousand pixels in the central field the statistical variation in intensity will by far exceed any effect due to field non-uniformity.

In the following we discuss measurements that have been done with a LSM 510 META on an AxioImager using a Plan-Neofluar 40x/1.3 oil objective lens (Carl Zeiss MicroImaging GmbH, Germany). The fluorescent thin layer sample has been provided by Prof. Brakenhoff. The LSM produced an image from which one can determine both the average analogue signal S_{A} and the standard deviation of the signal SD_{A}. Figure 3a shows a typical image and Fig. 3b the statistics from the selected region of interest (RoI). For most precise analysis it is best to collect a second image immediately following the first and to determine the standard deviation from the difference signal at each pixel (offset to avoid negative value). This substantially reduces errors in determination of SD due to inhomogeneities in the sample.

What has been discussed so far might seem almost trivial. A featureless sample has been imaged and the average signal and standard deviation from a region of interest determined. It is the very simplicity of the measurement and the featurelessness of the image that make it so valuable for use in performance characterisation, as we now show.

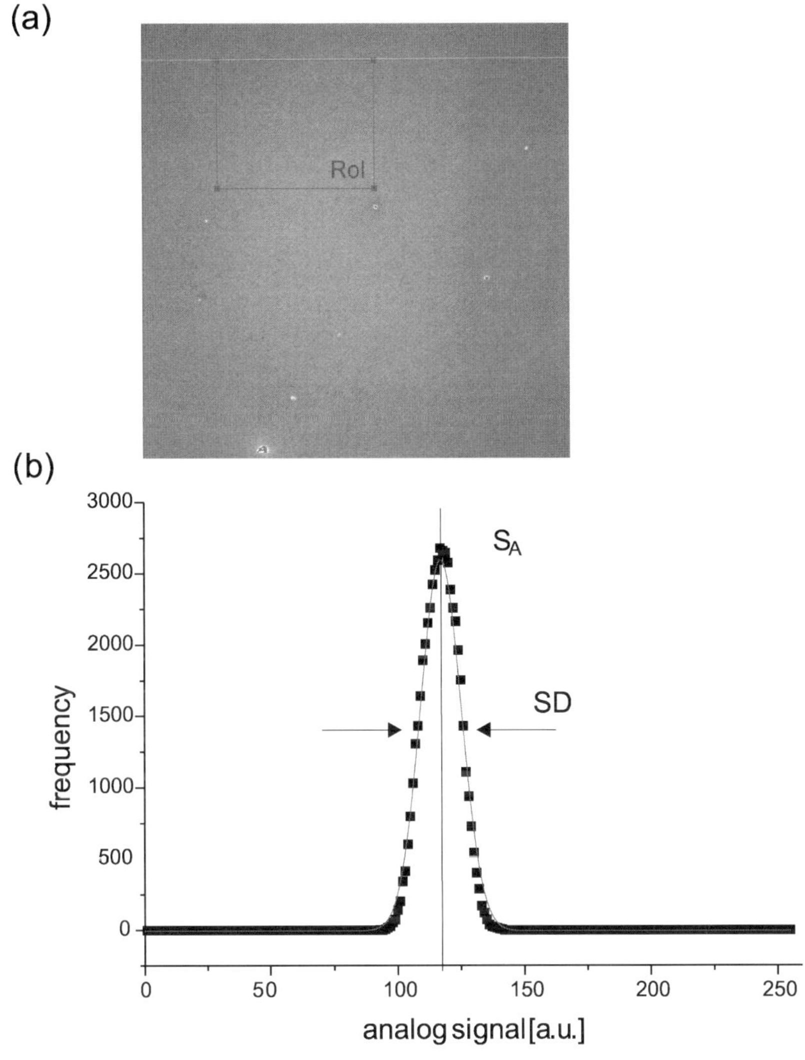

Fig. 3 **a** Fluorescence image of thin film sample with Region of Interest (RoI), **b** statistics showing the Gaussian distribution in the histogram plot of the fluorescent signal. Sample: homogeneous subresolution layer; Objective lens: Plan-Neofluar 40x/1.3 oil; Frame size: 512×512; Integration time: 1.6 μs; PMT voltage 565 V

Figures 4 and 5 show respectively the analogue signal S_A and the standard deviation squared SD_A^2 vs. laser power. Both plots are linear. For the S_A vs. power plot there is a slight residual signal at zero illumination intensity which is the analogue offset in the system. If the plot showed a negative offset then this would indicate incorrect set up since it would cause low signal regions of the image to appear black – apparently improving contrast but in

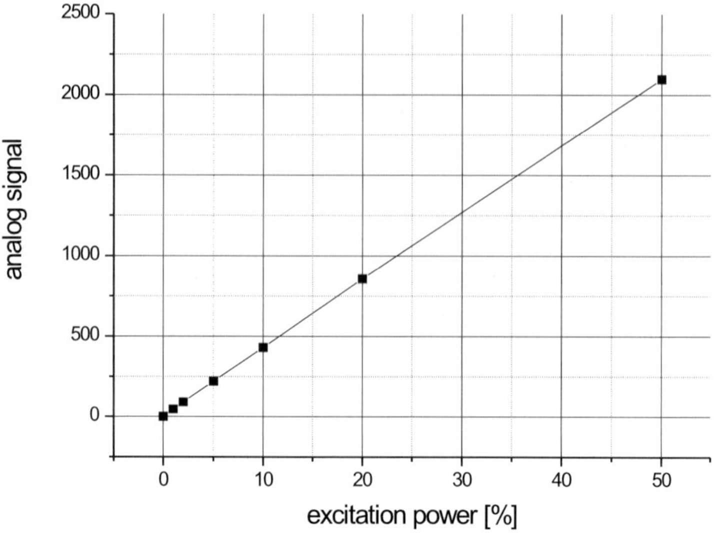

Fig. 4 Analog signal S_A vs. illumination intensity. The illumination intensity has been calibrated using a Plan-Neofluar 10x/0.3 lens and a powermeter Coherent Fieldmaster. The analog signal was measured by averaging over 512×512 pixels. Sample: homogeneous subresolution layer, Objective lens: Plan-Neofluar 40x/1.3 oil; Integration time: 1.6 μs; PMT voltage 565 V

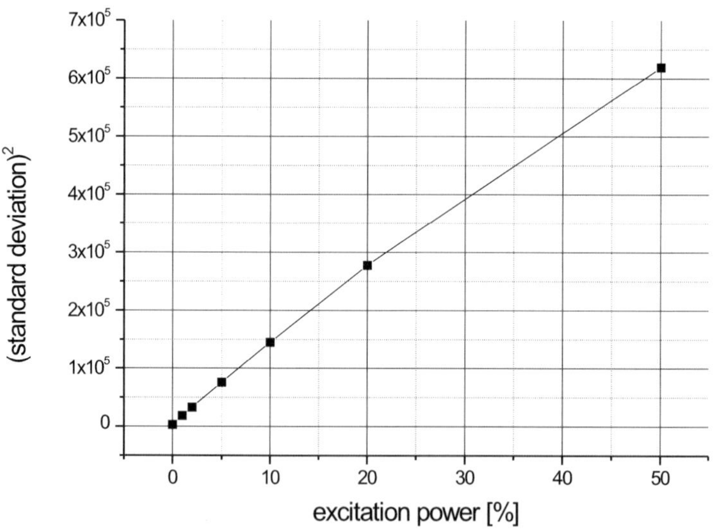

Fig. 5 Standard deviation squared SD_A^2 vs. illumination intensity. The illumination intensity has been calibrated using a Plan-Neofluar 10x/0.3 lens and a powermeter Coherent Fieldmaster. The standard deviation was calculated from a 512×512 image by fitting a Gaussian distribution in the histogram. Sample: homogeneous subresolution layer, Objective lens: Plan-Neofluar 40x/1.3 oil; Integration time: 1.6 μs; PMT voltage 565 V

reality misrepresenting the image. For the plot of SD_A^2 vs. power there is in this example, almost negligible residual SD_A^2 at zero laser intensity. This is a good diagnostic of residual noise in the system due principally to dark current from the detector or noise in the amplifier system. Already it can be seen that these simple measurements provide direct, accessible information about instrument performance.

The linear relationships of Figs. 4 and 5 are as expected. If they were not linear this would indicate either a non-linear response of the fluorescence probe (e.g., saturation) or a fault with the system. In the case of the relationship between S_A and illumination intensity this is routinely done by altering illumination intensity via the software interface, which assumes that the software control is linearised for the actual response of the AOTF or other intensity control device. If the response is not linearised this will show up as non-linearity in the S_A vs. intensity plot and would also appear as non-linearity in the S_A^2 versus intensity plot. A quick check for faults in this *set-up* linearisation is to use a reflective sample for the measurement.

Figure 6 shows the plot of S_A versus SD^2 for measurements made under the same conditions as for Figs. 4, 5. What further information does this plot provide about the system? The measurements presented here are for analogue detection of the signal as is common in LSM imaging systems. However, as discussed earlier, the fluorescence signal that the photomultiplier detects is

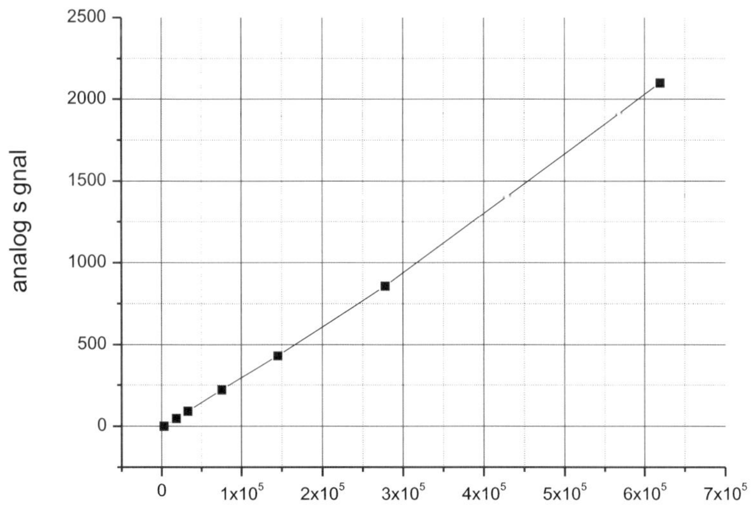

Fig. 6 Analog signal S_A vs. standard deviation squared SD_A^2. The standard deviation and the analog signal were calculated from a 512×512 image by fitting a Gaussian distribution in the histogram. Sample: homogeneous subresolution layer, Objective lens: Plan-Neofluar 40x/1.3 oil; Integration time: 1.6 μs; PMT voltage 565 V

a stream of photons. One can show that there is a direct relationship between the analogue signal and standard deviation, as plotted in Fig. 6, and the photon count signal that is initially detected.

In photon counting the standard deviation (SD_P) due to counting statistics is, to a good approximation, given by the following:

$$SD_\mathrm{P}^2 = S_\mathrm{P} \,. \tag{1}$$

Where S_P is the detected photon counts.

The photomultiplier and detection electronics amplifies this photon signal to produce a proportional analogue signal S_A as follows

$$S_\mathrm{A} = k{\cdot}S_\mathrm{P} \,. \tag{2}$$

Similarly the analogue standard deviation is given by

$$SD_\mathrm{A} = k^*{\cdot}SD_\mathrm{P} \,. \tag{3}$$

Where k^* differs from k only if there are additional sources of "statistical" noise introduced by the detection system such as multiplicative noise in the Photomultiplier.

From these simple equations one can deduce that

$$\frac{S_\mathrm{A}^2}{SD_\mathrm{A}^2} = \frac{k^2{\cdot}S_\mathrm{P}^2}{k^{*2}{\cdot}SD_\mathrm{P}^2} = \frac{k^2{\cdot}S_\mathrm{P}^2}{k^{*2}{\cdot}S_\mathrm{P}} = \left(\frac{k}{k^*}\right)^2 {\cdot}S_\mathrm{P}$$

or

$$\frac{S_\mathrm{A}^2}{SD_\mathrm{A}^2} = S_\mathrm{P}^* \,,$$

an equivalent photon signal.

What is valuable about this representation of the analogue data is that it shows a close equivalence between the analogue signal and the underlying photon count signal from which it derives. Indeed, if an instrument is able to operate both in photon count and analogue detection mode then the precise value of the factor $\frac{k^*}{k}$ can be determined. If there is no multiplicative noise $k^* = k$ and the factor is 1. If multiplicative noise is present then $k^* > k$ and the factor is < 1. This means that the photon count signal deduced from the analogue signal is less than the true photon count signal. If there are other sources of intensity dependent noise that are non-statistical – such as increased noise in the illumination at high intensity – then this would present itself as a non-linear relationship between S and SD^2. Thus with this plot, as in the previous figures a *linear* relationship is a key diagnostic of proper instrument performance.

There is one further point to note from Fig. 6. The slope of the straight line of S_A vs. SD_A^2 is the parameter required to directly convert the analogue signal to its equivalent photon signal as follows

$$S_p^* = \frac{S_A}{SD_A^2} \cdot S_A = c \cdot S_A ,$$

where c is the slope of the straight line plot S_A vs. SD_A^2.

The conversion factor c ideally should not depend on the gain of the detector system. Changing the gain does not alter the number of photons reaching the photomultiplier. In practice there will be some dependence of the parameter c on gain, particularly on the photomultiplier gain. At too low gain the detection efficiency of the photomultiplier may decrease and at high gain there may be additional noise from the PMT. Both these effects reduce the signal to noise ratio. At some intermediate PMT gain there should be a maximum signal to noise which, if possible, is the operating condition that should be chosen. Figure 8 shows S_p^* vs. PMT gain for a PMT. In this case S_p^* is nearly constant with PMT gain. Any departure from this on subsequent measurement would indicate a fault in the PMT.

A simple sample of thin film fluorescence is thus able to provide detailed information about the signal to noise (i.e., standard deviation) of the imaging system. Many of the key performance aspects of the instrument can be analyzed quickly, easily and both by the manufacturer, installation and service engineers and the user. If the sample can be produced reproducibly it offers the promise of being a simple tool for monitoring system performance over time.

Another example of the use of the tool is to measure signal and noise for images acquired with different integration time, either by collecting single scans with different pixel dwell times or by averaging a number of frames all taken with the same pixel dwell time. From counting statistics one would expect that four times the extended collection would give a twofold improvement in signal-to-noise, 16 times collection a fourfold improvement and so on. Figures 7a and 7b show that this is true over the range studied.

However, beyond a certain integration time, there will be no further improvement in signal to noise due to presence of other irreducible sources of noise such as fluctuations in the laser illumination intensity. The test sample can be a useful diagnostic tool in identifying changes in the residual noise of the system.

All of the characterization measurements so far discussed are "relative" measurements in which no attempt is made to deduce the absolute sensitivity – photons per mW of illumination intensity – of the instrument. Suppose that for the test set up the measured signal to noise with the test sample is found to have deteriorated since the previous measurement. The cause of this could equally likely be a reduction in illumination power for the given setting as a loss of collection efficiency. What both the manufacturer and the user require to know is in which part of the optical path the problem is. A loss of detection sensitivity is far more serious than a loss of illumination power.

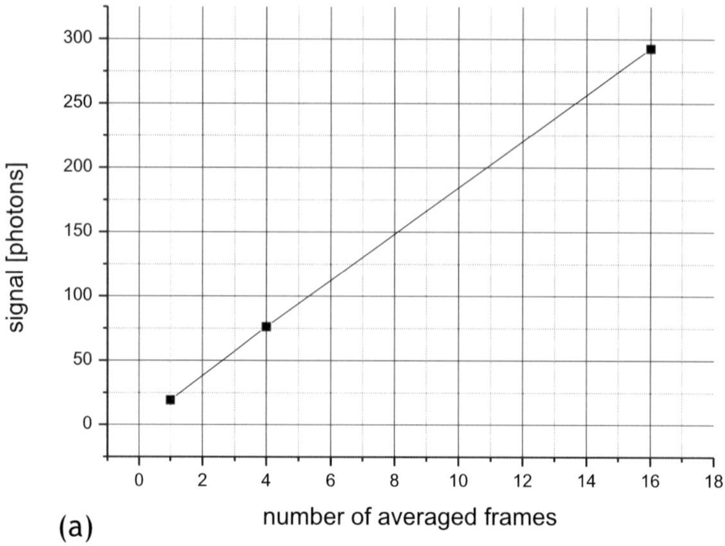

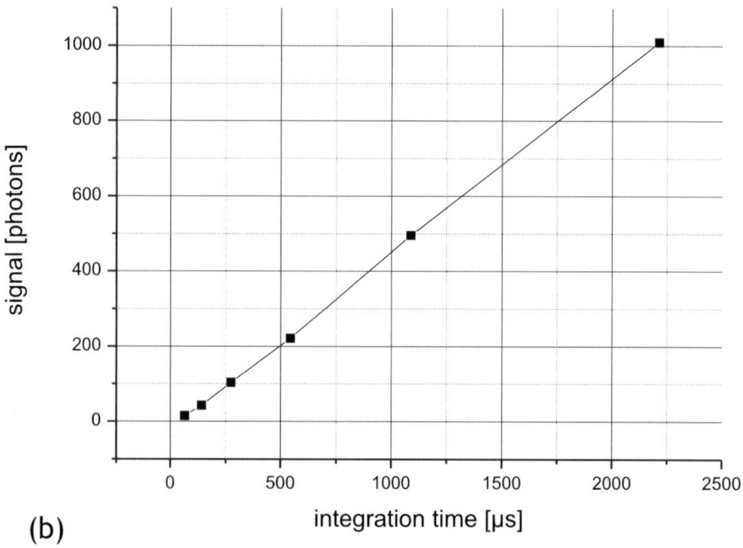

Fig. 7 **a** Signal S_p^* vs. number of averaged frames N; **b** Signal S_p^* vs. pixel dwell time. Sample: homogeneous subresolution layer, Objective lens: Plan-Neofluar 40x/1.3 oil; PMT voltage 565 V

This issue can be resolved if there is some independent measure of the illumination intensity at the sample. The obvious approach would be to use a power meter or a monitor diode that is integrated into the scanning module. However, in practice, it is surprisingly difficult to accurately measure illumi-

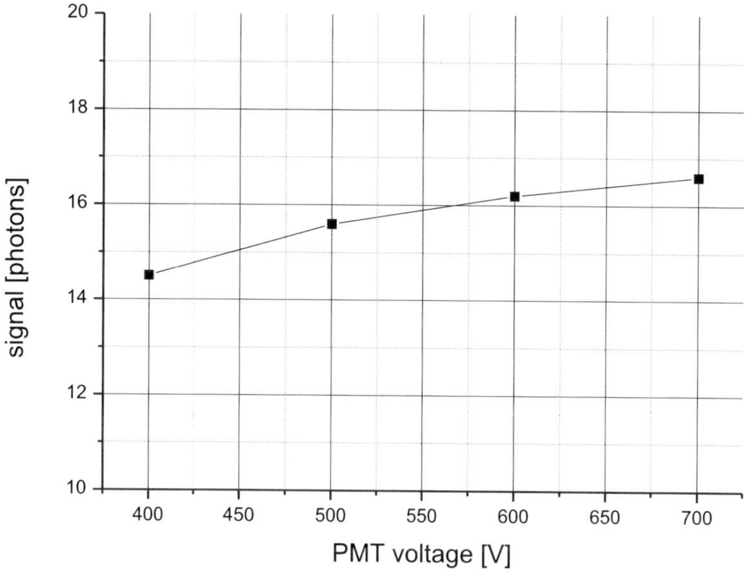

Fig. 8 Signal S_p^* vs. PMT voltage. Sample: homogeneous subresolution layer, Objective lens: Plan-Neofluar 40x/1.3 oil; Integration time: 1.6 μs

nation power at the output of an objective lens, especially at high NA. One approach is to mount a second identical objective lens on the opposite side of the sample and in this way couple light to a power meter. However this is not generally a practical method since it requires modification of the microscope stand to accommodate the second lens. A less conscientious approach would be to couple the power meter to the objective lens in a reproducible way and measure some unknown fraction of the illumination intensity. In this way measurements made at different times could be normalized. This approach is not unreasonable when monitoring performance of an individual instrument or instruments of the same type from one manufacturer but, even so, is prone to errors when used with very high NA objectives and could not be used with immersion lenses.

Brakenhoff [33] has proposed a different approach to monitoring illumination intensity by using a fluorescence sample with well characterized bleaching properties. The bleach rate itself can then be used as a direct measure of illumination power. This approach has the attraction, if practical problems can be overcome, of enabling the same thin film sample to be used to characterize both signal to noise and illumination intensity of a system. The advantage of such a sample is that it could be used to compare and characterize instruments of different design from different manufacturers and produce a single figure of merit. A large figure of merit would correspond to high sensitivity and low bleach rate – exactly what is required in biological imaging.

5 Conclusion

From all that has been discussed so far, it is clear that the performance of a fluorescence imaging system relies on synergy between a large number of components each susceptible to variation and disturbance of performance in different ways. However for both the manufacture, who must demonstrate and deliver a defined level of performance, and the user, who requires that this performance (or something close) be maintained consistently over a period of months and years, there is a requirement for characterisation tests that can assess global performance of the system without needing to assess the performance of individual components. Such a tool should at the same time help the engineer identify the specific cause of any reduced performance. Importantly the tests should allow performance to be measured independently of the particular application for which the system is being used and able to deliver, beyond argument, comparison of the performance of one system with another. Even better if the tests can compare the performance of systems of different design, possibly even from different manufacturers.

It is worth highlighting at this point that a manufacturer and his customer do not always agree about the results of performance characterisation tests. The service engineer may suspect the problem lies with poor sample preparation or some change in the user's imaging protocol. The users will prefer to dispute the performance of the system and defend the quality of the laboratory imaging protocol and method sample preparation. How is this confrontation to be avoided, since if it does occur there is risk of a lengthy "cold war" breaking out between manufacturer and user which is to the advantage of neither party? The answer almost certainly lies in the use of "arbitrated" samples for carrying out the performance tests. These are samples that are independently verified as to their performance characteristics and mode of use. Neither manufacture nor user should have any interest in their design although both may have contributed to the body of data that established their performance and consistency.

We can return now to our opening remarks about information content in a fluorescence image and the signal to noise as the key determinant of contrast. What the manufacturer requires are tests that assess the signal to noise performance of the system in a way that is accurate, repeatable, easy to understand and simple to perform both in the factory and in the field, by both a trained engineer responsible for manufacture or service of the system and a user responsible for ensuring that the system is performing consistently within specification.

Although there is yet no fully certified sample that meets this need we have shown that a thin film sample of uniform fluorescence can meet many of the requirements. It is easy to use and provides direct and easily understood information. The simplicity of obtaining an image devoid of all contrast except for statistical and other sources of noise allows the underlying performance of the instrument to reveal itself.

Appendix
Factors Affecting Signal and Noise in Confocal LSM

Table 2 Variability of the optical components of a confocal microscope (modified from Jim Pawley [24])

Instrumental variations: optics and their description [* Important especially in multi-photon microscopy]
Laser unit
• Power output stability: Usually noise and instability is < 1% but lasers can become much more unstable as they age.
• Efficiency of the optical coupling to the connecting fibre: Dust, misalignment or mechanical instability can be the source of random changes of 10–30%.
• Alignment and reflection characteristics of laser mirrors: Can be the source of long-term drift in laser output.
• Beam-pointing error/alignment: The location from which the laser light appears to emanate is determined by the laser mirrors. Instability here will show up as changes in brightness because changes in the apparent source position will change the efficiency of the optics coupling the laser light into the single-mode optical fibre used in most instruments.
• Repetition rate (only with pulsed lasers)*: Recognised repetition rate might deviate from triggered one due to setting of time-to-amplitude converter or offset.
• Pulse-width (only with pulsed lasers)*: The excitation efficiency in two-photon microscopy strongly relates on the pulse width. The smaller the pulse width the more peak power and thus more two-photon processes are triggered.
Objective lens
• Numerical aperture: Effects fraction of light emitted by specimen that can be collected. Ditto for light from laser.
• Objective magnification: Magnification is inversely related to the diameter of the objective lens entrance pupil. The objective will only function properly if the entire entrance pupil is filled with exciting laser light. Underfilling will reduce spatial resolution and hence peak intensity. Overfilling will cause some laser light to strike the metal mounting of the objective and be lost, also reducing the intensity in the spot.
• Cleanliness: Dirty optics produce much larger, dimmer spots.
• Transmission: The fraction of light incident on the objective that can be focussed into a spot on the other side. Varies with wavelength. Beware using older optics in IR or UV.

Table 2 (continued)

Instrumental variations: optics and their description
[* Important especially in multi-photon microscopy]

- Chromatic and spherical aberration:

Both make the spot bigger and vary with wavelength. Spherical also varies strongly with coverglass thickness and the refractive index of the immersion and embedding media.

- Diffraction/optical resolution:

Diffraction is the unavoidable limit to optical resolution. It effectively enlarges the image of objects smaller than the diffraction limit, making them appear dimmer than they should be.

Other optics

- Transmission:

Measure of the absence of absorption and reflectance losses in optical components, particularly: ND and/or bandpass filters, beamsplitters, and objectives. Also the transmission efficiency of Acousto-Optic Tunable Filters (AOTF) may drift over time mostly due to temperature effects.

- Reflections from air/glass interfaces:

Usually represent lost signal but may appear as bright spots, unrelated to specimen structure.

- Mirror reflectivity:

May be strong function of wavelength in the IR and UV and degrades with exposure to humidity and dust.

- Focus-plane position:

A feature slightly above of below the plane of focus will appear dimmer. When collecting 3D data, Nyquist sampling must also be practiced in the spacing of Z planes.

- Mechanical drift of stage:

Causes the plane of the object actually imaged to change with time.

Pinhole

- Size:

The detected signal is proportional to the square of the pinhole diameter. Usually set equal to the diameter of the Airy Disk at the plane of the pinhole.

- Alignment:

The image of the laser that is focused onto the specimen and then refocused back through the optical system should coincide with the centre of the pinhole.

Table 3 Variability of the non-optical components of a confocal microscope (modified from Jim Pawley [24])

Instrumental variations: others and their description
Scanning system
• Zoom magnification: This control determines the size of a pixel at the specimen. For Nyquist sampling, the pixel should be at least 2× smaller than the smallest features that you expect to see in your specimen. Assuming a Rayleigh criterion resolution of 200 nm, the pixels should be < 100 nm. Larger ones produce undersampling, reducing the recorded brightness of small features.
• Scan speed: The longer the dwell time on a particular pixel, the more signal will be detected and the less it will be distorted by Poisson noise. At high scan speeds (< 100 ns/pixel) signal from dyes with fluorescent decay constants that are longer than this dwell time can be reduced.
• Raster size: Together with the zoom magnification, the number of pixels along the edges of your raster will determine the pixel size. More pixels [1024×1024 vs. 512×512] makes undersampling less likely but means that one must either spend less time on each pixel [reducing the number of photons collected and increasing Poisson noise] or take more time to scan the larger image [possibly causing more bleaching]
• Geometrical distortion: Can be introduced by the optics or the scanning mirrors. Can result in discordance between the shape of the object and the image.
Detector: (PMT)
• Quantum efficiency (QE): The detected signal is directly proportional to QE. The effective QE of the PMTs used in most confocals drops from ~15% in the blue to ~4% in the red end of the spectrum.
• Response time: Most fluorescent signals can be amplified rapidly but detectors for others, such as transmembrane currents, respond only slowly, making slow scanning speeds necessary.
• PMT voltage: Determines the amplification of the PMT. An increase of 50 volts corresponds to a factor of ~2 more gain.
• PMT black level or brightness: This control permits the addition or subtraction of an arbitrary amount from the signal that is presented to the digitizer. Set so signal level in the darkest parts of the image is 5–10 digital units. Value is temperature dependent.
• Noise: In single point laser scanning microscopes the most commonly used detector is a photomultiplier. These detectors produce a "dark" noise due to thermally stimulated emission of photoelectrons from the photocathode. The dark current can be reduced by cooling the detector and is less for photomultipliers that are insensitive to red wavelengths.

Table 3 (continued)

Instrumental variations: others and their description
Digitization
• Linearity: The electronic signals presented to the digitizer of "8-bit" microscopes must be of a size to be recorded between 1 and 255. Because of statistical noise, > 10 and < 220 is safer.
• Digital conversion factor: The ratio between the number of photons detected and the number stored. Depends on PMT voltage and other electronic gain, but usually about 30 for "normal" specimens recorded on 8-bit instruments.

Table 4 Sample variability influencing a confocal microscope image (modified from Jim Pawley [24])

Sample variations and their description
Fluorophore
• Illumination wavelength: The best contrast between excitation of specific fluorescence (e.g. the dye) and non-specific fluorescence (e.g. autofluorescence) is commonly obtained at the excitation wavelength giving the maximum efficiency for fluorescence emission. Excitation at shorter wavelengths is often used when the dyes exhibit small Stokes shifts.
• Illumination intensity: Low intensity illumination at appropriate collection times prevent intensity saturation effects, e.g. departure of linearity in the relationship between illumination intensity and fluorescence signal. For a correctly set up system illumination intensity should lie within the range where both S and $(S/N)^2$ obey a linear relationship with illumination intensity.
• Fluorophore concentration: Within the Förster radius, commonly less than 10 nm, fluorophores transfer energy between each other and thus their light emission is altered (e.g. quenched). This effect is exploited in Förster Resonance Energy Transfer (FRET).
Sample
• Specimen & solvent: Both, specimen and solvent properties such as polarity and ion concentration influence the spectral properties of chromophores. Especially absorption and emission wavelength, extinction coefficient, and quantum yield are altered. Under the right conditions this may lead a complete loss of signal.
• Cleanliness: Dirty glass or plastic carriers induce aberrations and lead to blurred images.

Table 4 (continued)

Sample variations and their description
• Coverslip thickness: The least expensive optical component and the most likely to be carelessly chosen. Check with objective lens specifications and adjust correction collar. Standard cover glass thickness is 170±5 μm. Check each batch. • Immersion oil: Its refractive index must be exactly matched to the objective used. This may only occur over a small temperature range. Alternatively, it can be especially mixed.

Table 5 Environmental variations influencing confocal imaging (modified from Jim Pawley [24])

Environmental variations and their description [* Important especially in multi-photon microscopy]
Environment • Temperature: Changes in temperature influence the mobility (e.g. diffusion) of compounds within the sample. This aspect comes to the fore especially with live specimen and in Fluorescence Correlation Spectroscopy (FCS). Furthermore changes in temperature affect instrument performance through disturbance of the system alignment. Thus modern instruments are specified for an operating temperature range. • Incident light: Depending on the photostability of the specimen incident light, especially sunlight is harmful to the specimen. • Background light in room*: Incident light is much more a matter in two-photon microscopy than in confocal imaging due to missing pinhole. • Vibration: Vibration and stray EM fields can cause improper mirror deflections, resulting in distortions that may vary with time. The most prevalent source of vibration is air conditioning within the building which can induce a low frequency vibration within the entire structure. Fortunately it is relatively straightforward to isolate a system from such vibration, if necessary, using a stiffened table with anti-vibration supports. • Humidity: Humidity does not usually cause immediate effects on system performance. However over time in high humidity optical components can degrade.

References

1. Rost FWD (1991) Quantitative fluorescence microscopy. Cambridge University Press, Cambridge
2. Zwier JM, Van Rooij GJ, Hofstraat JW, Brakenhoff GJ (2004) J Microsc 216(Pt1):15
3. Zucker RM (2002) Microsc Tod 10(6):20
4. Zucker RM (2002) Microsc Tod 10(7):8
5. Hellmuth E, Mühlfriedel W (1996) Zeiss 1846–1905. Vom Atelier für Mechanik zum führenden Unternehmen des optischen Gerätebaus. Böhlau, Köln
6. Abbe E (1873) Arch Mikros Anat 9:413
7. McCarthy NJ, Evan GI (1998) Curr Top Dev Biol 36:259
8. Dunn KW, Mayor S, Myers JN, Maxfield FR (1994) FASEB J 8(9):573
9. Berland KM (2004) Methods Mol Biol 261:383
10. Hiraoka Y, Shimi T, Haraguchi T (2002) Cell Struct Funct 27:367
11. Dickinson ME, Bearman G, Tille S, Lansford R, Fraser SE (2001) Biotechniques 31(6):1272
12. Dickinson ME, Simbürger E, Zimmermann B, Waters CW, Fraser SE (2003) J Biomed Opt 8:329
13. Wouters FS, Verveer PJ, Bastiaens IH (2001) Trends Cell Biol (11):5
14. Kiyokawa E, Hara S, Nakamura T, Matsuda M (2006) Cancer Sci 97(1):8
15. Jares-Erijman EA, Jovin TM (2003) Nat Biotechnol 21(11):1387
16. Grunwald D, Cardoso MC, Leonhardt H, Buschmann V (2005) Curr Pharm Biotechnol 6(5):381
17. Kohl T, Schwille P (2005) Adv Biochem Eng Biotechnol 95:107
18. Gosch M, Rigler R (2005) Adv Drug Deliv Rev 57(1):169
19. Houtsmuller AB (2005) Adv Biochem Eng Biotechnol 95:177
20. Koster M, Frahm T, Hauser H (2005) Curr Opin Biotechnol 1:28
21. Mullineaux CW (2004) J Exp Bot 55(400):1207
22. Sprague BL, McNally JG (2005) Trends Cell Biol 15(2):84
23. Van Drogen F, Peter M (2001) Biol Cell 93(1–2):63
24. Pawley J (2000) BioTechniques 28(5):884
25. Zucker RM, Price O (2001) Cytometry 44(4):273
26. Henderson LO, Marti GE, Gaigalas A, Hannon WH, Vogt RF (1998) Cytometry 33:97
27. Schwartz A, Marti GE, Poon R, Gratama JW, Fernandez-Repollet E (1998) Cytometry 33:106
28. Schwartz A, Fernandez-Repollet E, Vogt R, Gratama JW (1996) Cytometry 26:22
29. Shapiro HM (1995) Practical Flow Cytometry. Wiley-Liss, New York
30. Chase ES, Hoffman RA (1998) Cytometry 33:267
31. Hoffman RA (2001) Methods in Cell Biology Vol. 63: Standardization and Quantitation in Flow Cytometry. Academic Press, New York
32. Wolf F, Geley S (2006) J Microsc 221(Pt1):72
33. Brakenhoff GJ, Wurpel GWH, Jalink K, Brocks L, Zwier JM (2005) J Microsc 219(Pt3):122
34. Stelzer EHK (1998) J Microsc 189(Pt1):15–24
35. Zucker RM (2005) Meth Mol Biol 319:77–136
36. Shotton DM (1995) Electronic light microscopy. Histochem Cell Biol 104:907–137
37. Sheppard CJR, Shotton DM (1997) Confocal Laser Scanning Microscopy. Springer, New York, p 9
38. Hell SW, Stelzer EHK (1995) Handbook of Biological Confocal Microscopy. Plenum Press, New York, pp 347–354

Springer Ser Fluoresc (2008) 6: 25–54
DOI 10.1007/4243_2008_029

Published online: 17 April 2008

Characterization and Calibration in Wide Field and Sectioned Fluorescence Microscopy SIPcharts

Fred Brakenhoff (✉) · Jurriaan Zwier

Swammerdam Institute for Life Sciences, University of Amsterdam, Kruislaan 316, 1098 SM Amsterdam, The Netherlands
brakenho@science.uva.nl

Abstract A fluorescence image calibration method is introduced based on the use of standardized uniformly fluorescing reference layers. Crucial to the approach is that these layers are highly uniform. It is demonstrated to be effective for the correction of non-uniform imaging characteristics across the image (shading correction) as well as for relating fluorescence intensities between images taken with different microscopes or imaging conditions. The approach can be used both in wide field or regular and sectioned (see the section on fluorescence microscopy).

In wide field it is shown that in addition the variation of the illumination intensity over the image can be determined on the basis of the uniform bleaching characteristics of the layers. This permits correction for the latter and makes bleach-rate-related imaging in wide field microscopy practical.

The significant potential of these layers for calibration in quantitative fluorescence microscopy is illustrated with a series of applications. The approach is also shown to be valuable for general microscope testing and characterization. Specifically in sectioning, specifically confocal, microscopy a set of parameters derived from through-focus datasets of such layers can be used to define a number of properties relevant to sectioned imaging. The main characteristics of a particular imaging situation can then be summarized in a sectioned imaging property chart (SIPchart), which turns out to be a very useful tool for characterizing the properties of particular sectioned imaging systems.

Keywords Confocal microscopy · Fluorescence microscopy · Fluorescence photo-bleaching · Image correction · SIPcharts · Sectioned imaging · Shading correction

Abbreviations

D(x,y)	Detection efficiency distribution
DPPC	Dipalmitoylphosphatidylcholine
F(x,y)	Fluorescer distribution
FRAP	Fluorescence Recovery after Photobleaching
FRET	Fluorescence Resonance Energy Transfer
I(x,y)	Illumination distribution
k(x,y)	Bleach rate distribution
LC	Liquid condensed
LE	Liquid expanded
NA	Numerical Aperture
NBDPC	NBD-phosphatidylcholine
P(x,y)	Product distribution
PSF	Point Spread Function
SIPchart	Sectioned Imaging Property chart
t_i (s)	Exposure time

1 Introduction

A fluorescence image calibration method is introduced based on the use of standardized uniformly fluorescing reference layers. Crucial to the approach is that these layers are highly uniform. It is demonstrated to be effective for the correction of non-uniform imaging characteristics across the image (shading correction) as well as for relating fluorescence intensities between images taken with different microscopes or imaging conditions. The approach can be used both in wide field or regular (Sect. 2) and sectioned (Sect. 3) fluorescence microscopy.

In wide field it is shown that in addition the variation of the illumination intensity over the image can be determined on the basis of the uniform bleaching characteristics of the layers. This permits correction for the latter and makes bleach-rate-related imaging in wide field microscopy practical.

The significant potential of these layers for calibration in quantitative fluorescence microscopy is illustrated with a series of applications. The approach is also shown to be valuable for general microscope testing and characterization.

Specifically, in sectioning microscopy, a set of parameters derived from through-focus datasets of such layers can be used to define a number of properties relevant to sectioned imaging. The main characteristics of a particular imaging situation can then be summarized in a sectioned imaging property chart, or SIPchart, which turns out to be a very useful tool for characterizing the properties of particular sectioned imaging systems.

2 Image Calibration in Wide Field Fluorescence Microscopy

2.1 Introducing Calibration in Wide Field Microscopy

For the purpose of this section on wide field imaging characterization the pixellated image P(x,y) – also called in this chapter the product distribution – of a fluorescence microscope can be described as:

$$P(x,y) = I(x,y) \cdot D(x,y) \cdot F(x,y) \cdot t_i(s) , \quad (1)$$

where I(x,y) is the illumination distribution over the image field of view, D(x,y) the detection efficiency distribution, F(x,y) the fluorescence distribution from pixel to pixel over the specimen, t_i (s) the image exposure time in seconds s, and x,y the image pixel coordinates.

In this section we address two types of fluorescence calibration:

1. Fluorescence of the fluorescence image intensity. This involves calibration at the level of the product I(x,y)·D(x,y) as needed for shading correction and image comparison.
2. Fluorescence of the variations in illumination intensity I(x,y) as required for the correction in bleach rate imaging.

The key to the approach is the use of fluorescent reference layers for the calibration that are both to a high degree *spatially uniform* as well as *reproducible*. In the presented procedure the fluorescence image is calibrated with the help of an image of the reference layer taken under identical imaging conditions as the image to be calibrated.

The work is partly a continuation of earlier work of our group [1, 2] and is related to the work done by Castleman [3] and Jericevic et al. [4]. The latter already showed that with a calibration layer spatial variation of the product of the illumination and detection pathways could be corrected. Ghauharali et al. [1] did obtain in addition separate illumination distributions by using a mono-exponential function for fitting the observed bleaching of their test layers. Fitting the bleaching characteristics using stretched exponential decay kinetics provides much better fits then with a mono-exponential function dependence.

Originally, we intended to develop two types of reference layers: one uniformly fluorescing, but non-bleaching for calibrating the product distribution P(x,y), and one uniformly bleaching to determine the illumination distribution. However, it turned out that the latter layers as developed could serve effectively both functions combined. While the bleaching was sufficiently slow to permit for fluorescence calibration with the first or second image of such a layer, it still showed enough bleaching over a finite time span to be practical for determining the illumination distribution from the bleaching dependence.

After illustrating the necessity for using stretched exponential fitting, we show that the fluorescence reference layers are suitable for the determination of both I(x,y) and D(x,y) in a range of intensities relevant to regular wide-field fluorescence microscopes. Subsequently, it is shown that the reference layers can be manufactured with narrow tolerances and with fluorescence and bleaching characteristics uniform within a few percent.

2.2 Bleach Kinetics

The excitation illumination distribution in a microscope image can be determined from the bleach behavior at each pixel point in a series of images taken as a function of exposure time. Ghauharali et al. [1, 2] have shown that with a suitable photo-bleachable test layer the distribution of both the excitation intensity and the detection efficiency over the image can be determined by this approach. Following up on their findings we set out to develop optimized calibration or reference layers which should show ideally mono-exponential irreversible photo-bleach kinetics with respect to the total irradiation dose of incident light. In practice, we found that none of the layers we produced did satisfy this requirement. Even at low dye concentrations where dye-dye interactions are minimized, still no mono-exponential decay could be observed in the layers produced by us. This does not come as a surprise, as it is known [5, 6] from polymer kinetics that in polymer films, dye molecules are subject to small differences in their environment affecting the local bleach rate. We found that by fitting the fluorescence bleaching with a stretched exponential function (Eq. 2) – often used to describe polymer kinetics – that good fits with small residuals can be obtained.

$$\mathrm{I_f}(t_\mathrm{b}) = \mathrm{C} + \mathrm{A}\exp\left((-\,\mathrm{k}t_\mathrm{b})^{\beta}\right) \; . \tag{2}$$

In Eq. 2, $\mathrm{I_f}$ (t_b) expresses the fluorescence intensity in counts, C the non-bleaching background fluorescence intensity, A the bleached fluorescence

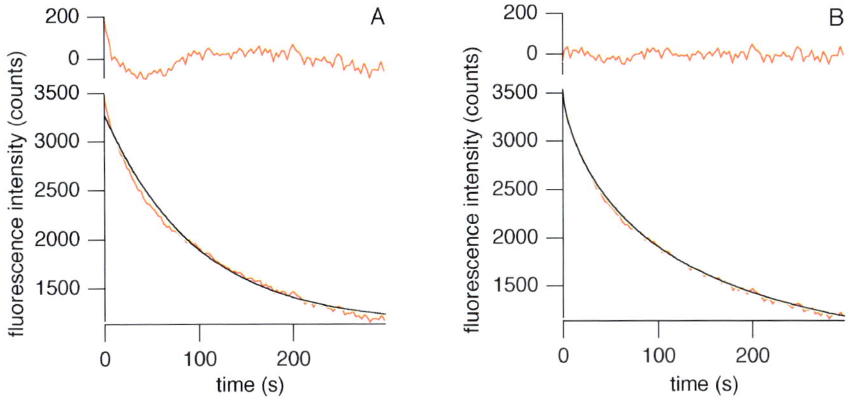

Fig. 1 Mono-exponential (**A**) and stretched exponential (**B**) fitting of the decay of fluorescence intensity measured for a single pixel with the residuals shown at the *top* of each figure

intensity, k the bleach rate, t_b the bleach exposure time, i.e., the time the layer is exposed to the illumination light and b the stretched exponential coefficient, which has a value between 0 and 1. Note that the stretched exponential function is equivalent to a mono-exponential function for $\beta = 1$. In an example on a bleach series from one pixel point, we see that the fit of the fluorescence bleaching behavior (Fig. 1) with the stretched exponential function shows a great improvement over a mono-exponential fit on the same data.

We also found (Fig. 2) that the bleach rate k obtained from the stretched exponential fitting procedure is linearly proportional to the illumination intensity within 2% over a range of excitation intensities relevant to regular arc-lamp fluorescence microscopy [1]. It is clear that such linearity is an absolute requirement for the successful application of this method for illumination calibration in practical microscopy.

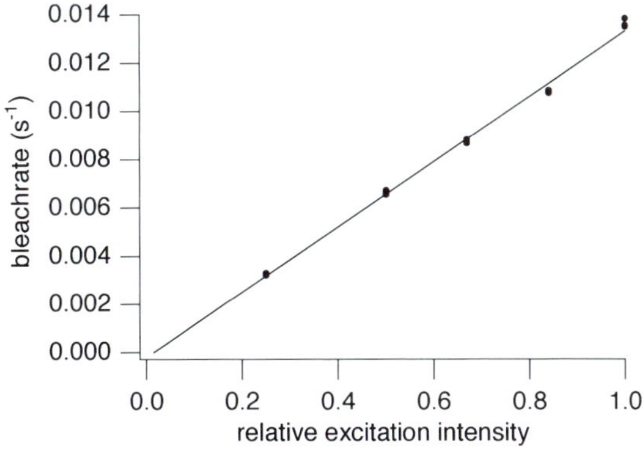

Fig. 2 Bleach rate *k* from the stretched exponential fitting procedure versus the relative excitation intensity, set by neutral density filters. 14 measurements are included

2.3 Fluorescence Reference Layer Development and Test Procedures

2.3.1 Preparation of Reference Layers

A fluorescence reference layer typically contains a fluorescent dye embedded in a uniform polymer film. For the irreversibly photo-bleaching dye we selected the well-known [7–9] highly fluorescing dye fluoresceine. It possesses suitable bleach sensitivity such that illumination calibration under typical specimen illumination conditions in an arc lamp equipped microscope can be done in a few minutes. Upon irradiation of fluoresceine in its absorption maximum, around 488 nm, an irreversible series of photo reactions takes place, leading to a change in the absorption spectrum, and therefore to a decrease in the fluorescence output, around 530 nm [7].

Since fluoresceine is water soluble, the polymer in which the fluoresceine is to be diluted has to be water soluble as well. Furthermore, the polymer solution should provide highly reproducible and well-defined layers after spinning. Polyvinylalcohols were identified as suitable polymer host layer material. Typically solutions were made comprising 0.01 wt % fluoresceine (Merck) in polyvinylalcohol (Aldrich, 87–89% hydrolyzed, MW 124 000–186 000), which were spin-coated (1250 rpm) on a 24×32 mm cover slide (Menzel), resulting in layers with a thickness – depending on the spin rate – between 150 and 200 nm and with each layer uniform in thickness within 5 nm. These layers were mounted and sealed with epoxyresin on a microscope slide (76×26 mm). Very reproducible layers could be obtained in this way. Due to the low concentration of fluoresceine we avoid intermolecular dye interactions as much as possible. As a result the fluorescence intensity from the layers is generally one order of magnitude lower then stained biological samples.

The layers are stored in the dark at room temperature and have been used more then one year after production, without any significant changes observed.

2.3.2 Instrumentation

Images were acquired with an Olympus BX60 fluorescence microscope equipped with a Photometrix Coolsnap fx digital camera. Excitation occurred with light from a Hg-arc lamp, which was filtered through an Olympus 41017-model UMF2 filter set, providing excitation at wavelengths between 451–490 nm light while transmitting fluorescence light to the camera between 491 and 540 nm. Measurements were carried out with an Olympus Ach 20x (NA = 0.4), or an Olympus UPlanFL 40x, (NA = 0.75) objective lens. Data collection and processing was done with IPLab Spectrum software from the Signal Analys Corporation with a custom written kernel added for the stretched exponential data fits. Spin coating of the layers was performed with a Delta 10TT system from BLE Laboratory Equipment.

2.3.3 Shading Correction and Microscope Calibration Procedure

An image in a fluorescence microscope (P(x,y)) can be described – see Sect. 2.1 – by:

$$P(x,y) = I(x,y) \cdot D(x,y) \cdot F(x,y) \cdot t_i(s) \,. \tag{3}$$

For characterization of the microscope imaging conditions we use an image $P_r(x,y)$ of the reference layer taken under identical imaging conditions as the fluorescence image to be calibrated:

$$P_r(x,y) = I(x,y) \cdot D(x,y) \cdot F_r(x,y) \cdot t_{ir}(s) \,. \tag{4}$$

By taking the ratio of both images a calibrated image $P_c(x,y)$ is obtained:

$$P_c(x,y) = \frac{P(x,y)}{P_r(x,y)} = \frac{F(x,y)}{F_r} \cdot \frac{t_i(s)}{t_{ir}(s)} \,, \tag{5}$$

where the pixel by pixel fluorescence is normalized in units of fluorescence with respect to the reference layer. We see that the actual imaging conditions described by $I(x,y) \cdot D(x,y)$ have dropped out. The fluorescence generation is assumed to be linear with respect to illumination intensity, i.e., only dose – $I(x,y) \cdot t_i\,(s)$ – dependent.

With the actual pixel by pixel imaging conditions removed in this image due to the division, the calibrated image $P_c(x,y)$ directly represents a shading corrected image.

For the same reason we have seen that fluorescence images taken under different imaging conditions, if no other factors play a role, can be directly quantitatively related to each other, as they are expressed in units of the standardized fluorescence of the reference layer.

2.3.4 Separation of I(x,y) and D(x,y)

As the reference layers – as shown below – possess highly spatially uniform bleaching characteristics it is in addition possible to obtain the specimen illumination distribution I(x,y) independently of the detection distribution D(x,y). This illumination distribution can be derived from the analysis of the bleaching behavior of the calibration layer. For this a time series of images is taken of the reference layer during which the layer is bleached down to about 30% of its starting fluorescence intensity. Using the stretched exponential bleach kinetics described in Sect. 1 we fit the bleach decay at each pixel

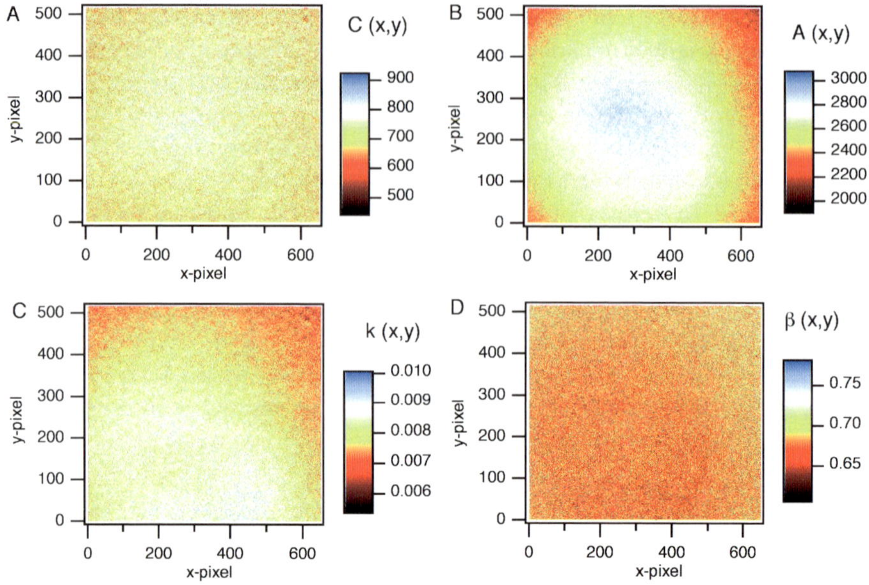

Fig. 3 Analysis of the fluorescence bleaching of a spatially uniform test layer by fitting (pixel by pixel) with a stretched exponential function. $I_f(t_b) = C + A \exp((-kt_b)^\beta)$: **A** C(x,y), **B** A(x,y), **C** k(x,y) and **D** β(x,y)

of this series of images with a stretched exponential (Eq. 2). The result of this operation can be represented as 4 images corresponding to the respective fitting parameters. A typical result obtained on our reference layers is shown in Fig. 3 with panel A the non-bleaching part of the fluorescence of the image C(x,y), panel B the bleached fluorescence intensity A(x,y), panel C the bleach rate k(x,y) and panel D the stretched exponential coefficient β(x,y). With $k(x,y) = k_0 \cdot I(x,y)$ over the relevant range of illumination intensities (Fig. 2) the illumination intensity distribution (I(x,y) can now be derived from the bleach rate image k(x,y) apart from a constant factor. k_0 is a bleach constant for the used bleaching material. Such an illumination distribution I(x,y) can

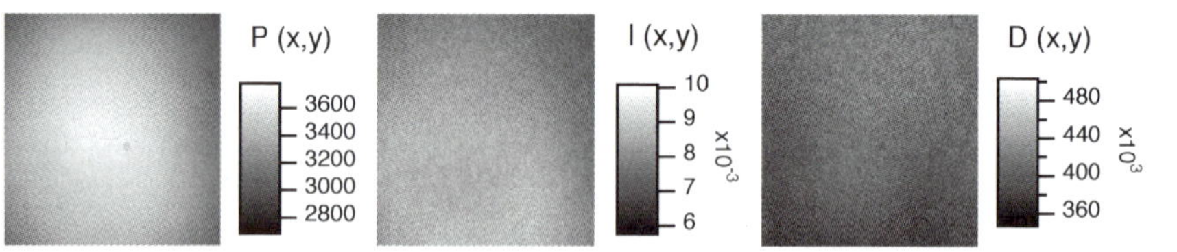

Fig. 4 The product distribution P(x,y) of the microscope (determined from the image at $t = 0$ of the test layer), divided by its illumination distribution I(x,y) (determined by the bleachrates k of the test layer), gives the detection sensitivity distribution D(x,y) of the microscope

be useful for determining the actual illumination conditions – such as alignment or uneven illumination in a microscope.

Dividing $P_r(x,y)$ by $I(x,y)$ obtained from the bleach procedure gives the detection sensitivity distribution, $D(x,y)$, of the microscope as is directly clear from Eq. 4. Figure 4 shows the results of the separation of $P(x,y)$ into $D(x,y)$ and $I(x,y)$. A remarkable feature in the $D(x,y)$ image is the appearance of dark spots solely in the detection distribution, which are due to irregularities such as dust particles in the detection pathway.

2.4
Calibration Layer Reproducibility and Uniformity

2.4.1
Uniformity of the Calibration Layer

2.4.1.1
Fluorescence

For application of the calibration procedures uniformity of the fluorescence and bleach properties across the layer are crucial. To determine if the reference layer is really spatially uniform, two fluorescence intensity images were taken at $t_b = 0$ at different spots on one reference layer (see Fig. 5g). To obtain such images, the layer is put into focus first using the diaphragm of the microscope, after which the layer is moved slightly with the light switched off. The measurement is started when the light is switched on.

The first image was then used as reference image $P_r(t_b = 0)$ – Fig. 5a – and the second – Fig. 5b – as the object image $P(t_b = 0)$. Then in the test for the layer uniformity the object image was "calibrated" by dividing it by the reference image resulting in the calibrated image Fig. 5c. If now both areas imaged are both uniform and show equal fluorescence, then in the histogram of pixel values of this calibrated image, we should see a narrow distribution with an average value of 1.

This "self" test using the reference layer itself is very effective because if the layer properties would not be uniform over the image area or would differ from location to location over the layer, then such differences would immediately show up as a broadening in the calibrated image histogram.

The images shown in Fig. 5 are in fact also an excellent illustration of the effectiveness of shading correction. The "uncalibrated or raw" reference layer images $P_r(t_b = 0$ and $P(t_b = 0)$ show in their respective histograms Fig. 5d (avg. 2210; fwhm 277) and Fig. 5e (avg. 2206; fwhm 292) intensity variations of up to 29% and relative standard deviation of ca. 5%. Correction leads to Fig. 5c with its corresponding histogram (Fig. 5f) (avg. 0.999; fwhm 0.038) with a clearly improved relative standard deviation of 1.5%. Furthermore the average value of the corrected image is close to 1.0, which is the value expected for a layer with identical fluorescence as its reference.

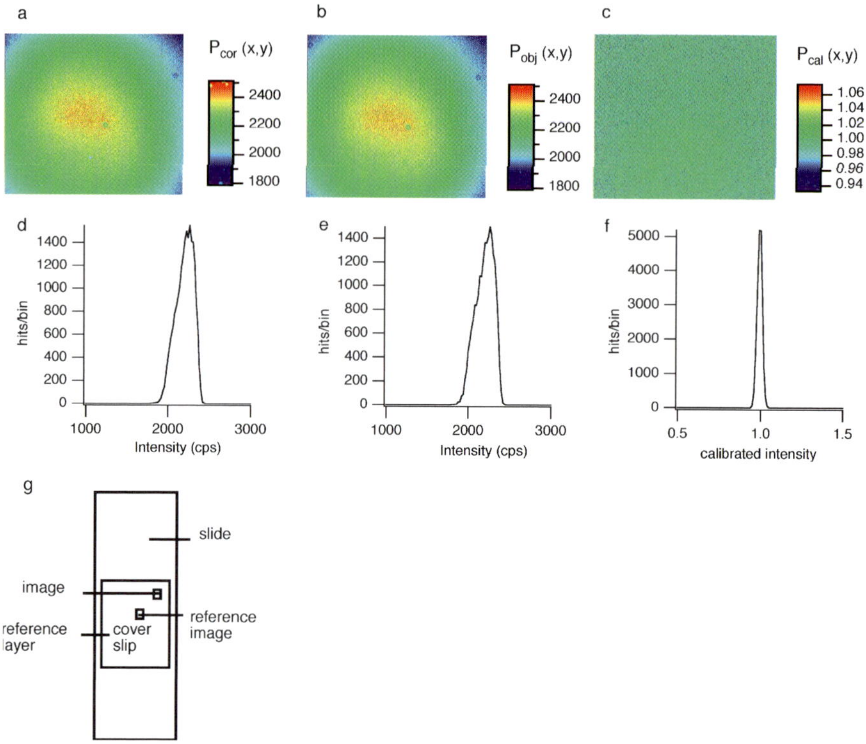

Fig. 5 The reference image (**a**) and object image (**b**) as used in the "self" test for layer uniformity with (**c**) the resulting calibrated image. **d,e** and **f** are the corresponding histograms of pixel intensity values of these images. **g** shows the configuration of the reference layer. See further text

2.4.1.2
Uniformity of Bleaching Characteristics

In a similar way as described above using the layer itself, the uniformity of the bleaching properties of the layer can be tested. The approach is correcting for the observed bleach rates in one location with the help of the illumination distribution data obtained at a second location of the layer. A narrow distribution in the bleachrate histogram in the illumination corrected bleachrate image then indicates that the bleach characteristics are indeed uniform over the layer.

A series of 100 images (515×630 pixels) was taken at identical time intervals of a reference layer. From these images an illumination distribution, $I_{cor}(x,y)$ (Fig. 6a) can be calculated as described in Sect. 2.3.4. This illumination distribution I_{cor} can now be divided by another illumination distribution, I_{obj} (Fig. 6b) obtained in a similar way at a different spot on the same reference layer or another reference layer. This results in a calibrated illumination distribution, I_{cal} (Fig. 6c).

Figure 6d–f show the histograms for the 3 images shown in Fig. 6a–c. I_{cal} is centered on 1.012 ± 0.023 (fwhm 0.054), whereas for a perfectly uniform test layer this value is expected to be one. The relative standard deviation of the uncorrected bleach rates from I_{obj} of 10.7%, after calibration is reduced to 2.3%.

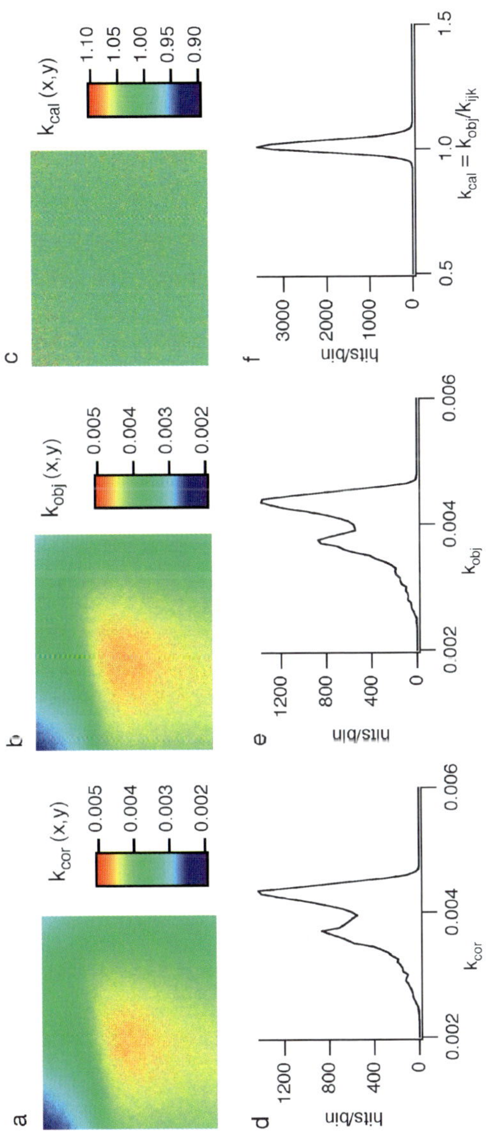

Fig. 6 Uniformity of reference layer bleach characteristics. **a** shows illumination distribution with which the bleach rate image (**b**) is corrected to obtain the uniform corrected bleachrate image (**c**). **d,e** and **f** are the corresponding histograms. See further text

2.4.2
Reproducibility of the Reference Layers

For a reference layer to be of practical use its properties should be reproducible from batch to batch during manufacturing. For a number of reference layers, prepared and measured under the same circumstances, the intensity at the onset of illumination and their respective bleach rate distribution have been measured. The results from the layers in one batch – prepared from the same fluorescer solution and under identical spinning and sealing conditions – are shown in Fig. 7A. For five samples, i.e., Fig. 7A(a–e), bleach rates with a relative standard deviation of 1.3% have been established, whereas their intensities at $t_b = 0$ have a relative standard deviation of 2.2%.

From batch to batch we observed very similar bleach properties in all properly sealed layers examined. Some variation in the absolute fluorescence intensities of the layers was observed both between batches and layers from

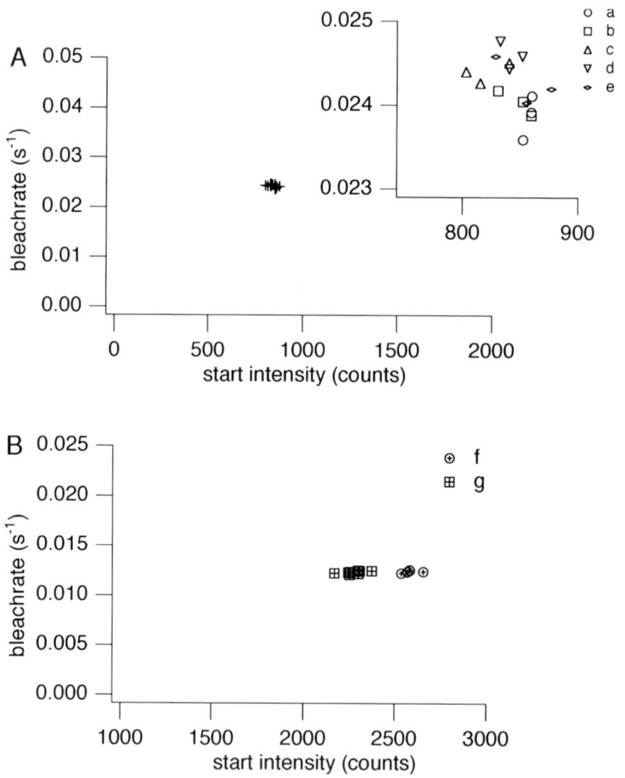

Fig. 7 **A** Bleach rate versus start intensity (emission at $t_b = 0$) at different locations in one test layer, inset enlargement of measurements *a*, *b*, *c*, *d* and *e*. **B** Bleach rate versus start intensity in layers from two different batches as indicated by *f* and *g*

one batch (Fig. 7B). The relative small variation in fluorescence observed is probably caused by fluorescer concentration variations from batch to batch and – within a batch – small variations in layer thickness due to spinning conditions. Some further optimization and calibration of the layer fluorescence against a common standard or in absolute terms – see below – can address this problem.

2.5
Application Examples in Wide Field Microscopy

2.5.1
Fluorescence Intensity

2.5.1.1
Shading Correction

In addition to the result presented in Sect. 2.4.1.1 we demonstrate the effectiveness of the shading correction procedure on a sample, which has an evenly distributed fluorophore concentration associated with recognizable morphological features. For this test liquid lipid monolayers of DPPC doped with the fluorophore NBDPC on a glass substrate were prepared. These monolayers give rise to two distinct morphological features: a liquid condensed

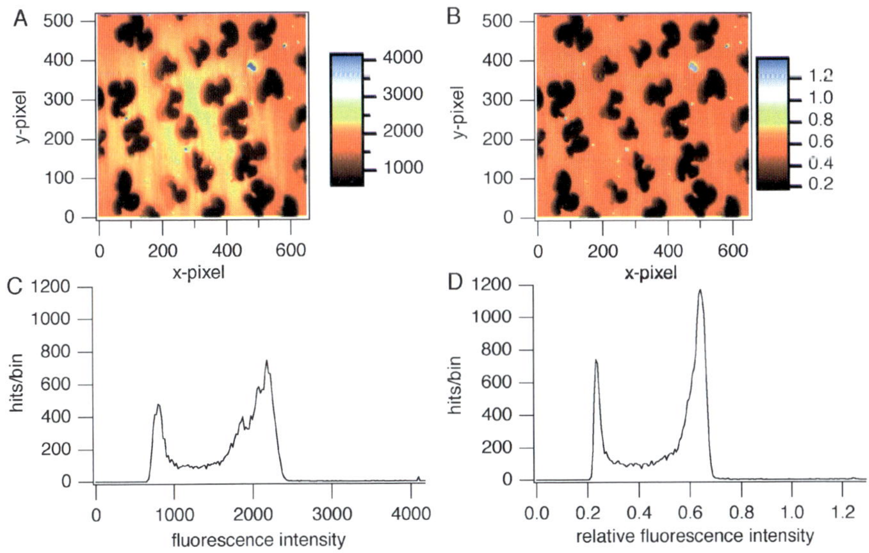

Fig. 8 False color fluorescence intensity images (**A,B**) and histograms (**C,D**) of DPPC monolayers doped with 4.4 mol% NBDPC, as obtained before (**A,C**) and after (**B,D**) shading correction

(LC) phase with low fluorescence intensity and a liquid expanded (LE) phase, characterized by higher fluorescence intensity [10, 11]. The uncorrected fluorescence image of these monolayers is shown in Fig. 8A while after correction with the reference layer image Fig. 8B is obtained. We observe after correction a much clearer association between respective regions of lower and higher fluorescence intensity and regions with LC and LE phases. The effect is also demonstrated in the histograms Figs. 8C and d of these images, where the distributions of associated with the LE and the LC phases are significantly better defined after shading correction than before.

2.5.1.2 Calibration of Microscope Conditions

It would be very valuable in fluorescence microscopy to be able to compare quantitative images taken at various imaging conditions. This is especially important as reproducing imaging conditions between microscopes – or even maintaining identical conditions in the same microscope over time – is difficult, if not impossible. When evaluating the possibilities of image calibration for comparing microscope conditions we found it to work well when comparing images obtained under similar NA conditions or different NA and similar object structure but not when both factors were different. A factor in this may be that the complexity of object structures – a flat layer vs. for instance cells of finite thickness in culture – affects the angles over which light is scattered. This may make the efficiency of fluorescence light collection NA dependent.

For the present we found it is useful to distinguish three different cases for evaluating the possibilities of image calibration as a function of microscope imaging condition:

(a) Comparing the imaging of objects in the imaging field with similar scattering properties and observed under different NA and magnification conditions.
(b) Idem with differently scattering objects but with identical NA and magnification and varying illumination conditions.
(c) Idem but with both differently scattering objects and different NA and magnification.

For demonstrating image calibration under uniform scattering we looked at images at different NA and magnification of liquid expanded and condensed lipid layers used above – see Sect. 2.5.1.1. These layers are basically flat and can be assumed to possess similar scattering properties over the whole image. We compared images obtained under 20× and 40× magnification. In order to compare images with these different magnifications a window which is about 1/4 of the total image in the 20× image was chosen, which exactly corresponds to the area covered by the 40× image. In Fig. 9 we see that

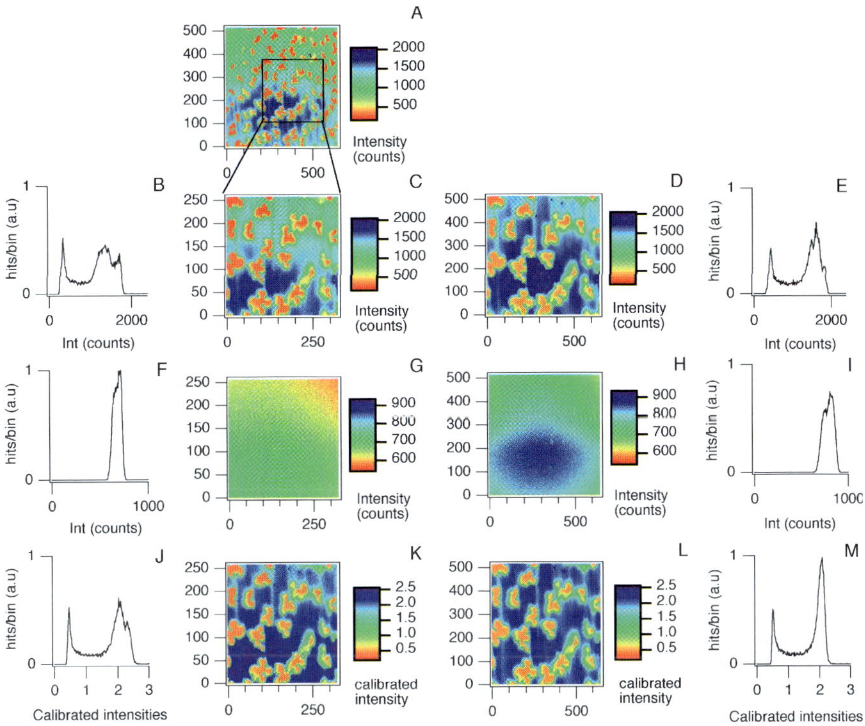

Fig. 9 Image calibration of fluorescence images of the labeled DPPC layer taken with 20×, N.A. = 0.40 (**A,C**) and 40×, NA = 0.75 (**D**) objectives respectively. For clarity the corresponding histograms are also given. Of the 20× image (**A**) the corresponding area viewed by the 40× lens (**C**) is shown as indicated by the sketch. With **C**, (**B**, histogram of **C**) and **D** (hist: **E**) serving as object images to be calibrated and **G** (hist: **F**) and **G** (hist: **H**) as reference images the calibrated 20× image **K** (hist: **J**) and 40× image **K** (hist: **L**) are obtained. From their false color representation it can be seen that the calibrated 20 and 40× images not only are shading corrected but also show closely similar calibrated intensities

the different intensity distributions in Fig. 9C and d after calibration (with Figs. 9G and H respectively) – Figs. 9K and L – show a nice correspondence and are also both shading corrected in the process.

Figure 10 shows the results of image calibration of images taken under strongly different illumination conditions, however at the same NA and magnification. These specimens are C3617 mouse cells transfected with GFP-GR (Glucocorticoid Receptor) [12]. Noteworthy is that in the ratio image h of the corrected images the non-bleaching background can be seen to have ratio values around 1 indicating good correlation between images after calibration. Due to some bleaching of the cells between the two images – image with objective 1 taken first – we see that in the ratio image h the cells show up somewhat brighter. The present result shows that in an object with some scattering and with very different product distributions – created here by on

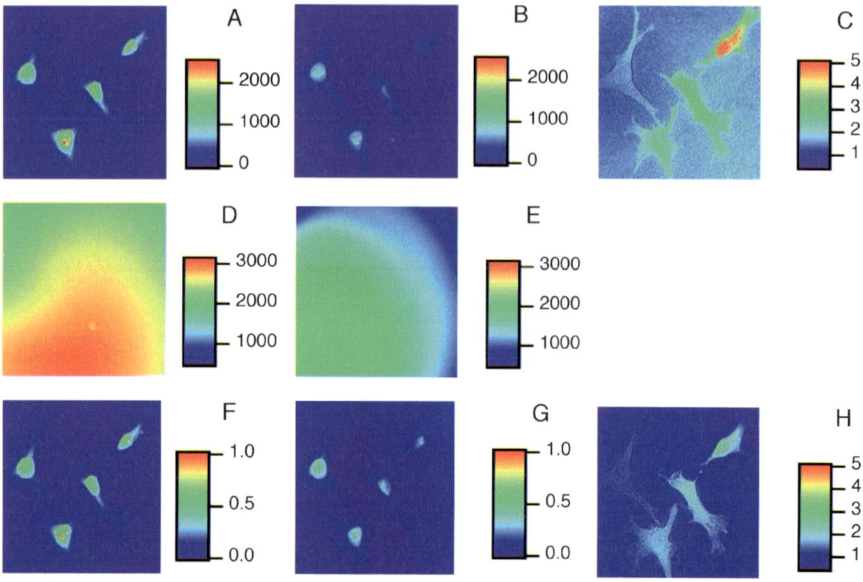

Fig. 10 Comparison imaging of C3617-mouse cells transfected with the GFP-glucocorticoid receptor using two different – but with the same NA – objectives, objective 1 and 2, and under different imaging or product distributions $P_1(x,y)$ and $P_2(x,y)$, respectively. **A** image taken with objective 1 and $P_1(x,y)$, **B** with objective 2, $P_2(x,y)$. The ratio image (**C**) (= a/b) shows very poor correlation between (**A**) and (**B**). After calibrating both images (**A**) and (**B**) with the respective product distributions **D** ($P_1(x,y)$) and **E** ($P_2(x,y)$) we see that the corrected images, **F** and **G** respectively, show much better correlation as also witnessed by the ratio image (**H**) (= f/g)

purpose disaligning the illumination conditions between the objective 1 and 2 images – still good image correlation can be achieved.

We found in preliminary experiments that on objects with finite scattering such as the cells used above, and observed under different NA and magnification conditions differences of up to 20 to 30% could be observed between the calibrated fluorescence of these objects, differences which could not be explained by bleaching. As indicated above these differences after calibration may be tentatively associated with the varying scattering properties of the structures imaged. A systematic exploration of this subject has not been done yet but we hope to address this issue at a later time.

2.5.2
Bleach Rate Imaging and Correction for Uneven Illumination

Bleach rate imaging becomes practical if the effect of uneven illumination – producing uneven bleaching over the image – can be corrected. We found during the imaging of the NBD chromophores present in the monolayers as

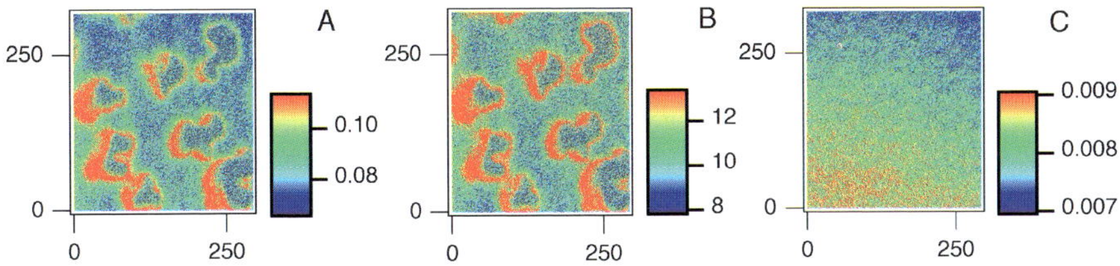

Fig. 11 False color images of the bleaching constants before (**A**) and after (**B**) illumination correction measured for DPPC monolayers doped with 4.4 mol % DPPC-NBD. **C** shows the illumination distribution used during the correction

described in Sect. 2.5.1.1, Fig. 8 that these are subject to substantial bleaching. Figure 11a shows the bleach constant k(x,y) image obtained by fitting with the stretched exponential fitting procedure an image bleach series of the central area shown in Fig. 8 corresponding to about 1/4 of the original image. After correction with the illumination distribution – Fig. 11C – we see in the corrected image Fig. 11B that both the LE and the LC phases bleach at a similar rate. However, in the phase coexistence region, the monolayer bleaches about 25% faster. While the underlying reason for this behavior is not fully clear – it could be associated with reduced ordering in the phase coexistence region – this result still shows that imaging in the bleach constant parameter can indicate new features in the image which would remain unnoticed otherwise.

3 Characterization of Sectioning Fluorescence Microscopy (3D) with Thin Uniform Fluorescent Layers: Sectioned Imaging Property or SIPcharts

3.1 Introducing Calibration in Sectioned Fluorescence Microscopy

Three-dimensional fluorescence microscopy has found widespread application in recent years, especially in molecular cell biology. Imaging in this type of microscopy is usually based on a series of sectioned images obtained by stepping the specimen through the focal region of a beam type scanning microscope. In most confocal or two-photon microscopes the signal at each lateral image position in a section is digitized and the data subsequently stored – together with the data of the other sections – as a 3D dataset. Ideally, the imaging properties should be identical over the imaging field. However, already at the inception of confocal microscopy it was realized that for instance the apparent fluorescent intensity in confocal imaging could vary significantly over the image field [13]. Also the actual confocal imaging conditions do vary significantly from microscope to microscope. In fact the actual sectioning properties of an instrument and the apparent image intensities are observed to depend sensitively on its optical properties related to the optics employed, and operator controlled factors like pinhole and alignment settings. The latter two factors especially cause uncertainty in reproducing settings with confidence making the comparison difficult of images obtained during different confocal sessions.

Up till now to our knowledge no reasonably easy to use and effective means are available for describing a particular imaging situation in 3D microscopy. Here we propose the use of thin uniformly fluorescing layers for characterizing the confocal or more general sectioning properties of a particular imaging situation. It has the specific advantages that it gives a good "feel" for the sectioning properties over the image field, is sensitive to small changes in the imaging conditions and possesses good signal to noise properties under regular imaging conditions, the latter because the fluorescence data from the thin layers can be binned to a substantial degree without loss of information on the lateral variation of measured sectioned imaging characteristics properties (see below). Its ease of use makes it feasible to use this method for routine determination and analysis of the 3D imaging properties as a function of parameters such as pinhole settings, alignment, and other parameters.

The method is based on the uniform fluorescent reference layers as utilized above for the calibration of regular wide-field fluorescence microscopy. Their uniform thickness and uniform fluorescence properties are also essential for the success of the presented method for 3D calibration. Schrader et al. [14] employed very thin – order of nms – fluorescent layers for monitoring the resolution in 4pi-microscopy. Their layers were neither aimed for use for general characterization of sectioning microscopy, nor specifically developed and tested for lateral uniformity.

3D datasets acquired by the deconvolution of non-scanned regular fluorescence images [15] can also, in principle, be characterized by the present approach: applications are restricted here to sectioned imaging obtained by the scanning approach.

It is to be noted that in the present approach only access is obtained to the axial imaging characteristics (or axial PSF, see below) but *not* the lateral variation of the point spread function (PSF) governing the imaging. While this constitutes a limitation on the presented method, we think that the axial PSF gives at least an excellent indication of the quality of a particular sectioned imaging system. Often the results will be more then sufficient for judging the relative imaging conditions between sessions or instruments with the ease of use and sensitivity of the method outweighing this limitation.

In Sect. 3.2, some basic aspects of confocal and 2-photon sectioned imaging by the scanning approach are described as an aid to the understanding of the sectioning imaging effects characterized by the presented method.

3.2
Imaging in Confocal and Two-Photon Scanning Microscopy

The image formation in confocal microscopy is governed by the confocal point spread function (PSF) formed by the product of the illumination distribution and detection sensitivity function distributions overlapping in specimen space. The former is given by the spatial distribution of the focused laser illumination while the latter refers to the spatial distribution of the probability that the fluorescence photons generated in the specimen by the focused laser excitation will in fact be detected and contribute to the imaging. Optically this distribution is represented by the back projection of the detection pinhole into specimen space.

Optimally the confocal PSF should be the product of ideal or diffraction limited illumination and detection distribution functions perfectly overlapping over the whole of the lateral imaging field both in the center as well as at the borders of the imaging field.

However, optical aberrations or alignment errors and often a combination of both may prevent this from being the case. For instance chromatic aberration in combination with off-axis aberration can cause relative walk-off of distributions, which were adjusted during alignment for optimal overlap in the center of the scanned image field. (Fig. 12). This then will result in a re-

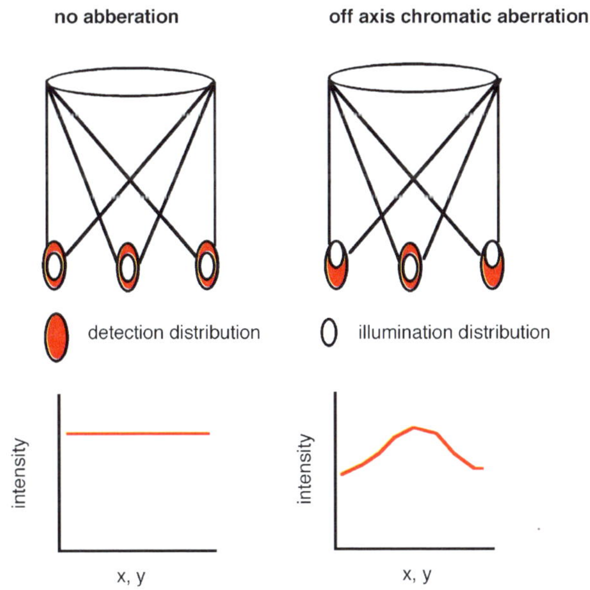

Fig. 12 Conceptual illustration of the walk-off due to chromatic aberration at off-axis scan-field positions between illumination and detection distributions and the resulting reduction in the detected confocal signal

duced confocal signal in the off-center regions. Also other parameters like the axial resolution may be similarly affected and often – see below – in an irregular manner over the imaging field.

In multi-photon microscopy the fluorescence generation in the specimen is proportional to the quadratic or higher power of the intensity of the focused excitation radiation in the microscope. Well focused, diffraction limited excitation distributions result in the highest multi-photon yield. This makes the fluorescence generation in this type of imaging sensitive to various on-axis and off-axis aberrations in the focusing of the excitation radiation during the scanned acquisition of a multi-photon image. As mostly no detection pinhole is employed, the situation on the signal collection side will be less critical.

A more extended treatment of both types of imaging has been written by Diaspro [16].

3.3 Sectioned Image Characterization, Principle and Analysis Parameters

3.3.1 Principle of the Method and Definition of the Axial PSF

The presented sectioned imaging characterization method utilizes a 3D image or data stack of a thin uniform fluorescence or reference layer, acquired through the standard 3D imaging routines as available in most confocal or two-photon microscopes. When the fluorescent reference layer is stepped through the confocal region in this routine the fluorescence signal at each lateral image point will track the axial dependence of the *laterally* integrated intensity of the confocal PSF, or "axial PSF," as further explained in Fig. 13.

It is essential in order to be able to measure the axial variations of the axial imaging properties with acceptable resolution that the layers used are reasonably thin with respect of to the dimensions of the axial point spread function. On the other hand a "too thin" layer will lead to lower signal to noise in the fluorescence data. With a typical axial PSF width under high NA conditions of around 700 nm we found that a layer thickness of the order of 100 nm proved a good compromise. The measured axial PSF will be in fact a convolution of the actual PSF. The increase in the apparent width due to the convolution of a layer of finite thickness will be approximately by a factor of $\sqrt{(1-(l/w)^2}$ with l the layer thickness and w the axial width of the PSF [17].

Similarly as wide field applications we have found that the fluorescent layers need to be laterally uniform to a high degree. Only then will the axial responses found at each x-y point do indeed represent a correct measurement of the axial PSF suitable for establishing the sectioned imaging characteristics at the various lateral points of the sectioned image. The layers, with a thick-

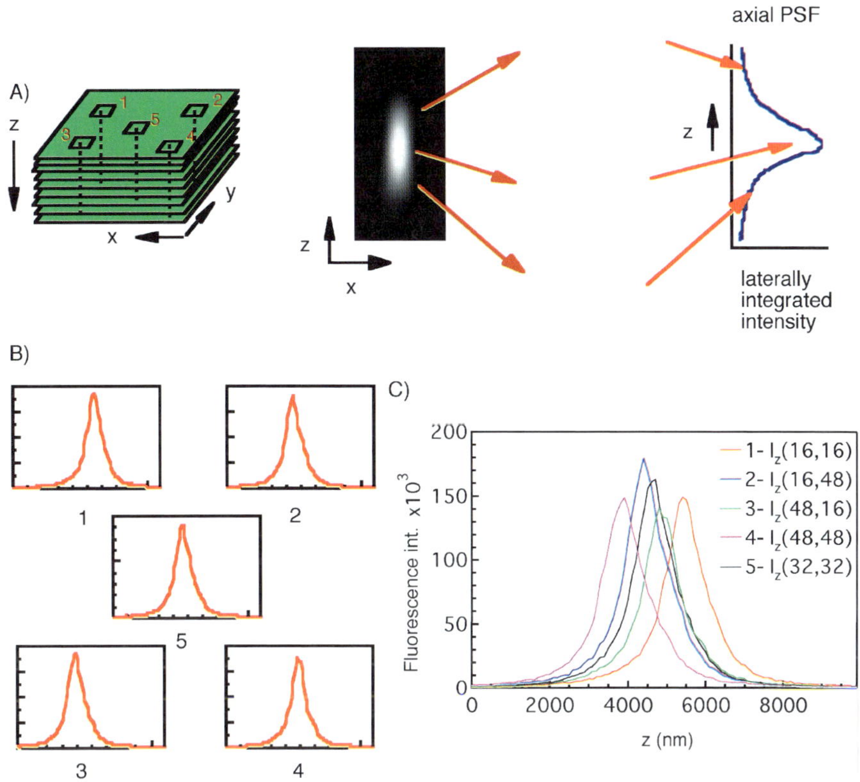

Fig. 13 The 3D image characterization is based on a 3D-data stack acquired by stepping a thin uniform fluorescence reference layer axially – i.e., along the *z*-axis – through the confocal region. As illustrated in (**A**) the set of values found at a particular x-y position as a function of z in the stack represent the axial variation of the laterally integrated PSF as sampled by the thin – about 100 nm – fluorescence reference layer. This set of values is called the axial PSF, the shape and amplitude of which will track the variations of the underlying PSF over the image scan field, as illustrated for field positions 1 to 5 in (**B**) and collected together in (**C**)

ness of ca 100 nm used for this application satisfy this condition with their fluorescence intensity and layer thickness uniformity similar to the ones described before [18]. With the layers sufficiently thin and uniform, the axial or z dependence of the fluorescence at each lateral image point in the 3D dataset of such a layer will in fact represent the axial PSF and can thus be used for characterizing the sectioned imaging at that point. Figure 13b and c show, as an example, taken from an actual measurement, the axial responses measured at 5 locations in the imaging field, showing that the actual axial PSF does vary over the imaging field. This is not unexpected in a beam scanning confocal instrument where the axial PSF may indeed be affected by off-axis optical aberrations in one form or another.

3.3.2
Analysis of the Axial PSF Properties

Various choices can be made to analyze these axial PSF responses in the terms of parameters. At present we have chosen the following, (see also Fig. 14):

I_{total} the total integrated intensity under the axial PSF response;
I_{max} the maximum fluorescence intensity found along the axial response;
Z_{max} the axial position at which the value of Imax is found;
fwhm the axial resolution as represented by the fwhm of the axial response;
skew s axial asymmetry of the axial PSF response.

For the purpose of this paper the skew s is defined as $s = (a - b)/(a + b)$ with a and b evaluated at the level of half maximum intensity of the axial PSF as indicated in Fig. 14. The sectioned imaging properties of a given system can conveniently be represented in a so-called sectioned imaging property chart or SIPchart (see Fig. 15a and b) based on the above parameters. As these parameters can be determined at each point in the lateral image field it is a logical step to represent the data in these charts in the form of color-coded images or maps. In addition the average and variation of the above parameters over the image can be calculated and are added as an inset in the respective color coded images. These values, summarized in a separate table, are useful for a numerical characterization of the imaging properties over the whole image field. The axial resolutions of the system, white against

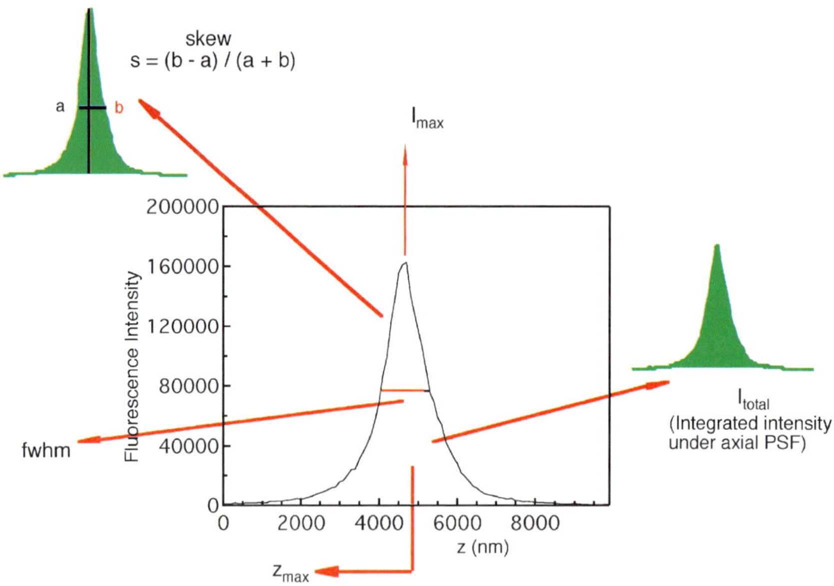

Fig. 14 Parameters for the characterization of the axial imaging characteristics of a sectioning microscope

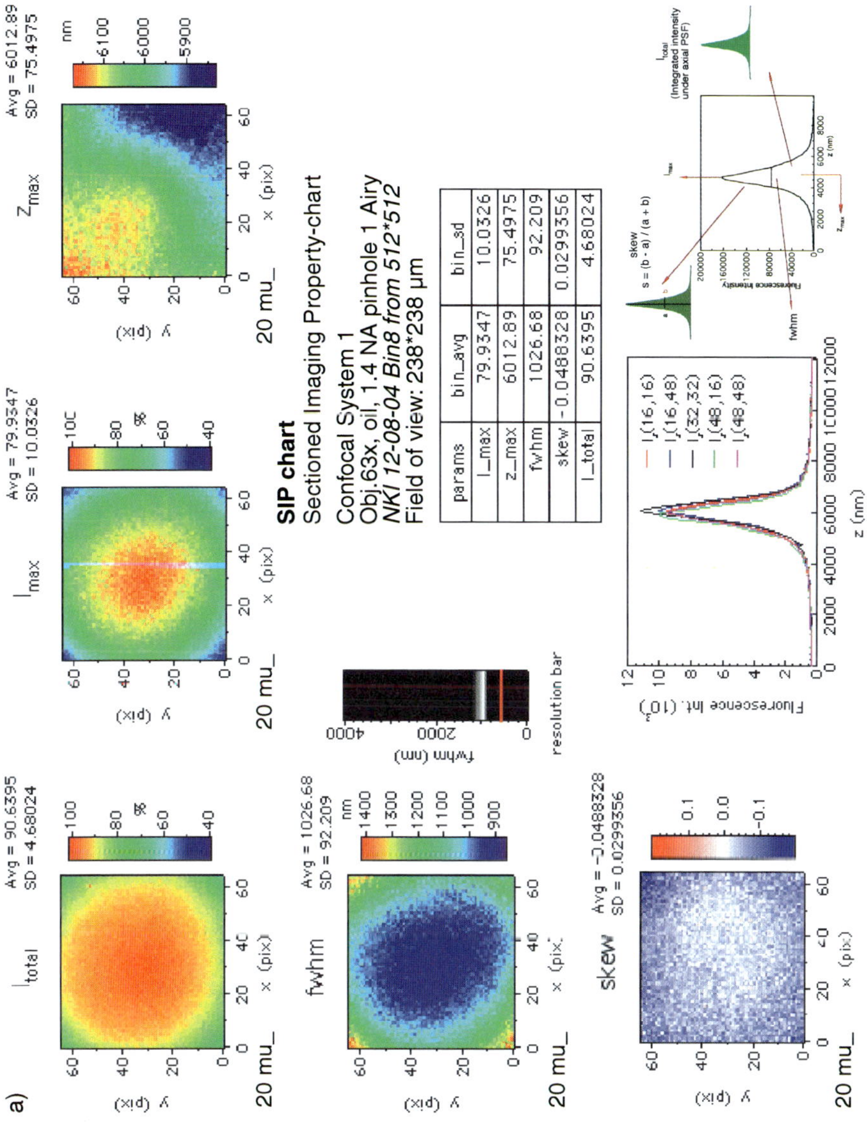

Fig. 15 (**a**) Sectioned Imaging Property charts or SIPcharts for two confocal microscope systems: SIPchart of confocal microscope system 1

the black background of the resolution bar, can be compared directly with the theoretical resolution – in red – to be expected at zero pinhole size and the numerical aperture NA of the used objective. Also included in the SIPchart are the actual axial responses measured in the center and 4 off-center locations. The SIPcharts shown are from an actual comparison of the sectioning conditions between 2 confocal systems as further discussed in the next sec-

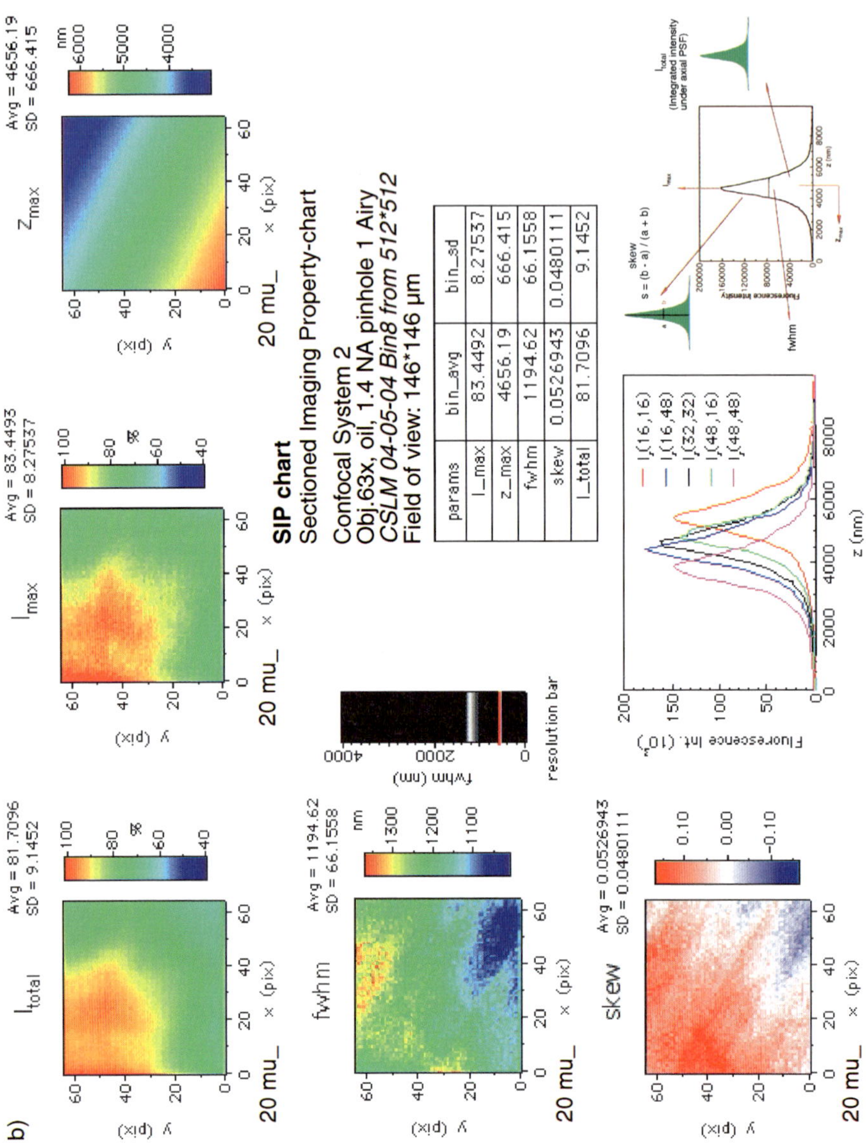

Fig. 15 (**b**) Sectioned Imaging Property charts or SIPcharts for two confocal microscope systems: idem of confocal microscope system 2. Both microscope systems are equipped with NA 1.4, 63× oil immersion lenses and operating at the same nominal pinhole setting of 1 Airy. See further text

tion. The color coded images are binned in this case to 64 by 64 from a set of images originally of 512 by 512 image points. As the various imaging properties can be assumed – and are in fact observed – to vary relatively slowly over the imaging field, this binning while improving signal to noise conditions does not cause any significant loss of information on the lateral variation of the represented parameters.

3.4 SIPcharts and 3D Imaging Assessment

The utility of SIPcharts for 3D image characterization is illustrated with an example based on SIPcharts taken from a comparison of two different confocal microscope systems 1 and 2 as presented in Fig. 15a and b, respectively. It should be noted that the point of this discussion is not to determine if one of the microscope systems is superior to the other, but to show that SIPcharts can be effective for evaluating and comparing their relative imaging properties. Both systems 1 and 2 are equipped with similar oil immersion lenses (63×, NA 1.4) and examined under similar settings for nominal pinhole (1 Airy) and zoom. For each measurement a 100 step z-scan with was made through focus. In both cases a zoom is chosen such that a – for this objective extended – scan field resulted: 238×238 μm for system 1, and 146×146 μm for system 2.

The panels I_{total} represent the integrated intensity along the axial response over the field and permit one to judge – together with the panels I_{max} – the degree to which the apparent fluorescent intensities in the confocal images are affected by not-optimal sectioned imaging. The I_{max} panel is useful to judge the maximum difference in apparent fluorescence in the separate sections due to microscope factors while integrated I_{total} panel has a similar function for extended depth or axially integrated images. With fluorescer distribution in the reference layers to a high degree laterally uniform, one would expect under ideal imaging conditions that the both the integrated I_{total} and the I_{max} images to show uniform fluorescence over the image field. That this is not the case is clear from a first glance at these panels. Looking in more detail it can be seen that the Imax panels of the SIPcharts of both systems show a variation in the maximum fluorescence intensity of 20% and 30%. For system 1 we see a maximum located around the center of the image field with the intensities falling off smoothly towards the edges. For system 2 a much more disordered, non-symmetrical, distribution over the image field of the axial PSF maxima is observed.

The Zmax panels show the axial positions at which the maxima shown in the Imax panels were found. In both cases we see that these are located in an approximately flat plane; however, these planes are not fully perpendicular to the optical axis but somewhat tilted by 300 nm (system 1) and 2600 nm (system 2), respectively, over the image field. The possible cause of these small tilts (up to 2% for system 2 over the image field) may be either a tilt of the specimen table with respect to the optical axis or an artifact connected to the optical scanning technique used.

Assuming that the observed fwhm values of the axial responses are close to and representative of the axial resolution then from the fwhm panels a good impression can be obtained of the resolution variations over the image field. We see that for system 1 areas with higher resolution correspond well with those with maximum intensities (I_{max}), as can be expected for a reasonable aberration free system. For system 2 this correspondence is not so clear-cut. In fact, the fwhm panel resembles the skew panel better then the Imax panel does. This suggests that aberrations in the latter system play an appreciable role in the image formation, as also witnessed by the much greater values for the skew and skew variation observed there. Also, comparing both systems, it is interesting to note that while in system 1 the average resolution is somewhat better than system 2, the opposite is the case for the resolution variations over the field. Thus system 2 has a more uniform resolution over the image field. The same is also the case for the fluorescence intensity variation as can be seen from the lower standard deviation of the Imax values for system 2. Of course, when making this judgement it should be noted that the imaging field shown of system 2 is appreciably smaller then the one of system 1.

The skew parameter s as defined in Fig. 14 is a parameter characterizing the first order asymmetry of the axial PSF and may be indicative of the presence of spherical or other optical aberrations. The severity or degree of aberration can and indeed often does vary over the image field. Comparing the data in the skew panels of the SIPgraphs of both systems very low skew values are seen close to 0 in the case of system 1 while in for system 2 a more irregular, somewhat striped pattern is seen with local skew values varying from 0.15 to –0.15.

The black resolution bar is useful to get an "at a glance" impression of the axial resolution and resolution variations of the system – the white band – in relation to the theoretically possible resolution – the red bar – at zero pinhole size. Finally in the SIPcharts the actual axial responses are given at 5 locations in the image field. With a binned image size of 64 by 64 the curve Iz(16,16) represents the axial response taken at point $x = 16$ and $y = 16$ etc. The availability – in the lower, middle panel of the SIPchart – of the actual responses is useful to recognize the presence of strong aberrations which sometimes cannot be effectively recognized from the fwhm and skew parameters values only.

System 1 and 2 represent two systems of major confocal manufacturers which were evaluated by the SIPchart method in the state we found them, including for instance, sub-optimal user alignment, etc. It is neither proper nor relevant for the purposes of this paper to further identify these systems, as the presented data are not necessarily representative of the imaging attainable with the instruments.

3.5 SIP-Charts, Analysis Examples, and Sensitivity

SIPcharts present a great amount of data to the researcher, which can serve subsequently as a convenient starting point for analysis of specific aspects of the imaging. SIPcharts and data extracted from them are effective and convenient tools for analyzing sectioned imaging conditions and are sensitive enough for tracking differences or changes in sectioned imaging conditions. Using the SIPcharts, various specific imaging aspects can easily be compared by directly extracting the applicable data/images from the SIPchart document for documenting the sectioning conditions under which, for instance, confocal images were acquired.

While for further examples of the use of SIPcharts in sectioning microscopy we refer to Brakenhoff et al. [19] we would like to include here one application illustrating the use of SIPcharts for tracking the influence of spectral conditions on sectioned imaging. Modern confocal microscopes can collect simultaneously or sequentially images at different excitation and detection wavelength settings. However, it is well known that image plane-shifts and other effects between imaging conditions may occur due to chromatic effects in the imaging. These can be documented very well with the help of the described procedures and the SIPchart representation. Figure 16A and B show the Z_{max} panels extracted from the SIPcharts of a microscope system acquired at two spectral settings: the first for excitation at 488 nm, using a detection band-pass filter of 503–530 nm and the second with a 543 nm excitation and 560–615 nm detection band-pass filter. The through-focus data stacks for both SIPcharts were obtained in one experimental run not changing the position of the reference layer (which in this case is based on a red fluorescing dye), only changing the filter settings. The full charts are shown to illustrate that many subtle differences may be noted between the imaging between both imaging conditions. Particularly important for work where data with a different spectral signature are correlated – as in the co-localization or FRET studies – is that not only the image planes between both conditions are found to be shifted with respect to each other but also that this shift is not uniform over the imaging field. Figure 16C, obtained by processing the Z_{max} data of these SIPcharts, illustrates this nicely.

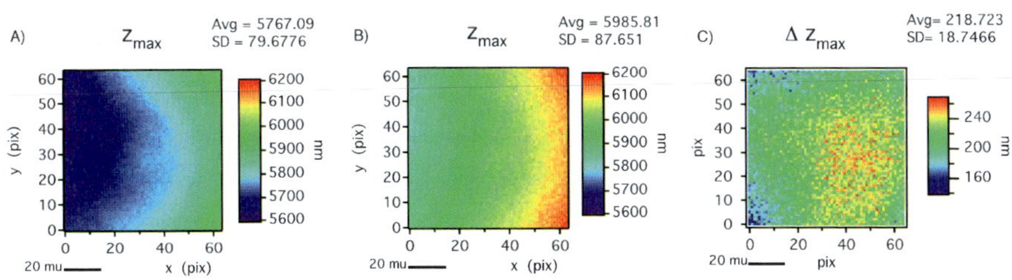

Fig. 16 Axial image plane position as a function of wavelength derived from SIPcharts of a multi-channel confocal microscope. Wavelength conditions: **A** excitation 488 nm, detection band-pass 505–530 nm, **B** excitation 543 nm, detection band-pass 560–615 nm. The data shown in panel (**A**) and (**B**) are represented on a common color scale. Panel **C** shows the axial height difference of the sectioned plane imaged at the two spectral conditions

4
Conclusions

We have shown in this chapter that using thin uniform fluorescent layers it is possible to do effective characterization and calibration in fluorescence microscopy. A major motivation behind effective calibration is that it would enable microscope users to derive quantitative specimen information from the primary fluorescence of their objects, to a first order independent of the microscope systems used. At present, quantitative data in microscopy are often determined by methods such as fluorescent life time, FRET [20] and FRAP [21], or ratiometric methods for measuring ion concentrations (Ca^{2+}, pH and others) [22].

Key to the presented approach is the availability of sufficiently uniform and reproducible layers as, for instance, produced here by spinning techniques.

The presented fluorescence reference layers may have significant value for characterizing microscope properties in general well beyond just their application in fluorescence microscopy quantification. For instance fluorescent yields under known illumination conditions allow microscope throughput or efficiency under various optical conditions to be assessed. Such illumination conditions can in fact be derived from layers with known uniform bleaching properties which were the basis of the bleach rate imaging demonstrated here. The bleach rate can serve as an environmental probe as the local bleach rate is known to be dependent on environment factors such as pH and molecular binding or as a proximity probe, the latter, for instance, through the mechanism that the mutual distance between excited molecule influences bleach probability [7, 23]. For the characterization and aligning of confocal microscopes presently different methods are employed. For instance, confocal microscopes are often aligned by maximizing the fluorescence yield from a slab of solid fluorescent material at the center of the image field. However no axial information becomes available for judging/optimizing sectioning conditions, possibly leading to sub-optimal instrument alignment. 3D imaging of fluorescent spheres can in principle give access to the full 3-dimensional point spread function, provided these beads are small in relation to the PSF. A limitation is that the small size and the hence limited number of fluophore molecules contained in these beads may make it difficult to obtain a sufficient fluorescence signal for accurate PSF determination before bleaching sets in. Also, we have found in practice that a substantial variation in apparent fluorescence between beads can be observed in many commercially available beads.

In contrast, thin uniform reference layers can provide axial PSF information at sufficiently low illumination conditions such that bleaching plays a minor role. Of course they do not provide access to the lateral PSF properties, but they do have the advantage that the laterally uniform layer fluorescence assures that fluorescence intensity variations related to instrumental properties are correctly mapped.

The SIPcharts, together with the underlying data could, in principle, be employed for correction purposes; correcting for the often observed variations in fluorescence intensity yield over the image field first comes to mind. We think the data contained in the total intensity image of the SIPchart can be used for an approximate first order correction.

Present day de-convolution algorithms are in general assuming a constant PSF over the image field. The data contained in the maps of the fwhm and skew variations in the SIPchart can be used to assess if this assumption is reasonably correct. Advanced de-convolution algorithms – incorporating PSF variations over the image field – can in principle be constructed. The presented skew and fwhm panels of the axial PSF can be a good starting point for such procedures.

Co-localization studies require accurate knowledge of relative axial positions of specimen elements imaged under different excitation and detection spectral conditions. Both on- and off- axis chromatic aberrations may cause shifts in the axial position at which these elements appear in the 3D image. By analyzing 3D datasets of the reference layer obtained at various wavelength conditions of a suitable reference layer we showed that the axial chromatic shift can be charted over the image field. We think that such shift data can be used for correction for such chromatic effects. At present lateral chromatic shifts cannot yet be tracked with the laterally uniform reference layers, but we are considering approaches to overcome this limitation.

We found – not shown here – that the presented characterization method can be very effective for the evaluation of the relative performance of otherwise identical microscope objectives when mounted on the same microscope.

The present work was mostly done using fluoresceine based uniform thin layers with an optimum excitation sensitivity around an excitation wavelength of 480 nm, but usable in a range from 430 to 490 nm. Layers suitable for any excitation and detection range are under development, with already promising results. In fact the data on chromatic effects on imaging – Fig. 16 – were acquired using a more red sensitive dye in the layer.

In this paper it is shown that the quantification and correction of fluorescence imaging can be successfully realized in wide field fluorescence imaging. Extension to sectioned microscopy imaging, a subject we are at present working on, seems to be feasible and the SIPchart representation of microscope system properties may be a good starting point for realizing this goal.

Acknowledgements We would like to thank Wijnand Takkenberg for assistance with confocal image acquisition, Mark Savenije for his help with processing the data and preparing the actual SIPcharts, Sjors Wurpel for the original development of the SIPchart analyses routine and finally Lauran Oomen, Lenny Brocks and Kees Jalink from the NKI Institute for extensive discussions and use of equipment. This work was supported by Stichting Technische Wetenschappen, Utrecht, The Netherlands under project no. ABI 4859.

References

1. Ghauharali RI, Hofstraat JW, Brakenhoff GJ (1998) J Microsc 192:99
2. Ghauharali RI, Brakenhoff GJ (2000) J Microsc 198:88
3. Castleman KR (1979) Digital Image Processing. Prentice-Hall, Englewood Cliffs, New Jersey
4. Jericevic Z, Wiese B, Bryan J, Smith LC (1989) In: Taylor DL, Wang Y (eds) Quantitative fluorescence imaging and spectroscopy, vol B. Academic Press Inc., San Diego, California
5. Arbe A, Colmenero J, Monkenbusch M, Richter D (1998) Phys Rev Lett 81:590
6. Levitus M, Talhavini M, Negri RM, Atvars TDZ, Aramendia PF (1997) J Phys Chem B 101:7680
7. Song L, Hennink EJ, Young IT, Tanke HJ (1995) Biophys J 68:2588
8. Talhavini M, Atvars TDZ (1998) J Photochem Photobiol A 114:65
9. Talhavini M, Atvars TDZ (1999) J Photochem Photobiol A 120:141
10. Leufgen KM, Rulle H, Benninghoven A, Sieber M, Galla H-J (1996) Langmuir 12:1708
11. Kagener V, Möhwald H, Dutta P (1999) Rev Mod Phys 71:770
12. Walker D, Htun H, Hager GL (1999) Methods 19:386
13. Sandison DR, Williams RM, Wells KS, Strickler J, Webb WW (1994) In: Pawley J (ed) Handbook of confocal microscopy. Plenum Press, New York
14. Schrader M, Hofmann UG, Hell SW (1998) J Microsc 191:135
15. Agard DA, Hiraoka Y, Shaw P, Sedat JW (1989) Methods Cell Biol 30:353
16. Diaspro A (2002) Confocal and two photon microscopy. Wiley-Liss, New York
17. Bracewell R (1965) The Fourier Transform and its Applications. McGraw-Hill Book Company, New York
18. Zwier JM, Rooij GJV, Hofstraat JW, Brakenhoff GJ (2004) J Microsc 216:15
19. Brakenhoff GJ, Wurpel GWH, Jalink K, Oomen L, Brocks L, Zwier JM (2005) J Microsc 219:122
20. Jares-Erijman EA, Jovin TM (2003) Nat Biotech 21:1387
21. Sprague BL, McNally JG (2005) Cell Biol 15:84
22. Fricker M, Runions J, Moore I (2006) Annu Rev Plant Biol 57:79
23. Ha T, Xu J (2003) Phys Rev Lett 90:223002

Springer Ser Fluoresc (2008) 6: 55–88
DOI 10.1007/4243_2008_027

Published online: 30 April 2008

Quantitative Fluorescence Microscopy: Considerations and Controls

Karl Garsha

Roper Bioscience Advanced Microimaging Group, 3440 E. Britannia Drive,
Tucson, AZ 85706, USA
kgarsha@roperscientific.com

Abstract It is important to have a working awareness of the many factors that can enhance, degrade or even distort the interpretation of quantitative data. Measurements in fluorescence microscopy may be discussed in the context of three major headings: (1) intensity, (2) spatial, and (3) temporal. The quantitative ability of instrumentation in each dimension is dependent on the performance characteristics of the instrument subsystems that contribute to the data gathering. In order for accurate and precise data to be recorded, not only must each subsystem perform well on its own merits, they must all be carefully orchestrated to work together in synergy. A number of basic considerations regarding the quantitative application of imaging instruments is outlined in the pages that follow; topics covered include detector technologies, illumination sources, optical limits, scan raster, specimen positioning and multi-channel acquisition.

Keywords Axial resolution · Bit-depth · CCD · Chromatic aberration · Dark noise · Dynamic range · EMCCD

Abbreviations

2D	Two-dimensional
3D	Three-dimensional
ADU	Analog/digital units
AOTF	Acousto-optical tunable filter
CCD	Charge-coupled device
CV	Coefficient of variation
EM	Electron multiplication
EMCCD	Electron multiplication charge-coupled device
F	Excess noise factor
FRET	Förster resonance energy transfer
FWHM	Full-width at half-maximum
ma	Mean pixel intensity
NA	Numerical aperture
PMT	Photomultiplier tube
QE	Quantum efficiency
sa	Standard deviation about mean pixel intensity
S/N	Signal-to-noise
λ	Wavelength
η	Refractive index

1 Introduction

Images can deliver an enormous psychological impact and so play a unique role in scientific communications. In spite of the convincing nature of image data, the weight of evidence reflecting such quantitative measures as protein concentration, co-localization, Förster resonance energy transfer (FRET), shifts in emission spectra, and identity of ambiguous fluorescent signals is dependent on both the methods used to acquire and analyze data as well as the accuracy and precision of a multitude of instrument functions. For today's advanced quantitative methods it becomes imperative to be aware of the many subtle factors that may contribute to measurement accuracy and precision.

There is an extremely rich variety of applications that fall under the topical heading of quantitative fluorescence microscopy. Much of the growth has been in the context of biological microscopy, and a number of factors have contributed to this recent explosion in popularity of fluorescence imaging. The development of molecular techniques and cloning technology has provided the prerequisite knowledge for deciphering the information content that drives living systems. The development of genetically encoded fluorescent proteins that can be used in living cells to provide a marker for genetic expression [1, 2] has revolutionized biology. Also significant are the development of advanced laser illumination, optical-sectioning technologies, efficient interference filters that permit unambiguous signal detection, real-time digital image capture and storage, precise and reliable instrument automation, along with effective computational processing and visualization of image data.

Today's researcher can directly monitor molecular interactions in living cells in multiple dimensions using fluorescence microscopy. Variables that can be quantified include lateral and axial spatial organization, shifts in the frequency distribution of fluorescent signal intensities [3–5], fluorescent decay lifetimes [6, 7], fluorescence polarization anisotropy [8, 9] as well as any temporal changes associated with these parameters. Sources of fluorescence contrast are under continual development and include such powerful tools as fluorescent chemicals for monitoring ion flux and membrane voltage potential [10, 11], genetically encoded fluorescent proteins that can act as targeted sensors of enzyme activity, ion flux, protein localization or gene expression [12, 13], second or third harmonic frequency conversion of incoming light by biological tissues [14, 15], and robust, functionalized semi-conductor nanocrystals of tunable emission wavelength [16].

The basic classes of instrumentation used for quantitative fluorescence microscopy include point-detection laser scanning microscopes as well as array detection scanning microscopes and array detection widefield illumination microscopes. Within each class is a plethora of technologies, each with its own strengths and weaknesses. To complicate matters, the term "quantitative microscopy" may be held to a wide range of interpretations. The level of rigor to which quantitative and semi-quantitative measurements are held may vary depending on the nature of the study.

A number of basic considerations regarding the quantitative application of imaging instruments is outlined in the pages that follow. It is important to have a working awareness of the many factors that can enhance, degrade or even distort the interpretation of quantitative data. Thus, it is helpful to confirm performance tolerances in the laboratory to ensure acceptable instrument performance and to assign a meaningful margin of error to the data generated.

Measurements in fluorescence microscopy may be discussed in the context of three major headings: (1) intensity (usually expressed as a function of spectral frequency), (2) spatial (three dimensions: x, y and z), and (3) temporal. Clearly, the quantitative ability of instrumentation in each dimension is highly dependent on the performance characteristics of the numerous instrument subsystems that may contribute to the data gathering. Oftentimes, meaningful data includes measurements from numerous dimensions (e.g., three spatial dimensions, one temporal dimension, three emission spectral scalars). Thus, in order for accurate and precise data to be recorded, not only must each subsystem perform well on its own merits, they must all be carefully orchestrated to work together in synergy.

2 The Intensity Dimension

2.1 Photon Detector Technologies

Intensity measurements are influenced by the image sensor as well as the illumination source. The detection subsystem chosen for a particular quantitative fluorescence application can have a profound impact on the feasibility of an experimental approach as well as the quality of the data acquired. The primary classes of detector commonly used in fluorescence microscopy fall into two major categories: (1) charge-coupled device (CCD) cameras, and (2) photomultiplier tubes (PMTs). CCD detectors are generally used on wide-field illumination systems because they are array detectors and can capture images of many points within the field of view simultaneously. CCDs are also found in line-scanning devices. PMT technology lends itself to point-scanning microscopes in which the signal for each pixel is sampled separately as the point-illumination rasters across the field of view. Each technology has different strengths and weaknesses with regard to signal quantification. A brief overview of general concepts, widely used technologies and corresponding considerations in the context of image sensors for quantitative fluorescence microscopy are outlined below.

Photon detection is a quantum mechanical event; because of this there is an inherent uncertainty in the number of photons actually registered for each pixel in a given exposure. This uncertainty follows a Poisson distribution. This noise due to quantum uncertainty, termed shot noise, is a physical limitation and cannot be rectified. In a Poisson distribution, the number of recorded events (photons) varies about the mean with a standard deviation equal to the square root of the mean. In other words, the standard deviation about 100 photons counted is 10. The uncertainty of the measurement in the latter example is 10%. If only 16 photons are counted, then the uncertainty is +/−4 photons or 25%. Thus, accumulating more photons reduces the relative contribution of shot noise to the uncertainty of the measurement.

The signal-to-noise (S/N) ratio is a measure of the data quality produced by an imaging system with a given sample. In other words, the S/N ratio is a figure of merit that relates the measured signal to the total system noise at each pixel. The signal-to-noise ratio has an inverse relationship with the uncertainty of the data, i.e., a high signal-to-noise ratio implies a low uncertainty with regard to the brightness levels recorded.

The S/N ratio is reflected in the intensity distribution of pixels corresponding to a constant signal level. Thus, the S/N ratio can be conveniently gauged in terms of a percentage known as the coefficient of variation or CV [17]:

$$CV = (sa/ma) \times 100\%\,, \tag{1}$$

where *ma* is the mean pixel intensity and *sa* is the standard deviation of the pixel intensities in the measured region.

The dynamic range of an imaging system is defined as the ratio of the largest single pixel intensity that can be quantified to the smallest measurable intensity that can be quantified. Dynamic range is a property of the detection system and is independent of experimental measurements. Imaging systems with higher dynamic range are able to quantitatively detect very dim and very bright pixels within a single image. Dynamic range is a figure of merit and carries no units. Dynamic range for imaging detectors is equal to the full-well capacity (number of photoelectrons that can be accumulated before the detector is saturated) divided by the total system noise of the detector. When the gain and/or bias offset are manipulated, such as on a confocal microscope or variable gain camera, the dynamic range of the detector is affected. This is because the full-well capacity is effectively reduced by increasing the off-chip digitization gain, while the noise level remains stable or may even increase. The bit-depth of the image may stay the same, however it is important to realize that the many brightness levels depicted may not correspond to actual variations in signal level due to molecular concentration, but rather may be an artifact of statistical variations (shot noise) in the signal. This phenomenon is most commonly encountered where very low levels of photons are contributing to the overall signal, as in point-scanning confocal images [18].

2.1.1 CCD Signal Detectors

The advantages offered by CCD detectors with regard to quantitative fluorescence imaging are several:

1. CCD technology provides an array of sensors that capture all pixel coordinates of a two-dimensional (2D) image simultaneously; this ensures that the data at each pixel is captured at a single time point. Furthermore, because the data is captured in parallel, many more photons are integrated for each pixel in the time taken to acquire a single frame than for a point-scanning PMT-based system [19]; this generally yields a higher S/N ratio for a given exposure time. As a general rule, because of the increased integration time for each pixel provided by parallel capture, much lower illumination intensities can be used to acquire an image in a given amount of time and this is beneficial to studies of living cells (where phototoxicity can be a major concern).

2. Modern CCDs have comparatively very high (some >90% across the visible spectrum) quantum efficiency as compared to PMT technology; again this can serve to reduce the exposure time for greater temporal resolution and for reduced phototoxicity/photobleaching.
3. CCDs have been developed for quantitative microscopy that have very high dynamic range. High dynamic range increases the range of brightness values that can be quantified.
4. The relationship between the number of photons integrated by any single pixel on the array and the electronic signal that is subsequently quantified is inherently very linear. CCD technology has reached a high stage of refinement; in many situations CCDs have superior qualities as imaging devices where the quantitation of intensity levels is of major concern.

There are a number of classical sources of noise that impact the signal-to-noise ratio achievable with a given CCD camera in the context of a given sample. The primary sources of noise with regard to CCD detection are shot noise (introduced above), read noise and dark noise. In order to arrive at the total noise value for a camera system, the individual noise components are added in quadrature:

$$Noise_{\mathrm{Total}} = \sqrt{(Noise_{\mathrm{Dark}})^2 + (Noise_{\mathrm{Read}})^2 + (Noise_{\mathrm{Shot}})^2}\,. \tag{2}$$

The following section discusses some of these potential sources of uncertainty in quantitative data acquired with a CCD camera.

The camera bias is the current charge on a CCD sensor and the associated electronic offset. A bias signal results from biasing the CCD offset at slightly above zero analog-to-digital unit (ADU) counts. In other words, the camera bias signal is an initial signal already on the CCD detector before an exposure is taken. The reason for the bias is to ensure that a high enough intensity such that a negative number does not get passed to the A/D converter; the A/D converter on a CCD camera can only process positive values. The bias voltage is artificially inserted after the data is read off of the CCD and before the data is received by the A/D converter. Generally, the bias is set at the factory and is stable over the lifetime of a camera. The bias level should be proportional to the bit-depth of a camera, e.g., a 12-bit camera (4095 brightness levels) may have a bias level of 50 to 70 ADUs, while a 16-bit (65 535 brightness levels) camera would have a bias level in the neighborhood of 500 ADUs. The camera bias can be determined by taking a readout of the CCD with zero exposure time. The average pixel value in the resulting image represents the bias offset of the camera.

The camera gain refers to the number of electrons that are assigned to each stepwise increase in the brightness value of a pixel, i.e., gain is a conversion factor that relates the number of electrons gathered in a pixel to digital numbers (ADUs). For example, an 8-bit image is capable of displaying 256 brightness values ranging from 0 to 255. If each pixel on the CCD sensor array is capable of gathering and holding 512 electrons before saturation, then we can say that the full-well capacity of each pixel is 512 *e* –. Full-well capacity is generally proportional to the physical size of the pixels on a CCD array. An 8-bit analog to digital converter can divide the full-well capacity into 256 brightness levels by assigning 2 *e* – to each analog-to-digital unit (ADU). The CCD gain can be estimated by dividing the full-well capacity by the bit-depth.

Intensities given in ADUs provide the user with a convenient method for comparing images and for comparing data generated by different camera systems. For camera systems with continuously variable user-modifiable gains, the exact same gain setting must be used between images for any meaningful comparison to be made between images. There is a single gain setting that optimizes the dynamic range of the camera to the digitization bit-depth. For dim samples, the gain can often be increased to permit easier visualization of the signal, but the dynamic range of the camera will be reduced. Gain should only be increased if a higher intensity is required and other conditions, such as exposure time, illumination level, and/or binning factor, cannot be changed.

Dark current is caused by spontaneous creation and accumulation of electrons in the pixel elements (storage wells) of a CCD. This constitutive accumulation of electrons in the storage wells is caused by thermal energy in the CCD. The rate at which electrons are liberated and stored in the pixel elements is temperature dependent, and the total number of electrons that will contribute to dark noise is a function of the integration time of the exposure at a given temperature. Thus, dark current is usually reported in *e* –/pixel/sec. Dark current noise follows a Poisson distribution. The dark noise is the square root of the dark current value. This is an important distinction: the dark current can be subtracted from an image, but the dark current noise (variability of the dark current) will remain. Thus, the best cameras should (and do) have very low dark current, and hence, very low dark current noise. The rate at which dark noise is accumulated is reduced by 50% for every 6.7 degrees Celsius reduction in temperature; for this reason, most research grade CCD cameras are deep cooled using peltier elements. CCD cameras intended for especially long integration times may even be cooled with liquid nitrogen. An image taken at a given exposure time with no light going to the camera will include the dark current as well as the bias offset. If a bias offset image is acquired (see above) and subtracted from an exposure taken with no light going to the camera, the average value of the difference between the two images will represent the dark current. If this average value is multiplied by the camera gain and divided by the exposure time (expressed in seconds), the result will be the dark current expressed in *e* –/pixel/sec. This value can be compared to a spec sheet provided by a camera manufacturer to confirm the dark current performance.

Read noise is the noise component that is attributed to the camera electronics during readout of the image. Most read noise is introduced in the output amplifier and preamplifier when the signal is boosted before analog to digital conversion. Careful electronic design and slow readout speeds can minimize the contribution of read noise. In general, a slow readout of a CCD array will generate low read noise, and higher read noise results from faster readout of the pixel values. At fast readout rates, under low light conditions, the readout noise may exceed the shot noise. Under such conditions we can say that the data is read noise limited.

The total read noise for a camera can be evaluated by taking two bias images (zero exposure time) and subtracting one from the other. The mean value of the resulting difference image is then multiplied by the camera gain and then multiplied by 0.707. This converts ADUs to electrons, and the result is the total system read noise expressed in $e-$. This value can then be compared to a camera spec sheet.

Linearity means that the relationship between the light level incident on the CCD and the signal that is digitized by the A/D converter has a linear relationship. In other words, when the light level is exactly doubled, the signal recorded by the camera should be exactly doubled as well. Well-designed CCD cameras typically have linearity deviations of less than 1% over the entire well.

Linearity can be measured using a stable light source by acquiring exposures at increasing integration times. Provided that a dark image (image taken at the given exposure time with no light incident on the CCD) is acquired and subtracted from each image of the illumination field, a plot of exposure time vs. signal intensity should yield a straight line.

A relatively recent development in CCD technology is the introduction of on-chip multiplication or electron multiplication (EM) gain technology. Such cameras are commonly referred to as EMCCD cameras. EM gain technology multiplies the photon-generated charge from each pixel on a CCD array to a level above the read noise; this permits signal detection at low light levels and high pixel readout rates, a regime that would not be possible using conventional CCD technology. This is because the read noise would dominate the image and cause excessive uncertainty in the data. This signal boosting process occurs before the charge reaches the on-chip readout amplifier, effectively reducing the read noise by the gain multiplication factor. This charge multiplication factor can be over $1000\times$. Despite clear advantages where high-speed, low-light imaging is concerned, EMCCD technology has some additional complexities that can impact the level of uncertainty in signal quantification.

The principle difference between a charge-multiplying CCD and a traditional CCD is the presence of a special extended serial readout register. This special serial register applies a high clock voltage. Because of this high clock voltage, the signal electrons generate secondary electrons with each clock cycle through an impact ionization process. The level of gain can be controlled by increasing or decreasing the clock voltage applied to move the electrons towards the readout register; the gain level is exponentially proportional to the voltage. The overall gain factor achieved through the impact ionization process can be greater than 1000×.

On-chip multiplication gain is useful only up to the point of overcoming read noise limitations. Traditional slow-scan CCDs with sufficiently low read noise achieve a better S/N ratio in the shot noise limited regime. The shot noise limited regime occurs at higher illumination levels or situations where longer exposures to accumulate many photons and/or slow readouts can be used. Because of this fact, EMCCDs have been developed with dual readout registers: an EM gain register as described above, and a conventional slow-scan readout register. Depending on the situation, the end user can select the appropriate readout technology on such cameras to permit maximum flexibility.

On-chip multiplication gain is a complex function of the probability of secondary electron generation and the number of pixels in the multiplication register. Mathematically, the gain function can be represented as:

$$Gain = (1 + g)^N , \quad (3)$$

where N is the number of pixels in the multiplication register and g is the probability of generating a secondary electron. The probability of secondary electron generation is dependent on the clock voltage being applied and ranges between 0.01 and 0.016. Because of the large number of pixels in the gain register, the total gain can be quite high even though the probability of secondary electron generation is relatively low.

On an EMCCD, the sharp inflections of the multiplication register clock waveform occasionally generate a secondary electron even if no primary electron is present. The probability of this phenomenon also increases slightly (along with the probability of secondary electron liberation by primary electrons) as temperature is decreased. This anomalous secondary electron creation is known as spurious charge. Spurious charge is usually added to the dark current to arrive at a total dark related signal in an EMCCD. Typically, a single spurious electron is generated for every 10 pixel transfers in the gain register; this yields a spurious charge value of 0.1 e –/pixel/frame. For example, for an EMCCD with 1.0 e –/pixel/sec dark current at a 30 ms exposure, the total dark related signal would be 0.133 e –/pixel/frame (0.033 e –/pixel/0.030 s dark current + 0.1 e –/pixel/frame spurious current).

Because EM gain is a probabilistic phenomenon, there is an inherent uncertainty associated with this form of gain. This uncertainty of secondary electron generation through the gain register is quantified by the excess noise factor (sometimes referred to as multiplicative noise). Experimental results show that the excess noise factor is between 1.0 and 1.4 for levels of on-chip multiplication gain as high as 1000× [20, 21]. In noise determinations, both the shot noise and the dark noise are multiplied by the EM gain and the excess noise factor. This additional uncertainty creates a situation analogous to the quantum efficiency of the EMCCD detector being reduced by approximately half [20, 21]. In order to make up for this limitation, the most effective EM-

CCD cameras use backthinned chips with very high (above 90%) quantum efficiency.

The determination of total system noise on EMCCDs is different than that for regular CCD technology. In an EMCCD, total system noise is given by:

$$Noise_{\text{Total}} = \sqrt{(Gain \times F \times Noise_{\text{Shot}})^2 + (Gain \times F \times Noise_{\text{Dark}})^2 + (Noise_{\text{Read}})^2}\,, \quad (4)$$

where *Gain* equals the system electron multiplication gain and F equals the excess noise factor. The corresponding signal to noise ratio is given by:

$$S/N = \frac{Signal \times Gain}{\sqrt{(Gain \times F \times Noise_{\text{Shot}})^2 + (Gain \times F \times Noise_{\text{Dark}})^2 + (Noise_{\text{Read}})^2}}\,. \quad (5)$$

2.1.2 PMT Detection

On laser scanning microscopes, only a few photons are generated in the time frame to sample a single pixel. These photons strike the photocathode of a photomultiplier tube (PMT); only a small fraction of these incident photons generate photoelectrons in turn. These photoelectrons can then be amplified by a factor of about 1 million by charge multiplication through the PMT dynodes. The signal emerging from the PMT is digitized under control of a clock signal that divides the time it takes for the laser to scan a single line of the image into the appropriate number of intervals for the number of pixels in each line.

Photomultiplier tubes (PMT) acquire light through a glass or quartz substrate covering a photocathode; the photocathode then releases electrons to be amplified by dynodes. The composition of the photocathode has a large influence on the spectral response, quantum efficiency, sensitivity, and dark current of a photomultiplier tube. Most photocathodes in the visible range are less than 30% efficient. Photomultiplier tubes have the advantage of a very high gain through the generation of secondary electrons by the dynodes, and very fast response. For this reason, PMTs are useful in the context of point-scanning devices, in which the number of photons generated per sampling period (pixel) is low, and the sampling periods are kept short to minimize acquisition time and photobleaching.

It should be noted that many factors might influence the signal as measured through the detector on a laser scanning microscope; in order to attribute any variation between measurements to the detection subsystem, all of these factors must be held constant [21]. Furthermore, the "noise" level in a confocal microscope has two fundamentally different major components: the noise level due to spontaneous generation of photoelectrons (dark noise), and photon (shot or Poisson) noise. Both dark noise and shot noise may become convolved with multiplicative noise (the excess noise factor due to uncertainty of amplification). PMTs may vary in their quantum efficiency (QE), gain response, and dark-count rate, even within the same model and manufacturing lot. PMT technology is primarily limited by contributions from shot noise and dark noise; read noise can generally be assumed to be negligible. Strategies for evaluating the relative contributions of dark noise and shot noise under standardized conditions are described below.

When assessing dark noise, stray light should be prevented from entering the detectors. Using standardized settings for PMT gain and offset, a single-scan image is collected; PMT noise will be displayed as single high-intensity pixels (or sometimes 2 or 3 bright pixels, always oriented in the direction of the scan line) scattered randomly throughout the image. Changing the confocal zoom setting will not alter the presence or size of these high-intensity noise pixels.

When imaging a fluorescent specimen of relatively low quantum yield, a great deal of non-uniformity is often evident within an image. This non-uniformity mainly reflects the statistical uncertainty inherent in the detection of any photon signal (shot noise). The histogram of the intensity levels within such an image provides an intuitive way to visually assess this noise level (a broader spread in intensities results in a broad histogram). For an accurate assessment, the signal source must be uniform, i.e., no visible features and no shading such as might be caused by signal loss at the edges of the field of view. This condition is more easily met by using a moderate level of zoom. The FWHM of this intensity histogram is a description of the noise in the image. The width of the distribution can be expressed quantitatively by calculating the coefficient of variation (CV) for the image as outlined previously.

It may be useful to measure and record CV values over a range of signal intensities. For instance, to describe the relationship between quantum yield and the sampling uncertainty at fixed laser power, a standardized laser power can be used to image increasingly dilute fluorochrome solutions; the detector gain may be raised proportionately to achieve a standardized mean intensity value. Mean pixel value and standard deviation about the mean are recorded for the dilution. These values can be used to calculate the CV for the image. Values for a range of dilutions can be recorded in this manner to provide a record of relative uncertainty over a range of quantum yields. An analogous test uses a sample having a standardized quantum yield and monitors CV as a function of the laser power [17]. For either test, all other variables such as scan speed, resolution, zoom, optics, dichroics and filters (alternatively beam splitter and spectral detection bandwidth settings), pinhole(s), beam expanders, PMT type, and laser power should be standardized and held constant [22].

The relationship between fluorescence quantum yield and the average PMT signal produced at a given gain voltage should be understood where quantitative studies are concerned. In this case, gain is set to a standardized level and left at that setting. Mean signal is next plotted as a function of either fluorochrome concentration (using a standardized laser power) or illumination intensity (using a standardized fluorochrome concentration). In either case, one would expect the relationship to be approximately linear except where: (1) fluorochrome concentration is at a point where quenching occurs, (2) there is chemical saturation of the fluorochrome solution, or (3) excessive laser power saturates the number of molecules that can be raised to an excited state. It is a good idea to ensure that the detector subsystem behaves as expected on a particular imaging system, and also to validate that the behavior is stable. This exercise also gives an idea of the dynamic range of the instrument in the context of a particular fluorophore.

2.1.3 Spectral Imaging Systems

Fluorescence microscopy segments differentially labeled features by virtue of the emission wavelength of the signaling moiety. For this reason it is important that quantitative instrumentation be able to accurately discriminate the spectral signatures of multiplexed probes; a variety of approaches are available, with the most sophisticated instruments being able to acquire high-resolution emission spectra from the sample at every pixel in an image. Evidence of the ability of such systems to solve problems such as distinguishing the identity of fluorochromes with overlapping emission profiles has been established [3, 4, 23–25]. It should be noted, however, that the integrity of such extrapolations is dependent on the accuracy and reliability of the underlying system [26–28].

Numerous approaches towards spectral imaging instrumentation are commercially available, both for point-scanning PMT-based systems and wide-field illumination systems equipped with CCD or EMCCD detectors. The advantages of a particular system will depend on the scope of applications an instrument is expected to handle. Strategies to evaluate the performance of such systems are of prime importance.

Three major aspects of the spectral imaging subsystem should be periodically evaluated: (1) the accuracy of the system in terms of the recorded location of spectral features, (2) the resolution in terms of the minimum bandwidth of spectral features that can be identified as discrete, and (3) the relationship of sensor efficiency with respect to wavelength.

As monochromatic light sources, the lasers installed on the system in question can be used as a convenient standard. In order to check the accuracy, resolution and wavelength response, multiple laser lines can be used simultaneously to provide a source of reflected light for measurement. An example of a spectral scan from a point-scanning system found to have miscalibrated spectral detectors by detecting reflected light from the laser lines is portrayed in Fig. 1. There are at least three potential advantages of using the instrument's integral laser illumination for spectral testing: (1) point source radiation provides a discrete cone of illumination so that both the lateral spectral–spatial resolution of the system, and the degree to which the pinhole

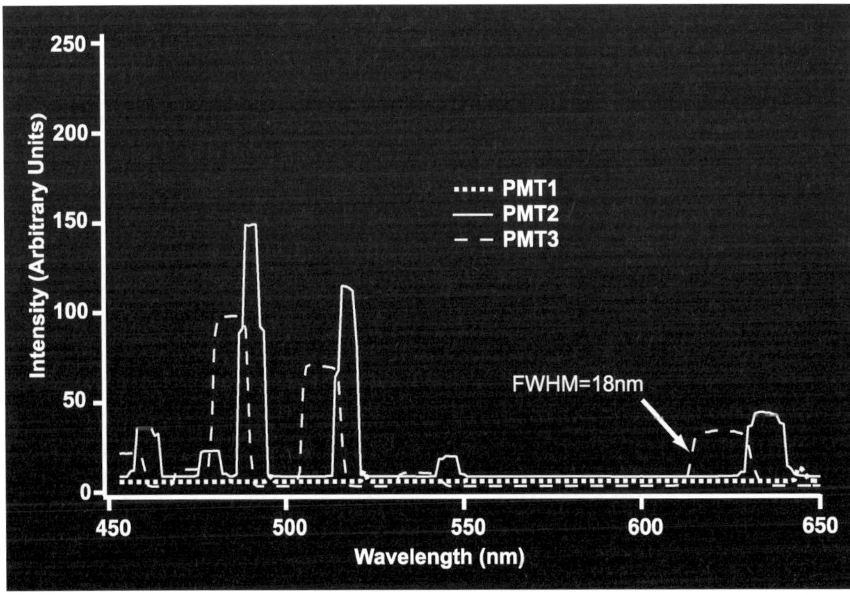

Fig. 1 Spectral scan over laser lines using backscattered light on malfunctioning point-scanning system. The wavelength selection mechanism for PMT 1 is inoperable, hence the low signal for that channel. The spectral reading for each channel is centered about a different wavelength and the spectral resolution approaches 18 nm in places. The laser lines corresponding to the peaks are 458 nm, 476 nm, 488 nm, 514 nm, 543 nm, and 633 nm

is effective in rejecting out-of-focus or scattered light from contaminating a spectral reading (axial spectral–spatial resolution) can be assayed, (2) the wavelength range between laser lines should have no signal, so baseline noise levels are easily evaluated, and (3) the power of individual spectral features (laser lines) can be easily measured and controlled.

An alternative approach is to use a calibration lamp standard in the manner described by Zucker and Lerner [28]. An ideal calibration lamp has numerous spectral features that form a sophisticated spectral fingerprint (Fig. 2); in theory, all instruments would be expected to reproduce the characteristic location, bandwidth and relative heights of peaks and valleys provided by the standard. The effect of sample aliasing on accuracy and precision is easily demonstrated using such a standard. This method isolates the detection subsystem from the integral illumination sources and may be useful where comparisons between different instruments are concerned, or where a system is not equipped with laser lines covering the full extent of the detection range. Traceable, well-characterized spectral calibration lamps based on elemental spectral peaks are available commercially. One cautionary note is that care should be taken when selecting a calibration lamp, however. It is a good idea to avoid lamps that depend on phosphors for some or all spectral

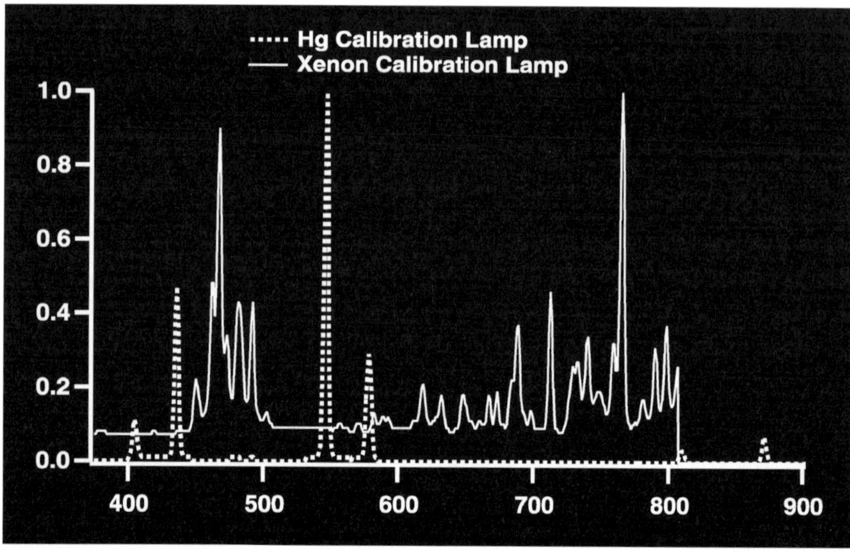

Fig. 2 Traces taken from a pure Hg calibration lamp (*dotted trace*) and a pure Xe calibration lamp using an EMCCD-based slit-scanning spectral imaging system. The Hg standard was used to calibrate the system because of the discrete peaks and high signal-to-noise ratio that can be achieved with this bright source. Once calibrated, a spectral recording of the output from a traceable Xe lamp standard (*solid trace*) was recorded and the peak locations noted. The peak locations for both lamps were determined to be consistent within 1 nm of the known values across the practical wavelength range for the instrument

features, the spectral emission profile of such fluorescent phosphors may be shifted due to contamination or temperature dependence.

2.2 Illumination

2.2.1 Broadband Arc Lamp Sources

High-quality fluorescence microscopy, and quantitative imaging in particular, is dependent upon stable and intense illumination sources. Most laboratory fluorescence microscopes rely upon mercury or xenon arc lamps for illumination. Unfortunately, traditional arc lamp sources have shortcomings that can compromise the integrity of quantitative measurements made in their context. Such light sources provide an intense broadband illumination source, but suffer from spatial and temporal instabilities [29]. For instance, the field illumination is not homogeneous and the intensity at the edges of the field can fall by a factor of two or more. The intensity distribution of conventional arc lamp sources for fluorescence microscopy is known to have hot spots that wander as a function of time; this in turn leads to spatial changes in the intensity of illumination at the image plane.

It is ideal to be able to reduce the impact of spatial and temporal inhomogeneity in illumination to below the limitations imposed by random shot noise; one proven strategy for accomplishing this is the use of light-guide delivery [29] to scramble the spatial variations in the raw lamp output. When properly implemented, light guide delivery delivers light with a very smooth intensity profile and has been demonstrated to provide greater than 100-fold improvement in spatial intensity variation over the field of view [29]. Remaining temporal intensity variations can be minimized using a closed loop approach in which the illumination intensity is sampled with a sensor and the lamp output is adjusted in real-time [29].

Recently, metal halide lamps for quantitative fluorescence microscopy have been introduced by the commercial sector. These modern broadband light sources simplify alignment, last many times longer than the traditional mercury arc lamps, have superior spectral outputs for imaging of green and red fluorophores, and have excellent stability over time. These optimized light sources are delivered through a liquid-filled fiber light guide to offer the advantages of light-guide scrambling.

2.2.2 Laser Illumination Sources

The consistency of results derived through the use of microscopy is dependent on predictable illumination levels. For this reason it is important to be able to confirm standardized illumination power levels at the specimen plane, and gauge the temporal stability of the illumination output.

Tracking long-term changes in power output requires the use of standardized settings. Because different objective optics exhibit different levels of transmission, it is important to consistently use a particular objective for power measurements. The same precautions apply in the context of widefield microscopy when measuring the power delivered from an arc lamp source. As a general rule, it makes sense to measure power through a dry objective of relatively low numerical aperture. This helps in that the power meter sensor can be located such that the angles of the incident rays are minimized and the cross-section of the cone of light exiting the objective is sampled in its entirety.

When measuring absolute average laser power it is important to keep in mind that readings taken during an active scan will usually fall well short of the actual value. This is because systems equipped with fast power modulation systems (such as acousto-optical tunable filter (AOTF) systems) may blank the laser beam during the flyback component of the scan raster, as well as during the short period between consecutive frames. Unless a facility is equipped with a relatively exotic high-speed chart recording power meter, the intermittent laser modulation will be averaged by the power meter, resulting in readings in the neighborhood of 1/2 to 2/3 of the actual value (this factor can fluctuate based on the relative speeds of the flyback and the speed of the forward raster portion of the scan cycle). On some systems, this can be alleviated through the use of a bi-directional scan at high zoom. In this type of scan, the sample is exposed during both the illumination phase of the scan and during the flyback. High zoom concentrates the laser power into a smaller area for easier gathering by the power sensor. The beam is usually blanked at the end of a frame for a short period of time, but the scan can be made slow enough that a reading can be obtained in the timeframe of a single scan using a slow dwell, high-resolution scan.

The method of determining laser power with the fewest contributing variables involves the use of "beam parking" or "point bleaching" features present on some platforms. In this procedure, software control provisions for bleaching a diffraction-limited point are used to designate a point in the center of the field of view, and a useful time of exposure (e.g., 30 s) is provided to the control software. The laser power is not modulated during beam parking; this is the ideal situation for power measurement. Several measurements should be taken to provide an idea of the precision of the data. In order to evaluate the laser line attenuation system, typically an AOTF, an extension of the basic power reading test may be used. Readings taken at different attenuation settings provide information on the linearity of the laser power response to AOTF gain, and they also provide data illustrating the amount of laser light that may be leaking past the AOTF when the attenuation is maximized.

Short-term temporal laser power stability can be measured with a power meter as well. An advantage of this approach is that the performance of laser power delivery system is isolated from problems related to the detection subsystem. The disadvantage to this approach is that rapid oscillations of laser output may be beyond the temporal resolution of the power meter.

Although less conclusive as a diagnostic measure, some may prefer to measure rapid power fluctuations by observing the impact of such fluctuations on a fluorescent sample under imaging conditions. For this test a standardized, stable fluorescent sample is used. In an effort to reduce contributions from photobleaching, the use of a freshly prepared, standardized dilution of fluorochrome in an index-matched solvent is helpful [24]. For those using inverted platforms, chambered coverslips with multiple wells work quite well as bulk fluorescent specimen holders. Carefully prepared concavity slides can be used as an alternative on upright microscopes. It is a good idea to centrifuge a test solution in order to remove particulates prior to its use as a standard. The microscope is focused into the bulk fluorescent standard near (but not at) the dye–coverslip interface. The procedure for recording data involves setting up a 2D time series with an appropriate acquisition interval and overall duration. Settings should be configured such that the laser power is standardized to a reasonable value, and the signal sensors should be adjusted such that the measured mean pixel intensity of the recorded field of view is about 3/4 of the available intensity maximum (e.g., 192 on a scale of 0 to 255) [20]. This helps to ensure that the PMT response to intensity variation is within the linear range. After acquiring the dataset, fluctuations in the fluorescent intensity can be expressed by plotting the mean pixel intensity as a function of time.

A caveat to the latter approach is that fluctuation of image intensity can be misleading as to the source; this is because recorded fluctuations may be due to a number of other subsystems (such as the PMT control board or spectral detector sliders on some platforms). Also, it may be tempting to use reflected light imaging instead of fluorescence in the protocol above. The intensity values recorded with the use of reflective samples can be exquisitely sensitive to slight movements of the z-positioning mechanisms [20]. For this reason, a movement in z that is only on the order of tens of nanometers can be wrongly assumed to represent fluctuations in power or efficiency of the detection system. Relatively thick fluorescent samples are more forgiving with regard to the latter problem.

2.2.3
Field Illumination

In order to perform experiments that presuppose a correlation between signal intensity and fluorochrome concentration, it is necessary to confirm an even illumination pattern across the field of view. This is easily accomplished using a fluorescent sea [30], fluorescent slides [31, 32] or fluorescent beads [33]. Alternatively, reflected light from a mirror standard may be used. For a sequence of increasing zoom levels, the fluorescence intensity near the coverslip interface is recorded for the field of view (FOV). In order for the distribution of intensity levels to be representative of the illumination intensity across the FOV, the highest and lowest pixel values should not exceed the dynamic range of the instrument, that is, bright pixels should not be saturated, and the lowest values should not drop below an intensity value of 0. To evaluate the data, a diagonal linear region of interest (ROI) is drawn from one corner to the opposite corner using image analysis software. A graph of intensity value as a function of position along the line is compiled (Fig. 3). When the field illumination is not uniform, it is possible that there is an alignment problem in the optical train [31]. An alternative explanation (in the context of point-scanning systems) would involve an uneven scan speed with relation to position.

Digital correction for residual uneven field illumination and calibration of brightness levels is possible in many cases [30, 34–36]. Such methods are based on the use of a uniformly fluorescent sample to record the field illumination properties (any flaws in the uniformity of detection will be convolved with flaws in the uniformity of illumination in the image of the reference sample). Classical field illumination correction involves dividing the image of

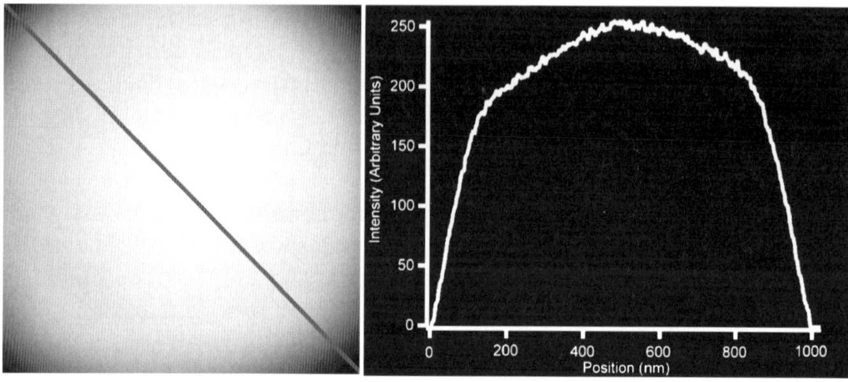

Fig. 3 Flatness of field. Deviation from an evenly illuminated field can be seen by plotting intensity as a function of position along the *diagonal line* drawn across the field of view. The intensity values have been normalized for purposes of illustration in this case

an experimental sample (with bias and dark current subtracted) by the image of a reference sample of even quantum yield across the field of view (again with the bias/dark current image subtracted first). The result of this ratio is multiplied by the mean intensity value of the difference between the reference sample image and the image representing dark current and camera bias. This method does not calibrate the intensity levels for comparison between imaging systems however [30]. In the manner presented by Zwier et al. [36], a fluorescent thin film composed of fluorescent polyvinyl alcohol polymer is spin coated onto coverslips to produce a very uniform fluorescent field with predictable bleaching characteristics. Such an approach permits correction for bleach-rate related imaging and provides methods for distinguishing the illumination distribution from the product of the illumination and detection pathways. Furthermore, such a standard sample can be used to characterize the imaging properties of an optical sectioning instrument [37]. In the methods introduced by Brakenhoff et al. [37], uniform fluorescent thin films (150–200 nm) are shown to permit standardized evaluation of axial resolution (see below), spherical aberration and off-axis chromatic aberrations in addition to illumination and signal collection uniformity.

3 The Spatial Dimension

3.1 Optical Limitations

Light has many fascinating properties, all of which may come into consideration during the design of sophisticated modern microscopes. Theoretical models that attempt to predict lateral and axial resolution of optical sectioning instruments under a variety of conditions have received much attention [38–49]. For a general approximation of what to expect with a given numerical aperture (NA), refractive index (η), and wavelength (λ), lateral resolution may be approximated by:

$$\frac{0.61 \times \lambda}{NA}, \tag{6}$$

and axial resolution by:

$$\frac{2 \times \lambda \times \eta}{NA^2}. \tag{7}$$

3.1.1 Sampling Frequency and the Nyquist Criterion

The resolvable detail in a digital image can be limited by the pixel sampling frequency of the detector. This phenomenon can be witnessed in situations where the physical size of the pixels on a CCD are larger than the smallest optically resolvable detail at the CCD, or when the number of pixels sampled for a scan line on a laser scanning instrument is inadequate. It has been demonstrated that the interval between the intensity measurements is less than half the period of highest frequency in the signal, and the original signal may be faithfully reconstructed from the digital values [18, 50, 51]. In other words, the pixel size at which spatial features are digitized must be at least half the size of the smallest optically resolvable unit at the image plane in order to take advantage of the optical resolution of a particular optics train. This concept is often referred to as the Nyquist criteria.

3.1.2 Lateral Resolution

In practice, the lateral resolution of a digital microscope is usually measured with either a subresolution point source standard or a special test slide [52, 53]. Subresolution point source standards can be made from a variety of sources. Fluorescent polystyrene beads measuring less than 200 nm, preferably less than 100 nm, can be purchased prelabeled with dyes suitable for measuring resolution at different wavelengths. When such standards are mounted in medium of the appropriate refractive index for an objective, a resolution test standard is created.

The lateral resolution is taken as the width of the intensity peak at 50% of maximum intensity on a plot of intensity vs. position for a linear region of interest taken through the center of the first-order intensity maximum (Fig. 4); this is known as the full-width at half-maximum (FWHM) value of the intensity profile. A projection through the z-axis of a volumetric dataset containing

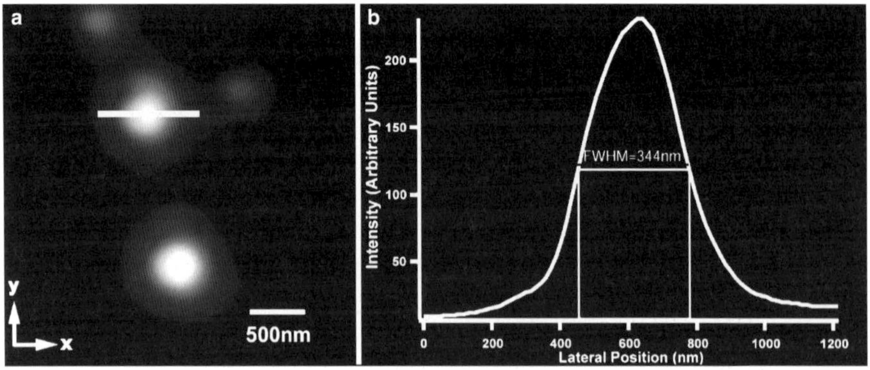

Fig. 4 Determining lateral (*xy*) resolution using fluorescent beads. The image in **a** represents the apparent lateral dimensions of a subresolution (100 nm) fluorescent bead taken on a laser scanning confocal microscope. The *line* is drawn through the center of the first-order intensity maximum and plotted with intensity as a function of position in the graph on the *right*. In this case the lateral resolution is taken as 344 nm. The objective used in this case was a 20× dry objective, *NA* 0.7, the excitation wavelength was 488 nm

a point source will ensure that the centroid of the intensity distribution is measured. The 3D diffraction pattern contained in such a dataset is known as a point-spread function (PSF) [39, 44, 46, 48, 49, 54]. Some drawbacks to using fluorescent polystyrene spheres include the fact that they may bleach under high zoom or high illumination intensity conditions, and the dye may leach into organic mountants (such as immersion oil) after a time.

Contrast and resolution are interrelated; optimal contrast is required to discern maximum resolution. Reflected light imaging is useful when values for ultimate resolution are desired or bleaching becomes a problem. This is because of the enormous contrast that can be created using reflected light with little concern for specimen degradation; very good signal to noise values can be achieved because of the proportionately large signal yield at a given level of illumination. An alternative point source standard for laser scanning instruments that can operate in reflected light mode makes use of colloidal gold and reflected light imaging.

3.1.3 Axial Resolution

Resolution with respect to the *z*-axis, or axial resolution, is a frequent concern held by users of optical sectioning instrumentation. When the instrument in question features an adjustable pinhole (or pinholes, as the case may be), it is prudent to take measurements at numerous pinhole settings all the way down to the smallest pinhole aperture (Fig. 5). By doing so, anomalies in the pinhole alignment can be discerned. Misaligned pinholes will yield poor results

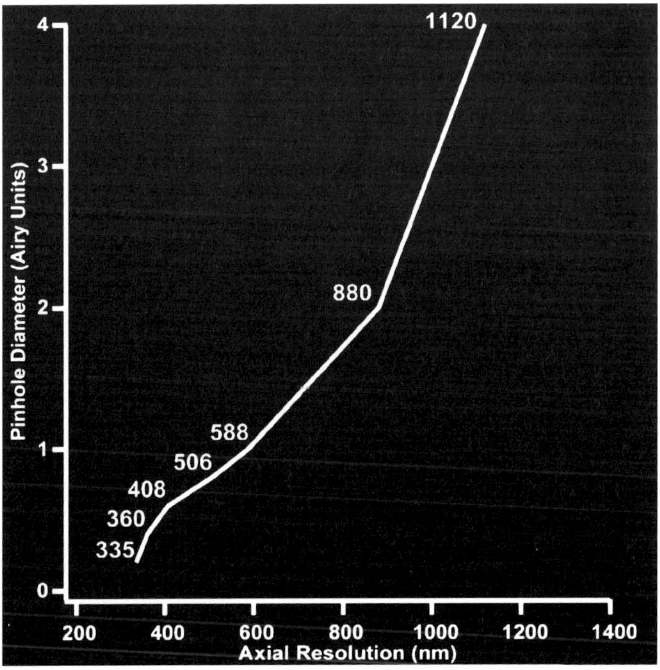

Fig. 5 Axial resolution as a function of pinhole diameter on a point laser scanning microscope. Measurements in this case were conducted using a front-face mirror standard with a 63× oil immersion objective at *NA* 1.32

for axial resolution, and the expected relationship among pinhole diameter, signal intensity, and axial resolution will likely be disturbed.

A variety of methods can be employed to judge *z*-resolution. The first is the use of subresolution fluorescent particles to generate a PSF [42, 54]. Similarly, a fluorescent thin film has been used in the context of measuring axial resolution using 4Pi microscopy [55]. Another approach involves measurement of the intensity component with respect to *z* of a mirror slide imaged with reflected light [33, 42, 56, 57]. Alternatively, a fluorescent sea [48, 54, 58], or a fluorescent plastic slide, may be used to provide a discontinuous fluorescent interface that is then imaged with respect to *z*.

In the interest of efficiency, it is possible to obtain both lateral and axial resolution values from a single high-resolution volumetric dataset containing a subresolution point source. This approach towards measuring axial resolution can utilize the same point source standards described above, and has the unique advantage that conditions can be adjusted to closely resemble the conditions under which fluorescent imaging is conducted. Once again, colloidal gold particles are highly reflective; they are a good option where bleaching or signal intensity proves to be a problem. The only major procedural difference from measuring lateral resolution is that measurement occurs along the *z*-axis. We must satisfy the Nyquist criteria: the resolution with respect to *z* hinges on the distance between sections (where a conventional *xyz* volume is acquired) or the distance between scan lines (where an *xzy* capable instrument is utilized). The *z*-resolution is taken as the FWHM of the axial intensity profile.

The front-face mirror test will generally yield more impressive results for axial resolution than can be achieved using subresolution particles [42]. Datasets are acquired in the same manner as above, and the axial resolution is taken as the FWHM of the first-order intensity peak (Fig. 6). The values

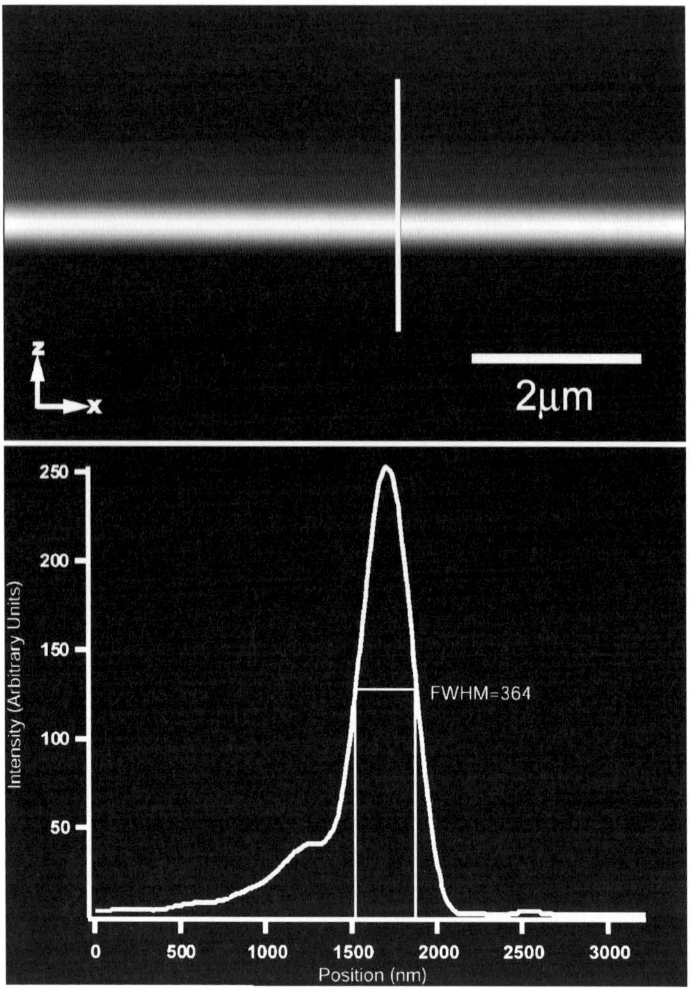

Fig. 6 Intensity profile through a front-face mirror standard using backscattered light on a point scanning confocal microscope. In this case, the FWHM is 364 nm. Measurements were conducted with a 63× oil immersion objective at *NA* 1.32, the incident wavelength was 488 nm, the confocal pinhole was at 0.5 airy units

obtained will also likely better reflect some of the theoretical models used to predict axial resolution on a laser scanning confocal microscope. It is unlikely, however, that users can expect equivalent axial resolution under conventional imaging conditions; this test is primarily useful as a comparison with theoretical models and an evaluation of the microscope against itself. Under practical biological imaging conditions, more often than not, slight refractive index mismatches, absorption, and scattering all tend to degrade both lateral and axial resolution. With samples that are relatively optically clear, spherical aberration and axial scaling due to refractive index mismatches can play a large role, thus, it is prudent to use water immersion optics for biological samples in an aqueous environment. Degradation of resolution and contrast due to spherical aberration, absorption and scattering tends to increase with depth of penetration into a given sample as a general rule.

3.1.4 Chromatic Registration

For multi-channel microscopy, it is important that the spatial registration of all channels can be confirmed. This is particularly true where experiments that seek to quantify co-localization and/or ratiometric measurements are concerned. Fortunately, there are simple tests that can evaluate the degree to which multi-channel registration may contribute to uncertainty in results. One effective approach is the use of a slide with subresolution calibration beads that emit at a variety of wavelengths. Probable sources of lateral chromatic registration error are found in situations where lasers delivered through separate fibers (or direct couplings) are used together, and/or when different optics such as dichroic beam splitters or beam expanders are used for separate channels in an automated sequential acquisition strategy. Lens aberrations or mismatched optics can be a source of lateral chromatic aberration. When testing laser lines that are delivered to the scan optics through separate couplings, it is advantageous if a multi-wavelength dichroic beam splitter can be used; this reduces the potential for ambiguity as to the source of any deviation (e.g., dichroic vs. laser-coupling alignment). By the same token, when beam splitter alignment or alignment of dichroics is suspect, it is advantageous to use laser lines that are delivered through the same fiber.

Axial chromatic aberration is best evaluated utilizing a front-face mirror test where reflected light imaging is possible. This test is very sensitive to chromatic aberration in the optical system, and even very small axial displacements of the focal plane between illumination wavelengths can be quantified [60–62]. Axial chromatic aberration is evaluated by plotting intensity as a function of z-position for multiple channels simultaneously (Fig. 7). A mismatch between the refractive index of the lens immersion and the specimen can exacerbate longitudinal chromatic aberration. Such discrepancies can be of major concern in the context of resolution-sensitive multi-channel experiments.

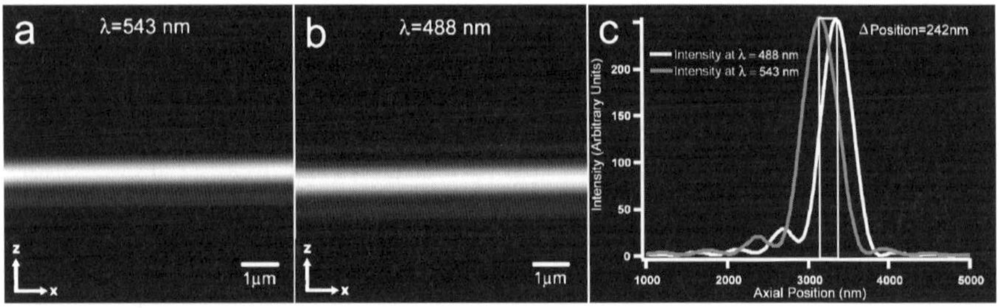

Fig. 7 Evaluating axial chromatic aberration. In this example, a mirror standard is imaged simultaneously in two channels; the peak-to-peak distance between the first-order intensity maximum for each respective channel is 242 nm (intensity plot on *right*). The laser wavelengths used to create this image were **a** 543 nm and **b** 488 nm, and a 63× *NA* 1.32 plan-apochromat, oil immersion lens was used

3.2 Scan Raster and Specimen Positioning

Accurate and precise spatial localization in three-dimensional (3D) space is a primary concern in quantitative fluorescence microscopy. In widefield systems with fixed arrays of image sensors covering the field of view, spatial measurements in the *xy*-plane are generally precise; calibration of the *xy*-axis using a camera of known sampling density is relatively straightforward [63]. Laser scanning systems are more complex because of the need to coordinate electromechanical movement with electrooptical sampling. Some approaches for verifying *xy* spatial calibrations are outlined below.

In the context of scanning instruments, spatial measurement and accurate morphometric classification rely on the accuracy and precision of the mechanisms used to move the focal volume through the specimen volume in the *x*, *y*, and *z* dimensions. From a practical standpoint, measures of the lateral and axial resolution on a digital imaging instrument are meaningless unless the accuracy and precision of pixel-to-pixel spacing in the *x*, *y*, and *z* can be verified. In the context of scanning instruments, non-uniform scan speed will also result in differential exposure of localized areas within the field of view [64]. This confounds accurate photometry and can result in increased phototoxicity when living specimens are being imaged.

3.2.1 Lateral Scanning Galvanometers

Standards for verifying lateral scan accuracy are easily purchased or fashioned from readily available components. A grid standard provides an efficient means of evaluating both the *x* and *y* scan rasters simultaneously. Examples of such standards include a 2000-mesh transmission electron microscope

(TEM) grid mounted in appropriate media. If a TEM grid (or other uncharacterized standard) is used, the accuracy and precision of center-to-center spacing between grid bars can be accurately determined on a properly calibrated widefield system. In order to do this, however, the possibility of lens aberrations should be taken into account. For this reason it is best to measure the center-to-center distances between grid bars using the central portion of the field of view and by moving each grid square into the position occupied by the square that was last measured. The objective used should be of sufficient numerical aperture and magnification, to permit diffraction-limited measurements of sufficient resolution to characterize the scanning performance of the resolution limit of the instrument in question. Variations on the use of a TEM grid mounted for transmitted or reflected light include mounting the TEM grid on a fluorescent plastic slide, or vacuum deposition of metal over a TEM grid on a coverslip to produce a negative grid (the resulting coverslip can be mounted to either a fluorescent plastic or glass slide), and the use of small optical slits or pinholes in place of a TEM grid.

Measurements of the center-to-center (side of one grid bar to the same side of the next) distance at different points within the field of view should be conducted for both *x* and *y* axes in conventional *xyz* imaging mode. It is important to realize the effect that undersampling can have on such measurements; sampling intervals for the scan resolution should satisfy the Nyquist criterion. A well-calibrated instrument will have accuracy within 1% of the known value and less than 1% variability between measurements. This should hold at both high and low zoom values, and across all scan speeds. An example of results from a well-performing instrument is depicted in Fig. 8. An example of noteworthy poor performance is documented in Fig. 9.

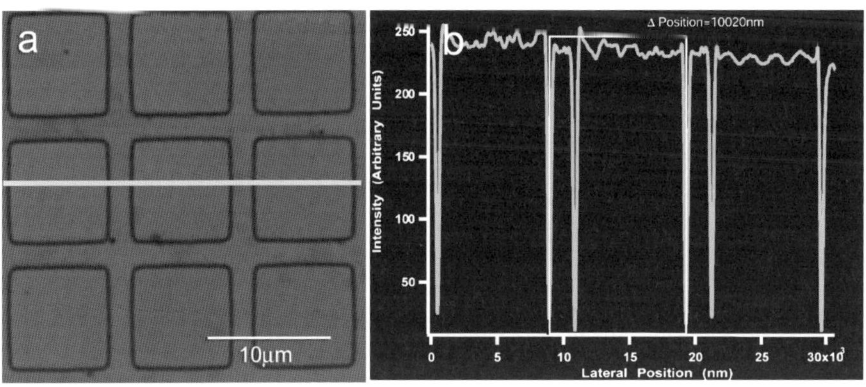

Fig. 8 Lateral scan accuracy and precision on a well-calibrated laser scanning instrument. **a** The standard in this case is a reflective etched silicon standard designed for reflected light microscopy. Each *square* is 10 μm per side. **b** A graph of intensity as a function of position on the *black line* in **a**. Quantitative measurements confirm the accuracy and precision of the *x* and *y*-galvanometers in this example

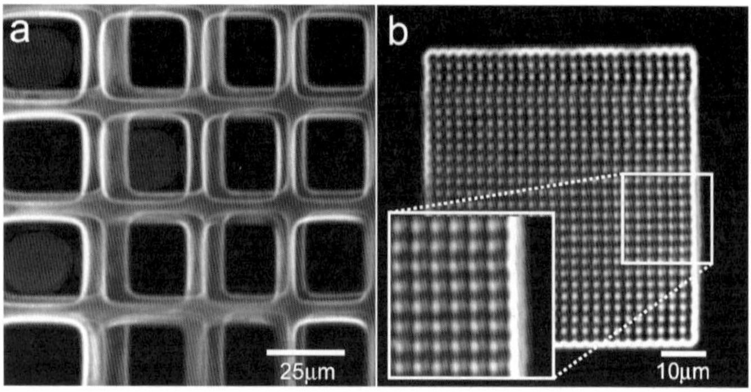

Fig. 9 Compromised lateral scan accuracy and precision on a laser scanning instrument. **a** Observation at low magnification indicates problems with both accuracy and precision between scans. Images from two scans taken in sequence are overlayed to show discrepancy between images of a stationary grid standard. **b** A scan taken at higher magnification on the same instrument. The standard in this case is milled into a coverslip surface with a focused ion beam (courtesy of Dr. Carlos Martinez). Again, the shifted overlay components of the image reflect the lack of precision (see *inset*). Wavering in the scan raster appears as distortion in the grid standard. The *square pattern* appears rectangular (narrower in the x-dimension) because of poor calibration of the x-scan galvanometer

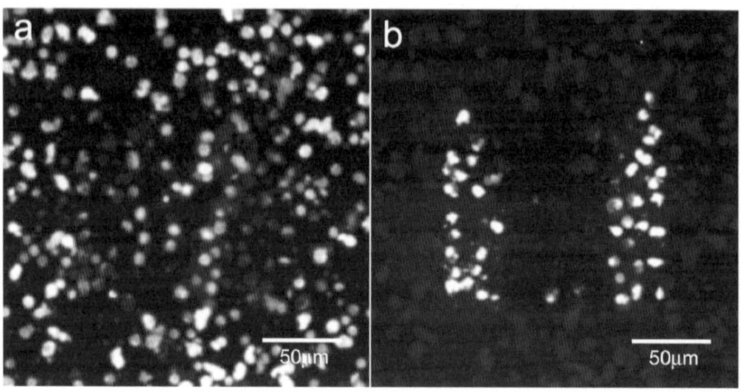

Fig. 10 Differential phototoxicity as a result of uneven scan speed across the field of view. In this image, metabolically active cells are stained with fluorescein diacetate (**a** green channel); the onset of propidium iodide (**b** red channel) indicates compromise of the plasma membrane associated with cell death. These images were captured at a lower zoom level such that the reason exposed during time-lapse is in the center of the field of view. The laser power was modulated by an AOTF to be attenuated during flyback so that the energy exposure per unit time should have been constant across a scan line. The speed of the scan raster in this case was deduced to be slightly slower at the edges than in the center of the field of view, over the course of a long time-lapse with frequent exposures, the impact of this slight deviation becomes increasingly significant

Some systems may be equipped with angular position sensor feedback mechanisms in the scanning galvanometers to control pixel sampling; this ensures accurate pixel-to-pixel spacing even when galvanometer speed is non-uniform. This measure does not prevent non-uniform image intensity and specimen damage, however (Fig. 10). Non-uniform exposure can be assayed using an easily bleached specimen such as Schott OG 530 glass [64] or a thin film of fluorescein applied to a coverslip surface. After repeated exposure (e.g., 10 scans) an image is recorded. Variations in intensity can be easily visualized on a plot of intensity as a function of position across the field of view in the direction of the scan. An image taken before repeated exposure should be used to control for possible intensity variations in the sample itself or the collection efficiency of the optics in different parts of the field of view. Non-uniform signal intensity in bleach-resistant specimens, such as a bulk dilution of fluorochrome in an index-matched solvent, can be quantified using the same type of plot. The pixel intensities from several sequential scans can be accumulated to get an idea of the additive effect of multiple exposures of the specimen to the scan raster.

3.2.2 Axial Focus Positioning

Evaluation of z-stepping performance requires that such measurements are performed under index-matched [42, 44, 46, 48, 64, 65] conditions. The materials in the optical path (standard and mountant) *must* match the design of the objective as closely as possible (Fig. 11). Dry objectives are designed to be free of spherical aberration under conditions at one z-position only: the sample coverslip interface. This is because the ratio of air to high index material in the optical path changes with depth into the sample. Axial measurements taken with dry objectives will not be accurate [64]. The disturbing disparity between z-axis measurements taken with a dry objective and an oil immersion objective are depicted in Fig. 12.

Boddeke et al. [66] describe a method for calibration of the automated z-axis of a widefield microscope that employs a test slide with a bar pattern mounted at an angle with respect to the object (xy) plane. In this method, one of the lines of the bar pattern (which lies perpendicular to the angle of the ramp) is brought into focus. The focus is determined objectively using an image processing algorithm based on a one-dimensional difference filter [67]. A movement of the tilted slide will cause a change in the lateral position of the in-focus line; this change is proportional to the tangent of the tilt angle. The z-axis motor step size is subsequently derived by determining the change in the in-focus z-position in the object plane as a result of a step in the z-direction.

An alternative approach to confirming z-calibration makes use of a carefully mounted 500-μm 90° microprism (Fig. 12). The microprism is a precision optical component and is manufactured to tight tolerances (angle toler-

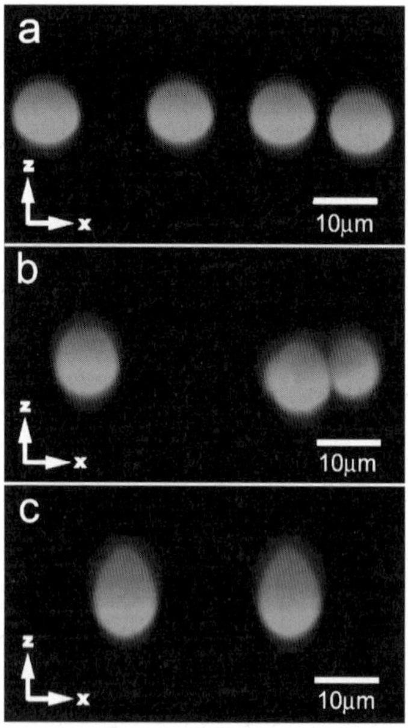

Fig. 11 The importance of refractive index matching with regard to measurements in the axial plane. **a** The 10.2 μm polystyrene spheres in immersion oil. The data maintain the correct aspect ratio, and accurate measurements in the *z*-axis are possible with an oil immersion lens. **b** The same beads in a classical 9 : 1 glycerol/PBS buffer mountant of $\eta = 1.43$ viewed with an oil immersion lens. Significant distortion in the *z*-axis is evident; this is due to contributions from spherical aberration as well as self-lensing. Accurate measurements with respect to *z* are not possible under these conditions. **c** The 10.2-μm polystyrene spheres in water ($\eta = 1.32$) viewed with an oil immersion lens. The distortion in the *z*-axis is exacerbated in accordance with Snell's law. The images in this case were acquired on an inverted microscope base, thus, the objective is located under the beads (below bottoms of images)

ance: +/−2 arc minutes; surface accuracy: 1/4λ). To prepare a *z*-calibration standard, the prism is mounted to a slide by the hypotenuse using a low-viscosity optical cement and the surface is rendered reflective through the use of a sputter coater or vacuum evaporator. Sections of wire measuring ~360 μm in diameter placed on two sides of the prism can be used to support a coverslip just above the prism apex. Optical cement (or another closely index-matched mountant) is drawn under the coverslip through capillary action by carefully feeding it to one of the open gaps between the coverslip and slide using a pipette. When the mountant is polymerized, such a sample provides an ideal and highly consistent standard for evaluating *z*-movement calibration using an oil immersion lens.

Measurement of *z*-movement calibration involves viewing the reflected light image of the prism from the side using an *xz* scan or by collecting a high-resolution *xyz* dataset and deriving the *xz* view. The divergence angle from

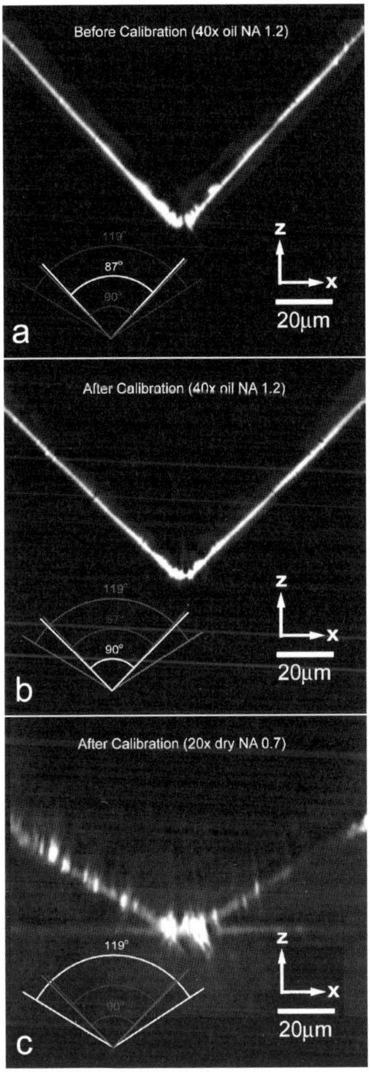

Fig. 12 Microprism standard for evaluation of z-movement calibration. Images **a** and **b** depict a 90° microprism standard [metal coated and mounted in acrylic ($\eta = 1.5$)] viewed in reflection mode using a 40× oil immersion objective (*NA* 1.2). **a** Represents the xz scan of a z-galvanometer stage before calibration; the prism angle measures 87°. This corresponds to a 12% discrepancy between the axial positions of the prism surface at the highest point within the field of view. **b** Depicts the same standard after adjustments were made to the calibration. The prism apex is determined to form the correct 90° angle. **c** Depicts the reflective microprism standard as viewed with a 20× dry (air immersion) lens (after calibration under index-matched conditions with oil immersion optics). Spherical aberration with dry optics is noteworthy, and the apparent depth of structures is fore-shortened up to 38% within the field of view. The coverslip/mountant interface is located just at the prism apex, barely visible in **c**

the apex should be 90° (Fig. 10b). If the prism angle is greater than 90°, then the z-movement mechanism is moving farther than it should (and vice versa); if the edges of the reflective interface are not straight, then it is likely that the speed of the z-movement is not constant through the range of motion.

In volumetric imaging or 2D time-lapse imaging it is important that the relationship between the position of the sample and the positioning of imaging optics is stable. In other words, when both the lens and the sample are expected to remain at their respective positions in z, they should; when one or the other is required to move to adjust the position of the focal plane in the sample, only the intended component should move. The tendency for the relationship between the nosepiece and the sample to remain stable can be assayed using the focus function approach described above [66], or, where reflected light imaging is an option, by using a mirror standard over a time course. Variations of this test can be used with fluorescence illumination; for instance, a test slide with subresolution fluorescent beads (on the order of 100 nm diameter) dried down to the coverslip surface and mounted in a semi-solid matrix of the appropriate refractive index can be used.

4 The Temporal Dimension

4.1 Multi-Channel Acquisition

The use of sequential acquisition of widefield images by means of automated electromechanical or manual filter wheels can introduce temporal artifacts in multi-channel images. For dynamic systems, molecules may drift from the area imaged by one pixel to an adjacent pixel in the time that it takes for the filter wheel to switch positions. An effective approach is to split the respective wavelength ranges using a dichroic mirror (or series of dichroic mirrors if more than two wavelengths are involved), and send the respective images to different areas of a single CCD camera (Fig. 13) or to multiple CCD cameras configured to capture in parallel. Such an approach alleviates the temporal distortion that can hamper high-speed multi-channel imaging on widefield systems and ensures that the images are obtained with the same capture parameters [68, 69]. A number of turn-key commercial solutions have become available for such applications.

4.2 Scan Raster Artifacts

Point-scanning devices sample each pixel in a 2D image in series, and it can be important to keep this in mind for dynamic imaging where temporal

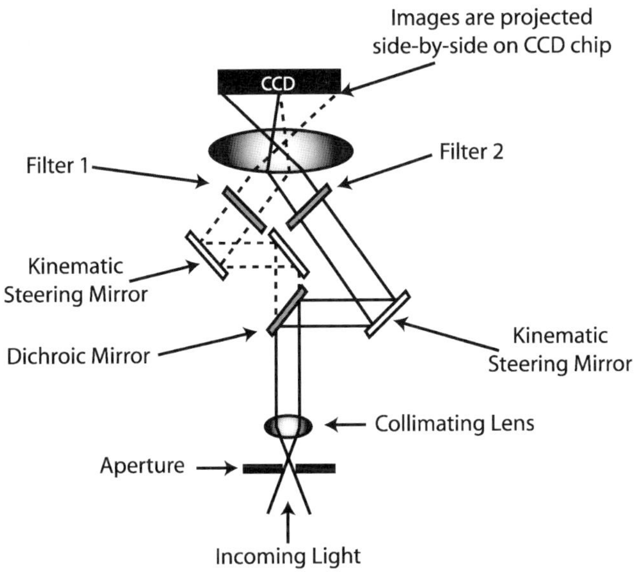

Fig. 13 Principle of operation for a dual-channel imager used with a single CCD. The light coming from the microscope base at the c-mount camera port is split into component wavelength channels using a dichroic mirror and the images for each wavelength are projected side-by-side on a CCD. This strategy can also be used to separate orthogonal polarization components and project the respective images side-by-side on a single CCD chip. Two noteworthy advantages of this strategy are that (1) kinematic mirrors can be used to adjust the image placement such that there is perfect alignment of the individual images with respect to one another, and (2) the images are acquired with perfect temporal registration. This is important for rapid dynamic events such as emission ratio imaging

resolution requirements approach the frame rate of acquisition. Each pixel is acquired at a different point in time, because of this, features at the end of the raster may represent events that occur at a later point in time from the features imaged at the beginning of the scan raster. When multiple channels are acquired, the use of multiple band dichroics in conjunction with rapid line-by-line sequential excitation will minimize the temporal shift between wavelength channels.

5 Discussion

Recent advances spanning only the past 20 to 30 years have fueled extremely rapid development and popular adoption of fluorescence microscopy as a practical and uniquely powerful tool for scientific discovery. The regular application of standard controls serves to demonstrate the outstanding capabilities of a well-maintained imaging system as well as provide meaningful information on the limitations of such equipment. Increasingly challenging applications for optical sectioning microscopy are rapidly evolving; however, the widespread adoption of rigorous methods for ensuring data integrity is still in its infancy. Powerful assays that measure such variables as relative concentration, co-localization, Förster resonance energy transfer, shifts in spectral emission, and the identity of ambiguous fluorescent signatures are dependent on the accuracy and precision of many interrelated instrument functions. In order to ensure quantitative repeatability, it is vitally important that data obtained in the context of a particular instrument have been confirmed to reflect the real-world situation.

To ensure maximum productivity, it is advantageous to identify anomalous instrument performance before the potential for grievous artifact or frank system failure becomes reality. Murphy's law dictates that there is a disproportionate chance such problems will be discovered at the onset of an important experiment or when results are needed in the face of looming deadlines. The considerations and controls in this manuscript are hoped to aid in realizing the goals of data integrity and peak performance for researchers using widefield and laser scanning microscopes in a variety of fluorescence applications.

References

1. Chalfie M, Tu Y, Euskirchen G, Ward WW, Prasher DC (1994) Science 263:802
2. Tsien RY (1998) Annu Rev Biochem 67:509
3. Dickinson ME, Bearman G, Tille S, Lansford R, Fraser SE (2001) Biotechniques 31:1272
4. Lansford R, Bearman G, Fraser SE (2001) J Biomed Opt 6:311
5. Zimmermann T, Rietdorf J, Pepperkok R (2003) FEBS Lett 546:87
6. Gadella TWJ, Jovin TM, Clegg RM (1993) Biophys Chem 48:221
7. Gratton E, Breusegem S, Sutin J, Ruan Q, Barry N (2003) J Biomed Opt 8:381
8. Rizzo MA, Piston DW (2005) Biophys J Biophys Lett L14
9. Mattheyses AL, Hoppe AD, Axelrod D (2004) Biophys J 87:2787
10. Loew LM (1996) Pure Appl Chem 68:1405
11. Dombeck DA, Blanchard-Desce M, Webb WW (2004) J Neurosci 24:999
12. Zhang J, Campbell RE, Ting AY, Tsien RY (2002) Nat Rev Mol Cell Biol 3:906
13. Shaner NC, Steinbach PA, Tsien RY (2005) Nat Meth 2 12:905
14. Müller M, Squier J, Brakenhoff GJ (1999) Proc SPIE 3605:92
15. Campagnola PJ, Wei M, Lewis A, Loew LM (1999) Biophys J 77:3341
16. Larson DR, Zipfel WR, Williams RM, Clark SW, Bruchez MP, Wise FW, Webb WW (2003) Science 300:1434
17. Zucker RM, Price OT (2001) Cytometry 44:295
18. Pawley J (2006) Points, pixels and gray levels: digitizing image data. In: Pawley J (ed) Handbook of Biological Confocal Microscopy, 3rd edn. Plenum, New York
19. Wang E, Babbey CM, Dunn KW (2005) J Microsc 218:148
20. Pawley J (2006) Fundamental limits in confocal microscopy. In: Pawley J (ed) Handbook of Biological Confocal Microscopy, 3rd edn. Plenum, New York

21. Art J (2006) Photon detectors for confocal microscopy. In: Pawley J (ed) Handbook of Biological Confocal Microscopy, 3rd edn. Plenum, New York
22. Pawley JB (2000) Biotechniques 28:884
23. LaMorte VJ, Zoumi A, Tromberg BJ (2003) J Biomed Opt 8:3:357
24. Huth U, Wieschollck A, Garini Y, Schubert R, Peschka-Süss R (2004) Cytometry A 57A:10
25. Hutter H (2004) J Microsc 215(2):213
26. Garini Y, Gil A, Bar-Am I, Cabib D, Katzir N (1999) Cytometry 35:214
27. Neher R, Neher E (2004) J Microsc 213(1):46
28. Zucker RM, Lerner JM (2004) Cytometry A 62A:8
29. Kam Z, Jones MO, Chen H, Agard DA, Sedat JW (1993) Bioimaging 1:71
30. Model M, Burkhardt J (2001) Cytometry 44:309
31. van den Doel LR, Klein AD, Ellenberger SL, Netten H, Boddeke FR, van Vliet LJ, Young IT (1998) Bioimaging 6:138
32. Zucker RM, Price OT (1999) Methods 18:447
33. Zucker RM, Price OT (2001) Cytometry 44:273
34. Jericevic Z, Wiese B, Bryan J, Smith LC (1989) Methods Cell Biol 30:47
35. Wilkinson MHF (1994) Comp Methods Prog Biomed 44:61
36. Zwier JM, van Rooij GJ, Hofstraat JW, Brakenhoff GJ (2004) J Microsc 216(1):15
37. Brakenhoff GJ, Wurpel GWH, Jalink K, Oomen L, Brocks L, Zwier JM (2005) J Microsc 219:122
38. McCutchen CW (1967) J Optic Soc Am 57:1190
39. Cox IJ, Sheppard CJR, Wilson T (1982) Optik 60:391
40. Hiraoka Y, Sedat JW, Agard DA (1990) Biophys J 57:325
41. Gu M, Sheppard CJR (1991) Optik 86:65
42. Hell S, Reiner G, Cremer C, Stelzer EHK (1993) J Microsc 169:391
43. Jacobsen H, Hell SW (1995) Bioimaging 3:39
44. Wilson T, Juškaitis R (1995) Bioimaging 3:35
45. Gu M (1996) Opt Lett 21:988
46. Scalettar BA, Swedlow JR, Sedat JW, Agard DA (1996) J Microsc 182:50
47. Young M (ed) Optics and Lasers. Springer, Berlin Heidelberg New York, p 181
48. Booth MJ, Wilson T (2001) J Biomed Opt 6:266
49. Cox G, Sheppard CJR (2004) Microsc Res Tech 63:18
50. Nyquist H (1928) Trans AIEE 47:617
51. Shannon CE (1949) Proc Inst Radio Eng 37:10
52. Centonze V, Pawley J (1995) Tutorial on practical confocal microsopy and the use of the confocal test specimen. In: Pawley J (ed) Handbook of Biological Confocal Microscopy, 2nd edn. Plenum, New York, p 549
53. Stark PRH, Rinko LJ, Larson DN (2003) J Microsc 212:307
54. Shaw PJ, Rawlins DJ (1991) J Microsc 163:151
55. Schrader M, Hofmann UG, Hell SW (1998) J Microsc 191:135
56. Visser TD, Brakenhoff GJ, Groen FCA (1991) Optik 87:39
57. Cox G (1999) Methods Mol Biol 122:357
58. Wijnaendts-van-Resandt RW, Marsman HJB, Kaplan R, Davoust J, Stelzer EHK, Stricker R (1985) J Microsc 138:29
59. Hell SW, Bahlmann K, Schrader M, Soini A, Malak H, Gryczynski I, Lakowicz JR (1996) J Biomed Opt 1:71
60. Akinyemi O, Boyde A, Browne MA, Hadravsky M, Petran M (1992) Scanning 14:136
61. Browne MA, Akinyemi O, Boyde A (1992) Scanning 14(3):145
62. Maly M, Boyde A (1994) Scanning 16:187

63. Mullikin J, van Vliet LJ, Netten H, Boddeke FR, van der Feltz GW, Young IT (1994) SPIE 2173:73
64. Carlsson K (1991) J Microsc 163:167
65. Visser TD, Oud JL, Brakenhoff GJ (1992) Optik 90:17
66. Boddeke FR, Van Vliet LJ, Young IT (1997) J Microsc 186:270
67. Boddeke FR, van Vliet LJ, Netten H, Young IT (1994) Bioimaging 2:193
68. Kinosita K, Itoh H, Ishiwata S, Hirano K, Nishizaka T, Hayakawa T (1991) J Cell Biol 115:67
69. Bastianello S, Ciubotaru C, Beltramello M, Mammano F (2004) Proc SPIE 5324:265

Springer Ser Fluoresc (2008) 6: 89–116
DOI 10.1007/4243_2008_028

Published online: 12 March 2008

Comparability of Fluorescence Microscopy Data and Need for Instrument Characterization of Spectral Scanning Microscopes

Katrin Hoffmann[1] · Ute Resch-Genger[1] (✉) · Roland Nitschke[2] (✉)

[1]Federal Institute for Materials Research and Testing (BAM), Richard-Willstaetter-Str. 11, 12489 Berlin, Germany
ute.resch@bam.de

[2]Life Imaging Center, Center for Systems Biology, Albert-Ludwigs-University Freiburg, Hauptstr. 1, 79104 Freiburg im Breisgau, Germany
roland.nitschke@biologie.uni-freiburg.de

Abstract The aim of this article is to illustrate the need for an improved quality assurance in fluorescence microscopy. From the instrument-side, this can be achieved by a better understanding, consideration, and regular control of the instrument-specific parameters and quantities affecting measured fluorescence signals. Particularly, the need for requirements on physical- and chemical-type instrument standards for the characterization and performance validation of spectral fluorescence microscopes (SFMs) is discussed and suitable systems are presented. Special emphasis is given to spectral fluorescence standards and to day-to-day intensity standards for SFMs. Fluorescence standards and well-characterized fluorescence microscopes are the first and essential steps towards the comparability and the understanding of the variability in fluorescence microscopy data in medical and life sciences. In addition, standards enable the distinction between instrument-specific variations and fluorescent label- or probe-related uncertainties as well as generally sample-related effects.

Keywords Calibration · Fluorescence microscopy · Microscopy standards · Spectral correction · Spectral imaging

Abbreviations

BiFC	Bimolecular fluorescence complementation
CLSM	Confocal laser scanning microscopy
DNA	Deoxyribonucleic acid
ELMI	European light microscopy initiative
EMBO	European molecular biology organization
EPA	Environmental protection agency
FDA	Food and drug administration
SFM	Spectral scanning fluorescence microscope
FRET	Förster or fluorescence resonance energy transfer
λ	Wavelength
MIDL	Multi-ion discharge lamp
NIST	National institute of standards and technology
NMI	National metrology institute
PMT	Photo-multiplier tube
SOP	Standard operation procedure
SRM	Standard reference material
XY	Lateral dimensions
3D	Three-dimensional
VIS	Visible
CCD	Charge coupled device
LED	Light emitting diode
NIR	Near infrared
UV	Ultraviolet

1 Introduction

Over the last decades, the use of fluorescence-based analytical techniques in areas such as bioanalysis, material sciences, environmental analysis, molecular genetics, cell biology, medical diagnostics, and drug screening [1–5] has been constantly growing. This is related to the unique potential of fluorescence methods in microscopy that enable a broad variety of sensing, screening, and imaging applications. These applications exploit the selectivity of fluorescence communication via different experimental parameters like excitation and emission wavelength, fluorescence intensity, fluorescence lifetime, and fluorescence (de)polarization and the unique sensitivity of fluorescence for monitoring and tracking of single molecules. This, in conjunction with spatial resolution, can provide a unique wealth of information like target-specific emission spectra or lifetimes at any point of the magnified image [5–9]. Moreover, using wide-field or confocal laser scanning microscopy, the theoretical limits of spatial resolution that are determined by the numerical aperture of the objective and the excitation wavelength can be realized in practice. In addition, labeling or probing of biological structures with fluorescent reporters allows their indirect visualization, even when their size is far below the optical resolution limit [10].

Because of the rapid advances in computer techniques and microscope equipment, fluorescence microscope techniques have developed into some of the most powerful and commonly used tools in life sciences. This has provided the basis for highly specialized imaging and non-imaging methods like Förster or fluorescence resonance energy transfer (FRET), bimolecular fluorescence complementation (BiFC), photoconversion as well as spectral unmixing. This trend has been further stimulated by the simultaneous development of a broad variety of (target-specific) fluorescent reporters for microscope applications like genetically encoded fluorescent proteins and caged chromophores [11–15] and the continuous improvement of these chromophores with respect to, for example, photostability and brilliance (the product of the molar absorption coefficient at the excitation wavelength and the fluorescence quantum yield) [16, 17]. At present, standard applications of fluorescence microscopy techniques are investigations of fixed (dead) samples like immunofluorescence studies and in situ hybridization measurements of DNA sequences. The most popular and rapidly developing field, however, is live cell imaging with measurements of the structure, organization, dynamics, and function of membranes, organelles, and other cellular structures or of biological active compounds, the determination of intracellular pH and physiologically important ions or second messengers as well as studies of protein structure and dynamics [5–8, 18].

At present, a broad variety of fluorescence microscope techniques do not specifically require high spectral resolution and quantification. However, in more and more fields of application, confocal and wide-field fluorescence microscopy have been developing from being only visualization techniques to a stage where quantification is becoming mandatory to be competitive with other techniques. Such tasks include, for example, the quantitative measurement of the concentration of an analyte or tracking of relative fluorescence intensities in 3D space over time (4D) [19, 20]. Simultaneously, applications are gaining in importance that demand an improved instrument (long-term) stability or at least tools for the correction of instrument drift as a prerequisite for long-term studies. Moreover, as a new trend in microscopy, an enhanced spectral resolution is increasingly valued. This can be, for example, exploited for the analysis of various targets in parallel via the simultaneous and independent interrogation of the label- or target-specific fluorescence parameter emission wavelength or color [21–23]. Such a multiplexed analysis can be performed also in the time domain (lifetime multiplexing) as well as principally in addition to spectral multiplexing or multi color imaging. Moreover, spectral information, i.e., emission spectra, can be employed for target identification [1]. To enable and encourage these trends, new commercial instruments have been emerging within the last years that offer an improved spectral resolution and thus, for example, the possibility to measure spatially resolved emission spectra. The increasing need for quantification, however, is still poorly met in most cases by commercial imaging systems that are designed mainly for high image quality (low background signals and image distortions, high light throughput, and good detection efficiency).

As a result of these emerging areas of application and the ever increasing complexity of the instrumentation used for confocal and high-end wide-field imaging microscopy, there is an urgent need for easy-to-use and adequately characterized instrument calibration and validation standards to improve the reliability and comparability of microscopy data. This cannot account for fluorescent label- or probe-related uncertainties like, for example, photoinduced changes (decomposition, photochemical conversion, blinking) or the micro-environment dependence of the spectroscopic properties of most dyes, sample-related uncertainties such as scattering or background fluorescence or for the heterogeneity of sample preparation procedures. Better and regular control and consideration of microscope parameters and quantities not only strongly reduce instrument-specific and measurement-related sources of uncertainty, yet present first steps towards the standardization of microscope techniques. Suitable calibration tools, that should be preferably supplied in conjunction with standard operation procedures (SOPs) or recommendations/guidelines for instrument characterization and performance validation, are similarly mandatory to pave the road for fluorescence microscope techniques in strongly regulated areas like, for example, medical diagnostics [24–31]. This, as well as laboratory accreditation, require proper documentation of instrument performance with for instance control charts. The consideration of instrument-specific effects and their separation from measured data is also a prerequisite for multi-laboratory studies and for the combination of data across instruments and laboratories.

To achieve this, in need are purpose-fit, robust, and easy-to-use instrument standards for the determination and control of all the relevant microscopy parameters and quantities (see Sect. 2). In addition, tools for the control and consideration of the day-to-day and long-term instrument performance that monitor the comparability and stability of the utmost important microscope parameters. Furthermore, often users of microscopes request fluorescence intensity standards [32][1] to compare the overall spectral sensitivity of microscopes. However, despite the widespread use of fluorescence techniques

[1] In fluorescence microscopy as well as in certain other fluorescence techniques, application-specific standards for certain fluorophores are also referred to as reference standards.

and the extensive literature dedicated to (micro)fluorescence measurements and potential standards [1–4, 33–35], at present, there exists only a limited number of truly reliable standards. This situation is further complicated by the very few number of recommendations on the performance of quantitative fluorescence measurements [36] and the characterization of fluorescence instrumentation [37–39, 44].

The aim of this work is to illustrate the general need for well characterized standards for wide-field and confocal fluorescence microscopy. As such microscope standards and quantitative microscopy are highlighted in other parts of this volume as well, we focus here on spectrally resolved microscope measurements and the influence of the spectral characteristics of microscopes on measured data. In addition, general requirements on suitable fluorescence microscopy standards are discussed without consideration of the specific demands of the multitude of fluorescence microscope techniques available. The aim is here to similarly pave the road for the understanding of the need for overall accepted quality criteria for standards.

2 Instrument-Specific Parameters and Quantities Affecting Spectral Measurements in Microscopy

Instrument-specific parameters and quantities affecting microscope measurements [40] include (1) the size of the illuminated and, therefore, also the bleached volume, (2) the spectral irradiance/excitation intensity reaching the sample, (3) the homogeneity of the sample illumination field, (4) the instrument's spatial (x,y) resolution, (5) the field flatness, z-distance, and z-resolution, (6) the spectral responsivity of the detection channel, and (7) the spectral resolution [1–4, 41, 42]. The resulting data and the final signal-to-noise ratio (S/N) of the recorded images are affected by all these parameters and by the spectroscopic features of the fluorophore(s) as well as by undesired background fluorescence. The latter can result from the sample, i.e., contaminations or unspecific binding of fluorescent reporters, the instrument's optics such as lenses, cover glasses, and optical filters [43], and from sample supports/containers like glass or polymer slides or the immersion medium. All these quantities together determine the signal size, the shape of the recorded spectra, and the limit of detection for a certain target.

Instrument-specific effects distorting otherwise sample-specific data are linked to the spectral irradiance at the sample position, the light collection properties and aberration correction of the microscope, and the spectral responsivity and sensitivity of the emission detection system, respectively. These quantities are wavelength- and polarization-dependent, and, due to the aging of optical and opto-electronical instrument components, also time-dependent [34, 35, 44]. Accordingly, the magnitude of these instrument-

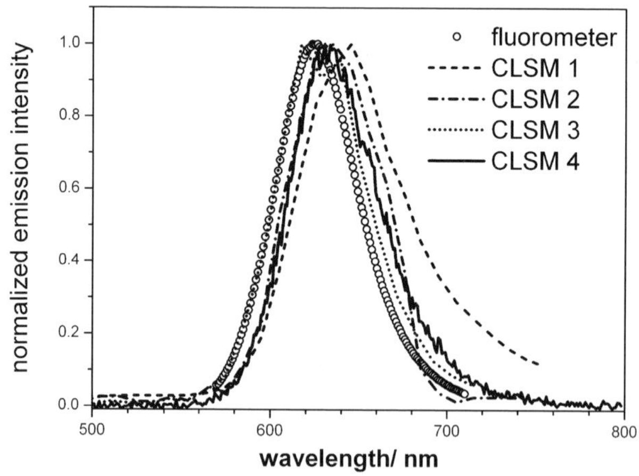

Fig. 1 Effect of the instrument-dependent spectral responsivity s(λ) of the emission channels of CLSMs on the measured emission spectrum of an exemplary chosen common organic dye: uncorrected emission spectra (*lines*) vs. emission spectrum (*symbols*), corrected for the instrument's spectral responsivity s(λ)

specific signal distortions depends on the spectral region of the dye's emission and the width of its emission band. These effects are exemplary illustrated in Fig. 1 for the (normalized) emission spectrum of an organic dye measured with common types of fluorescence microscopes. Similarly, instrument-specific spectral distortions can affect quantification if the emission spectra of the fluorophores to be quantified and the intensity standard do not match [35].

3
Comparability of Fluorescence Microscopy Data and Instrument Characterization

To improve the reliability of fluorescence microscopy data, the multitude of measurement-affecting instrument parameters addressed in the previous paragraph has to be characterized and taken into account. In the following sections, examples for commonly used microscopy standards for different microscopy parameters and quantities are introduced and the general need for quality criteria on instrument-type microscopy standards is discussed. In a second step, special emphasis is dedicated to standards suited for the determination of the wavelength accuracy, spectral resolution, and most important for the spectral correction of microscope data. In addition, potential day-to-day intensity standards for the correction of instrument drift and aging are discussed and approaches towards the determination of the linearity of microscope detection systems.

3.1
Commonly Used Microscope Standards

There is a broad variety of more or less suitable standards for fluorescence microscopy available. Such standards can be of physical or chemical nature. Classical examples for physical (transfer) standards[2] are calibrated lamps or detectors [35, 38]. Chemical standards are liquid, solid, or particle-type reference materials containing organic or inorganic chromophores. These reference materials are typically referred to as "fluorescence standards" for techniques like fluorescence microscopy. The majority of microscope standards in use are instrument calibration and instrument validation standards, see also the work by DeRose et al., 2008, in this volume. Instrument calibration standards enable the determination and correction of instrument bias to rule out instrumentation as a major source of variability and to yield instrument-independent fluorescence data. Instrument validation standards like day-to-day intensity standards represent tools for the periodic check of instrument performance. Both types of standards must not necessarily consider the spectroscopic properties of fluorescent samples. More application-specific fluorescence standards that are beyond the scope of this work are, for example, fluorescence intensity standards used typically for quantification. Such systems must accordingly take into account the spectroscopic properties of the fluorophore(s) to be quantified.

Table 1 summarizes typical microscopy standards and approaches to standards.

To check on instrument alignment, stability, and sensitivity as well as on spectral separation (e.g. spectral unmixing of different fluorophores) of both conventional microscopes as well as laser scanning systems, often multi-colored particles, labeled for instance with one fluorophore throughout and another fluorophore on the outer shell (see Fig. 2) are employed.

The performance of these particles is commonly limited by fluorophore photostability. Moreover, the bead mounting medium has to be matched to the bead refractive index and the microscope objective immersion media (oil, glycerin, water). Otherwise, refractive index mismatch can occur leading to distorted z-images and intensity artefacts due to lense effects of the beads. Fluorescent microspheres [39, 45–47] are also of widespread use as internal (added to the sample) or external (measured separately from the sample) fluorescence intensity standards to provide a relative reference intensity scale. One of the aims is here to account for instrument drift and day-to-day instrument variability thereby yielding comparable data on a single instrument basis.

Approaches to microscopy standards for the control of the (spatial) homogeneity of illumination and detection within a single field (shading or flat

[2] Physical or chemical transfer standards are terms used, for example, in radiometry and fluorescence spectroscopy for standards that are used to transfer a certain quantity like the spectral responsivity of a detection system to e.g. a physical scale representing for example a SI unit.

Table 1 Standards for fluorescence microscopy

Standard	Parameter	Commercially available	Refs.
Beads	Resolution x/y/z; emission spectra, intensity calibration	+	[6, 15, 26, 39, 46, 47]
Clondiag slide (fluorescent pattern)	Intensity calibration	+	[57]
Color slides (dye-doped polymers)	Excitation intensity at sample position; homogeneity of illumination; emission correction; intensity calibration; field flatness; linearity	+	[26, 54, 63–65, 76]
Detector slide/Power meter	Homogeneity of illumination; wavelength accuracy of excitation; power measurement	+	[26, 73, 75]
Fluorescent arrays	Intensity calibration	+	[84]
Fluorescent particles/cells in various matrices (wax; arrays, polymers)	Intensity calibration		[53, 55, 56]
Fluorescent solutions	Excitation intensity at sample position; emission correction, intensity calibration		[32, 48, 49, 52]
Glass (slide/bulk; broad band-emitter)	Excitation intensity at sample position; homogeneity of illumination; emission correction; intensity calibration; field flatness; linearity		[62, 66, 81] [86]*
Glass (slide/bulk; narrow band-emitter)	Excitation intensity at sample position; wavelength accuracy; emission correction; intensity calibration; field flatness; linearity		[62, 66, 81]
Light sources/lamps	Wavelengths accuracy; emission correction; linearity	+	[70, 71]
Meshes/grids	Resolution x/y	+	[6]

Table 1 (continued)

Standard	Parameter	Commercially available	Refs.
Mirrors	Linearity; wavelength accuracy	+	[6]
Quantum dots	Wavelength accuracy	+	[67, 68]
Richardson slide (specimen with defined micro-structure)	Resolution x/y	+	[6]
SipCharts (dye-doped polymer films)	Excitation intensity at sample position; homogeneity of illumination; intensity calibration; field flatness		[50, 51]
USAF target (specimen with defined micro-structure)	Resolution x/y	+	[6]

* Raman reference material.

Certain commercial equipment, instruments, or materials are identified herein to foster understanding. Such identification does not imply recommendation or endorsement by the Federal Institute for Materials Research and Testing, nor does it imply that the materials or equipment identified are necessarily the best available for the purpose.

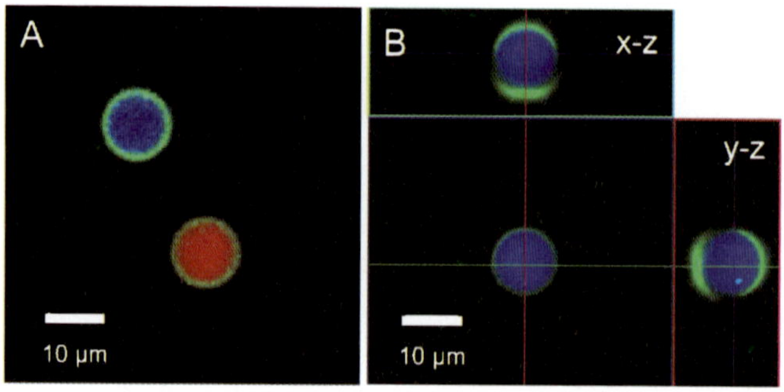

Fig. 2 **A** 3-channel overlay of Focal Check Beads (Molecular Probes Inc., Eugene, USA; F-7237 and F-7238) acquired sequentially with 364, 488, and 543 nm excitation. **B** The orthogonal x-z and y-z cross-section view from a 2-channel z-stack (excitation at 364 and 488 nm) shows a bead acting as an optical active element, which results in a distorted intensity profile of the green colored shell. Images were recorded on a confocal microscope LSM 510 Meta (Carl Zeiss MicroImaging GmbH, Jena, Germany) using a 63x/1.2 C-Apochromat water immersion lens

field correction) include microdroplets of fluorophore solutions [32], microcapillaries filled with dye solutions [33, 48, 49], and fluorophores immobilized within spin-coated polymers [50, 51]. Alternatives are based on concentrated dye solutions on regular slides [52], fixed fluorescent cells [53] and fluorescent polymers [54] as well as immobilized particle arrays [55] and wax films doped with fluorescent dyes [56]. Commonly used or suggested standards for the determination of parameters like the homogeneity of illumination, the spectral characteristics of fluorescence microscopes, and for the characterization of day-to-day instrument performance are slide-shaped supports with fluorescent coatings [57], glasses and polymers doped with inorganic metal ions [33, 58–62] or organic fluorophores [63–65] as well as inorganic ion-doped optical fibers [66].

Only recently, also organic and inorganic systems containing uniformly dispersed luminescent nanocrystals or so-called quantum dots at various concentrations have been suggested as potential fluorescence standards, for example, for parameters like wavelength accuracy and spectral resolution [67, 68]. Advantageous are here the—compared to organic fluorophores—generally improved photostability in combination with narrow symmetric emission bands and very broad absorption spectra providing a unique flexibility concerning the choice of the excitation wavelength. The suitability of such quantum dot-based materials as standards, however, requires further extensive research to overcome some inherent drawbacks such as photobrightening, i.e., the typically observed increase in fluorescence intensity upon illumination [68, 69], and blinking (critical for very short pixel dwell times).

Also physical-type standards can be used for microscope characterization. Simple physical standards are meshes and grids for testing the lateral resolution or better, more sophisticated micro-patterned microscopy test slides (Richardson; USAF, see Table 1) [6]. Other examples are well-characterized light sources [70] like a multi-ion discharge lamp (MIDL)-based calibration assembly [71], slide-shaped devices containing photodiodes for power and pulse length measurements at the sample position and a charge coupled device (CCD) detector for control of the wavelength accuracy of the excitation channel, and microscope test slides with a built-in light-emitting diode (LED) and different pinholes to mimic the emission characteristics of fluorescent cells [72] or equipped with intensity-adjustable LEDs of different color and a photodiode for internal and external calibration or power measurement [73–75].

3.2
Quality Criteria for Instrument-Type Microscopy Standards

Standards for the characterization of fluorescence microscopes have to meet specific demands in addition to the requirements imposed on standards for macroscopic fluorescence techniques (see for example, DeRose et al. and Resch-Genger et al., 2008, in this volume) [35, 38]. This is mainly related to the use of lasers as excitation light sources and the accordingly strongly enhanced spectral irradiances at the sample position as compared to, for example, steady-state spectrofluorometry. Furthermore, the high spatial resolution in fluorescence microscopy make stringent demands on the homogeneity of the fluorophore distribution within a standard. In addition, the size, the shape, and the physical robustness of standards gain in importance. Only dimension-adapted microscopy standards with a well-defined shape permit a correct intensity/volume relationship. Also, a standard should preferably have a surface sealed by a cover glass, that allows the use of liquid immersion media widely used in confocal microscopy, protects the calibration tool from dust, contaminations, and damage, and avoids optical aberrations as the optical design of almost all fluorescence microscope objectives is based on the use of a 170 μm cover glass. Finally, for applications such as the correction of instrument drift, for example, for long-term studies and the comparison of data between different microscopes, a highly stable and reproducible standard alignment is mandatory. This has to be considered in the design of the standard.

As derived from our experience with the development of spectral fluorescence standards for steady-state fluorometry and from tests with commercial microscopy standards, suitable microscopy standards must additionally meet the following general quality criteria to yield a reliable instrument characterization and minimize standard-related uncertainties. More application-specific requirements focusing on spectral and day-to-day intensity standards are discussed in the following sections.

Principally, standards must be adequately characterized with respect to all the parameters that can potentially influence their calibration-relevant properties. This includes the documentation of the measurement conditions used for the determination of the exploited features and if applicable, the dependence of these properties on, for example, temperature, excitation light intensity, and excitation wavelength. The spectroscopic properties of chromophore-based standards must be suitable for the desired application (i.e. excitable at typical laser wavelengths, very narrow spectra for the determination of the wavelength accuracy and spectral resolution, broad spectra for spectral correction; see also DeRose and Resch-Genger, 2008, in this volume) [35]. Accordingly, to facilitate the choice of standards, absorption and instrument-independent corrected fluorescence spectra should be provided. Suitable standards should be designed for straightforward use under routine measurement conditions (e.g. detector settings, measurement geometry etc.). This implies, for example, that their spectral radiances or emitted light intensities cover the range of the fluorescence intensities emitted by typical biological samples to avoid detector saturation or tedious attenuation procedures [35, 38]. This is a common problem for calibration lamps and, to a smaller extent, for physical wavelength standards. It can be also critical for certain highly doped plastic slides.

Suitable standards must be not only photochemically and thermally stable under ambient, i.e. microscope conditions using laser illumination over an adequate time period, but their long-term stability under typical usage conditions should be known and reported in a comprehensible way. However, sufficient photostability under (long-term) laser illumination is a rather stringent requirement as illustrated in Fig. 3 revealing the results from photostability tests of color slides [65, 76] that are frequently used in confocal and wide-field microscopy for spectrally resolved measurements, spectral calibration as well as for the adjustment or test of illumination homogeneity. The data shown in Fig. 3 were obtained from a time series of five images (DAPI Blue) or ten images (DV 488/519). Between recordings of the images, additional scans within defined regions of interest (ROIs) were run with maximum laser intensity to probe the bleaching characteristics of the calibration slides [35]. The dramatic effect of the photoinduced degradation of these polymer-based fluorescent slides is immediately evident from Fig. 3.

Chemical microscopy standards must reveal a homogeneous fluorophore distribution within the microscopic excitation volume. For solutions, this is commonly not a problem, yet dye homogeneity must be controlled in the case of solid standards. As typically, not single spots on a standard are measured and the reproducibility of the standard's alignment has to be considered as well, this criterion must be met at least for a sizable and clearly marked area of the standard if not for the whole system. An inhomogeneous chromophore distribution can result in a spatial dependence of the standard's calibration-relevant features and can affect, for example, the standard's day-to-day performance. From a practical point of view, critical for many fluorescence standards can be an improper geometry that renders already the mounting of the standard on the microscope table difficult and affects the

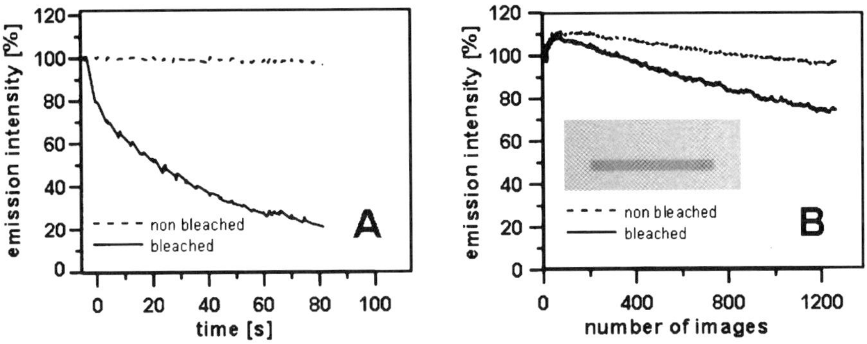

Fig. 3 Photostability test with two polymer-based calibration slides DAPI Blue (**A**) and DV 488/519 (**B**). The settings for the bleached ROI were 1000 scan iterations, pixel time 1.4 μs, total scanning time for the ROI ca. 2.5 s (DAPI Blue, excited at 364 nm) or 25 s (DV 488/519, excitation at 488 nm), respectively. *Dashed lines* represent the emission intensity over time within a nonbleached area. The time-dependent emission intensity within the marked ROI results in the bleached curves (*solid lines*)

identification of the appropriate focal plane for the calibration procedure and its reproducibility. This needs to be overcome by proper design facilitating standard alignment.

Eventually, the uncertainty of the standard's calibration-relevant properties including the procedure for its calculation is desired. This is, for example, a prerequisite for the traceability of the instrument characterization, and thus of microscopy data, to the relevant radiometric and physical scales and SI units like the spectral radiance and the spectral responsivity [37, 38, 44, 77, 78], see also Resch-Genger, 2008, in this volume. At present, this is only fulfilled by standards released or calibrated by National Metrology Institutes (NMIs) such as certain physical transfer standards and the spectral fluorescence standards presented in Sect. 3.5 [38, 79].

3.3 Instrument-Specific Correction Procedures Built Into Microscopes by Instrument Manufacturers

Instrument manufactures are becoming increasingly aware of the need for an improved instrument characterization and performance validation. Currently, manufactures of spectral CLSM try to improve the reliability and comparability of microscope data through instrument-specific correction procedures. To the best of our knowledge, this includes for example the adjustment of the wavelength accuracy exploiting built-in lasers (known spectral position; Leica, Zeiss) and the use of halogen light sources employed for conventional transmitted light illumination (Zeiss) as an internal standard for relative spectral sensitivity (gain matching of the PMTs). One instrument manufacturer (Nikon) implemented a calibration curve for the relative spectral responsivity of the instrument's detection channel into spectral CLSM. This correction curve was determined with a light source with calibrated spectral radiance (so-called spectral radiance transfer standard). Moreover, the currently widely used multi-anode PMT technology allows sensitivity correction on a per-channel basis (gain matching of the PMTs) and wavelength accuracy correction by movement of the sensor array.

These approaches may help to diagnose certain measurement errors on an instrument-specific and day-to-day basis. However, these standards and characterization procedures are currently used only to guarantee a defined performance level on a single instrument level. With a single exception, these procedures are not accessible to the microscope users. Moreover, the procedures and their evaluation are often not very well documented. Accordingly, these approaches do not represent substitutes for external calibration tools and internationally accepted standard procedures for instrument characterization. They do not improve the instrument-to-instrument comparability of microscope data and do not meet the requirements for the use of fluorescence microscopy in medical diagnosis, when quantification is needed.

3.4 Standards and Procedures for the Determination of the Wavelength Accuracy and Spectral Resolution

To meet the described trends in (spectrally resolved) fluorescence microscopy, particularly tools are desired to determine the spectral characteristics of fluorescence microscopes, to compare the wavelength-dependent sensitivity/spectral responsivity of different instruments, and to characterize microscope long-term stability [35]. Typically, control of the wavelength accuracy of the detectors is the first step towards the characterization of spectral fluorescence microscopes. Suitable (internal, i.e., instrument-integrated, or external) wavelength standards must reveal a multitude of very narrow emission bands or lines at known spectral positions within the wavelength region of interest typically the VIS/NIR. Accordingly, examples include lasers, atomic discharge lamps, and materials containing narrow band-emitters.

In CLSM, a spectral scan over the excitation laser lines using a mirror slide is often used to evaluate the wavelength accuracy (see Fig. 4, left panel, and Sect. 3.3). Principally ideal candidates for external wavelength standards with respect to their emission features are atomic discharge lamps displaying extremely narrow emission lines at well-known spectral band positions (including uncertainties) in the UV/VIS/NIR [38] and a multi-ion discharge lamp (MIDL) that contains mercury, argon, and inorganic fluorophores emitting a multitude of narrow and distinct bands [71]. Such lamps should be preferably dimension-adapted for microscope use. Disadvantageous can be here, however, the high spectral radiances of these

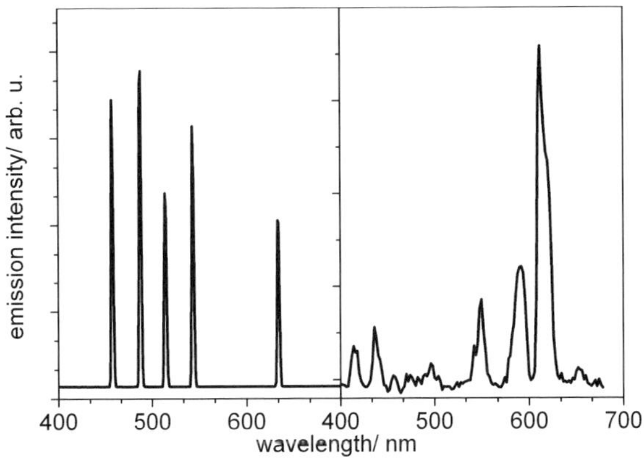

Fig. 4 Spectral position of different laser lines used for control of the wavelength accuracy of CLSMs (*left panel*) and emission spectrum of a fluorescent glass slide doped with a multitude of rare earth (RE) ions (*right panel*)

lamps that strongly exceed those of typical fluorescent biological samples and thus, cause problems with detector saturation. In addition, the alignment can be tedious, especially, if a high reproducibility is desired as mandatory, for example, for the comparison of intensity values between different measurements and instruments (see also Sect. 3.6). As the highest spectral resolution encountered in CLSM is about 2 nm for commercially available systems, dye-based wavelength standards exploiting chromophores or chromophore mixtures, that reveal several very narrow fluorescence bands, properly separated by at least 20 nm, within the UV/VIS/NIR spectral region, present an elegant alternative to such lamps. Such materials, that should be preferentially slide-shaped and easy to align, can be designed to display fluorescence intensities comparable to those of luminescent samples. The emitted intensities can be controlled via dopant concentrations. Examples are a dysprosium-activated yttrium aluminum garnet (DYAG) mounted in a cuvette-sized holder that is recommended as a wavelength standard for steady-state spectrofluorometry [80]. However, the dimensions of this material are not well suited for microscope applications, and to the best of our knowledge, no application as a wavelength standard in microscopy has been yet reported. Similarly suited are glasses doped with rare earth (RE) ions [81]. Such materials, the emission spectrum of which is shown in the right panel of Fig. 4, are currently tested and evaluated at BAM for use as (certified) wavelength standards for steady-state spectrofluorometry (cuvette-shaped materials) and for spectrally resolved microscopy (slide-shaped materials). Their very narrow bands can be also exploited for the determination of the instrument's spectral resolution and the intensity pattern can be related to the relative spectral responsivity of the emission detection system, see also Sect. 3.5.

3.5 Spectral Characteristics of Fluorescence Microscopes and Spectral Sensitivity

As the spectral characteristics of fluorescence microscopes distort measured data as illustrated in Sect. 3.1, the determination of the relative spectral responsivity of the microscope detection system is a prerequisite to comparable microscopy data. Moreover, the consideration of these quantities can decrease quantification uncertainties. Simultaneously, the regular determination of the spectral characteristics of fluorescence microscopes, especially of the spectral sensitivity of the emission channel, provides an elegant tool for instrument performance validation.

Determination of the relative spectral responsivity of the emission channel ($s(\lambda_{em})$) requires a source with a known wavelength dependence of the spectral radiance covering the currently most relevant spectral region of about 400 to 800 nm. The emission spectrum of this source should be preferably very broad and unstructured to minimize effects of spectral bandpass [38, 79]. This is fulfilled by (calibrated) light sources like tungsten ribbon lamps revealing an extremely broad emission spectrum [38, 44]. The use of such a physical transfer standard, however, is expensive (purchase and regular recalibration) and requires a certain background in optics as well as tedious attenuation procedures to perform the instrument characterization under routinely used measurement conditions and to simultaneously avoid detector saturation. Other problems can arise due to stray light, especially when the pinhole is opened above one airy unit. A more simple and straightforward approach are spectral fluorescence standards used as dye solutions. Corrected broad and unstructured emission spectra of a set of standard dyes F001 to F005 covering the spectral region from 300 to 770 nm have been only recently certified by BAM [38, 79, 82]. The liquid nature of these materials in conjunction with the very small overlap between absorption and emission spectra provide the basis for the very flexible use of these materials in a broad variety of measurement geometries and containers. For fluorescence microscopy, typically only the VIS standards F003-F005 are relevant revealing fluorescence emission spectra within the spectral region of ca. 400 to 770 nm. The fluorescence of these dyes can be excited with most of the commonly used standard laser lines or other typical light sources between 405 and 530 nm. F003-F005 have been tested by BAM, for example, for thermal and photochemical stability and fluorescence anisotropy [38, 79], as well as for the dependence of the shape of the emission spectra on the z-position and various excitation wavelengths. Similarly as calibrated light sources, these standards provide traceability to the spectral radiance scale. Whether other broad band-emitters presented in the works by DeRose et al. and Resch-Genger et al., 2008, in this volume, are also suitable for fluorescence microscopy remains to be tested [38, 60, 61].

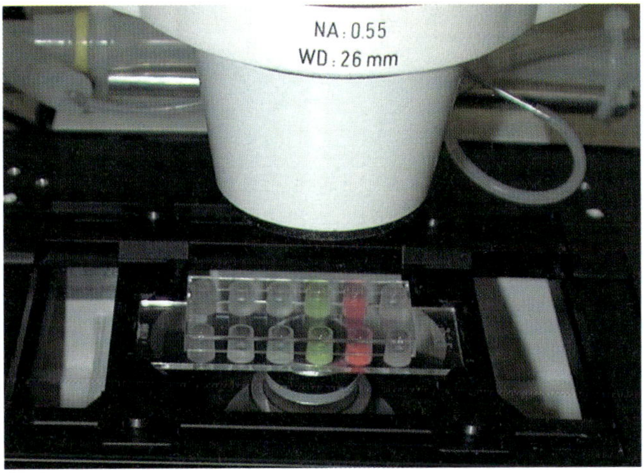

Fig. 5 Solutions of fluorescence standard dyes within the 30 µl channels of a microchannel microscopy slide (μ-Slide; ibidi GmbH, Germany)

To adapt this procedure developed for the characterization of spectrofluorometers to the determination of the spectral responsivity of confocal and widefield fluorescence microscopes, we chose a slide-type microchannel device with a defined optical path length [83], shown in Fig. 5, filled with (renewable) solutions of dyes F003 to F005. Moreover, the dye concentrations were varied to optimize measured signal intensities.

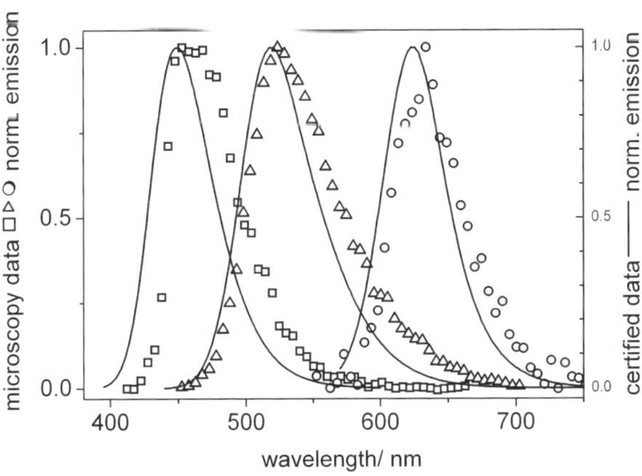

Fig. 6 Normalized uncorrected emission spectra of the spectral fluorescence standards F003 (*squares*), F004 (*triangles*), and F005 (*circles*) measured with a CLSM (*open symbols*, Leica TCSP, excitation at 405 nm) and the corresponding corrected certified spectra measured with a calibrated spectrofluorometer (*solid lines*). Evaluation of these data using a purpose-developed software (LINKCORR, BAM) yields the inverse relative spectral responsivity $s(\lambda^{-1})$ (see also the work by Resch-Genger et al., 2008, in this volume) of fluorescence measurement systems

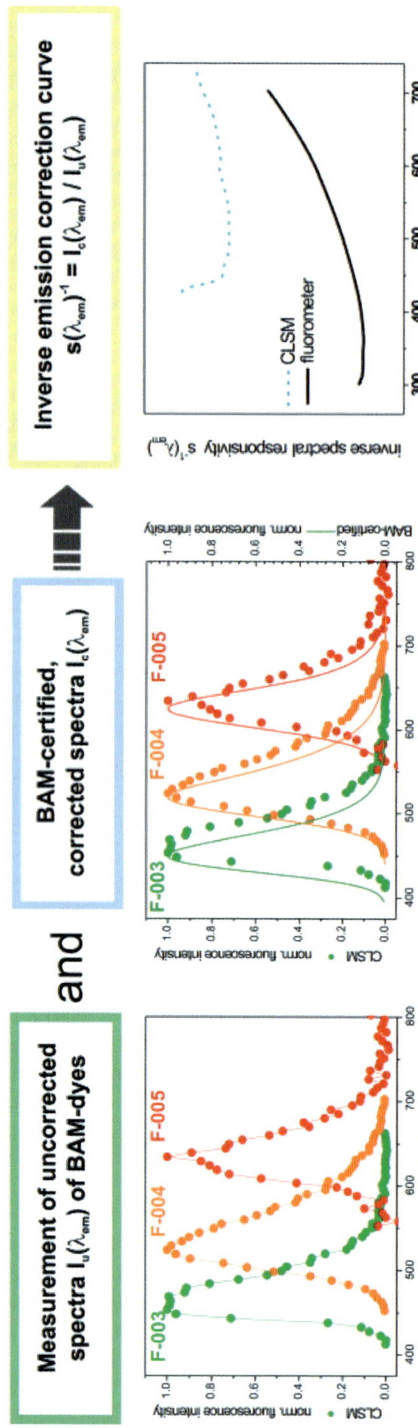
Measurement of uncorrected spectra I_u(λ_em) of BAM-dyes
and
BAM-certified, corrected spectra I_c(λ_em)
Inverse emission correction curve s(λ_em)^-1 = I_c(λ_em) / I_u(λ_em)
F-003
F-004
F-005
CLSM • norm. fluorescence intensity
BAM-certified — norm. fluorescence intensity
wavelength/ nm
inverse spectral responsivity s^-1(λ_em)
CLSM
fluorometer

◀ **Fig. 7** Determination of the inverse spectral responsivity of fluorescence measurement systems using the Calibration Kit "Spectral Fluorescence Standards" (working principle)

Figure 6 reveals the normalized uncorrected emission spectra of F003 to F005, measured with a CLSM, together with the corresponding BAM-certified spectra.

Evaluation of these data includes the calculation of the quotients of the corrected (here certified) and measured spectra for each dye ($I_c(\lambda_{em})$ divided by the instrument-dependent, uncorrected spectra ($I_u(\lambda_{em})$) and the subsequent statistically weighted combination of these wavelength-dependent quotients to a global correction curve according to the recently described working principle of the Calibration Kit (cf. Fig. 7) using the BAM-developed software *LINKCORR*. This yields the microscope's inverse relative spectral responsivity $s(\lambda^{-1})$ (see also Resch-Genger et al., 2008, in this volume) [38]. A prerequisite for the subsequent correction of measured data with this correction curve are similar settings of the measurement parameters like objective, magnification, excitation wavelength, laser intensity, PMT voltage, PMT off-set, size of the pinhole, and z-position.

Figure 8 displays the relative spectral responsivity $s(\lambda)$ of different commercial fluorescence measurement systems. These data were obtained according to the procedure illustrated in Fig. 7 using F003 to F005 in the case of the spectral CLSM. The relative spectral responsivity $s(\lambda)$ of the spectrofluorometer was obtained with a spectral radiance transfer standard and a white standard at the sample position following a procedure previously described by BAM [38].

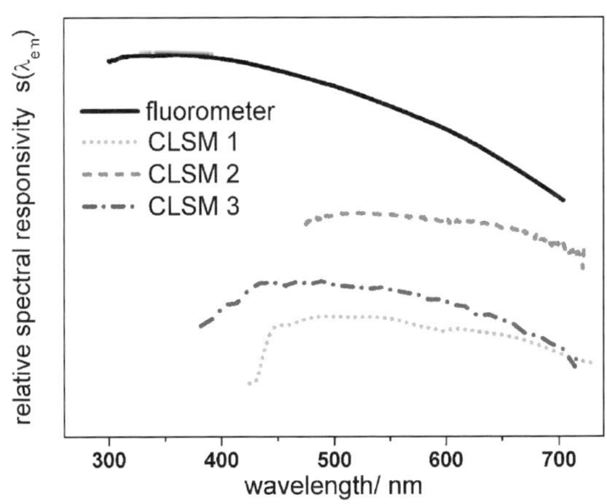

Fig. 8 Relative spectral responsivity $s(\lambda_{em})$ of different CLSMs (*broken lines*), determined with the *Spectral Fluorescence Standards* BAM-F003 to BAM-F005, as well as $s(\lambda_{em})$ of a spectrofluorometer, determined with a calibrated light source and a white standard following a previously described calibration procedure (*solid line*); [38]

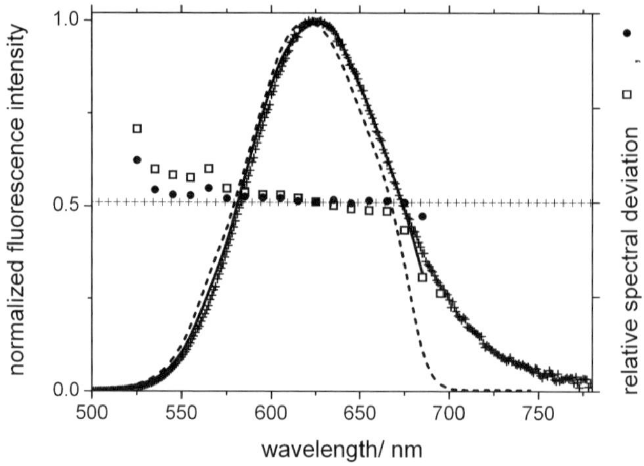

Fig. 9 Fluorescence emission spectrum of an organic dye measured with a CLSM (NIKON C1si) before (*dashed line*) and after (*solid line*) dye-based (F003 to F005) spectral correction. The corrected emission spectrum scattered line (–+–) obtained with a calibrated spectrofluorometer (SLM 8100, Spectronics Instruments) is given for comparison. Symbols represent relative spectral deviations of the uncorrected (*open squares*) and corrected (*solid circles*) spectra from the corrected emission spectrum obtained with the spectrofluorometer

To control the suitability of the dye-based determination of s(λ) also for spectral CLSM, the obtained corrected emission spectrum of a test dye was compared with the corrected emission spectra determined with a spectrofluorometer calibrated with physical transfer standards. This is exemplary revealed in Fig. 9 for measurements with the CLSM C1si from NIKON. The minimal relative spectral deviation of the two corrected emission spectra underlines the applicability of the dye-based calibration approach to fluorescence microscopy. It also illustrated the improvement in comparability of fluorescence data across instruments.

The presented dye-based approach to spectral correction is the first step towards the development of spectral fluorescence standards for microscopy and standardized calibration procedures. It can also contribute to a more reliable quantification. Whether the dye-filled calibration slide remains a single-use type approach or whether it can result in a standard with an adequate long-term stability depends on, for example, the leak tightness of the polymeric container and the photostability of the dyes.

3.6 Standards and Procedures for the Determination of the Day-to-Day Intensity and Instrument Long-Term Stability

Testing the instrument's day-to-day performance and long-term stability requires so-called instrument validation standards. Use of such standards is the prerequisite for the documentation of microscopy performance, as is mandatory, for example, in regulated areas or accredited laboratories and for the correction for these instrument-specific variabilities needed for long-term studies. Principally, the majority of instrument validation standards can also be applied as instrument-to-instrument intensity standards for the comparison of measured fluorescence intensities between instruments when measurement parameters are fixed. Whether such a comparison is possible at all for fluorescence microscopy, however, remains to be tested.

Generally, day-to-day intensity standards do not necessarily need to closely match routinely measured samples, yet should be measurable with typical instrument settings to guarantee the reliability of the instrument performance under routine measurement conditions. The most stringent requirements are a high reproducibility of the standard alignment and either a sufficient, well-characterized stability of the standard under applicable conditions, or, for single-use standards, an excellent reproducibility, preferably in combination with an assigned uncertainty. For fluorescence microscopy, especially the former, which requires a highly stable and reproducible position of the standard in relation to the objective, can be critical. Further prerequisites are known corrected spectra, if the standard's emission intensities need to be compared with those of other fluorophores or between instruments with different spectral bandpasses.

Candidates for day-to-day intensity standards are commonly either wavelength standards or standards for the determination of the instrument's spectral responsivity. In the case of physical standards such as "self-luminescent" lamps, exclusively the status and changes of the emission channels are determined. For chromophore-based systems requiring excitation as a prerequisite for the emission of light, both the excitation and the detection channel are measured simultaneously. As has been revealed in Sect. 3.4, the occasionally discussed use of calibration lamps as day-to-day intensity standards requires adaptation of the measurement geometry to microscope needs, guarantee of the reproducible lamp alignment, which can be very difficult to achieve, and rigorous testing of the lamps' short and long-term stability. Also, the integration of calibration routines using photo-diodes could be very helpful to regularly control excitation light output as typically exploited in steady-state spectrofluorometry (reference channel equipped with a photodiode built into many spectrofluorometers). Preferentially, the signal from this reference detector should be externally read out.

A straightforward chromophore-based approach to day-to-day intensity standards includes extremely robust inorganic crystal- or glass-based materials like metal ion-doped glasses [81] such as the RE ion-doped glasses shown in the right panel of Fig. 4 (see Sect. 3.4). Here, the intensity pattern provides a tool for the evaluation of changes in the relative spectral responsivity of the emission detection system and different peaks could be used for a relative intensity reference system, i.e. to link the fluorescence intensities of measured samples to an external, yet comparable reference. To avoid problems caused by the comparably long (species-specific) emission lifetimes of many RE-metal ions within the μs and ms time domain [1, 35], these doped glasses must be applied with a constant set of measurement parameters, at least for instruments with pulsed light sources and particularly for glasses containing chromophore mixtures. Currently, different glasses are tested by BAM for this application and protocols for their proper use as microscope standards are worked out. One approach to reliably position a slide made from such glasses is the integration of focus planes. The focal position within the slide must be well defined as it has to be taken into account for instrument characterization to assure highly repeatable measurement conditions and spectral performance can be also dependent on the sample depth or z-position within the sample (e.g. inner filter effects, local dependence of relative spectral sensitivity of PMTs etc.). In addition, different concepts for the integration of a cover glass into such a device are being tested.

Similar to the RE glasses, also the dye-filled microchannel device described in Sect. 3.5 can present a tool for tracing of aging-induced spectral effects in the emission channel and changes in the spectral sensitivity of microscopes at constant instrument settings like, for example, excitation wavelength, beam splitter configuration, PMT voltage, and alignment of the emission pinhole. Reliable positioning of this calibration device may be similarly achieved by the integration of focus planes.

3.7 Linearity of the Detection System

Instrument characterization as well as quantitative fluorometry both require the (previous) determination of the linear range of the detection system(s) within the commonly used wavelength region at application-relevant instrument settings. Per definition, there exist no linearity standards, but only methods, in combination with suitable materials, to measure the range of linearity of fluorescence instruments, their dynamic range, and the (instrument- and dye-specific) limit of detection [38].

The range of linearity of fluorescence instruments can be obtained by the controlled variation of the amount of light reaching the detector by physical or chemical means. Physical means are for instance different attenuators in combination with a light source. Suggested approaches include, for example, the already mentioned slide-shaped accessories with intensity adjustable LEDs of different colors and photodiodes for internal and external calibration and power measurements [72–75], see Table 1. The more simple chemical approach implies the variation of the concentration of a dye [38]. This can be for instance realized with a dilution series of standard solutions as recommended in ASTM E-578-01 [85]. We currently evaluate the potential of the spectral fluorescence standards BAM F003-F005 for the determination of the range of linearity of fluorescence detectors within the spectral region of 400 to 800 nm. The BAM dyes are especially well suited for this purpose due to the minimum overlap between absorption and emission that minimizes dye-specific inner filter effects [38, 79]. Exemplary, the very promising results from a dilution experiment with dye F004 and a NIKON C1si CLSM are displayed in Fig. 10. Here, the integral emission was measured at a defined z-position (excitation at 405 nm, unchanged measurement parameters) with dye solutions of different concentrations in the channels of a microscopy slide with six parallel channels (Ibidi GmbH, Germany) (cf. Fig. 5).

Another promising example for a solid system seems to be a recently introduced micro-slide system built for microarray scanners using 5 or 11 graded levels of two colors of fluorescent nanoparticles [84]. Its suitability for fluorescence microscope linearity calibration purposes has not been tested yet, but seems likely. However, as this slide is not sealed with a cover glass, it is sensitive to dust, contamination, damage and optical aberrations will occur with most objectives corrected for cover glass usage.

Although often underestimated, even by instrument manufacturers, the reliable determination of the linearity of the detection system for routinely used sets of parameters is of utmost importance for instrument characterization and quantification. Accordingly, in our opinion, internationally agreed methods/protocols in combination with suitable systems are needed for the determination of this quantity for fluorescence microscopy as well as for other fluorescence techniques [35]. This is the only way to eventually realize a reliable instrument characterization and quantitative microscopy.

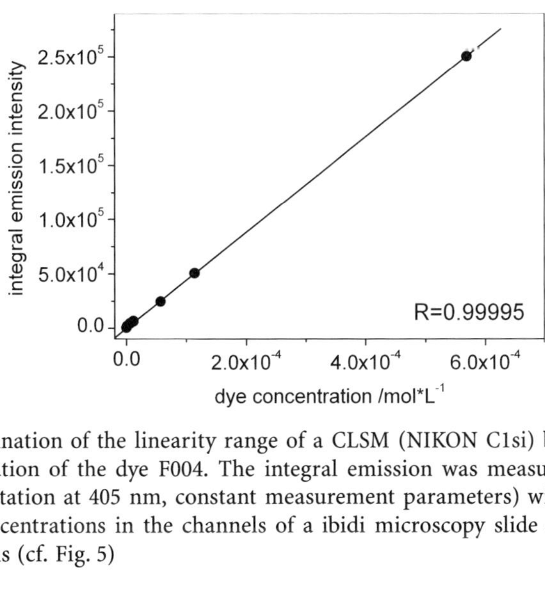

Fig. 10 Determination of the linearity range of a CLSM (NIKON C1si) by controlled dilution of a solution of the dye F004. The integral emission was measured at a defined z-position (excitation at 405 nm, constant measurement parameters) with dye solutions of different concentrations in the channels of a ibidi microscopy slide equipped with 6 parallel channels (cf. Fig. 5)

4
Conclusion, Future Requirements and Challenges

Aside from an enormous progress in fluorescence microscopy and the broad variety of fluorescence standards suggested, the suitability of many of these standards is still under debate, while the need for improved standards for the characterization of fluorescence microscopes and for quantification purposes becomes more and more obvious. Critical is here also the lack of internationally accepted protocols for instrument calibration, control of instrument specifications and performance validation as well as signal quantification. This all together hampers the acceptance of fluorescence microscope methods especially in strongly regulated areas such as medical diagnostics, where standardized instrument characterization is essential. Moreover, microscopists as well as spectroscopists need to become more aware of quality criteria for standards and reference materials as discussed in Sect. 3.2, that are common knowledge in other analytical techniques.

As derived herein (Sect. 3.4), attractive candidate materials for the design of easy-to-operate, solid standards for fluorescence microscopy are glasses doped with inorganic fluorophores with narrow emission bands or broad unstructured emission spectra. An alternative are microchannel devices that can be filled with many different chromophores as shown in Sect. 3.5. These standards enable the reliable characterization of the spectral characteristics of fluorescence microscopes and their performance validation under routine measurement conditions. Both approaches, that are currently evaluated by BAM, are developed to guarantee proper and reproducible standard alignment and incorporation of a cover slip. Another approach currently followed by the National Institute for Standards and Technology (NIST), the only National Metrology Institute (NMI) aside from BAM that develops and releases fluorescence standards, see DeRose et al., 2008, in this volume, includes slides with fluorescent coatings for the spectral region from ca. 450 nm to 900 nm. These materials also show little and well-characterized photodecomposition and a homogeneous distribution of fluorophores (see DeRose et al., 2008, in this volume, standards for the microarray area). This approach is currently being tested for standards for the microarray community.

The wide acceptance of microscopy standards in the microscope community in general requires not only easy-to-use systems, but also SOPs worked out for these standards and specific instruments. Such SOPs should include specific guidance software for standard use including standard measurement, data analysis, and data documentation. To guarantee the reliability of such standards and instrument characterization procedures, both should be tested for a broad variety of different microscopes by nonprofit organizations and individuals like NMIs, laboratories from regulating agencies such as the Environmental Protection Agency (EPA) and the Food and Drug Administration (FDA) and different core facilities and expert laboratories in Round Robin tests. The ultimate goal should be here to identify limitations of certain approaches and to provide the basis for the international acceptance of suitable materials and characterization procedures. The microscope standards should then be subsequently introduced to key users.

Because of the complexity of the instrumentation to be calibrated, in parallel to the development and evaluation of microscopy standards, a much better documentation of the instrument performance and the expected dependence on the operating conditions should be provided by instrument manufacturers. To simplify the choice of instrument to be purchased, a common and reliable catalog of instrument specifications for microscopes (determined using reported or even standardized procedures) would be helpful. In addition, a better training of microscope users is necessary to increase the awareness of the many factors influencing instrument performance and therefore, also the calibration procedures. Preferentially, such courses should be offered jointly by instrument manufactures and existing organizations of microscope users like ELMI, microscopical societies, or other scientific organizations like EMBO.

Future development of confocal instrumentation should ideally consider also the growing need for comparable and maybe eventually standardizable measurements in research, development, and diagnostics. Manufacturers are already trying to take this trend more into account by integrating internal calibration tools in their instruments. These approaches are promising, yet not openly accessible. Here, we would favor an open, standardized interface for calibration purposes. The definition of such an interface should be done in a joint effort by NMIs, regulatory agencies, instrument manufacturers, and expert core facilities.

Acknowledgements Financial support from the Federal Ministry of Economics and Technology (BMWi; grant VI A 2-17/03), the Federal Ministry of Education and Research (BMBF; grant 13N8848), and the Landesstiftung Baden-Württemberg (R.N.) is gratefully acknowledged. We express our gratitude to Dr. J. Rietdorf for critically reading the manuscript.

References

1. Lakowicz JR (2006) Principles of Fluorescence Spectroscopy, 3nd ed. Springer, Berlin Heidelberg New York
2. Lakowicz JR (ed) (1991–2006) Topics in Fluorescence Spectroscopy Series Vols 1–11. Plenum Press, New York and Springer, Berlin Heidelberg
3. Wolfbeis OS (ed) (2001–2005) Springer Series on Fluorescence, Methods and Applications Vols 1–3. Springer, Berlin Heidelberg
4. Schulman SG (ed) (1985–1993) Molecular Luminescence Spectroscopy Parts 1–3. Wiley Interscience, New York
5. Mason WT (1999) Fluorescent and Luminescent Probes for Biological Activity, 2nd edn. Academic Press, San Diego

6. Pawley JB (ed) (1995) Handbook of Biological Confocal Microscopy, 2nd ed. Kluwer Academic Publishers, New York
7. Inoue S (ed) (1986) Video Microscopy. Plenum Pub Corp, New York
8. Murphy DB (ed) (2001) Fundamentals of Light Microscopy and Electronic Imaging. Wiley-Liss, New York
9. Wang XF, Periasamy A, Herman B, Coleman DM (1992) Crit Rev Anal Chem 23:369
10. Ewers H, Smith AE, Sbalzarini IF, Lilie H, Koumoutsakos P, Helenius A (2005) Proc Natl Acad Sci USA 102:15110
11. Shaner NC, Steinbach PA, Tsien RY (2005) RY Nat Method 2:905
12. Tanke HJ, Dirks RW, Raap T (2005) Curr Opin Biotechnol 16:49
13. Zhang J, Campbell RE, Ting AY, Tsien RY (2002) Nat Rev 3:906
14. Sapsford KE, Berti L, Medintz IL (2006) Angew Chem Int Ed 45:4562
15. Haugland RP (ed) (2005) The Handbook: a Guide to Fluorescent Probes and Labeling Technologies, 10th ed. Invitrogen, Eugene, USA
16. Panchuk-Voloshina N, Haughland RP, Bishop-Stewart J, Bhalgal MK, Millard PJ, Mao F, Leung WY, Haughland RP (1999) J Histochem Cytochem 47:1179
17. Berlier JE, Rothe A, Buller B, Bradford JG, Gray DR, Filanoski BJ, Telford WG, Yue S, Liu J, Cheung CY, Chang W, Hirsch JD, Beechem JM, Haughland RP (2003) J Histochem Cytochem 51:1699
18. Andreeff M, Pinkel D (eds) (1999) Introduction to Fluorescence in Situ Hybridization: Principals and Clinical Applications. Wiley-Liss, New York
19. Dundr M, McNally JG, Cohen J, Misteli TJ (2002) Struct Biol 140:92
20. Dobrucki JW (2004) Methods in Cell Biology 75:41
21. Medintz IL, Uyeda HT, Goldman ER, Mattoussi H (2005) Nat Mater 4:435
22. Gao X, Cui Y, Levenson RM, Chung LWK, Nie S (2004) Nat Biotechnol 22:969
23. Dickinson ME, Bearman G, Lansford R, Fraser SE (2001) BioTechniques 31:1272
24. Zucker RM, Price OT (1999) Methods—A Companion to Methods in Emzymology 18:447
25. Centonze VE, Takahashi A, Casanova E, Herman BJ (2000) Histotechnol 23:229
26. Zucker RM (2006) Cytometry 69A:659
27. Zucker RM (2006) Cytometry 69A:677
28. Zwier JM, van Rooij GJ, Hofstraat JW, Brakenhoff GJ (2004) J Microscopy 216:15
29. Lerner JM, Zucker RM (2004) Cytometry 62A:8
30. Larkin JE, Frank BC, Gavras H, Sultana R, Quackenbush J (2005) Nat Methods 2:337
31. Weis BK, Members of the Toxicogenomics Research Consortium (2005) Nat Methods 2:352 (Standardizing global gene expression analysis between laboratories and across platforms)
32. Ploem JS (1970) Standards for Fluorescence Microscopy. In: Holborow EJ (ed) Stand Immunofluorescence Symp. Blackwell Sci Publ, Oxford, p 137
33. Velapoldi RA, Epstein MS (1989) Luminescence Applications in Biological, Chemical, Environmental and Hydrological Sciences. In: Goldberg MC (ed) ACS Symposium Series 383. ACS, Washington, DC, p 98
34. Niles WD, Cohen FS (1995) Rev Sci Instrum 66:3527
35. Resch-Genger U, Hoffmann K, Nietfeld W, Engel A, Neukammer J, Nitschke R, Ebert B, Macdonald R (2005) J Fluoresc 15:337
36. Marti GE, Vogt RF, Gaigalas AK, Hixson CS, Hoffman RA, Lenkai R, Magruder, LE, Purvis NB, Schwartz A, Shapiro HM, Waggoner A (2004) NCCLS Fluorescence Calibration and Quantitative Measurement of Fluorescence Intensity; Approved Guideline I/LA24-A, Vol 24, No. 26

37. EN ISO/IEC 17025: Good laboratory practice (GLP)/good manufacturing practice (GMP); ASTM International (2007) ASTM E578-07: Standard Test Method for Linearity of Fluorescence Measuring Systems; ASTM E579-04: Standard Test Method for Limit of Detection of Fluorescence of Quinine Sulfate in Solution; ASTM E388-04: Standard Test Method for Spectral Bandwidth and Wavelength Accuracy of Fluorescence Spectrometers
38. Resch-Genger U, Pfeifer D, Monte C, Pilz W, Hoffmann A, Spieles M, Hollandt J, Taubert RD, Schönenberger B, Nording P (2005) J Fluoresc 15:315
39. Zucker RM, Paul Rigby P, Clements I, Salmon W, Chua M (2007) Cytometry 71:174
40. Nitschke R (2004), Standardization and quantification in microscopy. Workshop AK PhotonicNet, Wetzlar
41. Eastman JW (1966) Appl Optics 5:1125
42. Galbraith W, Ryan KW, Gliksman N, Taylor DL, Waggoner AS (1989) Comput Med Imaging Graph 13:47
43. Erdogan T, Pradhan A, Mizrahi V (2003) Biophotonics Int 10:38
44. Hollandt J, Taubert DR, Seidel J, Resch-Genger U, Gugg-Helminger A, Pfeifer D, Monte C (2005) J Fluoresc 15:301
45. Haugland RP (ed) (2002) Handbook of Fluorescent Probes and Research Products, 9th ed. Molecular Probes, Eugene, OR, USA
46. Haaijman JJ, van Dalen JPR (1974) J Immunol Methods 5:359
47. Lockett SJ, Jacobson K, Herman B (1992) Anal Quant Cytol Histol 14:187
48. Sernetz M, Thaer A (1970) J Microscopy 91:43
49. Rost FWD (1991) Quantitative Fluorescence Microscopy. Cambridge University Press, Cambridge, UK, p 236
50. Zwier JM, Van Rooij GJ, Hofstraat JW, Brakenhoff GJ (2004) J Microscopy 216:15
51. Brakenhoff GJ, Wurpel GWH, Jalink K, Oomen L, Brocks L, Zwier M (2005) J Microscopy 219:122
52. Model MA, Burkhardt JK (2001) Cytometry 44:309
53. Sisken JE (1989) Methods Cell Biol 30:113
54. Turney SG, Culican SM, Lichtman JW (1996) J Neurosci Methods 64:199
55. Stevens PW, Kelso DM (2003) Anal Chem 75:1147
56. Jones AC, Millington M, Muhl J, De Freitas JM, Barton JS, Gregory G (2001) Meas Sci Techn 12:N23
57. Adelheim K, Emantraut E, Kaiser T, Tuchscheerer J (2002) 5:22
58. Kaplan DS, Picciolo GL (1989) J Clin Microbiol 27:442
59. Jongsma APM, Hijmans W, Ploem JS (1971) Histochemie 25:329
60. National Institute of Standards and Technology (2007) Certificate of analysis, Standard Reference Material 2940, Relative intensity correction standard for fluorescence spectroscopy: Orange emission
61. National Institute of Standards and Technology (2007) Certificate of analysis, Standard Reference Material 2941, Relative intensity correction standard for fluorescence spectroscopy: Green emission
62. Hoffmann K, Monte C, Pfeifer D, Resch-Genger U (2005) GIT Laboratory J 6:29
63. Starna (2007) Online catalog, Starna Scientific Limited. Hainault, UK http://www.starna.com; last visited: 08 February 2008
64. Weidner VR, Mavrodineanu R, Eckerle KL (1986) Appl Optics 25:832
65. Applied Precision, LLC (2007) Online catalog. Applied Precision, LLC, Issaquah, Washington, USA, http://www.appliedprecision.com; last visited: 08 February 2008
66. Velapoldi RA, Travis JC, Cassatt WA, Yap WT (1975) J Microscopy 103:293

67. Evident (2007) Online catalog. Evident Technologies Inc. Troy, USA, http://www.evidenttech.com; last visited: 08 February 2008
68. Knight A, Gaunt J, Davidson T, Chechnik V, Windsor S (2004) NPL Report DQL-AS 007, p 1
69. van Sark WGJHM, Frederix PLTM, van den Heuvel DJ, Gerritsen HC, Bol AA, van Lingen JNJ, de Mello CD, Meijerink A (2001) J Phys Chem B105:8281
70. Cho EH, Lockett SJ (2006) J Microscopy 223:15
71. Zucker RM, Lerner JM (2005) Microscopy Res Technique 68:307
72. Young IT (1983) Proc SPIE 387:326
73. Young IT, Garini Y, Vermolen BJ (2006) SPIE 6088:1
74. Young IT, Garini Y, Dietrich HRC (2004) SPIE 5324:208
75. APE GmbH (2007) Online catalog. APE GmbH, Berlin, Germany, http://www.ape-berlin.de; last visited: 08 February 2008
76. Chroma Technology Corp. (2007) Online catalog. Chroma Technology Corp., Rockingham, USA, http://www.chroma.com; last visited: 08 February 2008
77. Monte C, Resch-Genger U, Pfeifer D, Taubert RD, Hollandt J (2006) Metrologia 43:S89
78. Monte C, Pilz W, Resch-Genger U (2005) Proc SPIE 5880:588019-1
79. Pfeifer D, Hoffmann K, Hoffmann A, Monte C, Resch-Genger U (2006) J Fluoresc 16:581
80. Lifshitz IT, Meilman ML (1989) Sov J Opt Technol 55:487 (DYAG, Photon Technology International Inc, FA-2036)
81. Engel A, Ottermann C, Resch-Genger U, Spaeth J, Schweizer S, Selling J, Hoffmann K, Rupertus V (2006) SPIE Biophoton New Therap Frontiers 6191-36:259
82. Hoffmann K, Resch-Genger U, Nitschke R (2005) GIT Imaging Microscopy 3:18
83. ibidi GmbH (2007) Online catalog. Integrated BioDiagnostics GmbH. Martinsried, Germany, http://www.ibidi.de; last visited: 08 February 2008
84. Calslide I, Calslide II (2007) Capitalbio Corporation, Beijing, China, http://www.capitalbio.com; last visited: 08 February 2008
85. ASTM E 578-01 (2001) Linearity of Fluorescence Measuring System. In: Annual Book of ASTM Standards, Vol 03.06. American Society for Testing and Materials, Philadelphia, USA
86. Etz ES, Choquette SJ, Hurst WS (2005) Microchim Acta 149(3–4) 175

Springer Ser Fluoresc (2008) 6: 117–142
DOI 10.1007/4243_2008_030

Published online: 11 March 2008

Fluorescence Lifetime Imaging Microscopy: Quality Assessment and Standards

Alessandro Esposito[1,2] · Hans C. Gerritsen[3] · Fred S. Wouters[4,5] (✉)

[1]Laser Analytics Group, Department of Chemical Engineering, University of Cambridge, New Museums Site, Pembroke, Cambridge CB2 3RA, UK

[2]Physiological Laboratory, Department of Physiology, Development and Neuroscience, University of Cambridge, Downing Street, Cambridge CB2 3EG, UK

[3]Debye Institute, Utrecht University, PO Box 80 000, NL 3508 TA Utrecht, The Netherlands

[4]European Neuroscience Institute Göttingen and DFG Center for Molecular Physiology of the Brain (CMPB), Waldweg 33, 37073 Göttingen, Germany

[5]*Present address:*
Laboratory for Molecular and Cellular Systems,
Department of Neuro- and Sensory Physiology,
Institute for Physiology and Pathophysiology, University Medicine Göttingen,
Humboldtallee 23, 37073 Göttingen, Germany
fred.wouters@gwdg.de

Abstract The measurement of fluorophore lifetimes—the excited state duration—in the microscope provides unique quantitative information on the molecular environment and is therefore increasingly being used in cell biological questions. Perhaps the most popular use of fluorescence lifetime imaging (FLIM) is to measure Förster resonance energy transfer (FRET) to detect protein interactions, conformational changes, and activities in the context of the (living) cell. The analytical use of FLIM requires a detailed knowledge of the proper use and limitations of its different instrumental implementations, including platform standardization and calibration, and considerations regarding its optimization for increased throughput. The results obtained with FLIM are conditional on the quality of the data. Therefore, stringent data analysis assessment and analysis criteria have to be maintained in the imaging workflow. In particular, the issues of photobleaching and fluorophore saturation, and their effect and correction possibilities, are discussed. This chapter deals with the various aspects of FLIM that need to be taken into consideration when this powerful technique is to be used as an analytical tool in the life sciences.

Keywords Fluorescence lifetime · Förster resonance energy transfer · Quantitative microscopy

Abbreviations

CHO	Chinese hamster ovary cell
CSM	Immortalized rat nigrostriatal cell
DASPI	2-(*p*-Dimethylaminostyryl)-pyridylmethyl iodide
DBS	4-Dimethylamino-(4′-bromo)-stilbene
DCS	4-Dimethylamino-(4′-cyano)-stilbene
DFS	4-Dimethylamino-(4′-fluoro)-stilbene
DMS	4-Dimethylamino-(4′-oxymethyl)-stilbene
EGFP	Enhanced green fluorescent protein
FD	Frequency domain
FLIM	Fluorescence lifetime imaging microscopy
FRET	Förster resonance energy transfer
MCP	Multichannel plate
REACh2	Resonance energy-accepting chromoprotein
TCSPC	Time-correlated single-photon counting
TD	Time domain
YFP	Yellow fluorescent protein

1 Introduction

Fluorescence microscopy is a noninvasive technique that provides high contrast and spatial resolution. The revolutionary innovations of the past few decades in solid-state technologies, information technologies, and fluorescent staining techniques turned fluorescence microscopy into a highly flexible and quantitative analytical technique. With its many implementations and applications, fluorescence microscopy became one of the fundamental and most widely used techniques in the life, medical, and materials sciences. Furthermore, quantitative and multiparametric fluorescence imaging can nowadays be combined with the high throughput provided by sample handling robotics

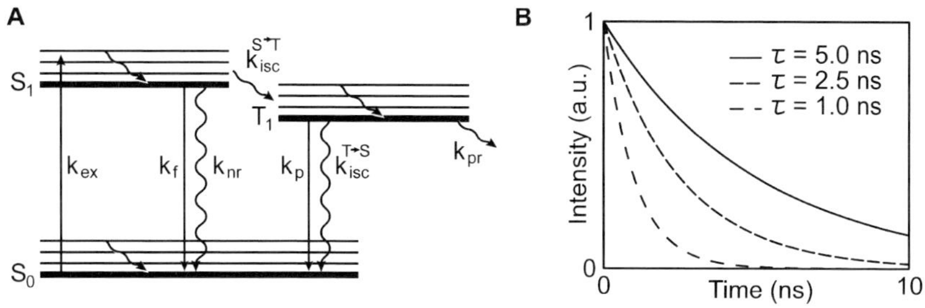

Fig. 1 Florescence lifetime. **A** Jablonski diagram of a fluorophore where the ground state (S_0) and the first singlet (S_1) and triplet excited states (T_1) are depicted. The transitions shown represent $S_0 \rightarrow S_1$ excitation (ex), $S_1 \rightarrow S_0$ fluorescence (f) decay, the nonradiative $S_1 \rightarrow S_0$ de-excitation intersystem crossing (isc) between singlet and triplet state ($S \rightarrow T$, $T \rightarrow S$), radiative triplet state depopulation (phosphorescence, p), the generation of photochemical reaction products (pr) from the triplet excited state, and the thermal relaxation within vibrational levels. **B** Competition between the de-excitation pathways defines the duration (lifetime) of the excited state and consequently gives rise to the fluorescence decays shown

and automation. Fluorescence lifetime imaging microscopy (FLIM) is an inherently quantitative technique that provides information on the biochemical environment of the fluorophore. Applied to biology, FLIM has provided a quantitative tool for the imaging of cellular biochemistry.

The fluorescence lifetime is the average time that a fluorophore spends in its excited state and depends on the transition rates of the singlet excited state (S_1) de-excitation pathways (Fig. 1A):

$$\tau = \left(k_f + k_{nr} + k_{isc}^{S \rightarrow T}\right)^{-1},$$

where k_f, k_{nr}, and $k_{isc}^{S \rightarrow T}$ are the radiative (fluorescence), nonradiative (internal conversion), and intersystem crossing transition rates (transition occurring between the singlet and triplet excited states $S \rightarrow T$), respectively. The quantum yield (Q) of a fluorophore is given by:

$$Q = k_f\left(k_f + k_{nr} + k_{isc}^{S \rightarrow T}\right)^{-1}.$$

It indicates the fraction of absorbed photons that will cause the emission of a fluorescence photon. Photon emission is a stochastic process and typically follows a normalized exponential decay distribution:

$$p(t) = \tau^{-1} e^{-t/\tau}.$$

The excited state lifetime can therefore be measured by resolving the fluorescence decays. Fluorescence lifetimes are commonly not longer than a few hundreds of nanoseconds, and the most frequently used fluorophores in the life sciences (e.g., Cy and Alexa dyes, Rhodamines, and fluorescent proteins) exhibit lifetimes between 1 and 5 ns (Fig. 1B).

Interactions between the fluorophore and its molecular environment may alter or add de-excitation pathways, causing the lifetime of the fluorophore excited state to change. For instance, fluorophores whose lifetime reports on pH, ion concentrations, and oxygen content in living cells and tissues have been described [23]. One of the most relevant applications for FLIM is the quantification of Förster resonance energy transfer (FRET [10]), a phenomenon commonly used for the detection of intermolecular interactions or molecular conformational changes. These changes are often exploited to construct FRET-based biosensors for specific cellular biochemical events.

FLIM has been implemented on wide-field [20, 31, 35, 47, 48] and laser-scanning microscopes [4, 6, 7, 12, 24, 38]. In recent years, its use has increased thanks to the availability of cost-effective pulsed or modulatable lasers and light-emitting diodes, and commercial FLIM systems (for instance, PicoQuant, Becker & Hickl, Nikon Instruments Europe, LaVision, LaVision BioTec, Lambert Instruments, HORIBA Jobin Yvon, Hamamatsu, ISS).

2
Förster Resonance Energy Transfer

FRET is the nonradiative transfer of energy (Fig. 2A) from a donor fluorophore to an acceptor chromophore through long-range dipole–dipole interactions. For this reason, the FRET efficiency, i.e., the fraction of energy transferred from a donor to an acceptor, strongly depends on the interchromophore distance (R, see Fig. 2B):

$$E = \frac{1}{1 + (R/R_0)^6} ,$$

where R_0, the Förster distance, is the distance at which 50% of energy is transferred. Typical FRET donor–acceptor pairs exhibit Förster distances of a few

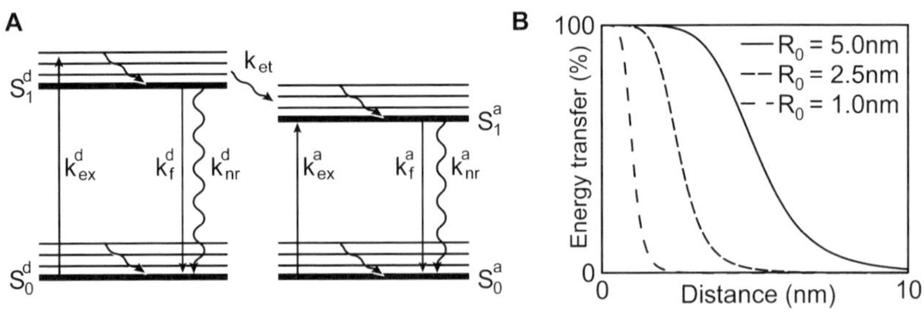

Fig. 2 Förster resonance energy transfer. **A** Jablonski diagram of a FRET pair, where the suffixes "d" and "a" indicate transition rates and energy levels for the donor and acceptor, respectively. K_{et} is the energy transfer rate. **B** Dependence of FRET efficiency on the interchromophore distance and Förster distance

nanometers and FRET generally does not occur at interchromophore distances exceeding 10 nm. FRET is thus sensitive to distances that are comparable to protein dimensions and can therefore report on direct intermolecular interactions.

The energy transfer efficiency increases from less than 2% to more than 98% when the intermolecular distance decreases only from twice to half of the Förster distance. FRET therefore provides a very high sensitivity for inter- or intramolecular distances, depending on whether the donor and acceptor label the same or different molecules, respectively.

The Förster distance depends on the refractive index of the medium (n), the donor quantum yield (Q), the overlap integral of the donor emission and acceptor absorption spectra (J), and an orientational factor (κ^2) according to [10]:

$$R_0 = (\kappa^2 Q J n^{-4})^{1/6} ,$$

where κ^2 is maximal ($\kappa^2 = 4$) only when the donor transition dipole and the acceptor absorption dipole are collinear. In all the other configurations, κ^2 is lower. A detailed description of the orientation factor can be found in [30, 42]. Briefly, when the two chromophores are free to rotate during the donor lifetime, κ^2 will be averaged over all possible orientations and will assume a value of 2/3. Otherwise, κ^2 can assume a value between 0 and 4, with a higher statistical probability for lower values.

FLIM provides a quantitative and robust technique for FRET imaging. The donor lifetime is reduced by FRET proportionally to the FRET efficiency [51]:

$$E = 1 - \tau/\tau_0$$

where τ_0 is the donor fluorescence lifetime in the absence of FRET.

3 Instrumentation and Techniques

The fluorescence lifetime can be measured in the time domain (TD) and in the frequency domain (FD) [30]. Both laser-scanning and wide-field systems for FD- and TD-FLIM have been developed. Laser scanning microscopes make use of detectors like photomultipliers and avalanche photodiodes, while wide-field systems are commonly built around a multichannel plate (MCP).

In the time domain, the sample is excited with a pulsed light source, commonly mode-locked femtosecond Ti:sapphire lasers or pulsed laser diodes with subnanosecond pulse widths. In the time domain, the two most frequently used techniques are time-correlated single-photon counting (TCSPC) and time gating. The former technique is based on the measurement of the arrival time of the first emitted photon relative to the excitation pulse. In time

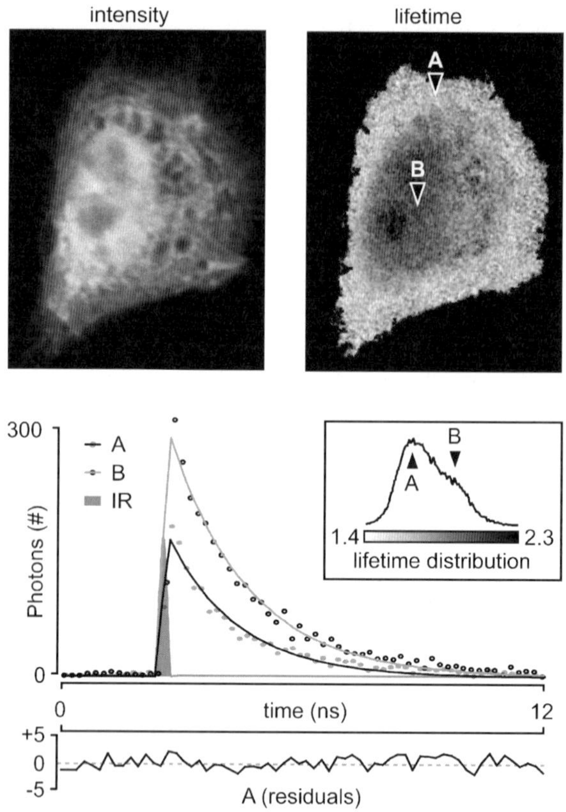

Fig. 3 Time-domain lifetime detection. Shown is a CSM cell coexpressing Cerulean:α-synuclein and YFP:tau. Their interaction is assessed by the measurement of the donor (Cerulean) lifetime. TD-FLIM provides the time resolution of fluorescence decays (*lower panel*). The experimental data (*gray dots*) can be fitted by exponential models (*black curve*) to estimate the fluorescence lifetime at each pixel location. Here, a single-exponential fit is performed, where more than 1000 photons were available at every pixel (binning 3×3). The randomly distributed residuals and the χ^2 equal to ~1.2 indicate that the single-exponential fit is sufficient for the representation of the fluorescence decay. To represent the multiexponential decay of the Cerulean fluorescent protein, higher counts are necessary. The lifetime map shows that the interaction of the two proteins occurs at the cellular periphery. The fluorescence decays in *regions marked A* and *B* are represented in the fit panel and in the lifetime distribution. The former confirms the bimodal lifetime distribution caused by the presence of pixels exhibiting FRET and regions in which the two proteins do not interact. The gray area in the *lower panel* shows the instrument response (IR)

gating, photons are counted by switching on different photon counters in adjacent time windows. Both techniques provide histograms of arrival times (Fig. 3) that can be fitted by appropriate physical models to estimate the fluorescence lifetime of the sample.

Figure 3 shows an image of a CHO cell expressing α-synuclein and the tau protein tagged with the fluorescent proteins Cerulean and YFP, respectively. The interaction of these two proteins, both involved in neurodegeneration, results in the reduction of the donor lifetime as a consequence of FRET. Regions in which α-synuclein and tau interact and areas where no interaction occurs become visible in the lifetime map of the cell and in the corresponding data fit. The images were acquired by a TCSPC system based on a TSC SP2 AOBS confocal microscope (Leica Microsystems Heidelberg GmbH, Mannheim, Germany) equipped with a Mira900 mode-locked femtosecond Ti:sapphire laser (Coherent Inc., Santa Clara, CA, USA), an MCP-PMT (R3809U-50 by Hamamatsu Photonics, Sunayama-cho, Japan), and an SPC830 acquisition board (Becker & Hickl GmbH, Berlin, Germany), as described in more detail in [17].

In the frequency domain, the sample is excited with a periodically modulated light source. Often, the light source is a continuous-wave laser that is sinusoidally modulated by feeding the light through an acousto-optical modulator. More recently, laser diodes and light-emitting diodes have been used. The finite fluorescence lifetime of the fluorophore injects a phase delay (ϕ) into the fluorescence emission and demodulates (m) its amplitude relative to the excitation light (Fig. 4):

$$m = (1 + \omega^2 \tau^2)^{-1/2}\,; \quad \phi = \arctan(\omega\tau)\,,$$

where ω is the modulation frequency of the excitation light. Therefore, the measurement of the phase delay and relative demodulation of the fluorescence emission provides two estimators for the fluorescence lifetime.

A photomultiplier tube with a modulated anodic potential is typically used in FD laser scanning systems. This enables the estimation of the lifetime by cross-correlation of the fluorescence emission and the excitation signal. Both homo- and heterodyning methods can be applied. In the first case, the detector is modulated at the same modulation frequency as the light source. In the second case, the detector is modulated at a slightly different modulation frequency; the phase and the demodulation information are here transferred to a harmonic signal at the difference between the excitation and detection frequencies. Detection in the frequency domain can also be conveniently performed by the use of lock-in amplifiers.

Figure 4 shows the average lifetime image (average of the modulation- and phase-lifetime estimations) of a CSM cell expressing α-synuclein and ubiquitin tagged with enhanced green fluorescent protein (EGFP) and a YFP-derived nonfluorescent chromoprotein (REACh2, REF), respectively. The fluorescence lifetime of the donor fluorophore is shortened when α-synuclein becomes ubiquitinated. Furthermore, this causes the lifetime heterogeneity to increase due to the presence of multiple fluorescence decays (i.e., donors undergoing FRET and noninteracting donors). The reduction of the lifetime

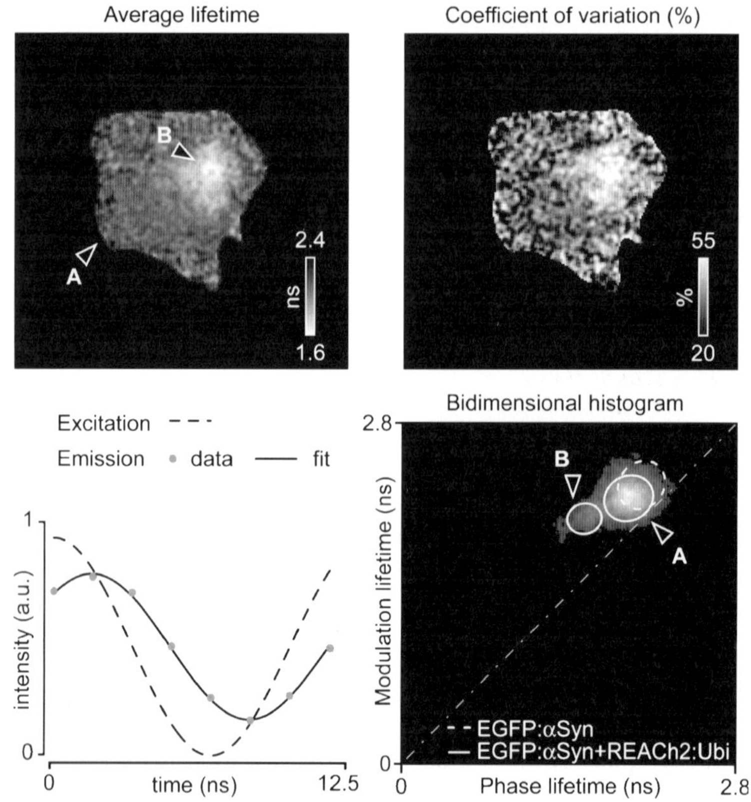

Fig. 4 Frequency domain detection. Shown is a CSM cell coexpressing EGFP:α-synuclein and REACh2:ubiquitin. The ubiquitination of α-synuclein is assessed by the measurement of donor (EGFP) lifetimes. In the frequency domain, the harmonic response of the fluorophore is measured to detect the relative phase delay and demodulation between the excitation light (*dashed line*) and the fluorescence emission (*gray dots*: experimental data, *black curve*: data fit). Here, the average response in the field of view is shown. The average lifetime of the cell expressing both proteins shows the presence of FRET in a cytoplasmic region of the cell. Lifetime heterogeneity is increased by FRET due to the presence of multiple-exponential decays. The plot of the modulation lifetime versus the phase lifetime estimation confirms a slight (∼5%) reduction of the lifetime over the entire cell (*solid ellipse, A*) relative to a control in which no REACh2:ubiquitin was expressed (*dashed ellipse*). Furthermore, the appearance of a pixel cluster (*solid ellipse, B*) at lower lifetime and with increased heterogeneity (shown by the increased distance from the diagonal of the plot) confirms the presence of localized FRET

and the increased heterogeneity can be visualized using bidimensional histograms. The image was acquired by an in-house developed FD-FLIM system built around a fully automated Axiovert200M microscope (Carl Zeiss Jena GmbH, Jena, Germany), equipped with an Innova 300C argon laser (Coherent Inc., Santa Clara, CA, USA) modulated by an acousto-optical modulator driven at ~80 MHz and a high-rate-imager MCP intensifier (Lavision GmbH, Göttingen, Germany), described in detail elsewhere [17].

Typical wide-field FLIM systems are based on the fluorescence detection by an MCP that can be gated or modulated at very high speeds (<1 ns rise time, ≫10 MHz modulation frequency). MCPs can be used in both the time and frequency domain and, depending on their operation, provide stacks of images that are collected at different time or phase delays. The fluorescence lifetime of the sample is computed from these images.

4 Fluorophore Standards

There is substantial interest in the specification of fluorescence standards that are useful for the calibration or benchmarking of lifetime imaging microscopes. However, many parameters can affect the intrinsic lifetime of a fluorophore. As mentioned before, the biochemical environment of the chromophore can significantly modify the lifetime of the excited state. Measurements with a precise composition of the solvent/buffer are therefore necessary. Fluorescence lifetimes can also be sensitive, for instance, to the pH and to the viscosity of the media, parameters that should therefore be controlled as well. Furthermore, the temperature dependence of the excited state lifetime is often overlooked. However, a temperature difference of only a few degrees may change the excited state lifetime of a nanosecond fluorophore by some hundreds of picoseconds. The stability of the temperature in laboratories where FLIM systems are operated is therefore of extreme importance.

A reflective or a scattering sample positioned at the focal plane provides useful zero-lifetime standards. However, optical filters and detector color effects may cause differences in the time/phase estimations measured at different spectral bands. Spectrally matched fluorophores would therefore constitute better samples for calibration.

Protocols have been suggested for the generation of lifetime gradients (see Table 1) that may serve for the calibration of systems at multiple lifetimes [9, 26]. For instance, quinine quenched with different concentrations of

Table 1 Lifetime gradients

Fluorophore	Additive	Lifetime range (ns)	Refs.
DASPI	Glycerol	0.05–0.30	[14]
Rhodamine 6G	KCl	0.5–4.1	[26]
Quinine	NaCl	0.189–18.9	[9]
γ-pyrenebutyrate	KI	18–115	[9]

sodium chloride provides a gradient of lifetimes ranging from a few hundreds of picoseconds up to ~20 ns. Similarly, γ-pyrenebutyrate can be quenched with potassium iodide to cover the range between ~20 and ~115 ns. Although useful, it is important to underline that quinine and γ-pyrenebutyrate exhibit simple single-exponential decays, whereas the quenched fluorophores possess a heterogeneous decay [30]. The contrast between Coumarin 314 (~3.3 ns) and DASPI (~0.3 ns) has also been used for the characterization of FLIM systems [14]; the fluorescence lifetime of DASPI can also be shortened by controlling the viscosity of the solvent to provide picosecond lifetime references (50–300 ps).

The measurement of lifetime heterogeneity is important for the understanding of fluorophore photophysics, or for the discrimination of different fluorescent moieties in the focal volume. Single-exponentially decaying fluorophores therefore play important roles as homogeneous fluorescence standards. A more commonly used fluorophore that provides a single-exponential decay is Rhodamine 6G with a decay of ~4.1 ns [26].

Short-lived fluorophores can also be used as fluorescence standards. An example is Rose Bengal which exhibits a single-exponential decay with a time constant of ~80 ps [39] and ~500 ps [30] in PBS or ethanol, respectively. A set of fluorophore derivatives of 4-dimethylamino-stilbene have also been characterized [32] to possess homogeneous fluorescence lifetimes in the range of ~60 and ~900 ps, and represent an interesting set of standards for multipoint system calibration and for the assessment of possible detector color effects.

Table 2 Single-exponential fluorescence lifetime standards

Fluorophore	ex/em (nm)	Solvent	Lifetime (ns)	Refs.
DCS	280–420/300–500 [a]	Cyclohexane	0.066	[32]
Rose Bengal	556/572	PBS	0.078	[39]
DBS	280–385/375–475 [a]	Cyclohexane	0.176	[32]
DASPI	400–550 [a]/580 [b]	Ethanol	0.274	[14]
DFS	280–375/375–450 [a]	Cyclohexane	0.328	[32]
Rose Bengal	556/572	Ethanol	0.519	[30]
DMS	280–375/375–475 [a]	Cyclohexane	0.88	[32]
Dimethyl-POPOP	300–400/390–560	Ethanol	1.45	[30]
Rhodamine B	540/625	Ethanol	2.88	[27]
Coumarin 314	436/460	Ethanol	3.32	[14]
Rhodamine 6G	525/555	H_2O	4.11	[26]
Quinine	250–350/450 [a]	0.1 N H_2SO_4	18.9	[9]
γ-Pyrenebutyrate	250–350/400 [a]	H_2O	115	[9]

[a] The single-exponential fluorescence decays have been shown to be constant over the reported spectral ranges.
[b] French PM and McGinty J (personal communication).

A list of single-exponentially decaying fluorophores and their references is presented in Table 2.

There is no particular need for FRET standards in lifetime microscopy, because once a system is properly calibrated for lifetime detection, FRET efficiencies will be detected quantitatively. However, the validation of new analytical methods may require standardized FRET samples. Fluorescent beads are probably the easiest model, where fluorophore can be conjugated to, for instance, (controlled-pore or solid) glass or polystyrene beads [18]. Using solutions of different fluorophores, such as Alexa 488 and Alexa 546, as donor and acceptor pairs at different concentrations, it is possible to obtain samples exhibiting defined FRET efficiencies. It is also possible to coat beads with fluorescent proteins and tag them by maleimide/succinimidyl-conjugated probes, e.g., green fluorescent protein conjugated to Cy3 [17]. Fluorescent proteins can also be engineered in tandem constructs with donor and acceptor present in the same molecule. This method is particularly useful for cellular imaging. Finally, we note the possibility to use DNA-conjugated fluorophores, where the DNA can be used as a fine subnanometer spacer between donor and acceptor [49]. Other polymers can also be used for these purposes [40].

5 System Calibration

TCSPC-based systems almost never require day-to-day calibration. However, fluorescence standards are useful for the characterization and benchmarking of the system. Particularly, single-exponential fluorescence standards are useful for testing the presence of excitation bleed-through on the detector, nonlinearities, and systematic errors caused by the instrumentation.

In general, a pulse picker is used to trigger the TCSPC electronics and if an appropriate delay is set, the fluorescence decay can be adequately stored in the board memory. Thanks to the relatively high time resolution of the currently available TCSPC electronic boards, the data fitting procedures can estimate the excitation instant with high precision [3]. Therefore, contrary to time-gated and frequency-domain systems, no routine calibration is required. However, TCSPC could require the determination of the system response in order to deconvolve the experimental data. As the detector response can exhibit color effects, it should be measured in the spectral range of the fluorescence emission.

Time-gated systems also require a correct time delay between the excitation pulse trigger and fluorescence signals [11]. In this case, however, the available number of time gates is typically less (2–8) than the number of time bins of a TCSPC (16–256). Time-gating systems therefore require the positioning of the first time gate immediately after the excitation pulse. An anticipated start of the gates has to be avoided. On the other hand, the injection of an initial delay may be beneficial to suppress the scattered light and short-lifetime fluorescence background, but at the cost of some detectable signal. In this case it is useful to calibrate the initial delay time relative to the trigger at every system startup by imaging a mirror or a (near-zero lifetime) fluorescence standard (Table 2). Furthermore, timing jitter of the electronics, detector response characteristics, and asymmetry between the time gates can require a thorough system characterization by the imaging of well-known standards.

It is advisable to characterize a fluorescence lifetime imaging microscope with different lifetime standards in order to minimize possible systematic errors. For instance, although a system may appear well-calibrated when imaging a fluorophore with a certain lifetime value, it may deviate in other lifetime ranges [26]. This may be caused by timing jitter of the electronics or mismatched time window widths. These two parameters cannot be both compensated when only a single-lifetime calibration is used. Gradients of fluorescence lifetimes (Table 1) may be useful in this regard, although the appearance of multiexponential decays in the presence of quenchers may necessitate the use of multiple distinct but individually single-exponentially decaying fluorophores (Table 2).

FD-FLIM apparatus requires more care during daily operation and calibration. In fact, phase-detection techniques are inherently and significantly sensitive to the initial phase of the system that needs to be carefully measured. An exponential decay, on the other hand, is scale-invariant and can be fitted without errors even in the presence of a different initial time delay. Therefore, a frequency-domain system requires the precise measurement of the initial phase and modulation of the excitation light at the sample plane before every use. This calibration requires periodic updates during experiments to compensate possible time and temperature drifts typically exhibited by the instrumentation.

Fluorescence lifetime heterogeneity can be resolved by single- or multifrequency measurements [17, 45, 46]. In any case, the lifetime heterogeneity is encoded in the differences between the apparent phase and modulation lifetimes. An error in their relative calibration can cause the heterogeneity to be measured incorrectly. Here, the use of single-exponential, well-characterized fluorescence lifetime standards (Table 2) is necessary.

A multipoint calibration has also been proposed to avoid systematic errors in the frequency domain [26]. It was shown that Rhodamine 6G quenched by potassium iodide may serve as a standardized lifetime gradient to detect systematic errors in the calibration parameters that are caused by differences between the optical paths that are used for the calibration and for the measurements.

Very often, the calibration procedure of a FD-FLIM instrument is performed sequentially to the measurement. In our laboratory, we automated the calibration process by positioning a low-efficiency scatterer in the motorized filter revolver of the microscope. The constant relative phase delay and demodulation between this internal reference and a reference at the sample plane are measured at the startup of the system, and the computer automatically performs regular calibrations without the need for sample handling. Recently, a calibration method for spectrally resolved lifetime measurements has been described [25] that permits (a fraction of) the excitation light to bleed through to the detector. The spectral separation of the excitation and emission light now provides a parallel and robust method for continuous system calibration.

6
Photon Efficiency and Photon Economy

The quality of a lifetime image depends on the number of detected photons, the instrumentation used, the adopted detection technique, and the data analysis. A figure of merit (F) can be defined [23] that describes the photon efficiency of lifetime detection by the ratio of the relative error of the lifetime estimation and of the fluorescence intensities:

$$F = (\sigma_\tau/\tau)N^{-1/2} \, ,$$

where σ_τ indicates the standard deviation of the lifetime and N the total number of counted photons. This figure of merit can be generalized, but is here given under the assumption that Poissonian noise dominates.

For $F = 1$, the lifetime estimator provides the highest achievable signal-to-noise ratio and can be defined *efficient*. For $F > 1$, the estimator is less efficient and therefore requires more photons to achieve similar results, i.e., F^2-fold more photons. The inverse of F is an indicator for the photon economy of lifetime detection.

The photon economy of many time- and frequency-domain lifetime detection systems has been extensively studied [8, 22, 23, 29, 36]. TCSPC and time-gated systems with multiple time gates (>16) can provide an F value equal to 1. In combination with pulsed excitation, frequency-domain detection also provides the same result.

For instance, with these efficient systems, a single-exponential decay can be resolved with a relative error of only 5% if at least 400 photons are collected. In the presence of Poissonian noise it is impossible to obtain a better result. A two-window time-gated system would require at least 900 photons and an MCP-based FD-FLIM system with a fully sine-modulated light source would require more than 10 000 photons to achieve the same error level. Resolving multiple-exponential decays necessarily requires higher fluorescence intensities.

Figure 5 shows the effect of the photon statistics on the quantification of lifetimes. Although it is possible to obtain decent intensity images at low

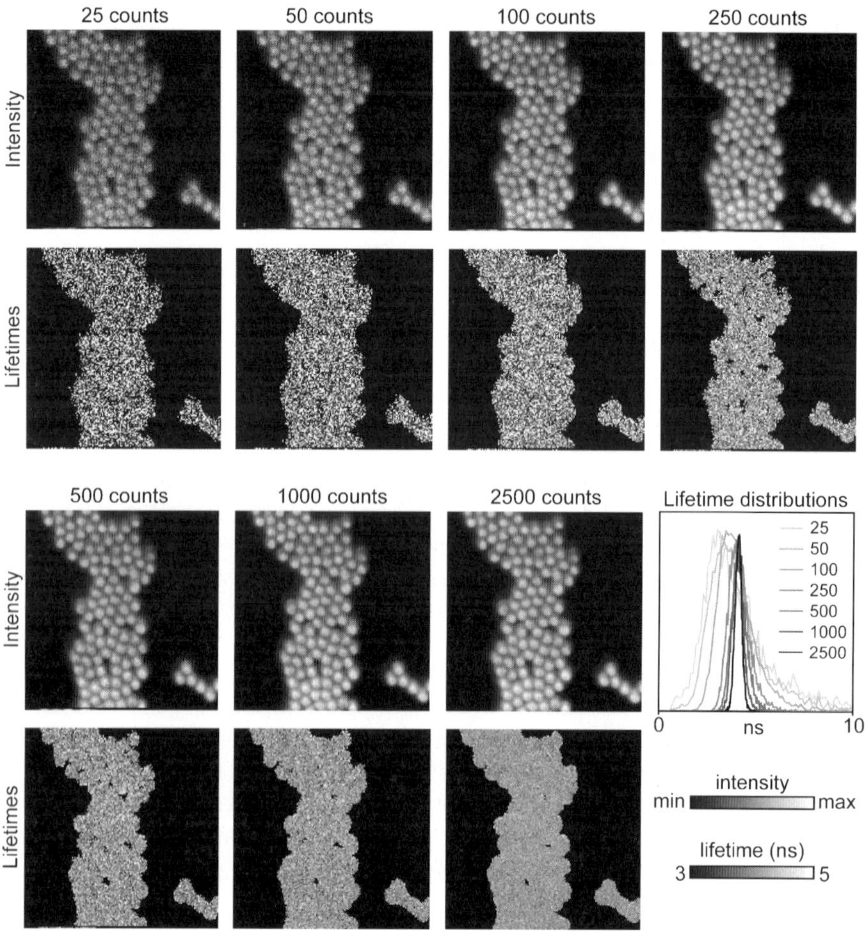

Fig. 5 Photostatistics and lifetime. Intensity (*upper*) and lifetime (*lower*) images of 2-μm-diameter Yellow Green beads (Invitrogen, Breda, The Netherlands) are shown at increasing photon counts: 25, 50, 100, 250, 500, 1000, and 2500. The images and lifetime distributions show that the lifetime estimation converges to the correct values and reaches acceptable signal-to-noise ratios only at counts higher than 250, while morphological information and contrast is still present at only 25 counts in the corresponding intensity images

photon counts (even less than 100), reliable lifetime quantification requires a higher number of counted photons (>250). At higher photon counts, the broadness of the fluorescence lifetime distributions decreases due to the reduced uncertainty of the measurement. This is apparent from a shift of the peak of the lifetime distribution. As a matter of fact, the minimization of χ^2 performed by the fitting routines may be trapped in local minima located around the initial fitting parameter values at lower counts. These artifacts are frequent with all the commercial software that we tested, and can be avoided

Table 3 FLIM photon-economy

FLIM technique	*F* value	Refs.
Time-correlated single-photon counting	< 1.1	[29]
Lock-in	< 1.1	[36]
Time gating (2 windows)	1.5	[11]
Time gating (8 windows)	1.2	[11]
Image intensifier—sine excitation[a]	10	[23]
Image intensifier (FD)	4.3	[36]

[a] Unless indicated otherwise, a Dirac pulse train is used for excitation.

by masking the pixels with low photon counts and careful inspection of the data for distribution peaks at the initial guess of the lifetime. We note that at low photon counts, χ^2 values will always be low and data masking by high χ^2 values would therefore not eliminate this artifact.

The images were acquired on a Nikon PCM2000 confocal microscope (Nikon Instruments Europe B.V., Badhoevedorp, The Netherlands) equipped with a 440-nm PicoQuant diode laser and a LIMO lifetime module described in detail elsewhere [43].

The photon economy depends on many factors, e.g., number of collected photons, number and width of time gates, timing jitter, excitation light profile, harmonic content of the detection system, and modulation depth of the light source. Parametric plots of *F* values versus such quantities are helpful for the optimization of the performance of a system. A summary of *F* values found in the literature for different systems is shown in Table 3 [8, 22, 23, 29, 36].

7
Acquisition Speed

The acquisition speed depends on the photon economy of the detection system. At *F* values higher than 1, the acquisition time would increase by a factor of F^2 to achieve a signal-to-noise ratio that is comparable to images obtained with an efficient system. In addition, many other factors also affect the acquisition time.

TCSPC can be relatively slow because of the dead time of electronics and detectors and the requirement that not more than a single photon should be detected per excitation pulse. Typically, TCSPC systems require exposure times from tens of seconds up to several minutes (even tens of minutes for precise FRET measurements [3]). Instead, time-gated scanning systems and lock-in imaging allow significantly faster operation. When implemented on scanning microscopes, time gating and lock-in imaging provide typical acquisition times ranging from 1 to 30 s depending on the pixel resolution of the scanned field of view and the brightness of the sample [22]. Under unfavorable conditions, some minutes of exposure time may still be required.

Wide-field imaging generally provides a faster performance than scanning-based microscopy. Wide-field FLIM is typically operated with exposure times from fractions of seconds to maximally $\sim$10 s. However, the speed of wide-field FLIM is limited by the common use of MCPs. These (FD/TD) MCP-based FLIM systems require the sequential acquisition of time-/phase-dependent images for lifetime estimation, thereby limiting the achievable acquisition speed ($<$1 Hz). In recent years, many architectures that are based on MCPs or solid-state detectors have provided the means for parallel imaging of the required information, offering instrumentation that is capable of fast lifetime imaging (20–100 Hz) for both time- and frequency-domain systems [1, 13, 16].

A further advantage of the full parallel imaging provided by such wide-field microscopes is that motion and photobleaching artifacts are cancelled so that fast lifetime changes or moving objects can be adequately imaged. As many cellular physiological reactions that are amenable for lifetime imaging are highly dynamic, these increases in speed represent important improvements for the emerging discipline of molecular physiology.

8
Data Quality Assessment and Analysis

The quality of FLIM results depends on the calibration and characteristics of the instrumentation, on the reproducibility of the results, and on the correct application of data analysis techniques. The retrieval of the quantitative information that is contained in the lifetime measurements demands that maximal care is taken in the correct use of the instrumentation, sample handling, photophysical models, and statistics.

The reproducibility of measurements depends on several factors. Reference [26] defines three classes of FLIM experiments:

- Type I, where lifetime differences are within an image. In this case, a pixel population or region of interest with fluorescence lifetimes significantly different from the average or from a reference area is detected. For example, in the case of localized posttranslational modifications such as ubiquitination (Fig. 4) [21].
- Type II, where a set of images exhibit diverse lifetimes. Here, image sets are compared to detect differences in lifetimes between different samples, for instance, when comparing protein–protein interactions (Fig. 3) among different protein mutants [51].

- Type III, where lifetime heterogeneity is measured. In this case, the intrinsic heterogeneity of the lifetime of a single fluorophore or the mixing of different fluorescent moieties (Fig. 4) can be exploited for the detection of physiologically relevant quantities [2].

Although originally described for FD-FLIM, this classification does not depend on the instrumentation used and can therefore also be used for the rationalization of quality assessment in the time domain.

Type I experiments are affected by heterogeneity in the calibration over the field of view. Scanning systems are usually not affected by these kinds of problems, but wide-field detectors may require a careful pixel-by-pixel calibration. The main factor that defines which lifetime differences are significant within a single image is the number of acquired photons. Variations that are comparable to or smaller than the errors predicted by the photon economy of the system are not relevant. Detectors that do not directly provide the readout of the number of counted photons can be calibrated to allow the estimation of the photon number [34]. At low photon counts, image binning is a commonly adopted solution. However, it is necessary to verify that the contribution of autofluorescence to the measurement remains negligible. When fluorescence signals are too dim, binning will cause the integration of a significant amount of autofluorescence. Border artifacts around fluorescent objects are common phenomena that are caused by this problem, and could easily be interpreted as membrane-localized biological effects by the inexperienced user.

Another possibility for errors is the saturation of the detector, which is simple to control and check with wide-field detectors, but more difficult to manage with scanning and single-photon counting systems. Here, pulse pile-up can cause apparent lifetime contrast in an image. Fluorescent control samples with a homogeneous lifetime but heterogeneous brightness can be imaged to verify the optimal imaging conditions. The photon count rates should always be lower than the maximum rate that is tolerated by the detector used in the system (10^4–10^7 counts per second, depending on the detector used).

Type II experiments are affected by more parameters. Time drifts of the system properties, temperature variations, and sample preparation are common sources of image-to-image variability. Laboratories that host FLIM systems require the tight control of room and sample temperature. It is furthermore advisable to carefully reproduce sample preparation protocols and to use identical stock solutions for samples that are to be compared. As described before, fluorescence lifetimes and FRET efficiencies are trivially dependent on differences in temperature, pH, viscosity, and refractive index. A high reproducibility can be achieved under these controlled conditions. It is advisable to regularly measure a control sample or a fluorescence standard to monitor potential instrumental drifts that can occur during and between experimental sessions. Frequency-domain systems usually require regular recalibration and its automation as described before significantly increases the ease of use of these systems.

The use of the correct model for the interpretation of the fluorescence decays is fundamental to the quality of FLIM measurements. Improper interpretation of course affects all types, I, II, and III, of experiments, but the latter more crucially depends on the choice of the model for the fitting of decays.

The general rule is to adopt the simplest possible model, such as a single-exponential decay, and to increase its complexity only when indicators of goodness of fit suggest statistical improvement. For example, with only 100 photons, data fits with models that are more complex than a single-exponential decay will not be able to represent the data with unambiguous fit parameters. The χ^2 value generally provides a good indicator for the quality of the data fit [30]. However, systematic errors in FLIM are not uncommon and χ^2 values higher than unity do not necessarily imply higher lifetime heterogeneity. In order to decide if a more complex model should be adopted, *F* statistics (not to be confused with the *F* value of the photon economy) can be used [30]. *F* statistics are based on the ratio of the χ^2 values of two models and return the confidence level for the assumption that one model is better than the other. In general, the inspection of the residuals offers the most direct way to estimate a deviation from the hypothesized models. Residuals should always be randomly distributed around zero. Deviations from this trend would suggest the need for a different model.

Multiple-exponential data fitting requires the collection of large numbers of photons for reproducible lifetime quantifications. There are alternative models that can fit the lifetime heterogeneity of the sample at the cost of the addition of only one extra parameter. In the past few years, stretched exponentials [33], power-like models [50], and moment analysis [17] methods were developed and adopted to overcome this problem. Biological samples do not always provide enough photons for more complex analyses. It is worth mentioning that global analysis of lifetime data [30, 45, 46] represents a robust approach that can be applied when the fluorescence lifetimes can be considered to be spatially, spectrally, or frequency invariant, depending on the dimension in which global analysis operates. The robustness of this statistical approach derives from its use of information present in the pixel ensemble rather than operating only at each single pixel location.

The comparison between different samples is commonly carried out by Student's test (*t* statistics) and the analysis of variance (ANOVA) when data are normally distributed. Type II experiments may provide a significant difference between samples even when the samples are actually not different. This can be avoided by alternating the acquisition order between samples when collecting data from samples that are to be compared, thus avoiding the acquisition of distinct data sets in sequential experimental sessions.

Cumulative histograms of the lifetime or FRET distributions computed on the entire dataset are useful for the comparison of fluorescence lifetime and FRET results. Probability density functions can be generated by normalization of the distributions to an integrated bin count of unity for comparison. For instance, the Kolmogorov–Smirnov test (*D* statistics), which makes use of the cumulative density functions, represents a robust assessment of statistical differences among samples that may also exhibit nonnormal distributions. It should be noted that biological samples often display nonnormal distributions due to the heterogeneous responses of different cell populations.

9 Photobleaching

As the fluorescence lifetime of a fluorophore is independent of its concentration, lifetime imaging is relatively insensitive toward photobleaching. However, photobleaching may generate fluorescent photoproducts that could contribute to the acquired images. The use of localized acceptor photobleaching [51] for the generation of internal non-FRET references in donor lifetime imaging may suffer from this problem. Furthermore, direct excitation of the acceptor or sensitized acceptor photobleaching, i.e., acceptor photobleaching due to energy transfer, can also result in the generation of a fluorescent product that may contaminate the donor fluorescence and, consequentially, lifetime.

For example, Texas Red can produce a blue-shifted fluorescent photoproduct that can alter the lifetime and FRET quantification, for instance, in combination with the donor Oregon Green 488. Analogously, photobleaching of YFP may generate a photoproduct with spectral properties similar to those of the cyan fluorescent protein, which could affect the lifetime determination of the latter [41]. These genetically encoded donor and acceptor fluorophores are very often used for FRET experiments in biology.

This is not the only deleterious effect of acceptor photobleaching. At relatively high FRET efficiencies, the donor quantum yield will significantly decrease. This would require longer exposure times or higher excitation intensities. If the acceptor is less photostable than the donor, it will undergo sensitized photobleaching and energy transfer will therefore be significantly underestimated [15]. Thus, photostable acceptors are crucial for reliable lifetime imaging. The underestimation of FRET is illustrated in Fig. 6, where the apparent FRET efficiency decreases at longer exposure times.

Donor photobleaching can affect techniques that require the acquisition of sequential images, as is typically the case for MCP-based systems. Here, photobleaching can alter the time- or phase-image stack which results in artifacts. However, robust techniques for the compensation of photobleaching exist. For example, an image stack can be acquired twice with a time-(phase-)reversed protocol [19]. Alternatively, the image stack can be acquired in a pseudorandom sequence [44]. In this way, photobleaching of up to 50% of the initial fluorescence can be accommodated without serious deleterious effects.

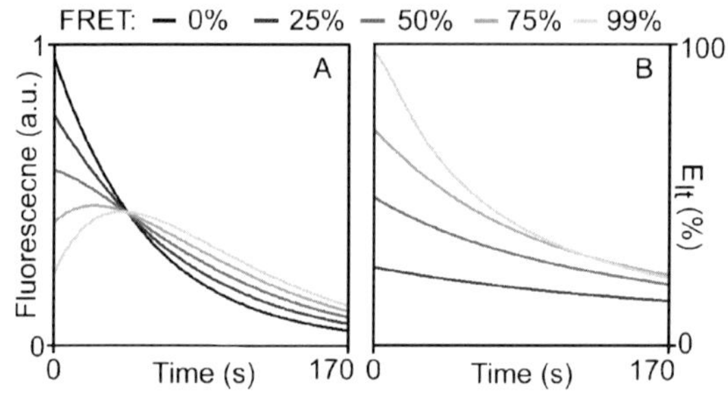

Fig. 6 Sensitized acceptor photobleaching. **A** Simulated donor fluorescence emission intensities over time during constant illumination. The time dependence of the emission is related to the combined photobleaching of both donor and acceptor. The simulation was performed with equal donor and acceptor photobleaching transition rates ($0.2\,s^{-1}$) and lifetimes (2.5 ns), and at donor excitation rates corresponding to ~10% of saturation (in the absence of energy transfer). Five energy transfer levels (0, 25, 50, 75, and 99%) are compared in the graphs. **B** The corresponding FRET efficiencies computed from simulated fluorescence lifetimes for the conditions shown in (**A**). The FRET efficiencies decrease with time due to sensitized acceptor photobleaching

The novel wide-field techniques that are based on parallel imaging are not affected by donor photobleaching [16] as long as the donor fluorescence remains higher than the autofluorescence of the sample.

10 Fluorophore Saturation

Fluorophore saturation is not difficult to achieve in scanning microscopes and may be easily overlooked. Fluorophore saturation can deteriorate the resolution of the microscope; furthermore, acceptor direct excitation at donor excitation wavelengths and energy transfer itself pump the acceptor in the excited state. When the acceptor excited state is saturated, FRET frustration, i.e., the inhibition of energy transfer due to the depletion of the acceptor ground state, occurs [15, 28]. This is particularly relevant for acceptor fluorophores with fluorescence or triplet state lifetimes that are longer than the donor singlet excited state lifetime. Therefore, the spectra and lifetimes of the FRET pair should be selected in order to minimize the generation of FRET-incompetent states that would cause FRET frustration. In general, the acceptor should exhibit a higher photostability and a shorter lifetime than the donor. Furthermore, the direct excitation of the acceptor by the donor excitation light source should be minimized by the most appropriate selection of excitation wavelength and donor Stoke's shift.

Fluorophore saturation causes a nonlinear dependence of the fluorescence emission on excitation light intensity. Frequency-domain techniques rely on the assumption that the fluorophore responds linearly to the excitation light. When this assumption is not satisfied, the harmonic content of the emission may be altered not only by the finite fluorescence lifetime of the fluorophore, but also by fluorophore saturation [15]. Fluorophore saturation, as used here, does not imply that the fluorophore ground state is fully depleted, i.e., a saturation of 100%. When sinusoidally modulated excitation light is used, nonlinear saturation-dependent effects can appear at saturation levels as low as 10%. On the other hand, when pulsed excitation sources are used, frequency-domain FLIM regains its robustness also in laser scanning systems where nonnegligible saturation levels may be present.

11 Fluorescent Stainings

Fluorescence lifetime imaging is also useful for the characterization of unstained samples by the detection of autofluorescence. However, FLIM is mostly used for the analysis of samples that were stained with exogenous fluorescent dyes [5]. Organic compounds for immunocytochemistry (e.g., Cy and Alexa dyes, Rhodamines, and fluoresceins), genetically encodable fluorophores (e.g., jellyfish and anthozoa fluorescent proteins and phycobilliproteins), tags and epitopes for site-directed in vivo labeling with synthetic dyes (e.g., FlAsH, ReAsH, Halo, and SnapTag), inorganic fluorophores (e.g., nano-diamonds, quantum dots, and other fluorescent colloidal particles), and cryptates provide a wide choice of labeling strategies and fluorophores.

Lifetime imaging can be used to read out fluorescent biosensors and for the quantification of protein–protein interactions or protein conformational change [51]. Biosensors and protein tags should minimize the disturbance of the physiological responses of a biological model to assure a reliable biological interpretation of the data. For instance, high-affinity calcium probes may alter the calcium homeostasis by competing with the endogenous buffering capability of the cell, protein overexpression can saturate binding partners, and protein tags can mislocalize the marked molecules. Furthermore, the introduction of electrostatic charges or hydrophobic regions in biomolecules may reduce physiological interactions or enhance oligomerization. Protein-tag-induced oligomerization has indeed been described for nonmonomerized fluorescent proteins [37]. Therefore, adequate (biological) control experiments are indispensable, not only for the assessment of the fluorescence lifetimes, but also for judging the potential alteration of biological responses.

When choosing fluorescent stainings for FRET experiments, one should always consider that FRET cannot be reliably detected at distances higher than twice the Förster radius or at a mutual orientation of the fluorophores that causes a low κ^2 orientation factor. Therefore, the molecular position of the tag and linkers may require optimization in order to detect significant FRET efficiencies. Finally, when deciding on the placing of donor and acceptor fluorophores on two interacting proteins, it should be considered that their interaction can only be optimally detected at a molecular excess of the acceptor-labeled protein in order to saturate the donor-labeled protein.

The selection of bright and photostable fluorophores, whose lifetime and spectra are in the optimal detection range of the instrumentation, is a basic issue in the selection of fluorescent stains. More subtle effects, however, can cause artifacts that, without proper controls and characterization of the fluorophore, may pass undetected.

Given the intrinsic high reliability, quantitative nature, and sensitivity of lifetime-based sensing, FLIM is expected to drive the current focus on "systems biology" approaches in the life sciences. That is, the requirement for numerical quantitative data on cellular events and processes for the formulation of mathematical models of larger integrative responses can ideally be fulfilled by FLIM. Furthermore, there is a current effort toward the generation of multiplexed fluorescent cellular biosensors, particularly those that are based on FRET, which allow the simultaneous observation of multiple events in single cells that is necessary to uncover causal connections in the complex cellular biochemical network. We will therefore likely see a large increase in the demand for accurate and efficient FLIM systems and, with that, a demand for even more stringent quality assessment tools and reliance on powerful statistical analysis methods.

The combination of molecular biology, which can provide genetically encodable tags and protein mutants for suitable controls, and a quantitative fluorescence imaging technique such FLIM provides a powerful tool set for the life and biomedical sciences.

12 Conclusions

Reliable and precise FLIM experiments require the characterization and calibration of the instrumentation, the use of experimental protocols to assure data reproducibility (e.g., constant temperature and sample preparation, alternation of acquisition of control and noncontrol data), the performance of controls for the validation of the biological model and for the characterization of the fluorescent tag, and a careful analysis of the data (see Table 4).

It is in the biological sciences, where FLIM is providing astounding results, that lifetime imaging requires the most meticulous and careful use. Lifetime heterogeneity due to the inherent properties of some fluorophores and to instrumental response variations sum up with the heterogeneity of the biological samples in the comparatively low throughput of typical microscopy systems. Therefore, careful calibration and experimental protocols are necessary to unveil subtle or heterogeneous biological effects. In this regard, the use of fast imaging instrumentation and automated systems may provide the higher throughput necessary for broader statistical samples. On the other hand, fluctuations in the instrumental response may cause the detection of false positive results which, however, can be avoided by appropriate protocols for routine calibration and sample measurement.

A collection of suitable single-exponential fluorescence lifetime standards (Table 2) is available for these purposes. Furthermore, multiple standards can be used to avoid or reduce systematic errors that cannot be corrected for by a single point calibration. Gradients of lifetimes can be created by the alteration of the chemico-physical properties of the solvent in a controlled

Table 4 Checklist

Sample preparation
1. Fluorophores should match the spectral and timing properties of the instrumentation.
2. Staining and, if needed, fixation and mounting should not alter the experimental interpretation of the data. Always use the same protocols and procedures.
3. Always prepare adequate (biological) sample controls.

System calibration
4. Characterize and calibrate the system on a regular basis.
5. Allow the temperature of the equipment, specimen, and laboratory room to stabilize.
6. Measure or estimate (see Table 3 and references) the *F* value of the used system.
7. Align/calibrate the system for the specific conditions of the experiment (before every use).

Data acquisition
8. Select an adequate field of view and acquisition speed for the investigated process.
9. Collect enough photons considering the model that should fit the data and the photon economy of the system.
10. In a two-photon system, long exposure times and lower excitation power could minimize photobleaching.
11. Check the autofluorescence background level of an unstained specimen. Sample brightness should be significantly higher than in unstained samples.
12. Keep the fluorophore photobleaching lower than 50% and ensure the remaining fluorescence dominates over the autofluorescence.
13. (For FRET experiments) Keep the acceptor photobleach low ($< 20\%$).
14. Avoid saturation of the detector, counting rates that exceed the detector capability, and pulse pile-up.
15. Alternate acquisition of different samples or at least check the stability of the system at the end of each experimental session.

Data analysis
16. Mask out pixels that do not contain sufficient photon counts.
17. Is the model correct? Check χ^2 and residuals. If a more complex model is needed, use *F* statistics.

Table 4 (continued)

18. When lifetime variations are present at the border of the objects, check if this is caused by wrong binning or insufficient photon counts. If control experiments imaged in the same conditions exhibit the same structures and an inhomogeneous lifetime, this is an imaging artifact.
19. (For FD-FLIM) Do the phase- and modulation-lifetime estimators provide similar information? The modulation lifetime should always be higher than the phase lifetime unless fluorescence is generated by a photochemical reaction at the excited state (e.g., acceptor fluorescence emission cross-talk). In other cases, the system may not be calibrated correctly.
20. When narrow lifetime distributions or lifetime values equal to the initial fit parameters are present, the fit did not converge to the actual lifetime value (not enough photons or wrong model).
21. Is there correlation between lifetimes and fluorescence intensities? Is this meaningful (higher FRET causes lower lifetimes and lower intensities) or the effect of photon statistics or pulse pile-up?
22. Compare adequate numbers of experiments, acquired in different experimental sessions (evaluate with t-test, ANOVA, Kolmogorov–Smironov test or other statistics).

Always

23. FLIM is a powerful tool, but good raw data are a prerequisite. Arbitrary editing of masks and imaging procedures that are not clearly declared in the protocols, or variation in sample preparation/data acquisition/data analysis from sample to sample, can change the scientific interpretation of the results!

manner. However, we note that the heterogeneity of such gradients should be carefully characterized.

In the past few decades, advanced fluorescence microscopy systems have become available to many research laboratories thanks to increasingly cost-effective and user-friendly technologies. In the near future, fluorescence microscopy will reach molecular resolution (<10 nm), will provide flexible and scalable systems (e.g., by the use of tunable laser supercontinuums, tunable optics, automation, and remote operation), and will provide quantitative multiparametric (intensity, lifetime, polarization, spectra, 4D spatio-temporal) imaging possibilities.

The necessity of fluorescence standards and standardized quality assessment protocols is therefore destined to grow. The development of standardized, stable, and easy-to-handle samples for the calibration of lifetime sensing, polarization detection, spectral resolution, and spatial resolution is therefore desirable.

Acknowledgements We thank Dr. Christoph Dohm for his cooperation on the study of α-synuclein, a collaborative project of which two images are here presented. We thank Prof. Paul M. French and Dr. James McGinty for providing spectral information about DASPI. This work was supported by the DFG Research Center for Molecular Physiology of the

Brain. The European Neuroscience Institute Göttingen (ENI-G) is jointly funded by the Göttingen University Medical School, the Max Planck Society, and Schering AG. FSW is further financed by the DFG Excellence Cluster 171 "Microscopy on the Nanometer Scale" and the project "FLI-Cam" in the BMBF Biophotonic III program. AE is funded by the Biotechnology and Biological Sciences Research Council (UK).

References

1. Agronskaia AV, Tertoolen L, Gerritsen HC (2003) J Phys D Appl Phys 36:1655
2. Barzda V, de Grauw CJ, Vroom J, Kleima FJ, van Grondelle R, van Amerongen H, Gerritsen HC (2001) Biophys J 81:538
3. Becker W, Bergmann A, Hink MA, Konig K, Benndorf K, Biskup C (2004) Microsc Res Tech 63:58
4. Bugiel I, König K, Wabnitz H (1989) Laser Life Sci 3:1
5. Bunt G, Wouters FS (2004) Int Rev Cytol 237:205
6. Buurman EP, Sanders R, Draaijer A, Gerritsen HC, Vanveen JJF, Houpt PM, Levine YK (1992) Scanning 14:155
7. Carlsson K, Liljeborg A (1997) J Microsc 185:37
8. Carlsson K, Philip JP (2002) Proc SPIE 4622:70
9. Chen RF (1974) Anal Biochem 57:593
10. Clegg RM (1996) Fluorescence resonance energy transfer. In: Wang XF, Herman B (eds) Fluorescence imaging spectroscopy and microscopy. Wiley, London, p 180
11. De Grauw CJ, Gerritsen HC (2001) Appl Spectrosc 55:670
12. Dong CY, So PT, French T, Gratton E (1995) Biophys J 69:2234
13. Elson DS, Munro I, Requejo-Isidro J, McGinty J, Dunsby C, Galletly N, Stamp GW, Neil MAA, Lever MJ, Kellett PA, Dymoke-Bradshaw A, Hares J, French PMW (2004) New J Phys 6:180
14. Elson DS, Siegel J, Webb SED, Leveque-Fort S, Lever MJ, French PMW, Lauritsen K, Wahl M, Erdmann R (2002) Opt Lett 27:1409
15. Esposito A (2006) PhD Dissertation, Utrecht University
16. Esposito A, Gerritsen HC, Lustenberger F, Oggier T, Wouters FS (2006) J Biomed Opt 11:34016
17. Esposito A, Gerritsen HC, Wouters FS (2005) Biophys J 89:4286
18. Forde TS, Hanley QS (2005) Photochem Photobiol Sci 4:609
19. Gadella TW Jr, Clegg RM, Jovin TM (1994) Bioimaging 2:139
20. Gadella TW Jr, Jovin TM, Clegg RM (1993) Biophys Chem 48:221
21. Ganesan S, Ameer-beg SM, Ng T, Vojnovic B, Wouters FS (2006) Proc Natl Acad Sci USA 103:4089
22. Gerritsen HC, Asselbergs MA, Agronskaia AV, Van Sark WG (2002) J Microsc 206:218
23. Gerritsen HC, Draaijer A, Van den Heuvel DJ, Agronskaia AV (2006) Fluorescence lifetime imaging in scanning microscopy. In: Pawley J (ed) Handbook of biological confocal microscopy. Plenum, New York, p 516
24. Ghiggino KP, Harris MR, Spizzirri PG (1992) Rev Sci Instrum 63:2999
25. Hanley QS, Ramkumar V (2005) Appl Spectrosc 59:261
26. Hanley QS, Subramaniam V, Arndt-Jovin DJ, Jovin TM (2001) Cytometry 43:248
27. Ivanova OV, Marcu L, Khoo MC (2005) Ann Biomed Eng 33:531
28. Jares-Erijman EA, Jovin TM (2003) Nat Biotechnol 21:1387
29. Kollner M, Wolfrum J (1992) Chem Phys Lett 200:199

30. Lakowicz JR (1999) Principles of fluorescence spectroscopy, 2nd edn. Kluwer Academic/Plenum, New York
31. Lakowicz JR, Berndt KW (1991) Rev Sci Instrum 62:1727
32. Lakowicz JR, Gryczynski I, Laczko G, Gloyna D (1991) J Fluoresc 1:87
33. Lee KCB, Siegel J, Webb SED, Leveque-Fort S, Cole MJ, Jones R, Dowling K, Lever MJ, French PMW (2001) Biophys J 81:1265
34. Lidke KA, Rieger B, Lidke DS, Jovin TM (2005) IEEE Trans Image Process 14:1237
35. Morgan CG, Mitchell AC, Murray JG (1990) Trans R Microsc Soc 1:463
36. Philip J, Carlsson K (2003) J Opt Soc Am A 20:368
37. Shaner NC, Steinbach PA, Tsien RY (2005) Nat Methods 2:905
38. So PTC, French T, Yu WM, Berland KM, Dong CY, Gratton E (1995) Bioimaging 3:49
39. Stiel H, Teuchner K, Paul A, Leupold D, Kochevar IE (1996) J Photochem Photobiol B 33:245
40. Stryer L, Haugland RP (1967) Proc Natl Acad Sci USA 58:719
41. Valentin G, Verheggen C, Piolot T, Neel H, Coppey-Moisan M, Bertrand E (2005) Nat Methods 2:801
42. Van der Meer BW (1999) Orientational aspects in pair energy transfer. In: Andrews DL, Demidov AA (eds) Resonance energy transfer. Wiley, London, p 151
43. Van der Oord CJ, De Grauw CJ, Gerritsen HC (2001) Proc SPIE 4252:119
44. Van Munster EB, Gadella TW Jr (2004) Cytometry 58A:185
45. Verveer PJ, Bastiaens PI (2003) J Microsc 209:1
46. Verveer PJ, Squire A, Bastiaens PI (2000) Biophys J 78:2127
47. Wang XF, Uchida T, Coleman DM, Minami S (1991) Appl Spectrosc 45:360
48. Wang XF, Uchida T, Minami S (1989) Appl Spectrosc 43:840
49. Watrob HM, Pan CP, Barkley MD (2003) J Am Chem Soc 125:7336
50. Wlodarczyk J, Kierdaszuk B (2003) Biophys J 85:589
51. Wouters FS, Verveer PJ, Bastiaens PI (2001) Trends Cell Biol 11:203

Part II
Single Molecule Spectroscopy

Springer Ser Fluoresc (2008) 6: 145–197
DOI 10.1007/4243_2008_032

Published online: 28 March 2008

State of the Art and Novel Trends in Fluorescence Correlation Spectroscopy

E. P. Petrov (✉) · P. Schwille

Biophysics, BIOTEC, Technische Universität Dresden, Tatzberg 47–51, 01307 Dresden, Germany
petrov@biotec.tu-dresden.de

Abstract This chapter presents a review of fluorescence correlation spectroscopy (FCS), an experimental technique with single-molecule sensitivity, which is based on the analysis of fluctuations of fluorescence intensity detected from a tiny volume. Correlation functions of fluorescence fluctuations can provide information on the translational and rotational diffusion of fluorophores, dynamics of singlet–triplet transitions, chemical reactions, flow, and active transport of fluorescent molecules. A detailed theoretical description of the fluorescence correlation and cross-correlation technique is followed by a discussion of various experimental aspects of FCS, including the choice of instrumentation and fluorophores, sample- and setup-related nonidealities and possible artifacts, statistical accuracy, and approaches to the analysis of FCS data. Additionally, some FCS application aspects are addressed, including the quantitative determination of translational diffusion coefficients and the use of FCS in studies of (bio)polymers and phospholipid membranes. The chapter concludes with an overview of both well-established and currently emerging varieties of FCS and related methods, including the use of two-photon excitation and the application of total internal reflection, nanoapertures, and stimulated emission depletion to confine the detection volume; the use of higher-order correlations; application of time-resolved and time-gated detection; multifocal and CCD-based FCS; and image correlation and scanning FCS techniques.

Keywords Anomalous diffusion · Brownian diffusion · Correlation function · Fluorescence correlation spectroscopy · Fluorescent proteins · Phospholipid membranes · Polymer dynamics

1 Introduction

Fluorescence correlation spectroscopy (FCS), also known as fluorescence fluctuation spectroscopy, is an experimental technique with single-molecule sensitivity based on studying fluctuations of fluorescence intensity detected from a tiny volume. Fluctuations of fluorescence take place as a result of fluctuations of the number and/or brightness of fluorescent particles in the detection volume with time (Fig. 1), e.g., as a result of their motion (diffusion, active transport) and/or photophysical or photochemical reactions.

The concept and basic principles of FCS were formulated by Magde, Elson, and Webb more than 30 years ago [1] and later successfully implemented [2–9] as a natural extension of the then well-developed field of dynamic light scattering (DLS) [10], especially of its applications to the study of number

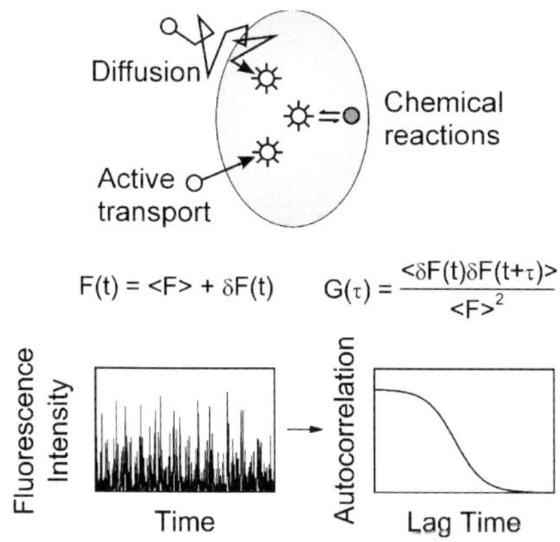

Fig. 1 Cartoon showing the concept of fluorescence correlation spectroscopy

fluctuations in dilute suspensions of colloidal particles [11, 12]. In the following two decades, FCS was established as a technique for the investigation of translational and rotational diffusion, active transport and flow, photophysical and photochemical transformations, chemical reactions, and molecular aggregation. A detailed account of this early stage of FCS development can be found in reviews [13–17]. However, in spite of the doubtless progress, the field stayed rather dormant, partly due to the lack of efficient hardware and software, and partly due to the more straightforward methods of steady-state and time-resolved fluorescence spectroscopy [18, 19] booming during this period.

It was the experimental demonstration of the single-molecule detection capabilities of FCS [20] in 1993 and the emergence in 1994 of the visionary paper by Eigen and Rigler [21], uncovering the huge potential of FCS for biophysical, biochemical, and molecular biological sciences, which resulted in

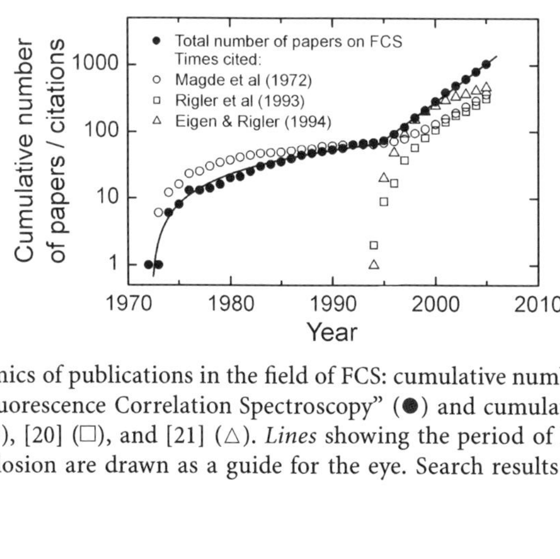

Fig. 2 The dynamics of publications in the field of FCS: cumulative number of papers with the keyword "Fluorescence Correlation Spectroscopy" (●) and cumulative number of citations of [1] (○), [20] (□), and [21] (△). *Lines* showing the period of linear growth and exponential explosion are drawn as a guide for the eye. Search results are from ISI Web of Knowledge^{SM}

an exponentially growing avalanche of FCS-based research in the last decade (Fig. 2; reviewed in [22–33]).

Over the decades, FCS has matured from a proof-of-principle concept to an important experimental method extensively used in a wide range of disciplines from photophysics to polymer physics to biophysical and biochemical sciences. In the present review, we give an outline of the present state of the art of FCS and discuss the novel trends in this experimental technique.

2 General Principles of FCS

2.1 Basic Concepts

The subject of FCS is the study of the character of time-dependent fluctuations of the fluorescence intensity $F(t) = \langle F \rangle + \delta F(t)$ collected from a tiny detection volume located in a sample which is considered to be in an equilibrium state. Generally, the fluorescence intensity can exhibit fluctuations due to fluorescent particles entering and leaving the detection volume as a result of Brownian motion or active transport, flow, or sample translation, or as a result of transient changes in the fluorescence brightness due to chemical reactions and photochemical or photophysical processes. Additionally, in cases where fluorophores are attached to objects exhibiting internal structural dynamics (e.g., macromolecules or biological membranes), the latter can substantially contribute to fluorescence fluctuations.

An efficient means to study intensity fluctuations of a stationary light source are intensity correlation functions [34, 35]. In particular, the expressions for the second-order (two-time) and third-order (three-time) intensity correlation functions are as follows:[1]

$$g_{ij}^{(2)}(\tau) = \frac{\big\langle F_i(t)F_j(t+\tau)\big\rangle}{\langle F_i\rangle\langle F_j\rangle}\,, \qquad \tau \geq 0\,, \tag{1}$$

$$g_{ijk}^{(3)}(\tau_1,\tau_2) = \frac{\big\langle F_i(t)F_j(t+\tau_1)F_k(t+\tau_1+\tau_2)\big\rangle}{\langle F_i\rangle\langle F_j\rangle\langle F_k\rangle}\,, \qquad \tau_1,\tau_2 \geq 0\,. \tag{2}$$

Angular brackets stand, generally, for ensemble averaging, which in FCS measurements on a sample under equilibrium conditions is replaced, under the assumption of ergodicity, by time averaging. If the fluorescence signals $F_{i,j,k}$ are recorded by the same detector, Eqs. 1 and 2 are referred to as autocorrelation functions (ACFs). In the case where the fluorescence signals $F_{i,j,k}$

[1] In some FCS-related papers, Eqs. 1 and 2 are referred to as the *first*- and *second*-order intensity correlation functions, respectively. We, however, prefer to use the standard definitions accepted in quantum optics and photon correlation spectroscopy [34, 35].

are recorded from different subvolumes within the sample, or/and at different spectral ranges using different detectors, the term cross-correlation function (CCF) is used.

The most common quantity reported and discussed in the context of FCS studies is the second-order correlation function of fluorescence intensity (Eq. 1) or the related correlation function of fluorescence intensity fluctuations:

$$G_{ij}(\tau) = g_{ij}^{(2)}(\tau) - 1 = \frac{\langle \delta F_i(t)\delta F_j(t+\tau)\rangle}{\langle F_i\rangle \langle F_j\rangle}, \tag{3}$$

which in what follows will be referred to as the FCS ACF. We first discuss some general properties of the FCS correlation function (Eq. 3), after which we consider in more detail some particular cases characteristic of typical FCS experiments.

Due to a random time separating the events of optical excitation and fluorescence emission by a dye molecule, the fluorescence signals from different molecules in the detection volume are completely incoherent. Therefore the fluorescence intensity detected at a particular time instant from a volume element located at a position $\boldsymbol{r}$ is proportional to the the position-dependent excitation intensity I, concentration of the fluorophore c in the fluorescent state, its excitation cross section σ_{exc} and fluorescence quantum yield ϕ, the overall fluorescence detection efficiency for the fluorophore κ, and the normalized fluorescence collection efficiency S. Additionally, in the case of polarized excitation and detection, the intensity of the detected fluorescence signal depends on the instantaneous orientation of the absorption and emission dipole moments with respect to the polarizer axes in the excitation and detection channels, as well as on the rotational correlation time and excited-state lifetime of the molecule. Therefore, provided that the parameters of the setup are not time-dependent, FCS measurements are sensitive to fluctuations of the concentration, photophysical properties, and orientation of fluorophores in the sample.

In most cases, though, the rotational diffusion can be neglected, and the detected fluorescence intensity is given by the expression:

$$F(t) = \int W(\boldsymbol{r}) q c(\boldsymbol{r}, t)\, \mathrm{d}V, \tag{4}$$

where the quantity $W(\boldsymbol{r}) = S(\boldsymbol{r})I(\boldsymbol{r})/I_0$ with $I_0 = \max(I(\boldsymbol{r}))$ describes the effective shape of the fluorescence detection volume (also known as the molecular detection efficiency profile), and $q = I_0\kappa\sigma_{\text{exc}}\phi$ is the so-called molecular brightness of the fluorophore.

If not mentioned explicitly, we will consider ideal solutions of point-like particles in an open volume, which means that the concentration fluctuations of different components of a system at the same time instant are independent and spatially uncorrelated, and their number fluctuations in a small open volume are governed by the Poisson statistic [36]:

$$\left\langle \delta c_i(\boldsymbol{r},0)\delta c_j(\boldsymbol{r}',0)\right\rangle = \bar{c}_i\delta_{ij}\delta(\boldsymbol{r}-\boldsymbol{r}')\,, \tag{5}$$

where $\bar{c}_i$ is the mean concentration of the i-th fluorescent component, δ_{ij} is the Kronecker delta, and $\delta(\,\cdot\,)$ is the Dirac delta function. Then the correlation function of concentration fluctuations of the i-th component is

$$\left\langle \delta c_i(\boldsymbol{r},t)\delta c_i(\boldsymbol{r}',0)\right\rangle = \bar{c}_i\Phi_i(\boldsymbol{r},\boldsymbol{r}',t)\,, \tag{6}$$

where $\Phi_i(\boldsymbol{r},\boldsymbol{r}',t)$ is the Green function describing its motion.

In this case for a single diffusing species the FCS ACF reads as follows:

$$G(\tau) = \frac{\iint W(\boldsymbol{r})\Phi(\boldsymbol{r},\boldsymbol{r}',t)W(\boldsymbol{r}')\,\mathrm{d}V\,\mathrm{d}V'}{\bar{c}\left(\int W(\boldsymbol{r})\,\mathrm{d}V\right)^2}\,. \tag{7}$$

The time zero limit of the correlation function is given by the inverse of the effective number of molecules in the detection volume:

$$G(0) = \frac{1}{\langle N\rangle}\,, \tag{8}$$

where the effective number of molecules is determined by the concentration of the fluorophore and the effective fluorescence detection volume:

$$\langle N\rangle = \bar{c}V_{\mathrm{eff}} \tag{9}$$

with $V_{\mathrm{eff}} = \left(\int W(\boldsymbol{r})\,\mathrm{d}V\right)^2 / \int W^2(\boldsymbol{r})\,\mathrm{d}V$.

In modern FCS, depending on the particular implementation, the size of the detection volume lies in the range of ~0.1–1.0 fl. This, along with practical limitations in reliable detection of the ACF, restricts the typical fluorophore concentrations to the range of ~0.1–100 nM, which corresponds to an average of a few (or even less than one) molecules in the detection volume. Thus, FCS can be classified as a single-molecule fluorescence technique.

The particular shape of the detection volume depends on the excitation–detection geometry and particular implementation of the technique. In most FCS experiments, confocal detection is used (as is shown in Fig. 3). In this case the fluorescence detection volume is considered to be well approximated by the 3D Gaussian ellipsoid [20]:

$$W(x,y,z) = \exp\left[-\frac{2(x^2+y^2)}{r_0^2} - \frac{2z^2}{z_0^2}\right], \tag{10}$$

whose z-axis coincides with the optical axis of the objective. This form of the fluorescence detection volume will be assumed in what follows, if not otherwise stated explicitly.

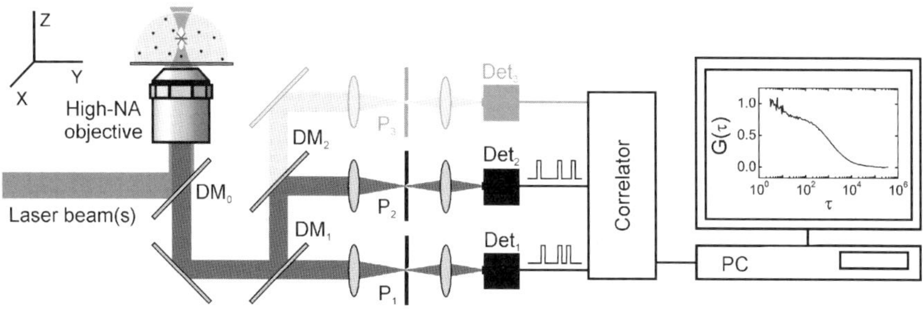

Fig. 3 Sketch of a typical confocal microscope-based setup for fluorescence (cross-)correlation spectroscopy. The optional third detection channel is shown in gray. DM_0 is a dichroic mirror separating the excitation and fluorescence emission; DM_1 and DM_2 are dichroic mirrors spectrally separating fluorescence into detection channels. In the two-channel configuration, DM_1 can be replaced by a beam splitter. Pinholes and detectors are denoted by P and Det, respectively

Within the Gaussian approximation (Eq. 10) of the detection volume shape, the effective number of molecules is given by $\langle N \rangle = \pi^{3/2} r_0^2 z_0 \bar{c}$. Thus, if the dimensions of the detection volume are known, the concentration of the diffusing fluorescent species can be determined from the amplitude of the FCS ACF.

Using the concept of the mean number of molecules in the detection volume, one can introduce the practical proxy to the molecular brightness, an important quantity in FCS experiments. This proxy, known as "counts per molecule per second" (CPM), is defined as $\mathrm{CPM} = \langle N \rangle^{-1} \langle F \rangle$ with the mean fluorescence intensity $\langle F \rangle$ measured in counts per second (the use of a single-photon counting detector is assumed).

In the presence of a noncorrelated background signal (due to detector dark counts and fluorescence or scattering background in the sample) with the detected mean intensity $\langle F_\mathrm{B} \rangle$, the amplitude of the FCS correlation function is reduced [4]:

$$G(0) = \frac{1}{\langle N \rangle} \left[1 - \frac{\langle F_\mathrm{B} \rangle}{\langle F \rangle + \langle F_\mathrm{B} \rangle} \right]^2 . \tag{11}$$

This effect should be accounted for when using FCS for estimating fluorophore concentrations and molecular brightnesses.

In most applications, FCS experiments are aimed at determining transport coefficients and reaction rate constants in systems under equilibrium conditions. Generally, concentration fluctuations in an n-component system of point-like particles under equilibrium conditions, whose dynamics is characterized by diffusion, chemical reactions, and active transport (or flow, or translation), are described by the following advective reaction–diffusion equation:

$$\partial_t c_i(\boldsymbol{r},t) = D_i \nabla^2 c_i(\boldsymbol{r},t) + \sum_{j=1}^{n} K_{ij} c_j(\boldsymbol{r},t) - \boldsymbol{v}_i(\boldsymbol{r}) \cdot \nabla c_i(\boldsymbol{r},t)\,, \tag{12}$$

where D_i and $\boldsymbol{v}_i$ are diffusion coefficients and flow velocities of the components, and K_{ij} are elements of the matrix of kinetics coefficients. In what follows we will consider several particular cases characteristic for typical applications of FCS, for which simple expressions for FCS correlation functions can be derived.

2.2 Correlation Functions in Typical FCS Experiments

2.2.1 Pure Diffusion

The simplest case of those described by Eq. 12 is pure diffusion. In this case the concentration fluctuations in a sample containing n independently diffusing fluorescent species are governed by the equation

$$\partial_t c_i(\boldsymbol{r},t) = D_i \nabla^2 c_i(\boldsymbol{r},t)\,. \tag{13}$$

The Green function for free diffusion of the i-th component in d dimensions ($d = 1, 2, 3$) is as follows (see, e.g., [37]):

$$\Phi_i(\boldsymbol{r},\boldsymbol{r}',t) = \frac{1}{(4\pi D_i t)^{d/2}} \exp\left[-\frac{(\boldsymbol{r}-\boldsymbol{r}')^2}{4D_i t}\right]. \tag{14}$$

By combining this expression with Eqs. 7 and 10 and evaluating the integrals, expressions for the FCS correlation functions can be obtained.

In particular, in the case of a one-component system, the FCS ACF takes the form

$$G(\tau) = \frac{1}{\langle N \rangle} G_{\mathrm{D}}(\tau)\,, \tag{15}$$

where $G_{\mathrm{D}}(\tau)$ is the normalized FCS ACF for free normal diffusion. For diffusion in 3D it is expressed as[2]

$$G_{\mathrm{D}}(\tau) = \frac{1}{\left(1+\tau/\tau_{\mathrm{D}}\right)\sqrt{1+f^{-2}\tau/\tau_{\mathrm{D}}}}\,, \tag{16}$$

where $f = z_0/r_0$ is the elongation of the Gaussian detection volume (Eq. 10).

[2] Here and in what follows we will use the subscript D to denote quantities related to diffusion.

In the case of 2D diffusion in the XY-plane (e.g., diffusion of a fluorescent probe in a phospholipid membrane lying in the focal plane of the objective),

$$G_{\mathrm{D}}(\tau) = \frac{1}{1+\tau/\tau_{\mathrm{D}}} \,. \tag{17}$$

In Eqs. 16 and 17

$$\tau_{\mathrm{D}} = \frac{r_0^2}{4D} \tag{18}$$

is the so-called diffusion time giving the characteristic decay scale of fluorescence fluctuations.

Since usually in experiments $f = 3.5–7$, the relation $G(\tau_{\mathrm{D}})\approx(1/2)G(0)$ can be used to obtain a quick estimate of the diffusion time τ_{D} in the case of pure single-component diffusion.

Thus, if the diffusion coefficient of the dye is known from independent experiments, the effective size of the detection volume and the mean concentration of fluorescent particles can be estimated by fitting the experimental correlation curve with Eq. 16 (or Eq. 17).[3] Once the effective detection volume has been characterized by measurements on a standard sample with the known diffusion coefficient, one can use FCS for determination of diffusion coefficients. Here we would like to emphasize that these diffusion coefficient measurements are relative and therefore necessarily require reporting of the diffusion coefficient of the reference substance employed.

In the case of an n-component mixture of noninteracting fluorescent particles characterized by molecular brightnesses q_i and mean concentrations $\bar{c}_i$, the correlation function of fluorescence fluctuations takes the following form:

$$G(\tau) = \frac{\sum_{i=1}^{n} q_i^2 \bar{c}_i G_{\mathrm{D}i}(\tau)}{V_{\mathrm{eff}}\left(\sum_{i=1}^{n} q_i \bar{c}_i\right)^2} = \frac{\sum_{i=1}^{n} q_i^2 Y_i G_{\mathrm{D}i}(\tau)}{\langle N_{\mathrm{total}}\rangle \left(\sum_{i=1}^{n} q_i Y_i\right)^2} \,, \tag{19}$$

where $Y_i = \bar{c}_i / \sum_{i=1}^{n} \bar{c}_i$ are the molar fractions of fluorescent components, and $\langle N_{\mathrm{total}}\rangle = V_{\mathrm{eff}} \sum_{i=1}^{n} \bar{c}_i$ is the total effective number of fluorescent particles in the detection volume.

Notice that in this case the inverse of the amplitude of the correlation function gives the apparent number of molecules in the detection volume

$$\langle N_{\mathrm{app}}\rangle = V_{\mathrm{eff}} \frac{\left(\sum_{i=1}^{n} q_i \bar{c}_i\right)^2}{\sum_{i=1}^{n} q_i^2 \bar{c}_i} = \langle N_{\mathrm{total}}\rangle \frac{\left(\sum_{i=1}^{n} q_i Y_i\right)^2}{\sum_{i=1}^{n} q_i^2 Y_i} \,, \tag{20}$$

which can differ substantially from $\langle N_{\mathrm{total}}\rangle$, and therefore this estimate should be used with extreme caution in the study of multicomponent systems.

Thus, FCS provides a means to study binding reactions under quasi-equilibrium conditions, provided the molecular brightnesses of the free and

[3] In reality other processes, such as, e.g., triplet blinking (see below), additionally contribute, as a rule, to the shape of the correlation curve and need to be taken into account.

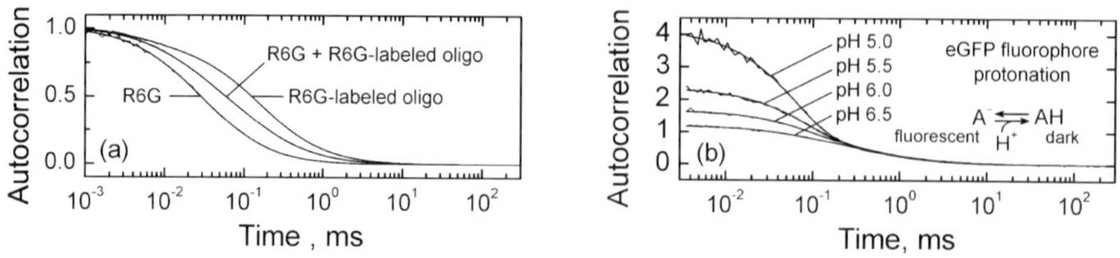

Fig. 4 Normalized FCS ACFs for **a** Rhodamine 6G dye, Rhodamine 6G-labeled oligonucleotide, and their 1 : 1 mixture freely diffusing in water ($\lambda_{\mathrm{exc}} = 532$ nm, $T = 298$ K, setup: Hamamatsu C9413) and **b** freely diffusing molecules of enhanced GFP (Clontech, Palo Alto, CA, USA) exhibiting pH-dependent blinking in a series of aqueous buffer solutions ($\lambda_{\mathrm{exc}} = 488$ nm, $T = 298$ K, home-built setup)

bound fluorescent labels are known. As a model example, Fig. 4a shows results for a free label, labeled oligonucleotide, and 1 : 1 mixture of the free and oligonucleotide-bound label. This approach was successfully used to follow slow irreversible binding reactions in solution [38, 39]. However, one should keep in mind that successful application of this approach requires a considerable difference in the diffusion coefficients of the bound and free fractions and their comparable contributions to the FCS correlation function [40].

2.2.2 Effects of Chemical Reactions

In the case where the dynamics of a system is determined by diffusion and chemical reactions, the concentration fluctuations are described by the following system of coupled reaction–diffusion equations:

$$\partial_t c_i(\boldsymbol{r}, t) = D_i \nabla^2 c_i(\boldsymbol{r}, t) + \sum_{j=1}^{n} K_{ij} c_j(\boldsymbol{r}, t) \,. \tag{21}$$

To be observed in an FCS experiment, the chemical reaction must satisfy at least one of the two following conditions: (1) the reaction should modify the transport properties of fluorescent species, and (2) fluorescence quantum yields or absorption cross sections should be modified as a result of reaction.

Although no simple solution of Eq. 21 can generally be obtained, the case of the two-state reversible chemical reaction

$$A \underset{k_{\mathrm{B}}}{\overset{k_{\mathrm{A}}}{\rightleftharpoons}} B \tag{22}$$

for freely diffusing species characterized by the identical diffusion coefficients $D_A = D_B = D$ can be solved exactly using, e.g., the technique described in [41]. By introducing the notation $\gamma_{XY} = \langle \delta c_X(\boldsymbol{r}, t) \delta c_Y(\boldsymbol{r}', t') \rangle$, $X, Y = A, B$ and defining the matrices $\boldsymbol{\Gamma} = [\gamma_{AA}\ \gamma_{BA}; \gamma_{AB}\ \gamma_{BB}]$ and $\boldsymbol{K} = [D\nabla^2 - k_{\mathrm{A}}\ k_{\mathrm{B}}; k_{\mathrm{A}}\ D\nabla^2 - k_{\mathrm{B}}]$, the correlation functions of concentration fluctuations are described by the following differential equation

$$\frac{\mathrm{d}}{\mathrm{d}t}\boldsymbol{\Gamma} = \boldsymbol{K}\boldsymbol{\Gamma}\,, \tag{23}$$

which in the case of freely diffusing particles can be solved, e.g., by applying the Fourier and Laplace transform techniques. Knowing the (cross-)correlation functions for the concentration, it is straightforward to obtain the corresponding correlation functions for fluorescence intensity fluctuations. By assuming different molecular brightnesses of the species in Eq. 22, the fluorescence correlation function for this system takes the following form:

$$G(\tau) = \frac{1}{\langle N_{\mathrm{app}}\rangle}\left[G_{AA}(\tau) + G_{BB}(\tau) + G_{AB}(\tau) + G_{BA}(\tau)\right], \tag{24}$$

where $G_{XX}(\tau) = \bar{c}_X^2 q_X^2[(\bar{c}_X q_X^2 + \bar{c}_Y q_Y^2)(\bar{c}_X + \bar{c}_Y)]^{-1}[1 + \bar{c}_Y\bar{c}_X^{-1}\exp(-k_{\mathrm{bl}}\tau)]G_{\mathrm{D}}(\tau)$ and $G_{XY}(\tau) = \bar{c}_X\bar{c}_Y q_X q_Y[(\bar{c}_X q_X + \bar{c}_Y q_Y)(\bar{c}_X + \bar{c}_Y)]^{-1}[1 - \exp(-k_{\mathrm{bl}}\tau)]G_{\mathrm{D}}(\tau)$, $X, Y = A, B$, with the blinking rate $k_{\mathrm{bl}} = k_{\mathrm{A}} + k_{\mathrm{B}}$, and, in agreement with Eq. 20, $\langle N_{\mathrm{app}}\rangle = V_{\mathrm{eff}}(q_A\bar{c}_A + q_B\bar{c}_B)^2/(q_A^2\bar{c}_A + q_B^2\bar{c}_B)$.

If one of the species is nonfluorescent, then Eq. 24 simplifies, and

$$G(\tau) = \frac{1}{\langle N_{\mathrm{total}}\rangle}\left[1 + \frac{\bar{c}_{\mathrm{dark}}}{\bar{c}_{\mathrm{bright}}}\exp(-k_{\mathrm{bl}}\tau)\right]G_{\mathrm{D}}(\tau)\,, \tag{25}$$

where $\bar{c}_{\mathrm{bright}}$ and $\bar{c}_{\mathrm{dark}}$ are mean concentrations of the fluorescent and dark species.

This situation can be experimentally observed, e.g., in FCS experiments with green fluorescent protein [42]. Its behavior can to a good approximation be described as a pseudo-first-order protonation–deprotonation reaction with the fluorescent deprotonated and nonfluorescent protonated forms [43, 44], which is illustrated in Fig. 4b.

The above model can be used to study small-scale (nanometer and subnanometer) conformational dynamics of biomacromolecules, provided that the conformational changes lead to strong variations in the molecular brightness of the fluorescent label attached (due to, e.g., nonradiative energy transfer or photoinduced electron transfer), so that the corresponding fluorescence fluctuations can be described in terms of the two-state on–off dynamics. For applications of this approach in FCS-based studies of conformational dynamics of DNA hairpins and proteins, see, e.g., [45–48].

2.2.3
Effects of Rotational Diffusion

In cases where the excitation light is linearly polarized and/or the detection is polarization sensitive, rotational diffusion of fluorescent particles leads to additional fluctuations in the detected fluorescence signal. For small dye molecules, the typical rotational dynamics in aqueous solutions at room temperature (typical conditions of an FCS experiment) takes place on the subnanosecond scale [49] and therefore virtually does not affect FCS correlation curves. However, if the fluorophore is rigidly attached to a larger particle, e.g., a biological macromolecule, the typical scale of rotational relaxation can shift from tens of nanoseconds to microseconds [18]. By this means, in the case of slowly rotating particles, FCS monitors rotational diffusion in the ground state, and therefore, in contrast to time-resolved fluorescence anisotropy measurements [18], rotational relaxation times accessible to FCS are not restricted to the scale of the excited-state lifetime of the molecule.

The theory describing the manifestation of rotational diffusion of fluorophores in FCS measurements with polarized excitation and/or detection was developed very soon after the appearance of FCS [5–7, 50] and was later extended to the case of driven rotational motion [51, 52]. The full theoretical description of polarized FCS is quite involved and accounts for the light polarization in the excitation and detection channels, mutual orientation of the absorption and emission dipole moments, the duration of the excited state, and the shape of the fluorescent particles [5, 6, 15]. In [6], approximate FCS ACFs for slow rotational diffusion of a spherical particle were obtained for different experimental geometries. In particular, in the case where the rotational correlation time is much longer than the excited-state lifetime, the rotational contribution to the FCS correlation function is expressed as follows [6, 50]:

$$G(\tau) = \frac{1}{\langle N \rangle}\left[1 + a_1 \exp(-6D_{\mathrm{R}}t) + a_2 \exp(-20D_{\mathrm{R}}t)\right]G_{\mathrm{D}}(\tau)\,, \tag{26}$$

where $G_{\mathrm{D}}(\tau)$ is the correlation function for free diffusion in 3D, D_{R} is the rotational diffusion coefficient, and the pre-exponential factors $a_{1,2}$ depend on the particular experimental geometry. In most cases the rotational contribution can be considered to be essentially single-exponential [53, 54].

Successful applications of polarized FCS include obtaining an estimate of the rotational correlation time of green fluorescent protein, being in a good agreement with time-resolved fluorescence anisotropy measurements [54]. Recently, polarized FCS was applied to study the rotational and translational diffusion of peptide-coated semiconductor nanorods [55]. One should keep in mind that, due to the use of high-NA objectives, polarized FCS measurements can be prone to experimental artifacts [56], which can still be minimized under certain conditions [57].

A related technique to study rotational dynamics which currently gains importance, especially in single-molecule fluorescence microscopy [58], is based on the study of correlation functions $C(\tau) = \langle P(t)P(t+\tau)\rangle$ of the fluorescence polarization $P = (F_{\parallel} - F_{\perp})/(F_{\parallel} + F_{\perp})$. Application of this approach is, however, far from trivial, which is illustrated by a recent retraction [59] of a series of experimental papers on slow single-molecule rotation dynamics. However, careful implementation of the method can provide valuable results on the rotational dynamics of single molecules immobilized in a matrix [60].

It has been recently shown [61] that application of the polarization correlation functions in cases where particles simultaneously exhibit both translational and rotational diffusion is an efficient means to eliminate the unwanted contribution of the translational motion, which otherwise can complicate determination of the rotational dynamics. The efficiency of a similar approach, but this time based on computing correlation functions of fluorescence anisotropy $r = (F_{\parallel} - F_{\perp})/(F_{\parallel} + 2F_{\perp})$, to study slow rotational motion of molecules was recently demonstrated [62].

It is known that the orientational photoselection of fluorophores becomes more efficient upon multiphoton excitation [18]. Therefore, an enhanced contribution of rotational dynamics to fluorescence fluctuations should be expected in FCS experiments with two-photon excitation.

2.2.4 Fast Fluorescence Dynamics: Triplet Blinking and Antibunching

The above expressions for the FCS correlation functions do not account for the quantum nature of the fluorescence emission by a single molecule. The proper account of quantum effects leads to anticorrelation of photocounts at short timescales, the phenomenon known as photon antibunching [34], which is a manifestation of the fact that a single quantum system cannot emit a light quantum immediately after undergoing a radiative transition. For the first time, the effect of the drop in the intensity ACF at short times was predicted as a by-product of the treatment of the rotational motion in FCS [5]; later, photon antibunching was independently predicted in a quantum electrodynamics analysis of single-atom resonance fluorescence [63]. The effect was first observed experimentally for sodium atoms [64], and later for dye molecules in solution at room temperature [65] and in a solid matrix at cryogenic temperatures [66].

On the other hand, the presence of the relatively long-lived triplet state, where the molecule can be "deposited" for certain periods of time and thus temporarily stops fluorescing, leads to the opposite effect of photon bunching, showing in the fluorescence ACF as an additional transient positive contribution at a timescale of the order of the triplet-state lifetime.

A unified description of population dynamics of a dye molecule accounting for the ground, first excited, and first triplet states (see scheme in Fig. 5) was developed in [67] based on optical Bloch equations. For dye molecules in solution at room temperature, the dynamics can be well approximated by a set of rate equations [67–69]. The model predicts photon antibunching at short times and photon bunching at longer times due to transitions to the nonfluorescent triplet state. Under the condition of $k_{21} \gg k_{23}, k_{31}$ (see scheme in Fig. 5) and with the assumption that photophysical processes and diffu-

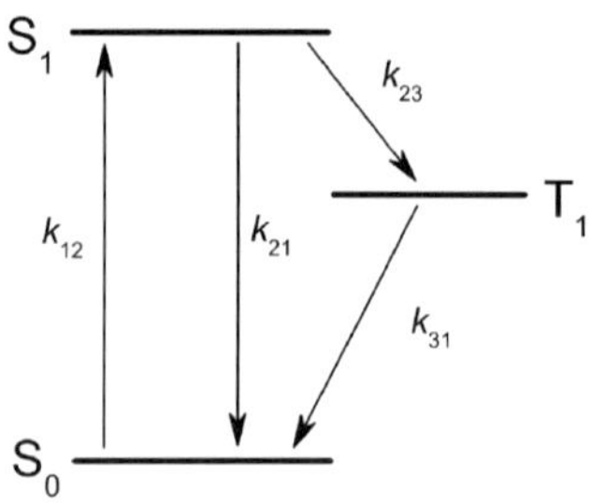

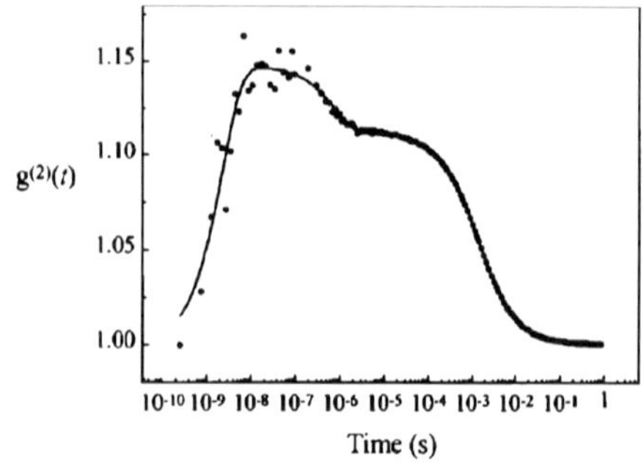

Fig. 5 *Left*: Simplified Jabłoński diagram of a fluorescent molecule. *Right*: Fluorescence intensity ACF for a Rhodamine 6G solution in a flat microcavity along with a fit according to Eq. 27. Reprinted with permission from [69]. © (1998) by the American Physical Society

sional motion of the molecules are independent, the following approximate expression for the FCS ACF can be obtained:

$$G(\tau) = \frac{1}{\langle N \rangle} \left\{ 1 + \frac{1}{m(1-T)} \left[-\exp\left(-\frac{\tau}{\tau_{\mathrm{A}}} \right) + T \exp\left(-\frac{\tau}{\tau_{\mathrm{T}}} \right) \right] \right\} G_{\mathrm{D}}(\tau) , \quad (27)$$

where $G_{\mathrm{D}}(\tau)$ is the correlation function for free diffusion, τ_{A} and τ_{T} are antibunching and triplet blinking times, $1/\tau_{\mathrm{A}} = k_{12} + k_{21}$, $1/\tau_{\mathrm{T}} = k_{31} + k_{12}k_{23}/(k_{12} + k_{21})$, $k_{12} = \sigma_{\mathrm{exc}} I$ (σ_{exc} is the excitation cross section, I is the excitation intensity), m is the number of fluorescent labels per particle [70], and $T = k_{12}k_{23}/[k_{12}(k_{23} + k_{31}) + k_{21}k_{31}]$ is the fractional population of the triplet state in the detection volume.

The possibility to measure the full correlation function of freely diffusing dye molecules showing the antibunching, triplet blinking, and diffusion components over the time range from picoseconds to seconds was demonstrated in experiments with a conventional correlator and cw excitation [69] (Fig. 5) and recently with pulsed excitation and time-resolved time-correlated detection [71].

At low excitation intensities the model agrees well with experimental observations, as is illustrated by Fig. 5. Several effects can lead to deviations from the simple expression (Eq. 27): due to the spatial dependence of the excitation power density, distributions of antibunching and triplet blinking times, rather than single-exponential terms, should be used [68, 69, 72]; at elevated excitation intensities further deviations from Eq. 27 should take place due to distortion of the concentration profile as a result of fluorescence saturation [68].

In most FCS studies, however, the time resolution is not high enough to observe the antibunching term, and, as a rule, diffusing particles are not multiply labeled. Therefore, the following expression is usually used to analyze FCS ACF data:

$$G(\tau) = \frac{1}{\langle N \rangle}\left[1 + \frac{T}{1-T}\exp\left(-\frac{\tau}{\tau_{\mathrm{T}}}\right)\right]G_{\mathrm{D}}(\tau) . \tag{28}$$

2.2.5
Effects of Molecular Binding via Two-Color Cross-Correlation

Investigation of binding reactions is extremely important in biochemical studies, and, in principle, can be carried out by following a change in the diffusion coefficient of fluorescent species, as has been discussed above. However, for proteins in the globular state the diffusion coefficient is approximately proportional to the cubic root of the molecular weight, and therefore at least an order of magnitude change in the molecular weight upon binding is required to observe the binding reaction in the fluorescence autocorrelation. The sensitivity and resolution can be enhanced dramatically by using so-called two-color fluorescence cross-correlation spectroscopy (FCCS). In this method, the sample contains two different labels with distinct excitation and fluorescence spectra, which are selectively excited by two overlapping laser beams; the fluorescence signal collected from the overlapping detection volumes is split into two spectral channels, intensities in which are monitored by two detectors whose outputs are cross-correlated (Fig. 3).

The two-color FCCS, which was inspired by varieties of the DLS technique [73, 74], was proposed in [75]. There, the theory of two-color FCCS was developed and the power of the approach was demonstrated in a study of correlated motion of interacting colloidal particles. The two-color FCCS method, however, did not receive considerable attention before Eigen and Rigler [21] pointed out its tremendous potential for biochemical studies on the single-molecule level. The power of this method in studies of diffusion and reactions in multicomponent solutions lies in the fact that tiny fractions of doubly labeled particles can be detected on the background of larger amounts of singly labeled particles, irrespective of whether their diffusion coefficients differ significantly or not. This was convincingly demonstrated in [76], after which two-color FCCS has become a tool of choice in studies of molecular interactions in biochemistry and cell biology.

As an example, consider a sample containing three types of particles: single-labeled ones carrying either a red or green label with the mean concentrations $\bar{c}_{\mathrm{r}}$ and $\bar{c}_{\mathrm{g}}$, respectively, and double-labeled ones carrying both labels, with the concentration $\bar{c}_{\mathrm{rg}}$. Assuming ideal conditions, namely, identical shape and position of the detection volumes corresponding to the red and green detection channels, their perfect spectral separation (no spectral cross-talk), and constant molecular brightnesses of the red and green fluorophores, irrespective of their binding conditions, one can find by using reasoning similar to that of Sect. 2.2.2 that the two-color CCF of fluorescence fluctuations in the red and green channels

$$G_{\times}(\tau) = \frac{\left\langle \delta F_{\mathrm{g}}(t)\delta F_{\mathrm{r}}(t+\tau)\right\rangle}{\langle F_{\mathrm{g}}\rangle\langle F_{\mathrm{r}}\rangle} \tag{29}$$

is characterized by the amplitude proportional to the concentration of the double-labeled species:

$$G_{\times}(\tau) = \frac{\bar{c}_{\mathrm{rg}}}{V_{\mathrm{eff}}(\bar{c}_{\mathrm{r}} + \bar{c}_{\mathrm{rg}})(\bar{c}_{\mathrm{g}} + \bar{c}_{\mathrm{rg}})} G_{\mathrm{Drg}}(\tau)\,, \tag{30}$$

where $G_{\mathrm{Drg}}(\tau)$ describes diffusion of the double-labeled species. The ACFs of the red and green species under these conditions are

$$G_{\mathrm{r,g}}(\tau) = \frac{\bar{c}_{\mathrm{r,g}} G_{\mathrm{Dr,g}}(\tau) + \bar{c}_{\mathrm{rg}} G_{\mathrm{Drg}}(\tau)}{V_{\mathrm{eff}}(\bar{c}_{\mathrm{r,g}} + \bar{c}_{\mathrm{rg}})^2}\,, \tag{31}$$

where $G_{\mathrm{Dr,g}}(\tau)$ describe diffusion of red and green species, respectively.

As a result, if the size of the detection volume is known, the absolute concentration of the double-labeled species can be determined:

$$\bar{c}_{\mathrm{rg}} = \frac{G_{\times}(0)}{V_{\mathrm{eff}} G_{\mathrm{r}}(0) G_{\mathrm{g}}(0)}\,. \tag{32}$$

The expressions above represent an ideal case of 1 : 1 ligand–receptor binding under ideal excitation and detection conditions. In reality, two-color FCCS measurements are prone to numerous artifacts, including different sizes of the two detection volumes, their relative displacement, cross-excitation of fluorophores, and spectral cross-talk of the two detection channels. The effect of the spectral cross-talk on cross-correlation amplitudes is extensively discussed in [77]. Technical aspects, including a discussion of the focal geometry, background and spectral cross-talk correction, as well as detailed treatment of the two-component binding equilibria as studied by FCCS, are presented in [78]. Practical strategies on quantitative correction for the spectral cross-talk are developed in [79]. Additionally, [79] presents a general theory of cross-correlation analysis for complex binding stoichiometries and changes in fluorescence efficiency, which was successfully applied in [80] to the cross-correlation analysis of enzyme–substrate interaction. A recent study [81] has shown the feasibility of FCCS with up to three different fluorophores using a single excitation wavelength, which allows one to study binding of three interacting partners.

The spectral cross-talk effects can be minimized in multicolor FCCS by interleaving the light sources exciting the different fluorophores and synchronizing the detection with the excitation, so that the source of each detected photon is known [82, 83]. The recently proposed optical designs for multicolor fluorescence cross-correlation based on the use of a grating [84] or a prism [85] should provide a more flexible selection of fluorescence detection ranges and thus more efficient multicolor cross-correlation experiments. For a review of recent advances in the application of FCCS in studies of living cells, see [33].

2.2.6 Effects of Flow and Active Transport

For the first time, the possibility of application of FCS to study systems with a laminar flow or uniform translation was explored in [9]. If a constant laminar flow is present in a sample with freely diffusing fluorescent particles, or the sample undergoes uniform translation, then fluctuations of the concentration of the fluorescent species are described by the advective diffusion equation

$$\partial_t c(\boldsymbol{r}, t) = D\nabla^2 c(\boldsymbol{r}, t) - \boldsymbol{v}(\boldsymbol{r}) \cdot \nabla c(\boldsymbol{r}, t)\,, \tag{33}$$

where $\boldsymbol{v}(\boldsymbol{r})$ is the vector of the flow or sample translation velocity. This equation should be solved with careful account of the relevant boundary conditions, especially when describing experiments in narrow microfluidic channels or capillaries. Far from channel boundaries, a free-space solution of Eq. 33 may be assumed:

$$\Phi(\boldsymbol{r}, \boldsymbol{r}', t) = \frac{1}{(4\pi Dt)^{3/2}} \exp\left[-\frac{(\boldsymbol{r} - \boldsymbol{r}' + \boldsymbol{v}t)^2}{4Dt}\right]. \tag{34}$$

The presence of the velocity-dependent factor in Eq. 34 modifies the FCS ACF compared to that for pure diffusion, which makes it possible to investigate simultaneously diffusion and flow. In particular, in the case of a uniform flow in the XY-plane with the velocity $v_\rho = (v_x^2 + v_y^2)^{1/2}$, the FCS ACF takes the following form:

$$G(\tau) = \frac{1}{\langle N \rangle} \exp\left[-\frac{(v_\rho \tau / r_0)^2}{1 + \tau/\tau_{\mathrm{D}}}\right] G_{\mathrm{D}}(\tau)\,, \tag{35}$$

where $G_{\mathrm{D}}(\tau)$ is the correlation function for free diffusion in 3D (Eq. 16). More general expressions involving flow or translation for a system undergoing diffusion in the presence of chemical reactions were considered in [86]. An extension of the technique employing a single laser spot with an elongated cross section to additionally measure the flow direction was proposed in [87] and later experimentally demonstrated in [88].

The more general technique of two-beam cross-correlation [89–91] utilizes two spatially separated laser foci and two corresponding detectors whose outputs are cross-correlated. The two-beam cross-correlation not only provides more reliable estimates of the diffusion coefficient and flow velocity, but additionally can determine the direction of flow. In particular, in the case of a constant flow with the velocity $\boldsymbol{v}$ and two identically shaped Gaussian ellipsoid detection volumes Eq. 10 displaced by the vector $\boldsymbol{R}$, the two-beam CCF takes the following form:

$$G_\times(\tau) = \frac{1}{\langle N \rangle} \exp\left\{ -\frac{\left[(\boldsymbol{v}_\rho \tau - \boldsymbol{R}_\rho)/r_0\right]^2}{1 + \tau/\tau_\mathrm{D}} - \frac{\left[(v_z \tau - R_z)/z_0\right]^2}{1 + (r_0/z_0)^2 \tau/\tau_\mathrm{D}} \right\} G_\mathrm{D}(\tau) , \qquad (36)$$

where $R_\rho = (R_x^2 + R_y^2)^{1/2}$. By setting $\boldsymbol{R} = \boldsymbol{0}$ one recovers the single-beam expression (Eq. 35). Notice that in the case where the separation of the laser foci is known, the two-beam cross-correlation can in principle be used to obtain absolute values of the transport coefficients.

Equation 36 implies that the two detection volumes are perfectly separated, so that detectors can "see" only fluorescence from the respective laser foci. In reality, a nonvanishing overlap of detection volumes (spatial cross-talk) leads to contamination of the cross-correlation curve by the so-called pseudo-autocorrelation, which is discussed in detail in [89, 91]. The effect of the spatial cross-talk between the detection volumes can be minimized by using pulsed optical excitation and introducing a delay for the excitation pulse in one of the beams by a time interval exceeding the fluorescence lifetime, as has been suggested in [92].

Fluorescence correlation and two-beam cross-correlation spectroscopy were successfully used to study transport [90, 92] and flow profiles [88, 93–95] in microfluidic structures, as well as chemical and photochemical reactions [96, 97]. Experimental aspects of the application of two-beam cross-correlation to flow measurements, including the limitations and artifacts of the technique, were recently reviewed in [98]. In spite of these impressive developments, experimental FCS-based studies of active transport in living cells still remain scarce [99].

2.2.7 Anomalous Diffusion

In certain cases, FCS measurements on systems presumably exhibiting inhomogeneity on the submicrometer scale produce correlation curves decaying slower than expected in the case of simple diffusion. During the last decade, it has become customary to explain this behavior by the presence of the so-called anomalous subdiffusion. Anomalous diffusion is a rather broad concept [100] generally related to diffusion processes where freely diffusing particles are characterized by the mean square displacement growing with time slower than in the case of normal diffusion, usually as $\sim t^\alpha$ with the exponent $0 < \alpha < 1$ reflecting the degree of diffusion "anomality". Subdiffusion was reported or discussed as a possible model in a number of FCS experiments with different systems, including biological membranes [101, 102], cell nuclei [103], cell cytoplasm [104, 105], gels [106], and crowded macromolecular environments [107, 108]. Recently, an alternative FCS-based approach to investigate anomalous diffusion by studying the dependence of the apparent diffusion coefficient as a function of the size of the detection volume has been proposed [109, 110].

In most FCS-related publications reporting anomalous diffusion, an assumption is made that the distribution of displacements of a diffusing particle is Gaussian with the mean square displacement growing with time as $\sim t^{\alpha}$. This assumption corresponds to the fractional Brownian motion (fBm) due to Mandelbrot and van Ness [111]. A few papers [103, 105] additionally mention the possibility to describe anomalous FCS curves using the continuous-time random walk (CTRW) fractional subdiffusion (FSD) [112]. Therefore, because of considerable attention to diffusion in disordered systems, understanding of the possible manifestations of these types of anomalous subdiffusion in FCS data is important for the correct interpretation of experimental results.

The fBm process is defined formally [111] by postulating that it is characterized by stationary and self-similar Gaussian increments having an infinite span of interdependence. Therefore, the Green function of the fBm in d dimensions is Gaussian by definition (Fig. 6):

$$\Phi_{\alpha}^{\mathrm{fBm}}(r,t) = \frac{1}{\left(4\pi D_{\alpha}^{\mathrm{fBm}} t^{\alpha}\right)^{d/2}} \exp\left[-\frac{r^2}{4D_{\alpha}^{\mathrm{fBm}} t^{\alpha}}\right], \tag{37}$$

where $D_{\alpha}^{\mathrm{fBm}}$ is the generalized anomalous diffusion coefficient dimensioned as $[\text{length}]^2[\text{time}]^{-\alpha}$. The variance of the distribution grows according to the power law $\langle r_{\mathrm{fBm}}^2(t)\rangle = 2dD_{\alpha}^{\mathrm{fBm}} t^{\alpha}$. The Gaussian form of the distribution of displacements for the fBm makes its use appealing due to the ease of related

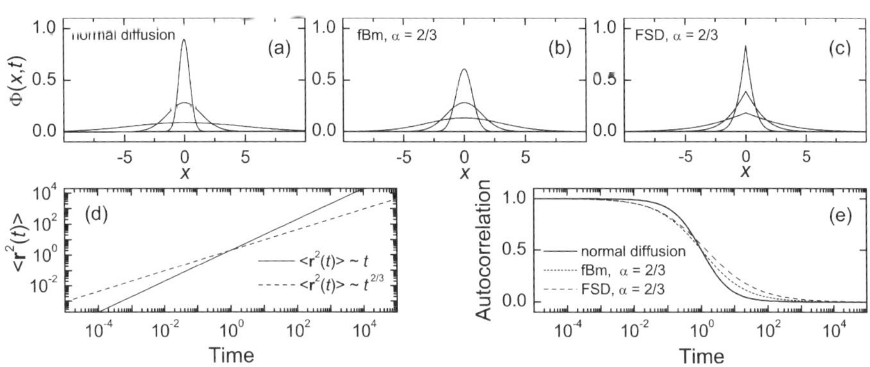

Fig. 6 Green functions of three diffusion types in dimensionless coordinates for three dimensionless time instants $t = 0.1$, 1, and 10: **a** normal (Fickian) diffusion ($D = 1$), **b** fractional Brownian motion ($\alpha = 2/3$, $D_{\alpha}^{\mathrm{fBm}} = 1$), and **c** fractional subdiffusion ($\alpha = 2/3$, $D_{\alpha}^{\mathrm{FSD}} = 1$). Panel **d** shows the corresponding mean square displacements for the normal diffusion (**a**) (——) and for the anomalous diffusion models (**b**,**c**) (- - -). Panel **e** shows the FCS ACFs for 2D particle motion according to normal diffusion (——), fBm with $\alpha = 2/3$ (- - -), and FSD with $\alpha = 2/3$ (– – –)

derivations: the normalized FCS ACF for fBm is expressed as follows:

$$G_{\mathrm{D}\alpha}^{\mathrm{fBm}}(\tau) = \frac{1}{\left(1+(\tau/\tau_{\mathrm{D}\alpha})^{\alpha}\right)\sqrt{1+f^{-2}(\tau/\tau_{\mathrm{D}\alpha})^{\alpha}}}\,, \tag{38}$$

where

$$\tau_{\mathrm{D}\alpha} \equiv \tau_{\mathrm{D}\alpha}\big(D_{\alpha}^{\mathrm{fBm}}\big) = \left(\frac{r_0^2}{4D_{\alpha}^{\mathrm{fBm}}}\right)^{1/\alpha}. \tag{39}$$

Unfortunately, the fBm model neither implies nor explains particular underlying physical mechanisms, since none of the known dynamical theories gives rise to fBm [37]. Additionally, attempts to "derive" the fBm Green function $\Phi_{\alpha}^{\mathrm{fBm}}(\boldsymbol{r},t)$ by "plugging in" a time-dependent diffusion coefficient into the diffusion equation (as is done, e.g., in [105, 108, 113–115]) are unphysical and misleading. For more detailes and discussion, see, e.g., [116].

Unlike fBm, another anomalous diffusion process leading to a power-law time dependence of $\langle \boldsymbol{r}^2(t)\rangle \sim t^{\alpha}$—the CTRW FSD—can be described on a physical basis [112] by assuming that the diffusing particle exhibiting a distribution of jump lengths $\lambda(x)$ with a finite variance can undergo transient trapping characterized by a waiting time distribution $w(t) \sim t^{-1-\alpha}$ with a diverging mean. The Green function of the FSD process is described by the fractional diffusion equation [112], which we write in the form

$$\partial_t \Phi_{\alpha}^{\mathrm{FSD}}(\boldsymbol{r},t) = \Gamma(1+\alpha) D_{\alpha}^{\mathrm{FSD}}\, {}_0\mathrm{D}_t^{1-\alpha} \nabla^2 \Phi_{\alpha}^{\mathrm{FSD}}(\boldsymbol{r},t)\,, \tag{40}$$

where $D_{\alpha}^{\mathrm{FSD}}$ is the FSD coefficient, ${}_0\mathrm{D}_t^{1-\alpha}$ is the Riemann–Liouville operator [117], and $\Gamma(\cdot)$ is the gamma function. This Green function is characterized by a power-law mean square displacement $\langle \boldsymbol{r}_{\mathrm{FSD}}^2(t)\rangle = 2dD_{\alpha}^{\mathrm{FSD}} t^{\alpha}$ [112]. With the use of the integral representation [118, 119] of $\Phi_{\alpha}^{\mathrm{FSD}}(\boldsymbol{r},t)$, we establish the following formal relation between the fBm and FSD:

$$\Phi_{\alpha}^{\mathrm{FSD}}(\boldsymbol{r},t) = \int_0^{\infty} A\big(D_{\alpha}^{\mathrm{fBm}},1\big)\Phi_{\alpha}^{\mathrm{fBm}}\big(\boldsymbol{r},t;D_{\alpha}^{\mathrm{fBm}}\big)\,\mathrm{d}\ln D_{\alpha}^{\mathrm{fBm}}\,, \tag{41}$$

where $A(s,t) = (t/(\alpha s))(D_{\alpha}^{\mathrm{FSD}}/s)^{1/\alpha} l_{\alpha}((D_{\alpha}^{\mathrm{FSD}}/s)^{1/\alpha}t)$, and $l_{\alpha}(\cdot)$ is the one-sided Lévy stable density [120]. Equation 41 clearly shows that the FSD process is characterized by a non-Gaussian distribution of displacements (Fig. 6c).

The effect of the non-Gaussian character of FSD clearly affects the corresponding FCS ACF, which can be expressed as follows:

$$G_{\mathrm{D}\alpha}^{\mathrm{FSD}}(\tau) = \int_0^{\infty} A\big(D_{\alpha}^{\mathrm{fBm}},1\big) G_{\mathrm{D}\alpha}^{\mathrm{fBm}}\Big(\tau;\tau_{\mathrm{D}\alpha}\big(D_{\alpha}^{\mathrm{fBm}}\big)\Big)\,\mathrm{d}\ln D_{\alpha}^{\mathrm{fBm}}\,. \tag{42}$$

While the fBm and FSD processes with the same α and $D_{\alpha}^{\mathrm{FSD}} = D_{\alpha}^{\mathrm{fBm}}$ give rise to identical mean square displacements $\langle \boldsymbol{r}^2(t)\rangle$, the corresponding FCS ACFs differ significantly (Fig. 6d and e).

Thus, the situation with the anomalous diffusion in FCS measurements is not as clear as it might seem from first sight. On the one hand, the fBm model leading to the Gaussian distribution of displacements (Eq. 37) and giving the simple expression for the FCS ACF (Eq. 38) is completely lacking a clear physical picture in the background, and therefore its choice as a basis for a model describing experimental data is hard to justify from the physical point of view. On the other hand, the CTRW FSD model, while having a clear and appealing physical basis, is characterized by a non-Gaussian distribution of displacements and leads to quite a different shape of the FCS ACF at the same value of α, showing a longer tail compared to the case of fBm. As a result, the physical insight gained from the empirical application of the concept of the anomalous diffusion to experimental FCS data may be quite limited, especially if no convincing microscopic origin of the deviation from the normal diffusion law is provided. Similar conclusions were drawn in a recent FCS study [121] of two-component membranes whose heterogencity is frequently claimed to result in subdiffusion of probe molecules [122]. In a recent FCS study of protein diffusion in an environment crowded by surfactant micelles [108], the two-component diffusion was favored over anomalous diffusion because of a more convincing physical picture behind the data. This shows that the anomalous diffusion model should be used with extreme caution and should always be tested against simpler physical alternatives.

3 Experimental Aspects

3.1 FCS Instrumentation

Numerous self-made FCS setups have been reported in the literature and share the general features depicted in Fig. 3. While the particular technical details of the experimental setup design are outside the scope of the present review, we still highlight some essential points related to the FCS instrumentation.

First of all, the need to provide a tightly focused laser beam and to collect the maximum amount of fluorescence emitted in a sample (which, as a rule, is close to water in its optical properties) requires the use of high numerical aperture (NA) water-immersion objectives designed for fluorescence microscopy, for example, C-Apochromat 40× NA 1.2 W from Zeiss or UApo 40× W/340 NA 1.15 from Olympus. Measurements in media with higher refractive indices will require an objective with a capability of refractive index correction, e.g., 40× Plan Neofluar from Zeiss. Most modern designs of FCS setups use fiber-coupled avalanche photodiodes (APDs) as fluorescence detectors, with the fiber entrance playing the role of a pinhole, as was first suggested in [123]. The use of APDs as photon detectors is motivated by their high quantum efficiency, which, however, comes at the expense of moderate time resolution, quite high dark count rate of the order of 250–300 s^{-1}, and rather high afterpulsing probability. On the contrary, photomultiplier tubes, which are very rarely used at present in FCS, mostly because of lower quantum efficiencies, may provide a higher time resolution, much lower afterpulsing probabilities, and dark count rates of just a few counts per second. A wide range of correlation hardware and software is currently available on the market, including the family of classical multiple tau correlators by ALV (Langen, Germany), the USB-based external digital correlators from correlator.com (Bridgewater, NJ), and novel data acquisition cards with sophisticated single photon counting modes which allow the extraction of complete information on the measured photon sequence [124–126].

A range of commercial research-grade FCS setups has become available during the last few years, including the ConfoCor2 [127] (recently upgraded to ConfoCor3) from Zeiss (Jena, Germany), which was the first company to offer an FCS system on the market, FCS2 from Leica (Wetzlar, Germany), MicroTime 200 from PicoQuant (Berlin, Germany), Alba FCS from ISS (Champaign, IL, USA), and FCS module C9413 from Hamamatsu Photonics (Japan). The first four setups combine a confocal microscopy system with fluorescence correlation/cross-correlation capabilities and are developed as user-friendly platforms mainly aimed at (but, of course, not restricted to) bio-applications. These setups include several lasers for excitation and several spectral detection channels with fiber-coupled APDs as detectors. (For additional options and advanced data acquisition methods implemented in specific setups, an interested reader is advised to consult the relevant producer's information.) By contrast, the compact Hamamatsu unit, featuring a solid-state laser for excitation, a Peltier-cooled photomultiplier tube with a very low afterpulsing as detector, and no microscopy capabilities, provides an affordable option for solution studies.

The large-scale FCS-based analysis of biomolecular interactions in drug discovery, research, and bioanalytics can be carried out using the Clarina™ II integrated platform by Evotec Technologies (Hamburg, Germany).

3.2 Fluorophores for FCS Studies

In principle, any fluorescent molecule can be used as a fluorophore in an FCS experiment, provided that the molecular brightness is high enough. However, the use of an organic fluorophore as a fluorescent label in biochemical studies imposes quite stringent requirements on the dye quantum yield, photostability, and chemistry. Presently, the most widespread fluorophores used in FCS are rhodamine and fluorescein derivatives, produced as the Alexa and Atto dye families (Molecular Probes and ATTO-Tec, respectively), which cover a wide range of wavelengths. Different types of functionalizations are available for specific labeling of proteins and DNA.

An inorganic alternative to autofluorescent proteins and fluorescent dyes are semiconductor nanocrystals, also known as quantum dots (QDs) [128, 129]. They are characterized by a broad absorption spectrum and relatively narrow symmetric luminescence band whose position is finely tuned by the size of the core of the QD (in the range from a few to ~ 10 nm). These features facilitate the use of QDs as labels in fluorescence cross-correlation investigations, including live cell studies [130]. An additional appealing feature of QDs is a high two-photon excitation cross section [131]. However, the relatively long lifetimes and complicated photophysics of QDs (see, e.g., [132–134]) still prevent their wide use as fluorescent labels in FCS. The behavior of QDs under the conditions of an FCS experiment is still not well understood, although significant progress has been made recently [135, 136]. Of particular interest are water-soluble QDs for biochemical applications. The solubility in water is achieved by surface modification, which leads to a significant increase in size compared to the naked QD, reaching as high as ~ 20 nm, and may lead to a reduction of fluorescence quantum yield. However, recent results [137] show that this problem can, at least to a certain extent, be resolved, and bright and stable water-soluble QDs with sub-10-nm diameter can be reliably produced.

Autofluorescent proteins [42, 138], which offer an important alternative to dyes and nanocrystals in cell research, include the green fluorescent protein discovered more than 30 years ago, as well as the more recently isolated cyan fluorescent protein, red-shifted green fluorescent protein [42], DsRed [139] with its monomeric mutants like mRFP1 [140], mTangerine, mStrawberry, and mCherry [141], and photoswitchable fluorescent proteins like Kaede [142] or Dronpa [143]. They are characterized by a particular arrangement of amino acid chains, which ensures formation of a naturally fluorescent group surrounded by a barrel-like structure. In applications, one should keep in mind that, as a rule, autofluorescent proteins are characterized by quite complicated photophysical behavior [43, 44, 144–149].

4 Nonidealities and Artifacts

4.1 Sample Nonidealities

4.1.1 Concentration Fluctuations in Nonideal Solutions

By monitoring concentration fluctuations of particles diffusing in and out of the detection volume, the FCS technique is sensitive to the mutual diffusion coefficient D_m [150] which describes relaxation of concentration gradients [37]. This contrasts FCS with the tracer-based techniques, like fluorescence recovery after photobleaching (FRAP) [151] and single-particle tracking (see, e.g., [152]), which provide information on the self-diffusion coefficient D_s describing the random walk of individual particles. At low concentrations, when interactions between diffusing particles can be neglected and the solution can be considered to be ideal, D_m tends to D_s. Thus, while at low concentrations FCS reports the self-diffusion coefficient, the mutual diffusion coefficient is measured by FCS in concentrated solutions. A related feature of the technique is that at higher concentrations, when the solute occupies a significant volume fraction of the solution, the particle number fluctuations in the detection volume are no longer described by the Poisson statistic. Instead, the fluctuations are determined by the isothermal osmotic compressibility [153, 154], and the amplitude of the correlation function in this case is $G(0) = k_B\Theta/[\langle N\rangle(\partial\Pi/\partial\rho)_\Theta]$, where k_B is the Boltzmann constant, Θ is the absolute temperature, $(\partial\Pi/\partial\rho)_\Theta$ is the isothermal osmotic compressibility, Π is the osmotic pressure, ρ is the number density of solute particles, and $\langle N\rangle$ is defined above (Eq. 9).

While these issues are of purely theoretical interest in the case of solutions of small molecules, they become of high importance in experiments with colloidal particles and macromolecules [75, 150, 153, 154] whose dynamic properties are strongly dependent on the concentration [155, 156]. Manifestations of these effects in FCS experiments have been demonstrated and successfully applied to study the concentration effects in diffusion of DNA [157, 158] and polystyrene macromolecules [159].

4.1.2
Effects of Fluorescence Saturation and Photobleaching

In deriving expressions for the FCS correlation function presented in Sect. 2, it was assumed that the fluorescence intensity is proportional to the intensity of excitation, and that the shape of the detection volume and the character of concentration fluctuations of fluorescent particles in the focal spot are independent of the excitation power. This assumption is valid, however, only for low excitation intensities, and, upon an increase in the laser power, the effects of fluorescence saturation and fluorophore photobleaching start to come into play.

Saturation of the fluorescence intensity with an increase in the excitation power leads to flattening of the fluorescence intensity profile around the laser focus. This increases the effective detection volume and produces a longer apparent diffusion time and larger apparent number of particles in the detection volume. The effects of fluorescence saturation in FCS experiments with a single-photon excitation were recently studied by several groups [160–162]. In [160], a simple saturation model was successfully used to explain the power dependence of the measured diffusion time and number of molecules in the detection volume. The study [161] revealed that the pulsed optical excitation has a stronger effect on the apparent diffusion time compared to excitation with a cw source with the same mean power. In the case of two-photon excitation, changes in the shape of FCS correlation curves with the laser power density were also discussed in the context of fluorescence saturation [163], although we believe that in this case the major role is rather played by photobleaching effects.

A fluorescent molecule diffusing in a focused laser beam has a nonzero probability to be photobleached—usually, irreversibly. As a result, fluctuations of the concentration of fluorescent molecules in this case are described by the following reaction–diffusion equation:

$$\partial_t c(\boldsymbol{r}, t) = D\nabla^2 c(\boldsymbol{r}, t) - B(\boldsymbol{r})c(\boldsymbol{r}, t)\,, \tag{43}$$

where the prefactor $B(\boldsymbol{r})$ in the sink term is the product of the photobleaching quantum yield, absorption cross section, and coordinate-dependent excitation intensity. Unfortunately, a solution of Eq. 43 cannot be obtained analytically for an arbitrary $B(\boldsymbol{r})$. In [164], an approximate solution to Eq. 43 for a low photobleaching rate was obtained using a path-integral approach and, in agreement with numerical simulations and experimental observations, revealed an accelerated decay of the FCS ACF due to photobleaching.

Photobleaching effects were studied experimentally for dye molecules freely diffusing in solution by FCS with single- [165, 166] and two-photon [163] excitation. Effects of excitation pulse duration and sequential excitation to higher-lying states upon two-photon excitation on the contributions of photobleaching and saturation in FCS measurements were considered [168]. Recently, a comprehensive analysis of the photobleaching effects in single-molecule measurements was carried out in [169].

In FCS studies of membrane dynamics, photobleaching becomes especially important as a result of slow diffusion of membrane-bound probes and their confinement to the 2D membrane surface, which prevents their easy escape from the laser focus. Photobleaching effects in FCS on membranes and in cells were recently studied in detail in [170, 171].

In an FCS experiment, fluorescence saturation and photobleaching are always present simultaneously. The net effect, however, depends on the combination of a number of factors, including the photophysical properties of the fluorophore, its diffusion coefficient, and the geometry of the system (Fig. 7). In particular, the saturation effects are likely to dominate in the case of freely diffusing dyes upon single-photon excitation, as exemplified by results for Alexa 488 in aqueous solution (Fig. 7a); similar behavior is observed for Rhodamine 6G (data not shown), which is in excellent agreement with the recent findings [160] but contrasts the earlier observations [165] (the reason for the latter discrepancy is still unclear). On the

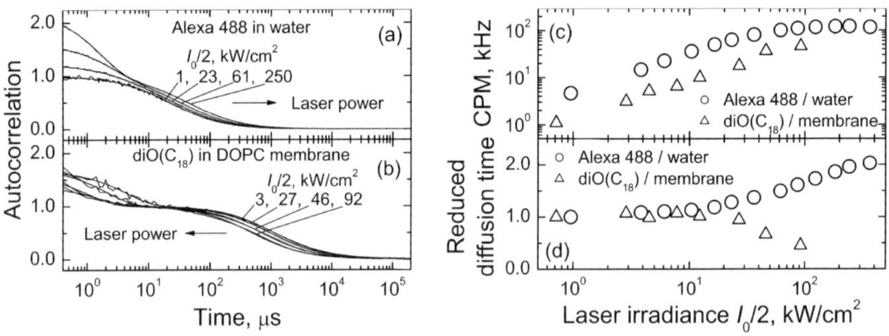

Fig. 7 Effect of the excitation laser power in FCS measurements on **a** freely diffusing Alexa Fluor 488 dye in water and **b** diOC$_{18}$(3) lipid probe undergoing two-dimensional diffusion in an L-alpha-dioleoylphosphatidylcholine bilayer membrane. Whereas in both systems the molecular brightness grows with the excitation power (**c**), the apparent diffusion time increases or decreases with the excitation power (**d**), depending on whether the fluorescence saturation or photobleaching is the dominating effect. The laser irradiance I_0 was estimated from the mean power P of the laser beam and the beam waist size: $I_0 = 2P/(\pi r_0^2)$ assuming $r_0 = 0.2\,\mu$m. $\lambda_{exc} = 488$ nm, $T = 298$ K. Setup: ConfoCor2 (Zeiss)

other hand, a fluorophore whose slow diffusional motion is confined to a 2D membrane has greater chances to be photobleached, which results in a faster decay of the correlation function at higher excitation intensities (Fig. 7b).

4.1.3 Effects of Optical Trapping

Tight focusing of the laser beam results in quite high laser power densities in the focal region which can, in principle, lead to (transient) trapping [172] of diffusing fluorescent particles and thus produce a slower decay of FCS autocorrelation curves. Whereas the early report on observation of optical trapping of dye molecules [173] is most likely a misinterpretation of fluorescence saturation effects at high excitation powers, the later observations for colloidal particles [174, 175] are plausible, as is confirmed by a recent theoretical study [176].

Recently, it has been shown that optical trapping in combination with FCS can be successfully exploited to study colloidal particles in a field of an off-resonant laser beam [177, 178]. In [179], a single IR laser beam was used to simultaneously create a trapping potential and provide two-photon excitation in FCS experiments with micrometer-sized multilamellar vesicles employed as biomimetic containers.

4.2
Nonidealities and Artifacts in the Experimental Setup

4.2.1
Real Shape of the Detection Volume

In the above discussion, it was assumed that the fluorescence detection volume can be well approximated by a 3D Gaussian ellipsoid (Eq. 10) (Fig. 8a). For the first time this model was introduced in [7] to describe FCS experiments with a nonconfocal detection. Although it was well understood that it should not hold in the case of confocal detection [180, 181], the 3D Gaussian model was put forward [20] as a good, simple approximation of the detection volume in confocal FCS, thus allowing simple expressions for FCS correlation functions to be obtained and providing an easy quantitative analysis of measurements.

The shape of the detection volume of the confocal microscopy system can be accurately described only by using the full electromagnetic treatment of the light field, and has been addressed in a number of papers both experimentally [182, 183] and theoretically [184, 185], including the effects of light polarization [186] and the stratified refractive index of the sample [187].

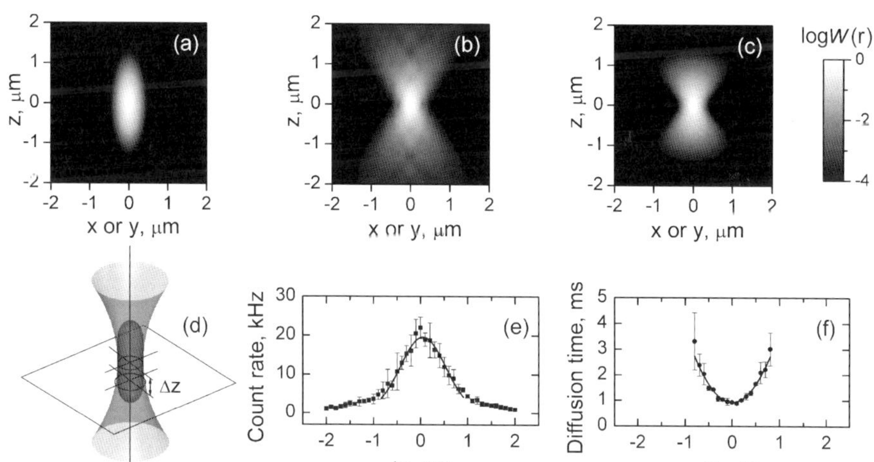

Fig. 8 Gray-scale images of the FCS detection volume: **a** 3D Gaussian approximation (Eq. 10); **b** calculated according to the theory [184] for parameters of the ConfoCor2 setup; **c** more realistic approximation (Eq. 44). Sketch **d** and results **e** and **f** of the detection volume "tomography" experiment demonstrating the non-Gaussian shape of the detection volume. *Solid curves* in **e** and **f** show Gaussian and quadratic fits, respectively. Sample: osmotically tensed giant unilamellar vesicles prepared from L-alpha-dioleoylphosphatidylcholine fluorescently labeled with $DiOC_{18}(3)$ lipid probe. $\lambda_{exc} = 488$ nm, $T = 298$ K. Setup: ConfoCor2 (Zeiss)

Computations carried out using the approach [183] for parameters of the ConfoCor2 setup [127] do indeed predict quite strong deviations from the simple 3D Gaussian model (Eq. 10), in particular considerable widening of the cross section at the tails (Fig. 8b). To verify the predictions of the theory, one can perform detection volume "tomography", e.g., by using a fluorescently labeled lipid bilayer parallel to the focal plane of the objective and performing FCS measurements at a set of its distances Δz from the focal plane.[4] In this case, while a decrease in the fluorescence intensity is expected with increasing $|\Delta z|$, the diffusion time τ_{D} of the probe molecules and the effective number of molecules $\langle N\rangle$ in the detection volume should grow with $|\Delta z|$. The experimental data in Fig. 8e and f are in good qualitative agreement with the expectations (data for $\langle N\rangle$ are not shown). This suggests the simple more realistic model for the detection efficiency profile:

$$W(x,y,z)=\frac{1}{1+z^2/h^2}\exp\left\{-\frac{2(x^2+y^2)}{r_0^2(1+z^2/h^2)}-\frac{2z^2}{z_0^2}\right\} \tag{44}$$

whose parameters can be determined from the measured z-dependences in the above tomography procedure. The gray-scale map of this approximation with parameters determined by the above procedure is shown in Fig. 8c.

The effects of the real shape of the fluorescence collection efficiency profile on deviations of the FCS correlation functions from the predictions of the simple model have been recently addressed in [189, 190]. Our practice shows that in the case of FCS experiments with probes freely diffusing in 3D, the influence of the non-Gaussian shape of $W(\boldsymbol{r})$ on deviation of the experimental results from the simple model (Eq. 16) are usually insignificant, the effects of fluorescence saturation and photobleaching at higher excitation powers being more pronounced. One should understand that fitting the FCS curve for a substance with a known diffusion coefficient to the model Eqs. 16 or 25 in fact does not allow one to determine the actual dimensions of the detection volume, but rather provides its effective description.

Due to the non-Gaussian shape of the detection efficiency profile, measurements of diffusion on phospholipid membranes can produce substantially underestimated diffusion coefficients if the membrane is displaced with respect to the focal plane of the objective, as is evident from Fig. 8 and has been demonstrated in [191]. To avoid this artifact, it is advised to carry out FCS measurements on membranes at the z-position corresponding to the minimum of the diffusion time (Fig. 8f).

Care must be taken to correctly account for the coverslip thickness by properly adjusting the correction ring of the objective and ensuring that the refractive index of the sample does not strongly deviate from that of water.

[4] In practice, it is easier to experiment with giant unilamellar vesicles [188] with radii of the order of 20 to 30 μm, whose curvature can be neglected compared to the size of the detection volume. To avoid artifacts due to membrane shape fluctuations, vesicles should be osmotically tensed.

Failing to provide these conditions can result in severe asymmetric distortion of the detection volume and thus produce hard to interpret data. In particular, the presence of the refractive index mismatch was shown to result in underestimation of the diffusion coefficient [190, 192]. For moderate refractive index mismatch, it was shown that the effect can be largely compensated by adjustment of the objective correction ring [193]. The use of an objective with immersion correction provides more reliable results in challenging FCS experiments in organic solvents, as has been demonstrated in [194].

4.2.2
Effects of Detector Afterpulsing and Dead Time

Two effects related to the properties of the fluorescence detector employed in an FCS setup need to be mentioned. One of these is known as detector afterpulsing. This effect is observed at low fluorescence intensities and takes place due to a nonzero probability of the detector to generate spurious pulses following a photon detection event. As a result, the measured ACF acquires artifactual short-lived components (Fig. 9) whose contribution increases with decreasing fluorescence intensity, which can hinder exact determination of the fast dynamics of fluorescent particles. The physics of detector afterpulsing is complicated and in the case of APD detection involves a number of different phenomena [195].

An experimental approach to eliminate the distorting effect of afterpulsing consists in splitting the fluorescence output into two channels and cross-correlating the outputs of two detectors, which was originally proposed in [196]. In this case, however, in addition to a more complicated setup, the fluorescence intensity on each of the detectors is reduced more than twice, which may be crucial in certain experiments. As an alternative, an approach to afterpulsing correction in ACFs which was first developed for DLS application [197, 198] was recently introduced in FCS practice [199]. It requires a separate determination of the detector afterpulsing pattern by performing an FCS measurement with a stabilized source of uncorrelated light. Recently,

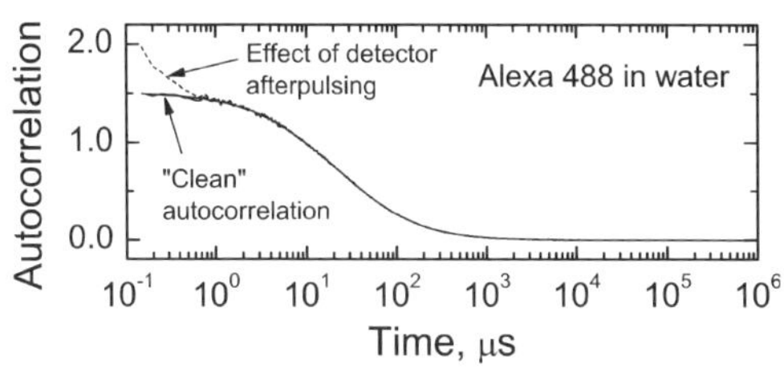

Fig. 9 Effect of the detector afterpulsing on the shape of the FCS curve and its elimination using the method described in [199]. Sample: Alexa 488 in water; $\lambda_{exc} = 488$ nm, $T = 298$ K. Setup: ConfoCor2 (Zeiss)

a method for elimination of the afterpulsing effects in FCS with pulsed excitation using time-correlated single photon counting (TCSPC) for separating photocounts due to fluorescence and detector afterpulsing was proposed in [200].

The other effect, detector dead time, distorts FCS correlation curves in the opposite way. Namely, it consists in appearance of a drop in the correlation function at short lag times and overall reduction of its amplitude. The dead time effect on correlation functions in FCS and DLS was studied theoretically in a number of works [4, 35, 201–203]. In a recent paper [204], a comparison of experimental data with different theoretical models for dead-time distortions [4, 201, 202] has shown that although the theory could not describe the observations quantitatively, computations according to the approach of [201] demonstrated a good qualitative agreement with data. The question of whether a more advanced theory (e.g., [203]) can provide a better agreement with observation still remains open. To date, the best approach to avoid the unwanted dead-time effects is to carry out FCS measurements at reasonably low fluorescence intensities, within the linearity range of the detector.

Since the dead-time effect is more characteristic for passively quenched APDs, today it is usually not a matter of concern in standard FCS measurements, which normally employ actively quenched APDs predominantly exhibiting the afterpulsing behavior (see, e.g., [195]). However, at higher count rates, the dead time of the whole electronic detection system may gradually come into play. The latter effect may become important, in particular in FCCS experiments with pulsed interleaved excitation [83].

5 Evaluation of Experimental FCS Data

5.1 Statistical Accuracy and Bias in FCS Data

Extracting trustworthy quantitative information from results of FCS measurements requires knowledge on the statistical accuracy of FCS data, which has some features specific for this particular experimental technique. To cover a wide as possible time range in a single experiment, FCS ACFs are usually recorded on a logarithmic lag timescale with quasi-exponentially growing channel widths, e.g., in the case when a multi-tau correlator [205] is used. As a result, the measured correlation functions show a nonuniform random noise whose amplitude is consistently higher at short lag times and decays toward the tail of the correlation curve. Therefore, a reliable data analysis requires knowledge of the lag time-dependent noise variance [206]. However, as opposed to, e.g., TCSPC experiments, where data are characterized by the

Poisson noise [19], the statistic of noise in the FCS data is much more complicated [4, 207–210].

The problem of the statistical accuracy of FCS measurements was addressed in [4] immediately after the introduction of the technique and was later considered in more detail [207–210]. The general conclusion of these studies is that the accuracy of the measured correlation curve dramatically increases with the molecular brightness of the fluorophore and improves with the duration of the measurement, though the particular dependences obtained in the above papers differ widely, depending on the assumptions made by the authors. Additionally, the statistical variations in the measured curve depend on the shape of the correlation function itself and on a particular implementation of the correlator architecture. Additionally, it should be pointed out that the noise in the correlation data is intrinsically correlated [211, 212]. This circumstance further complicates the data analysis, which is usually based on the assumption of statistical independence of the noise contribution at different lag times. In the absence of an easy-to-implement universal model of noise in FCS data, an estimate of the noise variance in correlation functions can be obtained from a set of repeated measurements on the same sample under identical conditions [209].

One should be aware that overly short measurement times can introduce a systematic distortion into the shape of the measured correlation function. The problem of bias in correlation functions due to the short sampling time has been known in statistics for many decades [213], and in the specific context of measurement of correlation functions with a digital correlator was addressed in the early 1970s [214]. The analysis of bias specific for the multiple-tau correlator was carried out in [211]. Recently, the accuracy and bias in the FCS data were analyzed theoretically in [210] in dependence on the molecular brightness and measurement time. (The reader, however, should be warned that this paper contains some unfortunate misprints in formulas.) Providing adequately long data collection times is especially crucial in measurements of ultraslow diffusion, e.g., in [215], or if exact measurements of the long tails of correlation functions of slowly diffusing particles are required, e.g., in [216].

Yet additional effects characteristic of multiple-tau correlators may contribute to shape distortion in measured correlation functions: the so-called triangular averaging [205] together with the exponential increase in the correlator channel width with increasing lag time can introduce distortions not only in oscillating correlation functions (which is quite obvious), but also in smoothly decaying correlation curves [217, 218].

The resolving power of FCS in determining diffusion coefficients was addressed in [40] and recently in [219]. In particular, it was found [40] that under the optimum experimental conditions fitting to a two-component model can reliably resolve two species with diffusion coefficients differing by at least a factor of 1.6. In [219], the optimum duration of the experiment necessary to guarantee the required accuracy of the diffusion coefficient estimate was studied for a range of molecular brightnesses as a function of the concentration and duration of the measurement.

5.2 FCS Data Analysis: Model-Based vs Inverse Problem Approach

The usual approach to experimental data analysis in FCS is a nonlinear least-squares fitting of experimental data using a model corresponding to the experimenter's assumptions on the system under investigation. The proper choice of a model is extremely important to obtain sensible and stable results. For example, in the case of fluorescent colloidal particles whose size cannot be neglected compared to the FCS detection volume, the particle size and shape should explicitly be included in the fitting model [220]. Another spectacular example is diffusion in narrow channels for which expressions for the FCS correlation functions should be derived by explicitly taking into account the presence of channel walls [221, 222]. Sometimes, sets of FCS curves are measured when studying the system behavior as a function of a particular control parameter. In this case, a global analysis of the whole set of data [223, 224] may provide a more accurate determination of the model parameters. As has been mentioned above, the use of the weighted least-squares fitting [206] with weights accounting for the noise structure in the correlation function is advantageous for obtaining reliable results.

An alternative approach to model-based analysis of FCS data is the recovery of (apparent) distributions of diffusion times $P(\tau_D)$ from the measured correlation curves. It is based on the representation of the diffusion part of the normalized correlation function via a distribution of diffusion times $P(\tau_D)$:

$$G(\tau) = \int_0^\infty P(\tau_D) G_D(\tau; \tau_D) \, d\tau_D \,, \tag{45}$$

where $G_D(\tau; \tau_D)$ is the FCS ACF for normal diffusion with the explicitly shown dependence on the diffusion time τ_D.

Reconstruction of the underlying distribution of diffusion times from experimental FCS data is an ill-posed inverse problem, and therefore a meaningful stable solution can be obtained only by applying regularization methods [225, 226]. The inverse-problem approach has come to the field of FCS from DLS and time-resolved fluorescence spectroscopy, where a number of linear and nonlinear regularization algorithms have been proposed and implemented during the last few decades (see, e.g., [227–232]).

An obvious application of this approach in FCS is simultaneous determination of the number of different freely diffusing species, their diffusion coefficients, and relative contributions. In this case, in accordance with Eq. 19, the distribution of diffusion times is a set of δ-functions

$$P(\tau_{\mathrm{D}}) = \sum_{i=1}^{n} P_i \delta(\tau - \tau_{\mathrm{D}i}) \,. \tag{46}$$

Upon successful inversion of the corresponding experimental data, this distribution will be reconstructed as a series of peaks, provided that the spacing of the diffusion times and ratios of contributions of components matches the data accuracy. Notice that in the inverse-problem approach, the resolvability of two discrete components will be worse than in the case of fitting to a two-component model, since, when solving an inverse problem, both the model and its parameters have to be determined from data. If the diffusion times of the components are closely spaced and/or the number of components is large, a (quasi-)continuous distribution of diffusion times will be reconstructed.

To date, the inverse-problem approach to the analysis of FCS data has been applied to problems involving heterogeneous systems, including determination of a distribution of fragment lengths in enzymatic DNA polymerization [233] and colloidal particle polydispersity analysis [234]. Other processes, including binding and diffusion on cell membranes [235–237] and effects of molecular crowding on protein diffusion [107], have also been addressed with the use of this approach. Simulation studies [238] have shown that the underlying distributions of diffusion times can be recovered from FCS data, provided that the noise in correlation functions is appropriately taken into account.

The fact that the inversion of the correlation function may produce a wide distribution of diffusion times does not necessary imply that the system represents a polydisperse ensemble of particles characterized by a distribution of sizes (a usual assumption in DLS measurements). To illustrate this point, we provide a few examples involving diffusion of a single species on a 2D surface. In this case, we rewrite Eq. 45 as

$$G(\tau) = \int_{0}^{\infty} \frac{p(\tau_{\mathrm{D}})}{1 + \tau/\tau_{\mathrm{D}}} \,\mathrm{d}\ln \tau_{\mathrm{D}} \,, \tag{47}$$

where $p(\tau_{\mathrm{D}}) = \tau_{\mathrm{D}} P(\tau_{\mathrm{D}})$ is the quantity usually computed by inverse-problem algorithms in the analysis of experimental data. As follows from the form of Eq. 47, $p(\tau_{\mathrm{D}})$ can be obtained using the Mellin transform technique [239].

Case 1. Anisotropic diffusion of fluorescent species in 2D:

$$G(\tau) = \frac{1}{\sqrt{(1 + \tau/\tau_{\mathrm{D}X})(1 + \tau/\tau_{\mathrm{D}Y})}} \,. \tag{48}$$

This situation can take place for diffusion of a marker on an anisotropic membrane in the XY-plane with diffusion coefficients $D_X \neq D_Y$ or, in the case of isotropic diffusion on a membrane making an angle with the XY-plane,

$\tau_{DX,Y} = r_0^2/4D_{X,Y}$; the limiting case $\min(D_X, D_Y) = 0$ corresponds to 1D diffusion (e.g., diffusion in a very narrow channel). Inversion of Eq. 48 produces the following apparent distribution of diffusion times (Fig. 10a):

$$p(\tau_D) = \begin{cases} \frac{1}{\pi\sqrt{(\tau_D/\tau_{min}-1)(1-\tau_D/\tau_{max})}}, & \tau_D \in (\tau_{min}, \tau_{max}), \\ 0, & \tau_D \notin (\tau_{min}, \tau_{max}), \end{cases} \tag{49}$$

where $\tau_{min} = \min(\tau_{DX}, \tau_{DY})$, $\tau_{max} = \max(\tau_{DX}, \tau_{DY})$.

Case 2. Mandelbrot–van Ness fractional Brownian motion (fBm) of fluorescent species in 2D. Inversion of the 2D variant of the correlation function $G(\tau) = G_{D\alpha}^{fBm}(\tau)$ (Eq. 38) leads to the following distribution (Fig. 10b):

$$\begin{aligned} p_\alpha^{fBm}(\tau_D) &\equiv p_\alpha^{fBm}(\tau_D; \tau_{D\alpha}) \\ &= \frac{1}{\pi} \frac{\sin(\alpha\pi)}{(\tau_{D\alpha}/\tau_D)^\alpha + 2\cos(\alpha\pi) + (\tau_D/\tau_{D\alpha})^\alpha} \end{aligned} \tag{50}$$

with $\tau_{D\alpha}$ defined in Eq. 39.

Case 3. Continuous-time random walk (CTRW) fractional subdiffusion (FSD) of fluorescent species in 2D. In this case, inversion of the correlation function $G(\tau) = G_{D\alpha}^{FSD}(\tau)$ (Eq. 42) leads to the following result (Fig. 10c):

$$p_\alpha^{FSD}(\tau_D) = \int_0^\infty A\left(D_\alpha^{fBm}, 1\right) p_\alpha^{fBm}\left(\tau_D; \tau_{D\alpha}\left(D_\alpha^{fBm}\right)\right) \mathrm{d}\ln D_\alpha^{fBm}, \tag{51}$$

where $A(D_\alpha, 1)$ was defined in Sect. 2.2.7 and p_α^{fBm} is given by Eq. 50.

Thus, a monodisperse system with a single fluorescent species characterized by a single type of diffusional behavior can indeed give rise to continuous distributions of diffusion times. Therefore, the mere fact that a distribution of diffusion times is obtained upon inversion of FCS data does not necessarily imply heterogeneity in the system under investigation.

It is remarkable that the maximum entropy inversion of the FCS data for protein diffusion in a crowded environment [107] produced a broad unimodal continuous distribution similar to the ones presented in Fig. 10b and c,

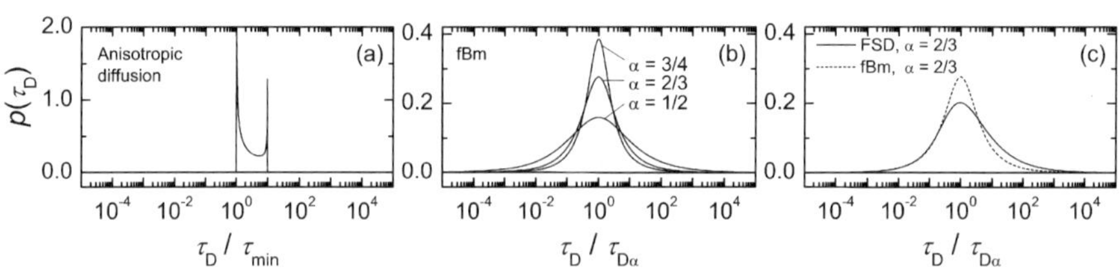

Fig. 10 Distributions of diffusion times for different types of 2D diffusion: **a** anisotropic diffusion with $D_{max}/D_{min} = \tau_{max}/\tau_{min} = 10$; **b** fractional Brownian motion for $\alpha = 1/2$, 2/3, and 3/4; **c** fractional subdiffusion with $\alpha = 2/3$ (——). For comparison, a distribution for fBm with $\alpha = 2/3$ is shown (- - -)

which speaks in favor of the conclusion of [107] on observing anomalous diffusion due to molecular crowding.

An alternative empirical approach to the distribution of diffusion times proposed in [240] consists in performing repetitive short measurements on one and the same system and histogramming the results of ACF fitting with a simple diffusion model. At present, it is unclear how the results of this method will compare with those based on the numerical solution of the inverse problem (Eq. 45), as well as with the actual distribution of diffusion coefficients in the system.

6 Some Application Aspects

In this section we will not consider specific FCS applications, which are mostly found in the field of biosciences, and on which a number of reviews have been published in the last few years [26–33]. Instead, we will focus on some application aspects related to the method itself, proper understanding of which is required for correct interpretation of FCS results.

6.1 Diffusion Coefficient Measurements with FCS

6.1.1 The Need for Fluorescent Standards with Reliable Consensus Values of Diffusion Coefficients

One of the usual goals of FCS experiments is the study of diffusion coefficients of freely diffusing fluorescent molecules or colloidal particles. However, the standard implementations of FCS methods do not provide absolute values of diffusion coefficients, since they require knowledge of the size of the detection volume which is hard to measure independently. Therefore, the most common way to measure the diffusion coefficient with FCS is to carry out relative measurements with a standard substance having a known diffusion coefficient.

To date, the majority of FCS papers focusing on determination of diffusion coefficients of fluorescent particles (see, e.g., [145, 241]) use an aqueous solution of Rhodamine 6G as a reference with the assumed diffusion coefficient of 2.8×10^{-6} cm^2 s^{-1} at 22 °C. However, in many instances either no reference for this value is provided, or it is cited indirectly via previously published FCS papers. Thus, a kind of FCS "folklore" was created, and the accuracy of the value cited so frequently is not questioned. In fact, this value can be traced back to one of the very first FCS papers published more than 30 years ago [3], where it was reported to have an error as high as 25%. What is less known is that the same paper gives another estimate for the diffusion coefficient of Rhodamine 6G in water, $(3.5 \pm 0.5) \times 10^{-6}$ cm^2 s^{-1}, this time based on a series of measurements for a range of laser beam waists. Thus, the value so frequently cited seems to be far from reliable, and standard values of diffusion coefficients, preferably obtained using an independent method not related to FCS, are required. In fact, these data can be found in the literature for a few dyes. For example, results of a microfluidic study of diffusion coefficients for some dye molecules in water at 25 °C are reported in [242], including Rhodamine 6G ($D^{25\,°C}_{\mathrm{Rh6G}} = (4.14 \pm 0.01) \times 10^{-6}$ cm^2 s^{-1}), Rhodamine B ($D^{25\,°C}_{\mathrm{RhB}} = (4.26 \pm 0.04) \times 10^{-6}$ cm^2 s^{-1}), and fluorescein ($D^{25\,°C}_{\mathrm{fluorescein}} = (4.25 \pm 0.01) \times 10^{-6}$ cm^2 s^{-1}). The latter value for fluorescein is in an excellent agreement with the previously published values [243, 244] obtained by methods not related to FCS.

Unfortunately, to the best of our knowledge, no such data exist for the fluorescent dye Alexa 488, an extremely popular fluorescent probe in the field of FCS. Our careful FCS measurements have shown [216] that its diffusion coefficient in water at 25 °C is, within experimental error, the same as that of Rhodamine 6G, i.e., $D^{25\,°C}_{\mathrm{Alexa488}} \approx 4.14 \times 10^{-6}$ cm^2 s^{-1}.

One should keep in mind that the viscosity of aqueous samples changes quite strongly with the temperature (e.g., upon an increase of the temperature from 20 to 25 °C, the viscosity of water decreases by ~11%), and the diffusion coefficients reported for a temperature T_0 should be recalculated for the temperature T_{exp} at which the experiment is carried out: $D_{T_{\mathrm{exp}}} = D_{T_0} T_{\mathrm{exp}} \eta_{T_0} / (T_0 \eta_{T_{\mathrm{exp}}})$, where D_T and η_T are the diffusion coefficient and viscosity of the solution at an absolute temperature T, respectively.

Thus, the quantitative FCS investigations of diffusion require a reliable database of independently measured diffusion coefficients for a set of standard dyes over a range of temperatures in water and aqueous solutions with compositions typical for biochemical and biophysical studies.

6.1.2 Absolute Diffusion Coefficient Measurements in FCS

As has been discussed above, extreme care should be taken when estimating diffusion coefficients from FCS measurements. Not only are the measured diffusion times prone to distortions due to fluorescence saturation and bleaching effects, but also the precise values of diffusion coefficients of dyes obtained in independent measurements are not available in abundance. Therefore, development of FCS techniques capable of delivering information on the diffusion coefficients with an internal reference is important.

The most natural approach in this case is to use spatial cross-correlation. The feasibility of two-beam cross-correlation for quantitative determination of diffusion coefficients was recently shown in [245, 246]. A variation of the technique employs spatial cross-correlation between concentric round and ring-shaped pinholes [247]. While the works [245, 247] are more of the proof-of-principle type, the method in [246] allows one to obtain quantitative results and thus may provide a basis for reference-free quantitative measurements of diffusion coefficients of fluorescent dye molecules. The diffusion coefficient for the Atto 655 dye ($(3.9 \pm 0.1) \times 10^{-6}\,\mathrm{cm}^2\,\mathrm{s}^{-1}$ at $25\,^{\circ}\mathrm{C}$) reported in [246] is in very good agreement with those of fluorescein and Rhodamine 6G obtained in [242], if the difference of the molecular sizes of the dyes is taken into account.

Studies of artificial and cell membranes can benefit from the two-foci scanning FCCS method [215], which has been recently demonstrated as a technique for measuring absolute values of diffusion coefficients of species undergoing slow diffusion in phospholipid membranes. This is especially important in view of the fact that the use of a dye freely diffusing in 3D as a diffusion standard for membrane measurements can introduce a systematic error due to the non-Gaussian shape of the detection volume.

Other FCS-related techniques incorporating an internal scale of reference, which are worth considering as alternative quantitative approaches to measure absolute diffusion coefficients, include circular-scanning FCS (see Sect. 7.6) with a known beam scanning radius,[5] as well as the varieties of FCS with patterned excitation, including those based on traveling [248] and stationary [249] interference patterns due to beam intersection, or due to reflection from a mirror [250, 251], and the closely related Fourier-imaging correlation spectroscopy technique [252].

6.2
FCS of "Soft" Systems with Internal Dynamics

Conformational fluctuations of polymers [155, 156] and membranes [253, 254] originate, as the Brownian motion, from thermal molecular motion and thus are an inherent feature of these objects. If a fluorescent label is bound to a macromolecule or membrane, conformational fluctuations of the latter (provided their scale is comparable to the size of the detection efficiency profile) can introduce additional fluctuations in the detected fluorescence intensity and thus modify the FCS correlation function.

A spectacular example of manifestation of the internal conformational dynamics in FCS data is presented in Fig. 11, which shows experimental results for single-end fluorescently labeled monodisperse fragments of double-stranded (ds) DNA [216]. We observe that the normalized FCS ACFs of dsDNA fragments show at short lag times a universal behavior, which extends to longer time ranges with increasing fragment length. Clearly, this is a manifestation of the conformational dynamics of dsDNA: while the center of mass of the macromolecule slowly diffuses through the laser focus, its conformation fluctuates,

[5] The feasibility of this approach has recently been successfully demonstrated [318]. In particular, the study confirmed our value for the diffusion coefficient of Alexa 488 in water (see previous section).

thus moving the fluorescent label in and out of the detection volume. This effect can be exploited to study the conformational dynamics of polymers in solution, including dsDNA [216, 255–257] and actin filaments [257], by means of the FCS technique. In this approach, the single-end labeling of monodisperse macromolecules makes the data analysis most straightforward. The approach allows one to obtain the diffusion coefficients (Fig. 11b) and polymer relaxation times (data not shown) of dsDNA fragments, which are in remarkable agreement with both theory [258] and independent non-FCS experiments [259–262]. Notice that an attempt to estimate diffusion coefficients by determining the lag time at which the FCS ACF decays to a half of its amplitude (see Sect. 2.2.1), while working reasonably well for dsDNA fragments up to 500 base pairs, completely fails for longer fragments. Although the effect of the intramolecular polymer dynamics on FCS data is somewhat reduced with increasing the labeling density of the macromolecule [257], it may still substantially affect FCS results, and therefore should be carefully taken into account in FCS experiments with (semi)flexible macromolecules whose size is comparable to that of the detection efficiency profile.

FCS measurements on artificial and cell membranes are usually carried out so that the membrane, which on the scale of the size of the laser focus can be considered to be flat, lies in the focal plane of the objective. Then, since the high-frequency spatial modes of membrane undulations are strongly damped [254], the undulations can be approximately considered as a random motion of a flat membrane up and down along the z-axis. If the mean position of the membrane lies in the focal plane, then due to the symmetry of the detection volume, the membrane undulations will lead to random drops in

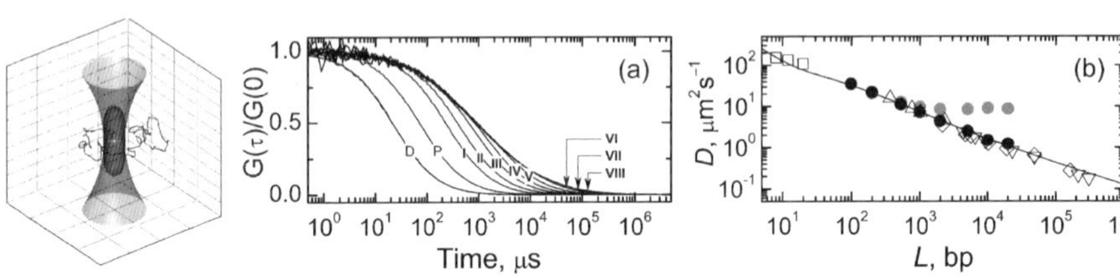

Fig. 11 **a** FCS autocorrelation functions of Alexa 488 single-end labeled dsDNA fragments with lengths of 0.1 (I), 0.2 (II), 0.5 (III), 1 (IV), 2 (V), 5 (VI), 10 (VII), and 20 kilobase pairs (VIII). For comparison, ACFs of freely diffusing Alexa 488 dye (D) and Alexa 488 labeled primer (P) are shown. **b** Diffusion coefficients of dsDNA fragments (corrected to 20 °C): (●) obtained using a semiflexible polymer model (for details, see [216]), and (●) estimated from lag times $\hat{\tau}_D$ corresponding to the condition $G(\hat{\tau}_D) = (1/2)G(0)$. For comparison, *open symbols* show diffusion coefficients of dsDNA fragments obtained by methods unrelated to FCS [259–262]. The *solid curve* shows a prediction of theory [254] for a semiflexible polymer with the persistence length of 50 nm. In determination of the diffusion coefficients, freely diffusing Alexa 488 dye, whose diffusion coefficient in water was found to be equal to that of Rhodamine 6G ($(4.14 \pm 0.01) \times 10^{-6}$ cm^2 s^{-1} at $T = 298$ K [242]), was used as a reference. $\lambda_{exc} = 488$ nm, $T = 298$ K. Setup: ConfoCor2 (Zeiss)

detected fluorescence signal. As a result, an additional component appears in the FCS curves (see Fig. 12), with the amplitude and timescale reflecting those of membrane undulations [263]. For example, giant unilamellar vesicles with diameters of ~20–30 μm may show undulations with amplitudes of ~0.2 μm at a timescale of ~100 ms, which can substantially contribute to the detected FCS ACF (Fig. 12). The contribution to the FCS ACF due to membrane undulations is independent of the concentration of dye molecules diffusing in the membrane, and therefore is more pronounced at higher labeling densities. Notice that, depending on its relative contribution, the overall effect of membrane undulations can be mistakenly interpreted as anomalous subdiffusion (Fig. 12b) or as an additional slow diffusion component (Fig. 12c and d).

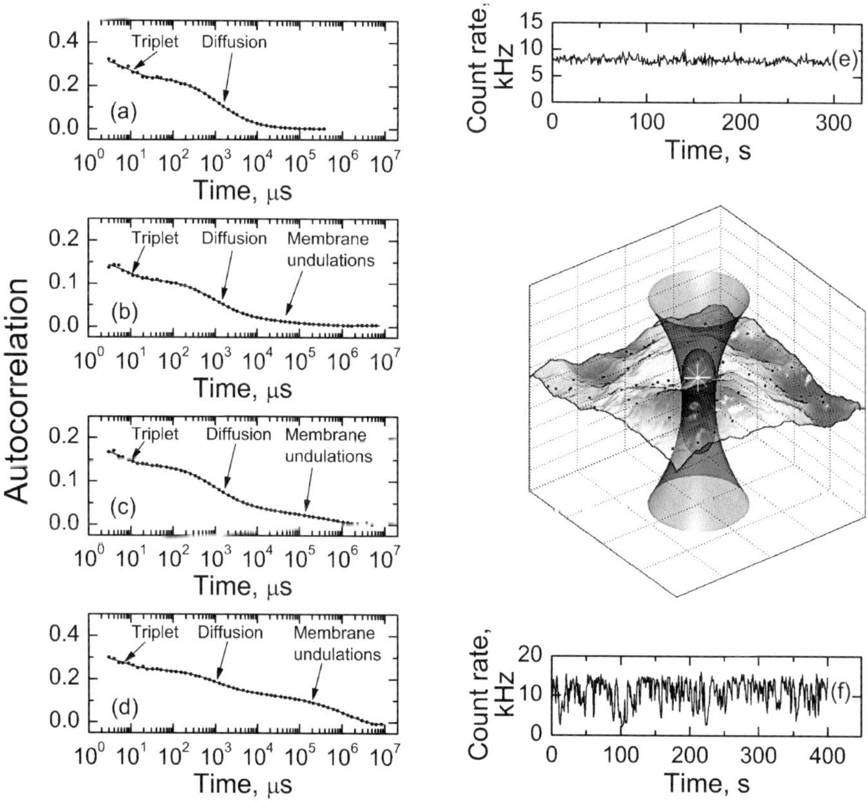

Fig. 12 FCS ACFs of the $DiOC_{18}(3)$ lipid probe in **a** osmotically tensed and **b–d** undulating L-alpha-dioleoylphosphatidylcholine giant unilamellar vesicles: (●) experimental data (every third data point is shown for clarity) and (——) fit accounting for diffusion and membrane undulations. The magnitude of undulations increases from **a** to **d**. Estimated root-mean-square membrane undulation amplitudes: 0.14 μm (**b**), 0.20 μm (**c**), and 0.33 μm (**d**). Panels **e** and **f** show the measured count rates for cases **a** and **d**, respectively. $\lambda_{exc} = 488$ nm, $T = 298$ K. Setup: ConfoCor2 (Zeiss)

Membrane undulations can also affect the results of FCS measurements in bulk solution if the detection volume is located in the vicinity of an undulating membrane, as has been shown in [264].

7
Varieties of FCS and Related Techniques

7.1
Two-Photon Excitation FCS

The use of two-photon excitation (TPE) in FCS studies was proposed in [265]. In this approach, as a result of the TPE probability being proportional to the square of the excitation intensity, the intensity of fluorescence excited by a tightly focused laser beam falls down rapidly along the z-direction (approximately as $\sim z^{-4}$), which provides a confined nearly Gaussian detection volume without the use of a pinhole. TPE FCS data on freely diffusing fluorophores can usually be successfully analyzed using the models described in Sect. 2 by assuming the 3D Gaussian shape of the detection volume, although the use of a more realistic approximation (a squared Gaussian–Lorentzian) for the detection volume shape provides more stable results [265]. TPE has proven to be very efficient in FCS studies on cells and tissues (see, e.g., [77, 102]): in spite of a higher photobleaching probability of the fluorophore in the detection volume, the overall bleaching probability over the whole cell is substantially reduced because the photobleaching region is confined to the two-photon focal spot. Additionally, the low absorption of tissues at the typical wavelengths used for TPE ($\sim$700–1000 nm) provides a deeper penetration of the excitation light into a tissue, and finally, Rayleigh and Raman scattering from the sample are excluded. For a recent review of TPE FCS applications, see [266, 267].

7.2
Confining the Detection Volume Using Total Internal Reflection, Nanoapertures, and Stimulated Emission Depletion

The idea to confine the detection volume in FCS measurements by using the effect of total internal reflection (TIR) was put forward and developed in [268]. In TIR-FCS, the sample is excited by an evanescent optical field exponentially decaying in the depth of the sample, as a result of total internal reflection of a beam impinging on a sample interface at a supercritical angle. This technique is an excellent tool to study surface binding reactions, as has been demonstrated in a number of studies (see, e.g., [269–271]). Recently, TIR-FCS was used to study the size dependence of protein diffusion close to membrane surfaces [272].

Successful applications of FCS require that just a few molecules are present on average in the detection volume, which, in the case of a standard confocal FCS setup with the diffraction-limited detection volume down to ~0.1 fl, requires nanomolar concentrations. However, for proper functioning of biological systems, micromolar concentrations are frequently required. Therefore, during the last few years, considerable efforts have been made to reduce FCS detection volumes.

Based on the idea of the objective-based TIR fluorescence setup [273], an objective-based TIR-FCS approach has recently been proposed [274, 275]. This technique employs epi-illumination through the periphery of a high-NA oil-immersion objective to create an evanescent field above the surface of the coverslide. The same objective is used to collect fluorescence, and additionally confines the observation volume, which can be approximately described as $W(\mathbf{r}) = \exp[-2(x^2 + y^2)/r_0^2 - z/h]$ [275], where $z = 0$ corresponds to the coverslide surface. Recently, a two-color cross-correlation extension of TIR-FCS was proposed and implemented [276]. The interpretation of the results on diffusing molecules can, however, be complicated not only due to possible interactions with the coverslide [275], but also due to a dependence of the diffusion coefficient on the distance from the surface [272].

Comparable or even better confinement of the detection volume can also be achieved by using subwavelength nanoapertures [277–279]. In this case, the effects of the electromagnetic mode structure are expected to enhance the spontaneous emission rate [280] and thus to considerably improve the detection efficiency.

Yet another approach to reduce the observation volume has been recently proposed in [281]. There, the stimulated emission depletion technique was successfully applied to obtain a subdiffraction FCS detection volume as small as ~20 al, thus providing an alternative method to study molecular dynamics at high concentrations.

7.3 Higher-Order FCS

The idea to study the higher-order fluorescence ACFs in the form

$$G_{mn}(\tau) = \frac{\langle \delta F^m(t+\tau)\delta F^n(t)\rangle - \langle \delta F^m(t)\rangle\langle \delta F^n(t)\rangle}{\langle F\rangle^{m+n}} \qquad (52)$$

was put forward in [282] and further developed in [180] (for a review, see [283]). While the use of the standard FCS ACF does not allow detection of the simultaneous presence of particles with different specific brightnesses (e.g., due to clustering), higher-order correlation functions provide such a possibility. One should keep in mind, however, that the use of higher-order correlations immediately reveals the non-Gaussian shape of the detection volume, which should be taken into account for adequate interpretation of experimental data [180]. The theory of the high-order correlation approach was further developed in [207]. There it was suggested that the more conventional form of the higher-order correlation function [34, 35], for example,

$$G_3(\tau_1, \tau_2) = \frac{\left\langle \delta F_i(t) \delta F_j(t + \tau_1) \delta F_k(t + \tau_1 + \tau_2) \right\rangle}{\langle F_i \rangle \langle F_j \rangle \langle F_k \rangle}, \quad \tau_1, \tau_2 \geq 0, \tag{53}$$

i.e., the form of the higher-order correlation function related to that of Eq. 2, is superior to the form of Eq. 52 since, while having an equally high information content, it does not contain an unwanted shot noise contribution. Additionally, the form of the higher-order correlation functions (Eq. 53) is more suitable for experiments employing single-photon detectors, where sequences of photon arrival times, rather than time-dependent fluorescence intensities, are measured.

Obtaining higher-order correlation functions from experimental data is a computationally intense task, and their use is presently hindered by the absence of fast dedicated hardware and software. However, further steps in the application of higher-order correlation functions have been taken. Recently, the three-color coincidence analysis involving the detection of the quantity

$$K_3 = g^{(3)}_{\mathrm{bgr}}(0, 0) = \frac{\left\langle F_{\mathrm{b}}(t) F_{\mathrm{g}}(t) F_{\mathrm{r}}(t) \right\rangle}{\langle F_{\mathrm{b}} \rangle \langle F_{\mathrm{g}} \rangle \langle F_{\mathrm{r}} \rangle}, \tag{54}$$

where subscripts b, g, and r indicate fluorescence intensities recorded by detectors sensitive in the blue, green, and red parts of the spectrum (Fig. 3), was demonstrated in [284] as a tool to detect formation and cleavage of molecular bonds in solution in real time.

Theoretical studies show a great potential in the application of higher-order correlation functions, in particular for discriminating biochemical reaction networks functioning in nonequilibrium steady states from the systems exhibiting equilibrium reaction kinetics [285].

7.4 FCS and FCCS with Time-Resolved and Time-Gated Detection

The use of pulsed laser excitation combined with time-resolved or time-gated fluorescence detection can substantially improve the information content of FCS data and thus enhance the capabilities of the method. The most general experimental implementation of this approach was recently presented in [71]. This is an advanced variety of the TCSPC technique allowing one to detect and register single photoelectron events with a picosecond accuracy over time periods of hours. Post-processing of sequences of photoelectron arrival times provides FCS and FCCS data spanning the lag time range from picoseconds to seconds, and makes it possible to additionally use the spectral and lifetime information to discriminate between different fluorophores.

A technically less demanding technique of time-resolved FCS [286] uses the pulsed excitation and time-resolved detection to discriminate between fluorophores based on their excited-state lifetimes. This allows one to obtain FCS correlation functions of different species using a single excitation wavelength and one detection channel, and thus presents an alternative to dual- or multicolor FCCS. A variety of this approach [200] uses the information on the excited-state lifetime of the fluorophore to eliminate the contribution of detector afterpulsing from FCS ACFs.

A simple, efficient alternative to the above time-resolved FCS methods is presented by the approaches to fluorescence cross-correlation based on pulsed interleaved excitation [82, 83]. In these approaches, several excitation sources are interleaved so that fluorescence excited by a light pulse decays completely before arrival of the next excitation pulse. The information on the excitation source for every photoelectron event is known, and hence the cross-talk between two spectral channels is eliminated. This approach is especially promising in cell applications of FCCS with autofluorescent proteins showing strong overlap of fluorescence spectra.

7.5 Multifocal and CCD-Based FCS Methods

Potential applications of FCS in high-throughput analysis require an increase in parallelism of experimental data collection. During the last few years, considerable progress has been made in this direction and a number of technical solutions have been proposed. In [287, 288] parallel multifocal FCS and two-color FCCS have been demonstrated by creating four independent detection volumes using a 2×2 fan-out diffractive optical element for excitation of fluorescence and 2×2 detector and correlator arrays for fluorescence detection. In [289] an alternative approach was used, where FCS correlation functions are quasi-simultaneously measured from several spots in the sample by sequentially directing the excitation beam to each of the spots in a cyclic manner over short periods of time using a Galvano mirror, and constructing the corresponding FCS correlation functions.

As an extension of the concept of detector arrays, a variety of FCS with charge-coupled device (CCD)-based detection was recently proposed. The CCD-based detection technique was first implemented in DLS [290–292], and only the limited sensitivity of CCDs prevented application of this experimental approach in FCS. The recent appearance of electron-multiplying CCDs with single-photon sensitivity resulted in a convincing demonstration of the possibility of CCD-based FCS [293, 294]. This has opened up exciting possibilities of flexible spatially resolved FCS, in particular on biological objects with complicated topology, especially in combination with diffractive optical elements to generate multifocal arrays with independently controlled positions of foci.

7.6
Image Correlation Spectroscopy and Scanning FCS

Prohibitively long acquisition times and effects of photobleaching prevent successful application of the standard FCS to systems exhibiting very slow motion of fluorescent particles. In this case, the use of the spatial, rather than temporal, correlation is advantageous. This approach is realized in the so-called image correlation spectroscopy (ICS) [295] and allows one to estimate the concentration of fluorescent particles and characterize their aggregation [296]. The potential of the higher-order correlation analysis in ICS was studied in [297]. The use of sequences of images taken at a number of time instants allows one to extract more information on the sample by adding the temporal dimension to ICS [298]. The limits of accuracy and dynamic range of ICS were recently studied in [299].

The generalization of the above approaches is spatio-temporal image correlation spectroscopy (STICS), where the spatio-temporal correlation function $r_{ab}(\varepsilon, \eta, \tau) = \langle \delta i_a(x, y, t)\delta i_b(x + \xi, y + \eta, t + \tau)\rangle / (\langle i_a\rangle_t \langle i_b\rangle_{t+\tau})$ of fluorescence fluctuations is investigated [300]. In addition to concentration and aggregation measurements, the STICS method provides information on slow diffusion as well as on velocities and directions of the flow. The accuracy and precision of the STICS method and correction of STICS data for photobleaching artifacts were recently studied in detail [301]. In another study, STICS was successfully employed to carry out a quantitative study of intracellular transport of nonviral gene delivery vectors (polyplexes) [302].

The term scanning FCS (SFCS) generally denotes the group of methods where a relative motion of the FCS detection volume and sample is implemented, and thus it occupies an intermediate position between the standard FCS and ICS. The approach was pioneered in [303], where rotation of a sample of slowly diffusing fluorescently labeled DNA molecules was used to achieve detectable fluctuations in the fluorescence signal and to determine the concentration and molecular weight of DNA. A variety of the method with linear motion of the sample was implemented in [304]. In more recent studies, scanning is achieved by moving the detection volume within a sample.

A line-scanning-based variety of SFCS was introduced in [305] to improve FCS capabilities in studies of slowly diffusing species on biomembranes with the goal of measuring aggregation. In [306], a commercial laser scanning microscope (LSM) was used to implement high-speed line-scanning SFCS with a small detection volume. A variety of two-focus sequential line scanning using an LSM was recently introduced in [215] to measure ultraslow diffusion in membranes.

The circular scanning version of SFCS was realized in [307] by rotating a tilted objective, and was further improved in [308] using an LSM with two-photon excitation. SFCS with circular scanning across the phospholipid membrane surface was used in [309] to study slow diffusion of a membrane-bound probe. The recently introduced position-sensitive circular scanning FCS technique [310] involves synchronization of data acquisition with the information on the beam position and as a result provides information on diffusion, flow velocity and direction, and immobilization of fluorophores.

More varieties of the above techniques can be found in the literature. In [311], scanning dual-color cross-correlation with line, circular, and random scanning was studied as a tool to assess colocalization of immobilized molecules. The raster image correlation spectroscopy (RICS) technique introduced in [312] uses raster scan images obtained using a commercial LSM to extract microsecond- and second-scale dynamics along the fast and slow scan axes, respectively, and thus takes an intermediate position between ICS and SFCS. For a detailed review of SFCS and ICS, see [313].

7.7 Other FCS-Related Techniques

Several FCS-related methods have been proposed in recent years which are based on detection of signals other than dye fluorescence, including X-ray fluorescence [314], Raman scattering [315], CARS [316], and the nonlinear optical response of colloidal particles [317]. Under certain circumstances, these methods may prove to be advantageous to conventional FCS by providing chemical selectivity without the need of fluorescent labeling.

8 Conclusions

More than 30 years from its inception, FCS today remains a dynamically developing field of intense research that is constantly expanding both its technical arsenal and range of applications. The highly interdisciplinary character of the technique and its applications thus calls for joint efforts of researchers working in the fields of molecular spectroscopy, optical instrument development, electronics and software engineering, biochemistry, and biophysics to further improve the accuracy and reliability of the technique and to search for new application areas.

Acknowledgements The authors gratefully acknowledge the support by the Deutsche Forschungsgemeinschaft within the Priority Program SPP 1164.

References

1. Magde D, Elson E, Webb WW (1972) Phys Rev Lett 29:705
2. Elson EL, Magde D (1974) Biopolymers 13:1
3. Magde D, Elson EL, Webb WW (1974) Biopolymers 13:29

4. Koppel DE (1974) Phys Rev A 10:1938
5. Ehrenberg M, Rigler R (1974) Chem Phys 4:390
6. Aragón SR, Pecora R (1975) Biopolymers 14:119
7. Aragón SR, Pecora R (1976) J Chem Phys 64:1791
8. Koppel DE, Axelrod D, Schlessinger J, Elson EL, Webb WW (1976) Biophys J 16:1315
9. Magde D, Webb WW, Elson EL (1978) Biopolymers 17:361
10. Berne BJ, Pecora R (1976) Dynamic light scattering. Wiley, New York
11. Schaefer DW, Berne BJ (1972) Phys Rev Lett 28:475
12. Schaefer DW (1973) Science 180:1293
13. Elson EL, Webb WW (1975) Annu Rev Biophys Bioeng 4:311
14. Magde D (1976) Q Rev Biophys 9:35
15. Ehrenberg M, Rigler R (1976) Q Rev Biophys 9:69
16. Elson EL (1985) Annu Rev Phys Chem 36:379
17. Thompson N (1991) In: Lakowicz JR (ed) Topics in fluorescence spectroscopy, vol 1: techniques. Plenum Press, New York, p 337
18. Lakowicz JR (1999) Principles of fluorescence spectroscopy, 2nd edn. Kluwer Academic/Plenum Press, New York
19. O'Connor DV, Phillips D (1984) Time-correlated single photon counting. Academic, London
20. Rigler R, Mets Ü, Widengren J, Kask P (1993) Eur Biophys J 22:169
21. Eigen M, Rigler R (1994) Proc Nat Acad Sci USA 91:5740
22. Rigler R, Elson E (eds) (2001) Fluorescence correlation spectroscopy: theory and applications. Springer, Berlin
23. Schwille P (2001) Cell Biochem Biophys 34:383
24. Hess ST, Huang S, Heikal AA, Webb WW (2002) Biochemistry 41:697
25. Widengren J, Mets Ü (2002) In: Zander C, Enderlein J, Keller RA (eds) Single molecule detection in solution. Wiley-VCH, Berlin, p 69
26. Thompson NL, Lieto AM, Allen NW (2002) Curr Opin Struct Biol 12:634
27. Krichevsky O, Bonnet G (2002) Rep Prog Phys 65:251
28. Bacia K, Schwille P (2003) Methods 29:74
29. Haustein E, Schwille P (2003) Methods 29:153
30. Gösch M, Rigler R (2005) Adv Drug Deliv Rev 57:169
31. Vukojević V, Pramanik A, Yakovleva T, Rigler R, Terenius L, Bakalkin G (2005) Cell Mol Life Sci 62:535
32. Kahya N, Schwille P (2006) Mol Membr Biol 23:29
33. Bacia K, Kim SA, Schwille P (2006) Nat Methods 3:83
34. Loudon R (2000) The quantum theory of light, 3rd edn. Oxford University Press, Oxford
35. Saleh B (1978) Photoelectron statistics. Springer, Berlin
36. Landau LD, Lifshitz EM (1980) Statistical physics, part 1, 3rd edn. Butterworth-Heinemann, Oxford
37. Mazo RM (2002) Brownian motion: fluctuations, dynamics and applications. Oxford University Press, Oxford
38. Kinjo M, Rigler R (1995) Nucleic Acids Res 23:1795
39. Schwille P, Oehlenschläger F, Walter NG (1996) Biochemistry 35:10182
40. Meseth U, Wohland T, Rigler R, Vogel H (1999) Biophys J 76:1619
41. Gardiner CW (2004) Handbook of stochastic methods for physics, chemistry and the natural sciences, 3rd edn, Chap 8. Springer, Berlin
42. Tsien RY (1998) Annu Rev Biochem 67:509
43. Haupts U, Maiti S, Schwille P, Webb WW (1998) Proc Nat Acad Sci USA 95:13573

44. Widengren J, Terry B, Rigler R (1999) Chem Phys 249:259
45. Bonnet G, Krichevsky O, Libchaber A (1998) Proc Nat Acad Sci USA 95:8602
46. Chattopadhyay K, Saffarian S, Elson EL, Frieden C (2002) Proc Nat Acad Sci USA 99:14171
47. Neuweiler H, Doose S, Sauer M (2005) Proc Nat Acad Sci USA 102:16650
48. Kim J, Doose S, Neuweiler H, Sauer M (2006) Nucl Acid Res 34:2516
49. Kawski A (1993) CR Anal Chem 23:459
50. Yardley JT, Specht LT (1976) Chem Phys Lett 37:543
51. Hoshikawa H, Asai H (1985) Biophys Chem 22:167
52. Scalettar BA, Klein MP, Hearst JE (1987) Biopolymers 26:1287
53. Kask P, Piksarv P, Mets Ü, Pooga M, Lippmaa E (1987) Eur Biophys J 14:257
54. Widengren J, Mets Ü, Rigler R (1999) Chem Phys 250:171
55. Tsay JM, Doose S, Weiss S (2006) J Am Chem Soc 128:1639
56. Hinze G, Diezemann G, Basché T (2004) Phys Rev Lett 93:203001
57. Wei CYJ, Kim YH, Darst RK, Rossky PJ, Vanden Bout DA (2005) Phys Rev Lett 95:173001
58. Rosenberg SA, Quinlan ME, Forkey JN, Goldman YE (2005) Acc Chem Res 38:583
59. Vanden Bout DA, Deschenes LA (2006) Science 312:195
60. Uji-i H, Melnikov SM, Deres A, Bergamini G, De Schryver F, Herrmann A, Müllen K, Enderlein J, Hofkens J (2006) Polymer 47:2511
61. Barcellona ML, Gammon S, Hazlett T, Digman MA, Gratton E (2004) Microsc Res Tech 65:205
62. Zondervan R, Kulzer F, van der Meer H, Disselhorst JAJM, Orrit M (2006) Biophys J 90:2958
63. Carmichael HJ, Walls DF (1976) J Phys B 9:L43
64. Kimble HJ, Dagenais M, Mandel L (1977) Phys Rev Lett 39:691
65. Kask P, Piksarv P, Mets Ü (1985) Eur Biophys J 12:163
66. Basché T, Moerner WE, Orrit M, Talon H (1992) Phys Rev Lett 69:1516
67. Bernard J, Fleury L, Talon H, Orrit M (1993) J Chem Phys 98:850
68. Widengren J, Mets Ü, Rigler R (1995) J Phys Chem 99:13368
69. Kitson SC, Jonsson P, Rarity JG, Tapster PR (1998) Phys Rev A 58:620
70. Mets Ü (2001) In: Rigler R, Elson E (eds) Fluorescence correlation spectroscopy: theory and applications. Springer, Berlin, p 346
71. Felekyan S, Kühnemuth R, Kudryavtsev V, Sandhagen C, Becker W, Seidel CAM (2005) Rev Sci Instrum 76:083104
72. Mets Ü, Widengren J, Rigler R (1997) Chem Phys 218:191
73. Griffin WG, Pusey PN (1979) Phys Rev Lett 43:1100
74. Kam Z, Rigler R (1982) Biophys J 39:7
75. Rička J, Binkert T (1989) Phys Rev A 39:2646
76. Schwille P, Meyer-Almes F-J, Rigler R (1997) Biophys J 72:1878
77. Heinze KG, Koltermann A, Schwille P (2000) Proc Nat Acad Sci USA 97:10377
78. Weidemann T, Wachsmuth M, Tewes M, Rippe K, Langowski J (2002) Single Mol 3:49
79. Bacia K (2005) PhD Thesis, Technische Universität Dresden
80. Kim S, Heinze KG, Bacia K, Waxham MN, Schwille P (2005) Biophys J 88:4319
81. Hwang LC, Gösch M, Lasser T, Wohland T (2006) Biophys J 91:715
82. Thews E, Gerken M, Eckert R, Zäpfel J, Tietz C, Wrachtrup J (2005) Biophys J 89:2069
83. Müller BK, Zaychikov E, Bräuchle C, Lamb DC (2005) Biophys J 89:3508
84. Burkhardt M, Heinze KG, Schwille P (2005) Opt Lett 30:2266
85. Hwang LC, Leutenegger M, Gösch M, Lasser T, Rigler P, Meier W, Wohland T (2006) Opt Lett 31:1310

86. Palmer AG III, Thompson NL (1987) Biophys J 51:339
87. Asai H (1980) Japan J Appl Phys 19:2279
88. Lenne P-F, Colombo D, Giovannini H, Rigneault H (2002) Single Mol 4:194
89. Brinkmeier M (1996) PhD Thesis, Technische Universität Braunschweig
90. Brinkmeier M, Dörre K, Stephan J, Eigen M (1999) Anal Chem 71:609
91. Brinkmeier M (2001) In: Rigler R, Elson E (eds) Fluorescence correlation spectroscopy: theory and applications. Springer, Berlin, p 379
92. Dittrich PS, Schwille P (2002) Anal Chem 74:4472
93. Gösch M, Blom H, Holm J, Heino T, Rigler R (2000) Anal Chem 72:3260
94. Lumma D, Best A, Gansen A, Feuillebois F, Rädler JO, Vinogradova OI (2003) Phys Rev E 67:056313
95. Kuricheti KK, Buschmann V, Weston KD (2004) Appl Spectrosc 58:1180
96. Dittrich PS, Müller B, Schwille P (2004) Phys Chem Chem Phys 6:4416
97. Dittrich PS, Schäfer SP, Schwille P (2005) Biophys J 89:3446
98. Brister PC, Kuricheti KK, Buschmann V, Weston KD (2005) Lab Chip 5:785
99. Köhler RH, Schwille P, Webb WW, Hanson MR (2000) J Cell Sci 113:3921
100. Bouchaud J-P, Georges A (1990) Phys Rep 195:127
101. Schwille P, Korlach J, Webb WW (1999) Cytometry 36:176
102. Schwille P, Haupts U, Maiti S, Webb WW (1999) Biophys J 77:2251
103. Wachsmuth M, Waldeck W, Langowski J (2000) J Mol Biol 298:677
104. Weiss M, Hashimoto H, Nilsson T (2003) Biophys J 84:4043
105. Weiss M, Elsner M, Kartberg F, Nilsson T (2004) Biophys J 87:3518
106. Fatin-Rouge N, Starchev K, Buffle J (2004) Biophys J 86:2710
107. Banks DS, Fradin C (2005) Biophys J 89:2960
108. Szymański J, Patkowski A, Gapiński J, Wilk A, Hołyst R (2006) J Phys Chem B 110:7367
109. Masuda A, Ushida K, Riken TO (2005) Biophys J 88:3584
110. Wawrezinieck L, Rigneault H, Marguet D, Lenne P-F (2005) Biophys J 89:4029
111. Mandelbrot BB, van Ness JW (1968) SIAM Rev 10:422
112. Metzler R, Klafter J (2000) Phys Rep 339:1
113. Wang KG, Lung CW (1990) Phys Lett A 151:119
114. Feder TJ, Brust-Mascher I, Slattery JP, Baird B, Webb WW (1996) Biophys J 70:2767
115. Lutz E (2001) Phys Rev E 64:051106
116. Lim SC, Muniandy SV (2002) Phys Rev E 66:021114
117. Hilfer R (ed) (1999) Applications of fractional calculus in physics. World Scientific, Singapore
118. Saichev AI, Zaslavsky GM (1997) Chaos 7:753
119. Barkai E (2001) Phys Rev E 63:046118
120. Feller W (1970) An introduction to probability theory and its applications, vol 2. Wiley, New York
121. Hac AE, Seeger HM, Fidorra M, Heimburg T (2005) Biophys J 88:317
122. Schütz GJ, Schindler H, Schmidt T (1997) Biophys J 73:1073
123. Schneider J, Rička J, Binkert T (1988) Rev Sci Instrum 59:588
124. Eid JS, Müller JD, Gratton E (2000) Rev Sci Instrum 71:361
125. Benda A, Hof M, Wahl M, Patting M, Erdmann R, Kapusta P (2005) Rev Sci Instrum 76:033106
126. Becker W, Bergmann A, Haustein E, Petrasek Z, Schwille P, Biskup C, Kelbauskas L, Benndorf K, Klöcker N, Anhut T, Riemann I, König K (2006) Microsc Res Tech 69:186
127. Weisshart K, Jüngel V, Briddon SJ (2004) Curr Pharm Biotechnol 5:135

128. Woggon U (1996) Optical properties of semiconductor quantum dots. Springer, Berlin
129. Gaponenko SV (1998) Optical properties of semiconductor nanocrystals. Cambridge University Press, Cambridge
130. Michalet X, Pinaud FF, Bentolila LA, Tsay JM, Doose S, Li JJ, Sundaresan G, Wu AM, Gambhir SS, Weiss S (2005) Science 307:538
131. Larson DR, Zipfel WR, Williams RM, Clark SW, Bruchez MP, Wise FW, Webb WW (2003) Science 300:1434
132. Kapitonov AM, Stupak AP, Gaponenko SV, Petrov EP, Rogach AL, Eychmüller A (1999) J Phys Chem B 103:10109
133. Zenkevich E, Cichos F, Shulga A, Petrov EP, Blaudeck T, von Borczyskowski C (2005) J Phys Chem B 109:8679
134. Petrov EP, Cichos F, von Borczyskowski C (2006) J Luminesc 119–120:412
135. Doose S, Tsay JM, Pinaud F, Weiss S (2005) Anal Chem 77:2235
136. Yao J, Larson DR, Vishwasrao HD, Zipfel WR, Webb WW (2005) Proc Nat Acad Sci USA 102:14284
137. Jin T, Fujii F, Sakata H, Tamura M, Kinjo M (2005) Chem Commun 2829
138. Zhang J, Campbell RE, Ting AY, Tsien RY (2002) Nat Rev Mol Cell Biol 3:906
139. Baird GS, Zacharias DA, Tsien RY (2000) Proc Nat Acad Sci USA 97:11984
140. Campbell RE, Tour O, Palmer AE, Steinbach PA, Baird GS, Zacharias DA, Tsien RY (2002) Proc Nat Acad Sci USA 99:7877
141. Shaner NC, Campbell RE, Steinbach PA, Giepmans BNG, Palmer AE, Tsien RY (2004) Nat Biotechnol 22:1567
142. Ando R, Hama H, Yamamoto-Hino M, Mizuno H, Miyawaki A (2002) Proc Nat Acad Sci USA 99:12651
143. Ando R, Mizuno H, Miyawaki A (2004) Science 306:1370
144. Schwille P, Kummer S, Heikal AA, Moerner WE, Webb WW (2000) Proc Nat Acad Sci USA 97:151
145. Schenk A, Ivanchenko S, Röcker C, Wiedenmann J, Nienhaus GU (2004) Biophys J 86:384
146. Hess ST, Heikal AA, Webb WW (2004) J Phys Chem B 108:10138
147. Cotlet M, Goodwin PM, Waldo GS, Werner JH (2006) ChemPhysChem 7:250
148. Schäfer SP, Dittrich PS, Petrov EP, Schwille P (2006) Microsc Res Tech 69:210
149. Dedecker P, Hotta J, Ando R, Miyawaki A, Engelborghs Y, Hofkens J (2006) Biophys J 91:L45
150. Phillies GDJ (1975) Biopolymers 14:499
151. Axelrod D, Koppel DE, Schlessinger J, Elson E, Webb WW (1976) Biophys J 16:1055
152. Saxton MJ, Jacobson K (1997) Annu Rev Biophys Biomol Struct 26:373
153. Abney JR, Scalettar BA, Hackenbrock CR (1990) Biophys J 58:261
154. Abney JR, Scalettar BA (1995) Biophys Chem 57:27
155. Doi M, Edwards SF (1986) The theory of polymer dynamics. Oxford University Press, Oxford
156. Teraoka I (2002) Polymer solutions: an introduction to physical properties. Wiley, New York
157. Scalettar BA, Hearst JE, Klein MP (1989) Macromolecules 22:4550
158. Wilk A, Gapinski J, Patkowski A, Pecora R (2004) J Chem Phys 121:10794
159. Liu R, Gao X, Adams J, Oppermann W (2005) Macromolecules 38:8845
160. Nishimura G, Kinjo M (2004) Anal Chem 76:1963
161. Gregor I, Patra D, Enderlein J (2005) ChemPhysChem 6:164
162. Nagy A, Wu J, Berland KM (2005) J Biomed Opt 10:044015

163. Nagy A, Wu J, Berland KM (2005) Biophys J 89:2077
164. Enderlein J (1996) Phys Lett A 221:427
165. Widengren J, Rigler R (1996) Bioimaging 4:149
166. Eggeling C, Widengren J, Rigler R, Seidel CAM (1998) Anal Chem 70:2651
167. Dittrich PS, Schwille P (2001) Appl Phys B 73:829
168. Niesner R, Roth W, Gericke K-H (2004) ChemPhysChem 5:678
169. Eggeling C, Widengren J, Brand L, Schaffer J, Felekyan S, Seidel CAM (2006) J Phys Chem A 110:2979
170. Delon A, Usson Y, Derouard J, Biben T, Souchier C (2006) Biophys J 90:2548
171. Widengren J, Thyberg P (2005) Cytometry A 68:101
172. Ashkin A (1997) Proc Nat Acad Sci USA 94:4853
173. Osborne MA, Balasubramanian S, Furey WS, Klenerman D (1998) J Phys Chem B 102:3160
174. Ding Y, Meng FB, Chen B, Ma H, Lei J, Chen DY (2001) Acta Phys Sin 50:2269
175. Chirico G, Fumagalli C, Baldini G (2002) J Phys Chem B 106:2508
176. Meng F, Ma H (2005) J Phys Chem B 109:5580
177. Hosokawa C, Yoshikawa H, Masuhara H (2005) Phys Rev E 72:021408
178. Hosokawa C, Yoshikawa H, Masuhara H (2006) Jpn J Appl Phys 45:L453
179. Haustein E (2006) PhD Thesis, Technische Universität Dresden
180. Palmer AG III, Thompson NL (1989) Appl Opt 28:1214
181. Qian H, Elson EL (1991) Appl Opt 30:1185
182. Shaw PJ, Rawlins DJ (1991) J Microsc 163:151
183. Juškaitis R, Wilson T (1998) J Microsc 189:8
184. Visser TD, Wiersma SH (1994) J Opt Soc Am A 11:599
185. Török P, Wilson T (1997) Opt Commun 137:127
186. Dorn R, Quabis S, Leuchs G (2003) J Mod Opt 50:1917
187. Haeberlé O (2004) Opt Commun 235:1
188. Dimova R, Aranda S, Bezlyepkina N, Nikolov V, Riske KA, Lipowsky R (2006) J Phys Condens Matter 18:S1151
189. Hess ST, Webb WW (2002) Biophys J 83:2300
190. Enderlein J, Gregor I, Patra D, Dertinger T, Kaupp UB (2005) ChemPhysChem 6:2324
191. Benda A, Beneš M, Mareček V, Lhotský A, Hermens WT, Hof M (2003) Langmuir 19:4120
192. Enderlein J, Gregor I, Patra D, Fitter J (2004) Curr Pharm Biotechnol 5:155
193. Chattopadhyay K, Saffarian S, Elson EL, Frieden C (2005) Biophys J 88:1413
194. Zettl H, Häfner W, Böker A, Schmalz H, Lanzendörfer M, Müller AHE, Krausch G (2004) Macromolecules 37:1917
195. Dravins D, Faria D, Nilsson B (2000) Proc SPIE 4008:298
196. Burstyn HC, Sengers JV (1983) Phys Rev A 27:1071
197. Burstyn HC (1980) Rev Sci Instrum 51:1431
198. Campbell L (1992) Rev Sci Instrum 63:5794
199. Zhao M, Jin L, Chen B, Ding Y, Ma H, Chen D (2003) Appl Opt 42:4031
200. Enderlein J, Gregor I (2005) Rev Sci Instrum 76:033102
201. Schätzel K (1986) Appl Phys B 41:95
202. Palmer AG III, Thompson NL (1989) Rev Sci Instrum 60:624
203. Apanasovich VV, Paltsev SV (1995) J Opt Soc Am B 12:1550
204. Nishimura G, Kinjo M (2005) Appl Opt 44:3458
205. Schätzel K, Drewel M, Stimac S (1988) J Mod Opt 35:711
206. Bevington P, Robinson DK (2003) Data reduction and error analysis for the physical sciences, 3rd edn. McGraw-Hill, New York

207. Qian H (1990) Biophys Chem 38:49
208. Kask P, Günther R, Axhausen P (1997) Eur Biophys J 25:163
209. Wohland T, Rigler R, Vogel H (2001) Biophys J 80:2087
210. Saffarian S, Elson EL (2003) Biophys J 84:2030
211. Schätzel K (1990) Quantum Opt 2:287
212. Harrison D, Fisch MR (1996) Langmuir 12:6691
213. Marriott FHC, Pope JA (1954) Biometrika 41:390
214. Jakeman E, Pike ER, Swain S (1971) J Phys A 4:517
215. Ries J, Schwille P (2006) Biophys J 91:1915
216. Petrov EP, Ohrt T, Winkler RG, Schwille P (2006) Phys Rev Lett 97:258101
217. Phillies GDJ (1996) Rev Sci Instrum 67:3423
218. Kojro Z, Riede A, Schubert M, Grill W (1999) Rev Sci Instrum 70:4487
219. Enderlein J, Gregor I, Patra D, Fitter J (2005) J Fluoresc 15:415
220. Starchev K, Zhang J, Buffle J (1998) J Colloid Interface Sci 203:189
221. Gennerich A, Schild D (2000) Biophys J 79:3294
222. Gennerich A, Schild D (2002) Biophys J 83:510
223. Widengren J, Schwille P (2000) J Phys Chem A 104:6416
224. Skakun VV, Hink MA, Digris AV, Engel R, Novikov EG, Apanasovich VV, Visser AJWG (2005) Eur Biophys J 34:323
225. Tikhonov AN, Arsenin VY (1977) Solutions of ill-posed problems. Winston, Washington
226. Hansen PC (1998) Rank-deficient and discrete ill-posed problems: numerical aspects of linear inversion. SIAM, Philadelphia
227. Provencher SW (1982) Comput Phys Commun 27:213
228. Livesey AK, Brochon JC (1987) Biophys J 52:693
229. Siemiarczuk A, Wagner B, Ware WR (1990) J Phys Chem 94:1661
230. Langowski J, Bryan R (1991) Macromolecules 24:6346
231. Gakamsky DM, Goldin AS, Petrov EP, Rubinov AN (1992) Biophys Chem 44:47
232. Petrov EP, Kruchenok JV, Rubinov AN (1999) J Fluoresc 9:111
233. Björling S, Kinjo M, Földes Papp Z, Hagman E, Thyberg P, Rigler R (1998) Biochemistry 37:12971
234. Starchev K, Buffle J, Pérez E (1999) J Colloid Interface Sci 213:479
235. Rigler R, Pramanik A, Jonasson P, Kratz G, Jansson OT, Nygren P-Å, Ståhl S, Ekberg K, Johansson B-L, Uhlén S, Uhlén M, Jörnvall H, Wahren J (1999) Proc Nat Acad Sci USA 96:13318
236. Pramanik A, Olsson M, Langel Ü, Bartfai T, Rigler R (2001) Biochemistry 40:10839
237. Módos K, Galántai R, Bárdos-Nagy I, Wachsmuth M, Tóth K, Fidy J, Langowski J (2004) Eur Biophys J 33:59
238. Sengupta P, Garai K, Balaji J, Periasamy N, Maiti S (2003) Biophys J 84:1977
239. Davies B (2002) Integral transforms and their applications, 3rd edn. Springer, New York
240. Jankevics H, Prummer M, Izewska P, Pick H, Leufgen K, Vogel H (2005) Biochemistry 44:11676
241. Krouglova T, Vercammen J, Engelborghs Y (2004) Biophys J 87:2635
242. Culbertson CT, Jacobson SC, Ramsey JM (2002) Talanta 56:365
243. Moore AW Jr, Jorgenson JW (1993) Anal Chem 65:3550
244. Paul PH, Garguilo MG, Rakestraw DJ (1998) Anal Chem 70:2459
245. Jaffiol R, Blancquaert Y, Delon A, Derouard J (2006) Appl Opt 45:1225
246. Dertinger T, Pacheco V, von der Hocht I, Hartmann R, Gregor I, Enderlein J (2007) Chem Phys Chem 8:433

247. Blancquaert Y, Delon A, Derouard J, Jaffiol R (2006) Proc SPIE 6191:61910B
248. Hattori M, Shimizu H, Yokoyama H (1996) Rev Sci Instrum 67:4064
249. Hansen RL, Zhu XR, Harris JM (1998) Anal Chem 70:1281
250. Asai H, Ando T (1976) J Phys Soc Japan 50:1527
251. Rigneault H, Lenne P-F (2003) J Opt Soc Am B 20:2203
252. Knowles MK, Grassman TJ, Marcus AH (2000) Phys Rev Lett 85:2837
253. Brochard F, Lennon JF (1975) J Phys France 36:1035
254. Granek R (1997) J Phys II France 7:1761
255. Lumma D, Keller S, Vilgis T, Rädler JO (2003) Phys Rev Lett 90:218301
256. Shusterman R, Alon S, Gavrinyov T, Krichevsky O (2004) Phys Rev Lett 92:048303
257. Winkler RG, Keller S, Rädler JO (2006) Phys Rev E 73:041919
258. Harnau L, Winkler RG, Reineker P (1996) J Chem Phys 104:6355
259. Eimer W, Pecora R (1991) J Chem Phys 94:2324
260. Sorlie SS, Pecora R (1990) Macromolecules 23:487
261. Smith DE, Perkins TT, Chu S (1996) Macromolecules 29:1372
262. Nkodo AE, Garnier JM, Tinland B, Ren H, Desruisseaux C, McCormick LC, Drouin G, Slater GW (2001) Electrophoresis 22:2424
263. Petrov EP, Schwille P (2005) Biophys J 88:524A
264. Fradin C, Abu-Arish A, Granek R, Elbaum M (2003) Biophys J 84:2005
265. Berland KM, So PTC, Gratton E (1995) Biophys J 68:694
266. Diaspro A, Chirico G, Collini M (2005) Q Rev Biophys 38:97
267. Sánchez SA, Gratton E (2005) Acc Chem Res 38:469
268. Thompson NL, Burghardt TP, Axelrod D (1981) Biophys J 33:435
269. Starr TE, Thompson NL (2001) Biophys J 80:1575
270. Lieto AM, Cush RC, Thompson NL (2003) Biophys J 85:3294
271. Lieto AM, Thompson NL (2004) Biophys J 87:1268
272. Pero JK, Haas EM, Thompson NL (2006) J Phys Chem B 110:10910
273. Axelrod D, Hellen EH, Fulbright RM (1992) In: Lakowicz JR (ed) Topics in fluorescence spectroscopy, vol 3. Plenum Press, New York, p 289
274. Ruckstuhl T, Seeger S (2004) Opt Lett 29:569
275. Hassler K, Leutenegger M, Rigler P, Rao R, Rigler R, Gösch M, Lasser T (2005) Opt Express 13:7415
276. Leutenegger M, Blom H, Widengren J, Eggeling C, Gösch M, Leitgeb RA, Lasser T (2006) J Biomed Opt 11:040502
277. Levene MJ, Korlach J, Turner SW, Foquet M, Craighead HG, Webb WW (2003) Science 299:682
278. Wenger J, Rigneault H, Dintinger H, Marguet D, Lenne P-F (2006) J Biol Phys 32:SN1
279. Samiee KT, Moran-Mirabal JM, Cheung YK, Craighead HG (2006) Biophys J 90:3288
280. Rigneault H, Capoulade J, Dintinger J, Wenger J, Bonod N, Popov E, Ebbesen TW, Lenne P-F (2005) Phys Rev Lett 95:117401
281. Kastrup L, Blom H, Eggeling C, Hell SW (2005) Phys Rev Lett 94:178104
282. Palmer AG III, Thompson NL (1987) Biophys J 52:257
283. Thompson NL, Mitchell JL (2001) In: Rigler R, Elson E (eds) Fluorescence correlation spectroscopy: theory and applications. Springer, Berlin, p 438
284. Heinze KG, Jahnz M, Schwille P (2004) Biophys J 86:506
285. Qian H, Elson EL (2004) Proc Nat Acad Sci USA 101:2828
286. Böhmer M, Wahl M, Rahn H-J, Erdmann R, Enderlein J (2002) Chem Phys Lett 353:439
287. Gösch M, Serov A, Anhut T, Lasser T, Rochas A, Besse PA, Popovic RS, Blom H, Rigler R (2004) J Biomed Opt 9:913

288. Gösch M, Blom H, Anderegg S, Korn K, Thyberg P, Wells M, Lasser T, Rigler R (2005) J Biomed Opt 10:054008
289. Takahashi Y, Sawada R, Ishibashi K, Mikuni S, Kinjo M (2005) Curr Pharm Biotechnol 6:159
290. Wong APY, Wiltzius P (1993) Rev Sci Instrum 64:2547
291. Cipelletti L, Weitz DA (1999) Rev Sci Instrum 70:3214
292. Viasnoff V, Lequeux F, Pine DJ (2002) Rev Sci Instrum 73:2336
293. Kannan B, Har JY, Liu P, Maruyama I, Ding JL, Wohland T (2006) Anal Chem 78:3444
294. Burkhardt M, Schwille P (2006) Opt Express 14:5013
295. Petersen NO, Hoddelius PL, Wiseman PW, Seger O, Magnusson KE (1993) Biophys J 65:1135
296. Brown CM, Petersen NO (1998) J Cell Sci 111:271
297. Vanden Broek W, Huang Z, Thompson NL (1999) J Fluoresc 9:313
298. Wiseman PW, Squier JA, Ellisman MH, Wilson KR (2000) J Microsc 200:14
299. Costantino S, Comeau JWD, Kolin DL, Wiseman PW (2005) Biophys J 89:1251
300. Hebert B, Costantino S, Wiseman PW (2005) Biophys J 88:3601
301. Kolin DL, Costantino S, Wiseman PW (2006) Biophys J 90:628
302. Kulkarni RP, Wu DD, Davis ME, Fraser SE (2005) Proc Nat Acad Sci USA 102:7523
303. Weissman M, Schindler H, Feher G (1976) Proc Nat Acad Sci USA 73:2776
304. Nicoli DF, Briggs J, Elings VB (1980) Proc Nat Acad Sci USA 77:4904
305. Petersen NO (1986) Biophys J 49:809
306. Koppel DE, Morgan F, Cowan AE, Carson JH (1994) Biophys J 66:502
307. Meyer T, Schindler H (1988) Biophys J 54:983
308. Berland KM, So PTC, Chen Y, Mantulin WW, Gratton E (1996) Biophys J 71:410
309. Ruan QQ, Cheng MA, Levi M, Gratton E, Mantulin WW (2004) Biophys J 87:1260
310. Skinner JP, Chen Y, Müller JD (2005) Biophys J 89:1288
311. Amediek A, Haustein E, Scherfeld D, Schwille P (2002) Single Mol 3:201
312. Digman MA, Brown CM, Sengupta P, Wiseman PW, Horwitz AR, Gratton E (2005) Biophys J 89:1317
313. Petrášek Z, Schwille P (2008) In: Rigler R, Vogel H (eds) Single molecules and nanotechnology. Springer, Berlin, p 83
314. Wang J, Sood AK, Satyam PV, Feng Y, Wu X-Z, Cai Z, Yun W, Sinha SK (1998) Phys Rev Lett 80:1110
315. Schrof W, Klingler JF, Rozouvan S, Horn D (1998) Phys Rev E 57:R2523
316. Hellerer T, Schiller A, Jung G, Zumbusch A (2002) ChemPhysChem 7:630
317. Brehm G, Sauer G, Fritz N, Schneider S, Zaitsev S (2005) J Mol Struct 735-736:85
318. Petrášek Z, Schwille P (2008) Biophys J 94:1437

Springer Ser Fluoresc (2008) 6: 199–212
DOI 10.1007/4243_2008_051

Published online: 5 April 2008

Single Molecule Spectroscopy: Instrumentation and Multiparameter Detection

Volker Buschmann[1] (✉) · Felix Koberling[1] · Benjamin Schuler[2] (✉) · Frank Hillger[2] · Daniel Nettels[2]

[1]PicoQuant GmbH, Rudower Chaussee 29, 12489 Berlin, Germany
info@picoquant.com

[2]Universität Zürich, Biochemisches Institut, Winterthurerstrasse 190, 8057 Zürich, Switzerland
schuler@bioc.unizh.ch

Abstract Single molecule spectroscopy has become a standard tool in analytics. It is possible not only to detect single molecule fluorescence, but to characterize individual fluorophores by their excitation and emission spectra, quantum yield, anisotropy, and fluorescence lifetime. Dedicated setups based on pulsed excitation, confocal detection, and intelligent data processing allow to simultaneously record and analyze several properties of a single dye fluorescence. A substantial advantage of single molecule measurements is the detection and characterization of subpopulations even in heterogeneous mixtures. A popular example is the analysis of macromolecules labeled with fluorescent donor and acceptor molecules capable of Förster resonance energy transfer (FRET). As an example, we show a multiparameter analysis of single molecule measurements of recombinant bovine liver rhodanese labeled with Alexa 488 and Alexa 594. Native and unfolded states can be clearly resolved and exhibit distinct FRET efficiencies, donor lifetimes, and anisotropies. In FRET measurements, pulsed interleaved excitation can be applied to distinguish molecules with and without fluorescent acceptor, as illustrated on a peptide labeled with Alexa 555 and Alexa 647.

Keywords Förster resonance energy transfer (FRET) · Multiparameter analysis · Pulsed interleaved excitation (PIE) · Single molecule detection (SMD)

1 Introduction

Since its beginnings in the 1990s, single molecule spectroscopy has become a standard technique in analytics [1, 2]. Most approaches today are based on fluorescence spectroscopy; a technique which allows to characterize and identify certain fluorophores [3], to resolve different subpopulations [4] and to determine the transition kinetics between different conformational states on the single molecule level, e.g., in the course of an enzymatic reaction [5] or in protein folding [6].

Single molecule fluorescence measurements can be performed in various ways. For example, in order to study slow conformational dynamics [5, 7, 8], the sample is usually immobilized on a surface. This procedure leads to an increased observation time of the single molecule, and the observation time is then only limited by photobleaching. For these measurements, a careful interpretation of the data is needed as interactions with the matrix (surface) have to be excluded in order to ensure native behavior of the observed sample.

For molecules freely diffusing in solution, the observation time is limited by the diffusion through the detection volume in the range of milliseconds. However, the accessibility of dynamics on short time scales with optimum time resolution, the absence of surface interactions, and the ease with which these experiments can be performed make them very popular.

2 Accessible Parameters

In single molecule fluorescence studies, five parameters are accessible in order to characterize a single fluorophore [9]: Excitation and emission spectra, quantum yield, anisotropy, and fluorescence lifetime. The measurement of each property requires dedicated instrumentation (Table 1).

For measuring the excitation spectrum, the wavelength of the excitation light needs to be varied [10]. Those measurements are, therefore, performed on immobilized samples which can be observed over a longer period of time.

Information about the emission spectrum is obtained more easily. In the simplest case, the emission is separated by a dichroic mirror and monitored

Table 1 Approaches to study different fluorescence characteristics in single molecule measurements

Chromophore properties	Instrumentation
Excitation spectrum	Excitation sources with different wavelengths
Emission spectrum	Spectral separation of emission
Fluorescence quantum yield	Quantitative measure of the signal intensity
Fluorescence anisotropy	Polarization separation of the emission
Fluorescence lifetime	Pulsed excitation, time-correlated single photon counting

by two detectors simultaneously. The fraction of light falling on one detector is called fractional intensity and it is characteristic for every fluorophore in a given setup. Changes in the emission spectrum result in changes of the fractional intensity [3]. This method is well suited for measurements in solution as it provides information about spectral shifts in a comparably short time and with a relatively small number of photons. Recently, even the recording of complete single molecule emission spectra in aqueous solution has been reported [11].

The quantum yield is the fundamental parameter that determines the photon emission probability for each excitation cycle. In practice, however, not the quantum yield, but the molecular brightness is measured, as this information is more easily accessible. The molecular brightness is the average photon count rate which can be detected from a single fluorescent molecule. It depends on the excitation spectrum, wavelength and intensity, the detection efficiency of the system, and the quantum yield of the fluorophore. This parameter can be extracted from a single molecule trace and is specific to the measurement conditions used [12].

Anisotropy measurements provide information about the rotational mobility of the fluorophore. While completely immobilized molecules in general have fixed absorption and emission dipoles, molecules in solution rotate in a similar time range as the fluorescence lifetime, resulting in fast depolarization of the fluorescence signal. The steady state anisotropy, r, is calculated by the polarized fluorescence signal, which is oriented parallel ($f_{\parallel}(t)$) and perpendicular ($f_{\perp}(t)$) to the linearly polarized laser excitation

$$r = \frac{\int_T \mathrm{d}t \left[f_{\parallel}(t) - f_{\perp}(t)\right]}{\int_T \mathrm{d}t \left[f_{\parallel}(t) + 2f_{\perp}(t)\right]} . \tag{1}$$

In order to measure anisotropy, the fluorescence needs to be split on two detectors using a polarizing beam splitter. However, excitation and emission polarization orientation is changed in the microscope objective lens and a correction factor needs to be introduced [13]. Then the ratio of the fluorescence intensity registered by the two detectors provides information about the anisotropy.

Analogous to the spectral properties, the fluorescence lifetime is characteristic for every fluorophore. However, changes in the local environment, such as viscosity, solvent polarity, or quencher molecules, may change the fluorescence lifetime. Thus, the fluorescence lifetime can also be used to gain information about the environment. In terms of instrumentation, lifetime measurements on single molecules are performed using pulsed excitation and time-correlated single photon counting, as described in Chap. 3.

Even more parameters can be extracted with sufficiently photostable immobilized molecules, such as absorption and emission dipole orientation, anisotropy decay time, intersystem crossing quantum yield, and triplet state lifetime [14].

The range of information obtainable from a single molecule experiment becomes even broader if macromolecules that are specifically labeled with two fluorophores capable of Förster resonance energy transfer (FRET) [15] are studied. The transfer efficiency, E, between donor and acceptor chromophore depends on the distance, R, between the fluorophores, as shown in Eq. 2. Therefore, FRET has been extensively used as a "spectroscopic ruler" for distance measurements on the nm scale in ensemble as in single molecule applications [16–18].

$$E = \frac{1}{\left[1 + (R/R_0)^6\right]} \tag{2}$$

The Förster distance,R_0, depends on the spectral characteristics of donor and acceptor dye. In general, it is in the range of 1–10 nm. For an intensity based determination of the FRET efficiency, the fluorescence of donor and acceptor need to be spectrally separated and monitored by two detectors, similar as for detection of spectral fluctuations. The transfer efficiency is then related to the number of photons in the donor channel (n_D) and the acceptor channel (n_A), according to

$$E = \frac{n_A}{\left(n_A + n_D\right)}, \tag{3}$$

where n_A and n_D need to be corrected for the quantum yield of the dyes, spectral bleedthrough, direct acceptor excitation, and detection efficiencies at the different wavelengths [19, 20].

While some parameters such as the emission spectrum and the anisotropy are independent, others are not. For example, the molecular brightness of a fluorophore decreases if a quencher molecule gets into its proximity. If quenching is achieved via the excited state of the fluorophore (dynamic quenching), the fluorescence lifetime also decreases. Such quenching mechanisms have been exploited, e.g., in the development of DNA hairpins for detection of unlabeled DNA sequences [21].

These dependencies are even more important for measurements of single FRET-pairs. The donor lifetime is directly connected to the FRET efficiency via Eq. 4, where τ_D is the donor lifetime in the absence of the acceptor and τ_{DA} in the presence of the acceptor:

$$E = 1 - \frac{\tau_{DA}}{\tau_D}. \tag{4}$$

Anisotropy and FRET efficiency also influence each other. If the acceptor is excited not by linearly polarized light, but via energy transfer, the anisotropy drops significantly. As a consequence, the donor behavior, rather than the acceptor behavior, is typically analyzed in anisotropy measurements. On the other hand, the relative orientation and rotational mobility of donor and acceptor influence the Förster distance, R_0, which characterizes the distance at which the energy transfer between donor and acceptor is 50%. As the rotational mobility of a fluorophore is reflected in its anisotropy, the anisotropy of donor and acceptor may also have an influence on the FRET efficiency.

It is obvious that information from single molecule signals increases with the number of parameters recorded, and it is, therefore, beneficial to record as many parameters as possible. Using multiparameter analysis of events from single freely diffusing molecules in solution, up to 16 different species could be distinguished [9].

3
Instrumentation

While single molecules have been observed freely diffusing in aqueous solution by wide field microscopy [22] using charge-coupled device (CCD) cameras, this technique is in general less suited for multiparameter analysis – mostly because the CCD cameras lack the high time resolution necessary for fast lifetime determination.

Confocal imaging has, therefore, become the method of choice for such measurements. Confocal microscopes can be equipped with the optics and electronics for recording virtually all aspects of fluorescence dynamics of microscopic samples.

In a confocal experiment, a laser beam is focused onto the sample, e.g. into an aqueous solution. The fluorescence is spatially filtered through a pinhole before being directed to the detector(s), leading to an observation volume in the femtoliter regime. Molecules crossing through the detection volume can then be identified by an increase of the detector count rate. Such fluorescence bursts can be identified by setting a simple intensity threshold [23] or, as an alternative, by decreasing interphoton times during such a passage [24, 25]. The passage time for most molecules in aqueous solution spreads from several tens of microseconds to a few milliseconds, depending on the exact size of the observation volume, and the diffusion coefficient of the observed molecule.

The number of photons that can be collected from a photon burst depends greatly on the optical setup and the detection efficiency of the detector. Typical values range from 20 to 200 photons. Single photon avalanche diodes (SPADs) have, therefore, become the detectors of choice for single molecule experiments, as they have a detection efficiency of up to 70%. Furthermore, they are fast enough to be used in single photon counting experiments for lifetime determination and can be selected for low dark counts. A disadvantage of these detectors is the small active area in the range of 100 μm

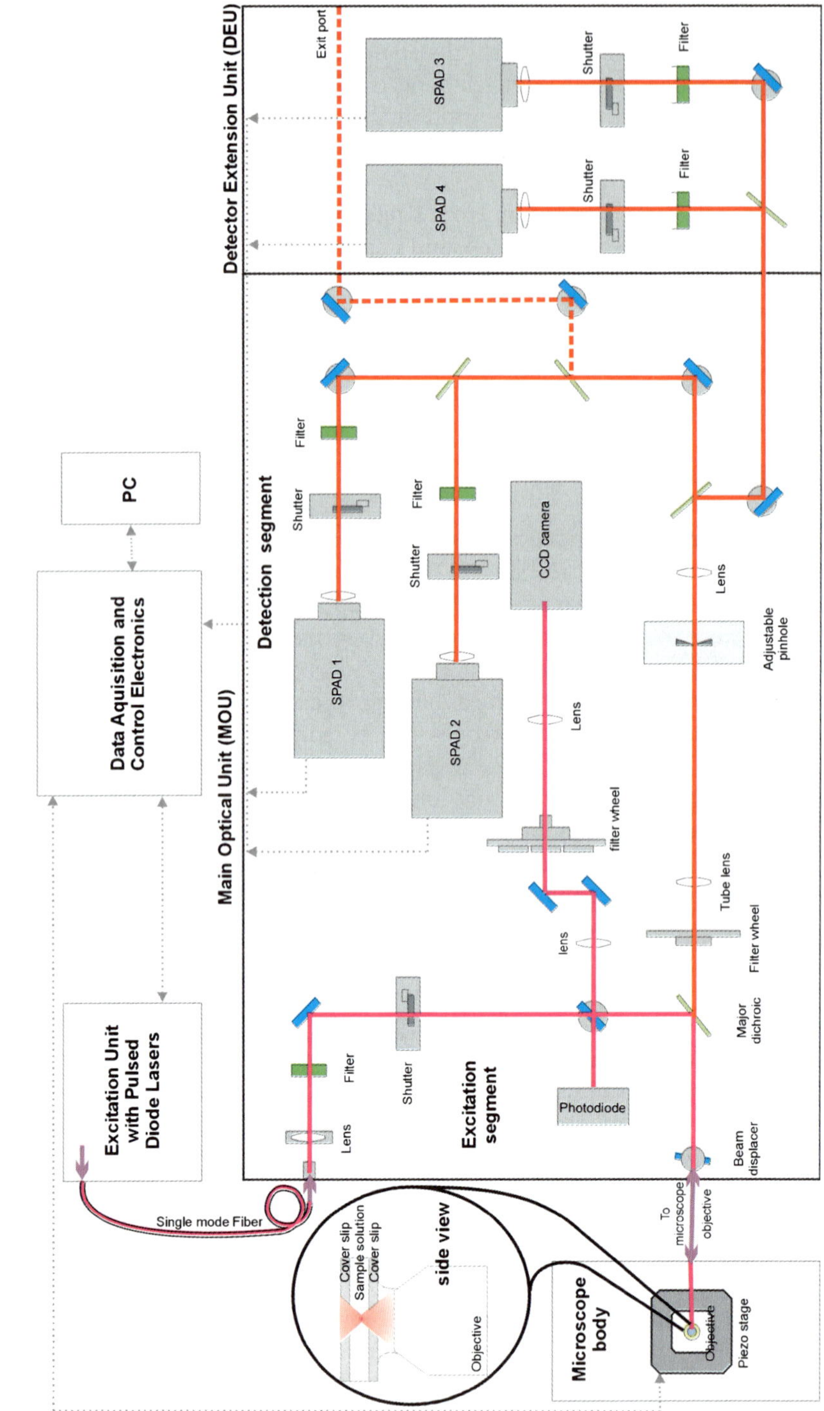

Detector Extension Unit (DEU)
Exit port
SPAD 3
SPAD 4
Shutter
Filter
Main Optical Unit (MOU)
Detection segment
SPAD 1
SPAD 2
CCD camera
Lens
Adjustable pinhole
filter wheel
Tube lens
Filter wheel
Major dichroic
lens
Photodiode
Excitation segment
Beam displacer
PC
Data Aquisition and Control Electronics
Excitation Unit with Pulsed Diode Lasers
Single mode Fiber
To microscope objective
side view
Cover slip
Sample solution
Cover slip
Objective
Microscope body
Piezo stage

◀ **Fig. 1** Scheme of the components of the confocal microscope used for the measurements, with detailed view on the main optical unit (MOU)

diameter. The sample fluorescence must, therefore, be accurately focused on such a detector, and a perfect alignment of the instrument is crucial for optimal performance, especially as the time resolution of some SPADs decreases, if the detected light is not focused to the center of the active area.

In general, every confocal setup consists of five basic parts: excitation unit, microscope body, detection unit, data acquisition and control electronics, and software. A scheme of the setup used for the measurements described here (MicroTime 200, PicoQuant, Germany) is shown in Fig. 1. Parts of the excitation unit and the detection unit are included in the main optical unit (MOU).

In nearly all single molecule experiments, lasers are used as excitation sources. For a multiparameter analysis, pulsed lasers have to be used instead of cw lasers, as only pulsed excitation allows lifetime determination via time – correlated single photon counting (TCSPC). Popular excitation sources are Ti:Sa lasers for two photon excitation and pulsed diode lasers for one photon excitation. Ti:Sa lasers work at a fixed repetition rate of typically 80 MHz and provide a tunable wavelength in the range of approx. 750 to 950 nm as well as extremely short pulses in the range of $\sim$150 fs. As the efficiency of two photon excitation depends on the pulse width, such a laser beam is usually directed into the microscope directly without an optical fiber that would broaden the laser pulse. However, latest research provides pulse shape modules, which compensate for these effects [26]. In contrast to tunable Ti:Sa laser systems, pulsed diode lasers emit at fixed wavelengths, but the repetition rate can be freely varied. They are available for many different wavelengths, and laser pulses can be as short as 70 ps. For multi-color excitation, different laser beams have to be overlaid in space and time, which is best achieved by coupling of all used lasers into a polarization maintaining single mode fiber. The laser beam is then directed into the objective via the major dichroic beam splitter.

The excitation unit of the confocal microscope used in the experiments described here additionally contains a beam splitter to direct excitation light on a photodiode for monitoring the excitation intensity as well as a camera that monitors the light reflected from the cover slip for focus control.

The sample is placed above the objective, which is the most important part of the microscope, in this case an IX 71 inverse microscope (Olympus, Hamburg, Germany). As single molecule detection requires efficient photon collection, objectives with high numerical aperture have to be used. A piezo-scanning stage with exact positioning capabilities is used to enable measurements on immobilized samples.

In the confocal detection unit, which is connected to the microscope body via an external exit port, the fluorescence is spatially filtered through a pinhole and directed to the detectors. Multiparameter recording requires use of several detectors – the combination of spectral and polarization imaging, for example, requires four detectors and three additional beam splitters for spectral and polarization splitting. High-quality filters are also used in front of each detector to efficiently reject scattered light. The detectors used in our setup are SPCM-AQR-14 SPADs (Perkin Elmer, Fremont, USA), which exhibit less than 50 dark counts per second. A detailed view on the main excitation and detection path is given in Fig. 1.

The data storage and handling is crucial for multiparameter measurements. For measurements in solution, as performed here, one must be able to measure the photon arrival density (in time) at a time resolution of typically microseconds in order to identify photon bursts from freely diffusing single molecules passing through the detection volume. For lifetime measurements, the relative times between the laser excitation and the corresponding fluorescence photon arrival times at the detector have to be measured additionally with a resolution down to a few picoseconds. For lifetime determination, these delay times of photons from a single burst have to be histogrammed and fitted (see Chap. 3). Furthermore, all detected photons need to be assigned to the corresponding detectors in order to provide the full information. In principle it is possible to collect all photon data on a TCSPC board – this approach is fast but limited in terms of memory capacity. Therefore, a continuous processing and real time storage on the PC is a far more flexible approach [27, 28]. For the measurements presented here, a TCSPC board (TimeHarp 200, PicoQuant, Germany) is utilized. The Time-Tagged Time-Resolved (TTTR) measurement mode of this highly integrated PCI-plug-in-board records all information needed for each photon as described above and allows to perform different measurement tasks based on one single data format without sacrificing any information available from each single photon. Thus, all measured data can be handled in a standardized and yet flexible way.

4 Data Analysis in Multiparameter Measurements

As an example for multiparameter analysis, single molecule measurements of a recombinant bovine liver rhodanese [29, 30] variant are presented. The rhodanese variant contains two surface-exposed cysteine residues (K135C, K174C) which were labeled with a FRET donor/acceptor dye pair (Alexa 488/Alexa 594, Molecular Probes). The labeled mutant rhodanese was mixed with unlabeled wild type rhodanese and unfolded in 4 M guanidinium chloride (GdmCl). Refolding and aggregation were initiated by 10-fold dilution into GdmCl-free buffer. After five minutes, the sample was further diluted to a total protein concentration of 50 nM. For excitation, a 470 nm laser diode with a repetition rate of 40 MHz (LDH, PicoQuant, Germany) was used. Bursts were identified using the interphoton time Δt and the total number of photons per burst n as thresholds (Δt<100 μs, n>50), including raw data corrections, as described previously [19].

Using four detectors, the information about lifetime, FRET-efficiency, and anisotropy can be extracted. The FRET efficiency is generally plotted in a transfer efficiency histogram of the identified bursts, as shown in Fig. 2 (top). The efficiency histogram reveals three subpopulations: while the first population shows no transfer at all, the other two populations show clearly distinct FRET efficiencies. The FRET efficiency of 0.7 corresponds to natively folded rhodanese molecules. The fraction of molecules with a FRET efficiency of approx. 0.4 was assigned to non-native rhodanese molecules located in aggregates. The peak showing no transfer could be due to several reasons, including: incomplete labeling, acceptor photobleaching, or a fraction of molecules with a conformation with very low transfer efficiency.

Further information is available from the middle and bottom panels of Fig. 2. In the middle panel, the donor anisotropy is plotted vs. the transfer efficiency. The anisotropy of the donor dye(s) in each burst was calculated

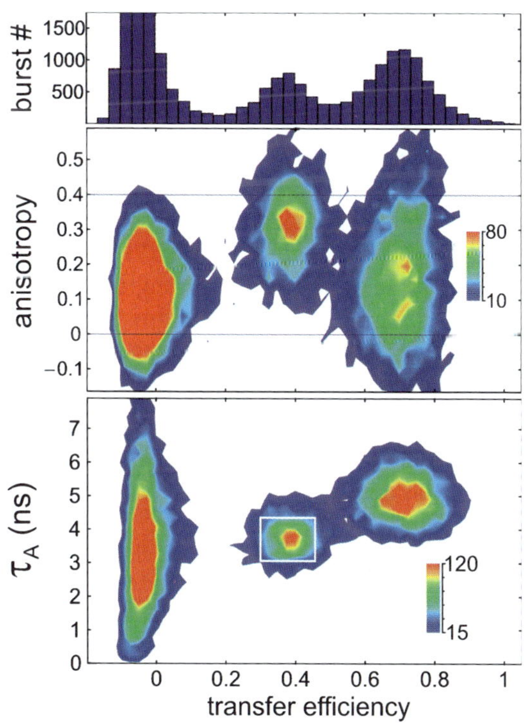

Fig. 2 Multiparameter analysis of rhodanese molecules labeled with Alexa 488 and Alexa 594. *Top*: Transfer efficiency histogram. *Middle*: Two-dimensional histogram of donor anisotropy vs. transfer efficiency. *Bottom*: Two-dimensional histogram of acceptor lifetime vs. transfer efficiency

from the number of photons detected in each of the two donor channels, corresponding to horizontally and vertically polarized emission, according to the procedure described in [13]. The donor anisotropy of the subpopulation characterized by a FRET efficiency of $E \approx 0.4$ is significantly higher than that of the subpopulation with $E \approx 0.7$. As expected, the aggregates thus show a higher anisotropy than the correctly folded rhodanese molecules. The anisotropy distribution of the subpopulation at $E \approx 0.7$ is significantly broader because of the smaller number of donor photons emitted at higher transfer efficiencies. Finally, the lower panel of Fig. 2 shows the acceptor lifetime vs. the FRET efficiency. This illustrates the sensitivity of the acceptor dye for its molecular environment: its average fluorescence lifetime is significantly lower in aggregated compared to natively folded molecules. The zero FRET efficiency peak shows a broad lifetime distribution due to the few counts in the acceptor channel.

In multiparameter analysis, it is of great importance that, after separation of different subpopulations, data from an individual population can be combined to accurately determine an average property with high accuracy. As a single burst contains only 20–200 photons distributed over different detectors, every property is only obtained with a relatively high uncertainty. The degree of uncertainty is directly represented in the distributions of the data in the 2D-plots in Fig. 2, where the data scatter quite significantly about their average value. The low number of photon counts complicates fluorescence lifetime determination, because the fitting procedure requires a good signal-to-noise ratio for a result with low errors. An alternative is, therefore, to combine data from all bursts of a certain subpopulation, which allows accurate fitting of the data, as shown in the TCSPC histogram in Fig. 3, which uses the photons of all bursts in the white rectangle in Fig. 2. As can be seen, this procedure combines more than 2000 photons in the maximum channel, which can be fitted with a much higher accuracy than data from single bursts. This procedure is especially useful in cases where two subpopulations are

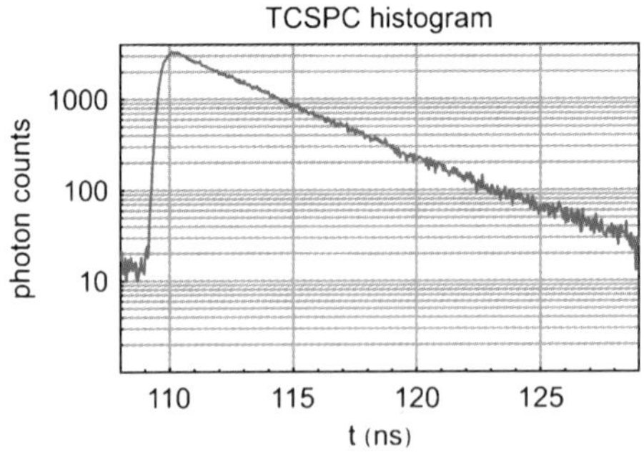

Fig. 3 TCSPC histogram calculated from bursts within the *square* in the lower part of Fig. 2

not completely separated in a multiparameter analysis plot. In this case, the combined fluorescence decay of all events would need to be fitted with a multiexponential decay, an analysis that requires a large number of photons to be reliable.

Analysis of FRET measurements is often complicated by the fact that not all molecules studied contain an active acceptor dye. Especially, a subpopulation with a low FRET efficiency is difficult to discriminate against a subpopulation with only the donor present. For correct detection of these subpopulations, donor and acceptor dye are directly excited alternately with two excitation wavelengths. Only if both donor and acceptor are present, a fluorescence signal is generated after excitation with either wavelength. Two techniques, named pulsed interleaved excitation (PIE) and alternating light excitation (ALEX), have been developed for identifying bursts with complete FRET pairs [31, 32]. The difference between PIE and ALEX is in the way the alternating excitation is achieved. For ALEX, electrooptical modulators (EOM) are utilized to switch the excitation wavelength in the μs time range, while for PIE, alternating donor and acceptor excitation laser pulses are generated, and, thus, the switching occurs in the nanosecond time regime. The advantage of PIE over ALEX is the possibility to use data from a single molecule experiment also for other methods like FCS or fluorescent lifetime analysis. In ALEX, switching of the excitation in the microsecond regime interferes with photophysical processes that take place in a similar time frame. Recently, nanosecond (ns)-ALEX was also performed [33].

The scheme in Fig. 4 shows the principle of PIE FRET. Four different data streams are generated in a PIE experiment. First, the donor is excited, and donor and acceptor fluorescence are probed by two spectrally separated detectors. Afterwards, the acceptor is probed, and the fluorescence response is again measured in the donor and the acceptor channel. If the spectral separation of the fluorescence is efficient, the donor channel should contain only background photons.

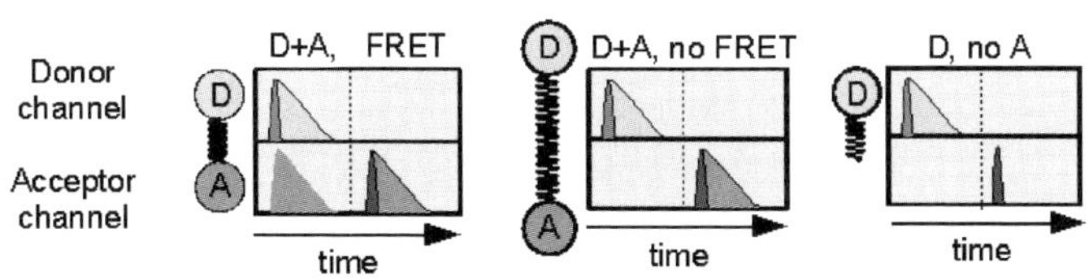

Fig. 4 Concept of PIE for identifying different subpopulations. The donor and the acceptor are excited consecutively, and the fluorescence is monitored in the donor and the acceptor channel. If donor and acceptor are present, a fluorescence response is detected after each excitation. If the donor/acceptor pair undergoes FRET, fluorescence after donor excitation is also measured in the acceptor channel (*left*), while this signal is missing in case when no energy transfer occurs (*middle*). If only the donor is present, no signal is detected after acceptor excitation (*right*)

As an example, a peptide consisting of a cysteine and a glycine separated by 12 proline residues was labeled on the N-terminus with Alexa Fluor 647 as acceptor and at the C terminus via the cysteine with Alexa Fluor 555. A pulsed laser at 532 nm (PicoTA, PicoQuant, Germany) and a pulsed diode laser at 638 nm (LDH, PicoQuant, Germany) were used as excitation sources. A time delay of 12.8 ns between donor and acceptor excitation was applied to allow separation of the fluorescence signals originating from each excitation. To demonstrate the effect of PIE, a FRET efficiency histogram was first calculated using all identified photon bursts, i.e., molecules lacking an active acceptor dye were also included. The resulting FRET efficiency histogram is shown in Fig. 5 (left) and is dominated by a maximum close to an efficiency of 0.9. The width of the distribution is due to shot noise and other, currently unidentified, sources of broadening. Additionally, a peak at low FRET efficiencies, the zero efficiency peak, can also be observed. This peak is most likely due to molecules lacking an active acceptor dye. Using the information of the PIE-FRET measurement now allows to exclude the bursts from molecules without fluorescent acceptor and to recalculate the FRET efficiency histogram. The result of this calculation is shown in Fig. 5 (right). As can be seen, the peak exhibiting zero efficiency is completely removed (Note that in contrast to Fig. 2, the FRET efficiency here is calculated without any corrections, such as bleedthrough or direct excitation).

Although the example shown here has a high FRET efficiency, it is obvious that the power of this method is the identification of subpopulations showing a low FRET efficiency, because otherwise they cannot be distinguished from populations without fluorescent acceptor.

In this chapter, we have shown which parameters can be measured in single molecule experiments and how combining the available data in multi-

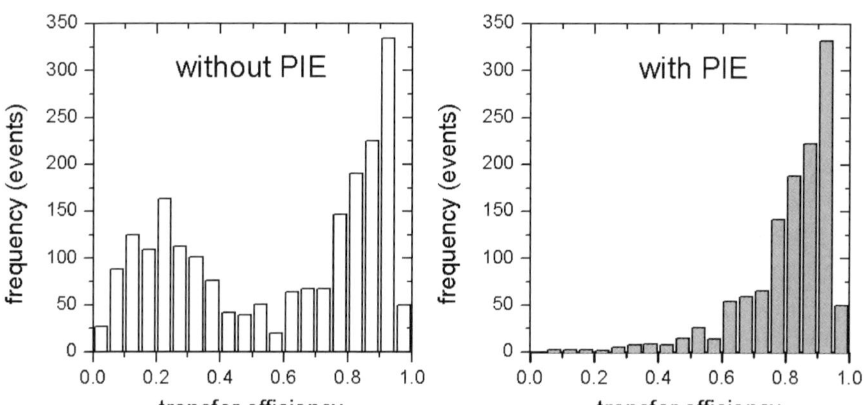

Fig. 5 FRET efficiencies taking all burst events into account (*left*) and using PIE for selecting only events arising from bursts with a fluorescent acceptor present (*right*)

parameter experiments can help to extract information from single molecule experiments that cannot be resolved by other techniques. Additionally, the technique can be efficiently combined with other methods, such as FCS, because the full information for all photons is recorded. High sensitivity and minimal perturbation of the system, without the need of performing complicated separation steps, make it an ideal tool for analytical applications and biophysics.

Acknowledgements This work was supported by grants from the Volkswagen Stiftung and the Schweizerische Nationalfonds (to B.S.) and the BMBF (grant 13N8850).

References

1. Moerner WE (2002) J Phys Chem B 106:910
2. Weiss S (1999) Science 283:1676
3. Herten DP, Tinnefeld P, Sauer M (2000) Appl Phys B 71:765
4. Deniz A, Laurence TA, Beligere GS, Dahan M, Martin AB, Chemla DS, Dawson PE, Shultz PG, Weiss S (2000) Proc Nat Acad Sci USA 97:5179
5. Rueda D, Bokinsky G, Rhodes MM, Rust MJ, Zhuang X, Walter NG (2004) PNAS 101:10066
6. Schuler B (2005) Chem Phys Chem 6:1206
7. McKinney AA, Declais AC, Lilley DM, Ha T (2003) Nat Struct Biol 10:73
8. Rhoades E, Cohen M, Schuler B, Haran G (2004) J Am Chem Soc 126:14686
9. Widengren J, Kudryavtsev V, Antonik M, Berger S, Gorken M, Seidel CAM (2006) Anal Chem 78:2039
10. Tamarat P, Maali A, Lounis B, Orrit M (2000) J Phys Chem A 104:1
11. Ma Y, Shortreed MR, Yeung ES (2000) Anal Chem 72:4640
12. Kask P, Palo K, Ullmann D, Gall K (1999) Proc Nat Acad Sci 96:13756
13. Schaffer J, Volkmer A, Eggeling C, Subramaniam V, Striker G, Seidel CAM (1999) J Phys Chem A 103:00331
14. Prummer M, Sick B, Renn A, Wild U (2004) Anal Chem 76:1633
15. Förster T (1948) Annalen der Physik 2:55
16. Stryer L, Haugland LP (1967) Proc Nat Acad Sci USA 58:719
17. Deniz AA, Dahan M, Grunwell JR, Ha T, Faulhaber AE, Chemla DS, Weiss S, Schultz PG (1999) Proc Nat Acad Sci USA 96:3670, Ha T (2001) Methods 25:78
18. Schuler B, Lipman AE, Steinbach PJ, Kumke M, Eaton WA (2005) Proc Nat Acad Sci USA 102:2754
19. Schuler B (2006) Application of Single Molecule Förster Resonance Energy Transfer to Protein Folding. In: Bai Y, Nussinov R (eds) Protein Folding Protocols, Vol. 366. Humana Press, Totowa, New Jersey
20. Deniz AA, Laurence TA, Dahan M, Chemla DS, Schultz PG, Weiss S (2001) Annu Rev Phys Chem 52:233–53
21. Knemeyer JP, Marmé N, Sauer M (2000) Anal Chem 72:3717
22. Grünwald D, Hoekstra A, Dange T, Buschmann V, Kubitscheck U (2006) Chem Phys Chem 7:1
23. Zander C, Sauer M, Drexhage KH, Ko DS, Schultz A, Wolfrum J, Brand L, Eggeling C, Seidel CAM (1996) Appl Phys B 63:517

24. Enderlein J, Robbins DL, Ambrose WP, Keller RA (1998) J Phys Chem A 102:6089
25. Eggeling C, Fries JR, Brand L, Günther R, Seidel CAM (1998) Proc Nat Soc Sci USA 95:1556
26. Wolleschensky R, Dickinson ME, Fraser SE (2002) Group-Velocity Disperion and Fiber Delivery in Multiphoton Laser Scanning Microscopy. In: Diaspro A (ed) Confocal nd Two-Photon Microscopy. Foundations, Applications and Advances, Wiley-Liss, New York
27. Wahl M, Erdmann R, Lauritsen K, Rahn HJ (1998) Proc SPIE 3259:173
28. Böhmer M, Pampaloni F, Wahl M, Rahn HJ, Erdmann R, Enderlein J (2001) Rev Sci Instrum 72:4145
29. Horowitz PM, Simon D (1986) J Biol Chem 261:13887
30. Mendoza JA, Rogers E, Lorimer GH, Horowitz PM (1991) J Biol Chem 266:13587
31. Rüttinger S, MacDonald R, Krämer B, Koberling F, Roos M, Hildt E (2006) J Biomed Opt 11:024012
32. Lee NK, Kapanidis AN, Wang Y, Michalet X, Mukhopadhyay J, Ebright RH, Weiss S (2005) Biophys J 88:2939
33. Laurence TA, Kong X, Jäger M, Weiss S (2005) Proc Nat Soc Sci USA 102:17348

Part III
Fluorescence-Based Microarray Technology: Applications, Future Trends, and Need for Standardization

Springer Ser Fluoresc (2008) 6: 215–237
DOI 10.1007/4243_2008_036

Published online: 6 March 2008

DNA Microarrays: Applications, Future Trends, and the Need for Standardization

Sige Zou[1] · Hua-Jun He[2] · Yaping Zong[3] · Leming Shi[4] · Lili Wang[2] (✉)

[1]Laboratory of Experimental Gerontology, National Institute on Aging, National Institutes of Health (NIH), 5600 Nathan Shock Drive, Baltimore, MD 21224, USA

[2]Biochemical Science Division, National Institute of Standards and Technology (NIST), 100 Bureau Drive MS 8312, Gaithersburg, Maryland 20899-8312, USA
lili.wang@nist.gov

[3]Full Moon Biosystems, 754 N. Pastoria Avenue, Sunnyvale, CA 94085, USA

[4]National Center for Toxicological Research, US Food and Drug Administration (FDA), 3900 NCTR Road, Jefferson, AR 72079, USA

Abstract Microarray technology allows high-throughput analysis of tens or even hundreds of thousands of genes in a single experiment, and has been applied broadly to address a wide variety of questions in basic and applied sciences. The applications of the microarray technology range from gene-expression profiling and genotyping to DNA–protein interactions and genome sequencing, and the list of applications keeps growing, especially when combined with other technologies such as proteomic technologies. However, many steps are involved in each microarray experiment and a number of microarray platforms exist, each with its unique features. The challenge is how to standardize the methods and materials to allow intra- and inter-comparison of microarray data collected in the same or different set of experiments. With the challenge in mind, we address the need for standardization for each step of the microarray experiments with emphasis on quality control of array fabrication and scanner calibration and verification. The proposed standards are designed for checking the quality of mRNA, fabricating slides, hybridization, and collecting, analyzing, and storing data. By implementing standards for each step of the microarray process, the full potential of the microarray technology will be realized, especially in the area of disease diagnosis and drug development.

Keywords Bead-based arrays · cDNA arrays · Expression profiling · External RNA control · Genotyping · Oligonucleotide arrays · Scanner performance validation

Abbreviations

bp	Base pair
CCD	Charged-coupled device
cDNA	Complementary deoxyribonucleic acid
CGH	Comparative genomic hybridization
ChIP	Chromatin immunoprecipitation
DLBCL	Diffuse large B-cell lymphoma
EST	Expressed sequence tags
EB	Ethidium bromide
MAQC	Microarray quality control
MIAME	Minimal information about a microarray experiment
MIAPE	Minimal information about a proteomics experiment
NIST	National Institute of Standards and Technology
NRO	Nuclear run-on
ORFs	Open reading frames
PCR	Polymerase chain reaction
PMT	Photo-multiplier tube
RNA	Ribonucleic acid
aRNA	Antisense RNA
mRNA	Messenger RNA
SNP	Single nucleotide polymorphisms
TF	Transcription factor

1 Introduction

Genomic sequences have been determined for a number of organisms from single-cell organisms to human beings [1–8]. For the human genome alone, at least 20 000 genes have been predicted to be expressed in human cells [4]. This enormous amount of sequence information has presented tremendous (but challenging) opportunities for basic research to study biological processes at the genomic level and for business such as disease diagnosis and drug development. The microarray technology responds to this genomic challenge allowing researchers to simultaneously analyze tens of thousands of genes or samples at the genome level in a single experiment [9]. One of the first genome-sized microarrays with complementary DNA (cDNA) clones was introduced by Schena et al. shortly after the genomic sequence of the baker yeast *Saccharomyces cerevisiae* was initiated by an international sequencing consortium. This was followed by introduction of oligonucleotide arrays by Affymetrix, Inc. (Santa Clara, CA) [3, 10]. Since then, the development and applications of the microarray technology have exploded and now has become a popular and almost standard molecular biology tool. The typical applications of microarray technology include genome-wide gene-expression profiling and large-scale detection of single nucleotide polymorphisms (SNPs) [11] (Fig. 1).

Many types of microarrays have been developed for various applications. The major type is the DNA microarray that employs the basic nucleotide hybridization principle, in which complementary sequences bind to each other [12]. This technology consists of two key components: the probe sequences immobilized on the solid substrates, and the target sequences, which are used to hybridize with the immobilized probes. In a typical gene-expression study, each individual probe represents a unique gene in the genome of interest and the targets are derived from messenger RNA (mRNA) isolated from cells, tissues, or whole organisms. Hybridization between probes and targets allows quantification of molecules in the targets. In this chapter we will outline the microarray technology and its applications and then we will discuss future trends, concerns, and standardization of this technology.

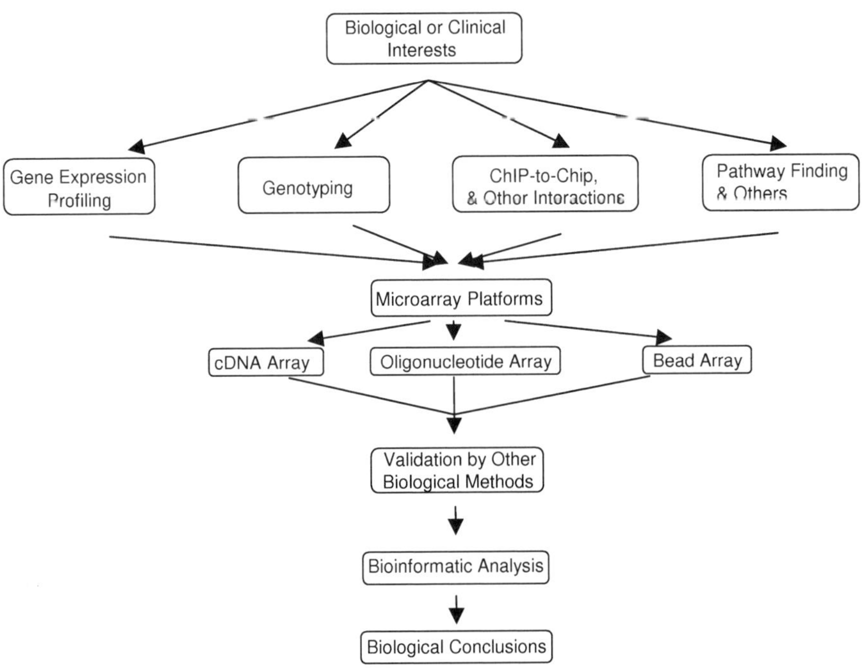

Fig. 1 Overview of the microarray technology and its applications

2
Microarray Technology

A typical DNA microarrary experiment is conducted through a series of steps using a number of materials that could significantly impact the success of an experiment (Fig. 2). One such critical component is the arrays containing a number of orderly positioned probes. The probes are either synthesized single-stranded short (≤25 mer) or long (50–70 mer) oligonucleotides, or long double-stranded DNAs (>100 bp), the latter of which are normally generated from a library of expression clones (EST) or gene-specific primers in polymerase chain reaction (PCR) amplification reaction [10, 13–15]. Several approaches have been employed using these probes for at least three major types of the DNA microarrays: cDNA arrays, oligonucleotide arrays, and bead-based arrays, each having its own unique features. Once the arrays are made, target molecules labeled with fluorescent dyes or biotins are hybridized to the probes on the arrays. The labeling is normally achieved by directly incorporating fluorescent dyes into the target molecules or by indirect labeling, the latter of which is done by first incorporating modified nucleotides, such as amino allyl-dUTP or biotin conjugated dCTP, followed by coupling the fluorescent dye to them. After hybridization, the fluorescent intensities of the dyes on each probe are recorded by an imaging system. Microarray data are then subjected to bioinformatic analysis to extract out biological information.

There are three major approaches for array fabrication. One is the printing (or spotting) method, in which the PCR products or pre-synthesized oligonucleotides are deposited on a glass surface by touching pins or ink-jet nozzles with a high-precision robot (Fig. 3A). To facilitate the attachment of probes, glass slides are modified by coating with poly-L-lysine or other chemical modifications, such as silanes possessing amine or epoxy groups [10, 16–18]. Drawbacks to the pin-spotting method include the difficulty in controlling the amount of DNA deposited on the slides and the inconsistency in spot morphologies related to the properties of slide surfaces and printing buffers and sample concentrations. These two features of spotted arrays may generate huge systematic errors downstream in the microarray experiments. However, it is still comparatively inexpensive to generate arrays with spotting methods, and the arrays can be of high quality if sufficient care is taken in the printing steps. In our hands, the chemical modifications of glass slides and types of buffers were also critical for high-quality printing.

The second method to generate arrays is in situ synthesis, in which short or long oligonucleotides are directly synthesized in situ on a glass surface. These technologies have been primarily developed by Affymetrix, Inc. and Agilent Technologies, Inc. (Palo Alto, CA) (Fig. 3B). Affymetrix has adapted a photolithographic technology commonly used in the semiconductor industry to first manufacture high-density arrays with hundreds of thousands of oligonucleotide probes synthesized in situ base-by-base on a glass sur-

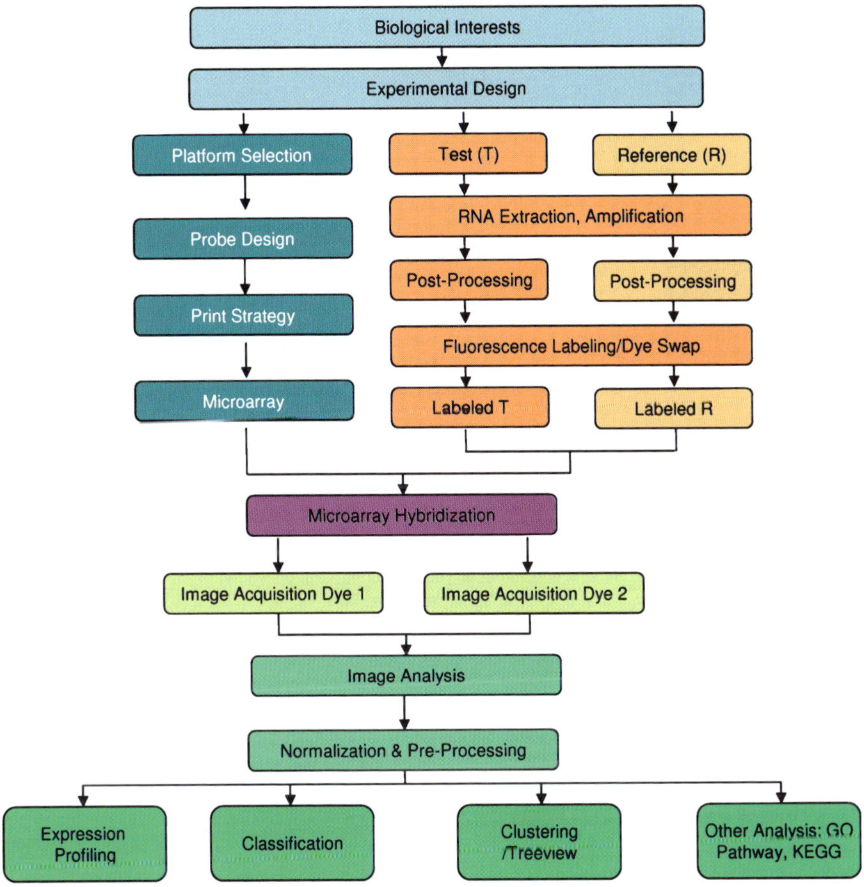

Fig. 2 Schematic of a typical microarray experiment design. It begins with biological interests and experimental design. The samples include one serving as a test sample and another as a reference (control sample). The RNAs are extracted, reverse-transcripted and sometimes amplified, followed by labeling with different fluorescent dyes, and co-hybridized to a microarray. In some cases, a dye-swap is performed to minimize the fluorescent dye effect. For the array, one can select a suitable platform, probes for the targets, and a print strategy. After hybridization, microarrays are scanned to acquire fluorescent images. Image analysis is then performed to obtain the raw signal data for each spot. Data with poor quality are filtered out, and high-quality data are collected for further analyses. Depending on the aim of the study, one can infer statistical significance of expression profiling, perform classification, clustering/treeview analysis or carry out analysis such as GO, pathway and KEGG. Note that the data from all the steps should be complied according to standards, for example, minimum information about a microarray experiment (e.g., MIAME), and be archived properly

face of about 1.6 cm^2 [13]. Due to the coupling efficiency in each step of oligonucleotide synthesis, the probe length on the Affymetrix chip is always limited to 25 bases to ensure that sufficient full-length probe is present in each

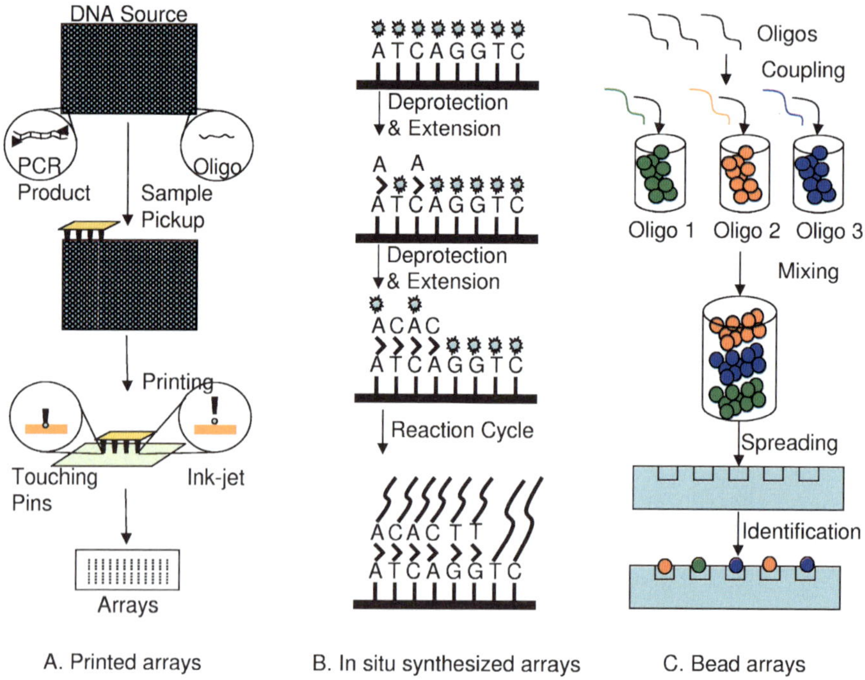

Fig. 3 Fabrication of arrays. **A** Printed arrays. PCR products or oligonucleotides are printed on the glass substrates either by pins touching the slides or by the ink-jet technology. **B** In situ synthesis of oligos on the glass substrates. Oligos are synthesized base-by-base by in situ phosphoramidite solid-phase chemistry (Affymetrix and Agilent). **C** Bead arrays. Oligos are coupled to microbeads and randomly distributed on a slide with etched wells. The identity of each bead type is determined by a decoding method before being used for array experiments

spot. On the other hand, Agilent has developed another method to synthesize oligonucleotides in situ on an activated glass surface. It combines the ink-jet technology commonly used in the laser printer industry and solid-phase phosphoramidite chemistry to perform an iterative oligonucleotide synthesis loop [19]. The efficiency of the phosphoramidite synthesis permits manufacturing arrays with long oligonucleotides (typically 60 mers), which is proven to increase hybridization sensitivity compared to arrays with short oligonucleotides when used in gene-expression profiling experiments.

Recently, Illumina Inc. (San Diego, CA) has invented another microarray platform, called BeadArray technology, to manufacture arrays [14]. In this technology, oligonucleotides of any desired and feasible length are first individually immobilized on to derived microscopic beads (~3 μm in size) with bead identifiers in a bulk process. Meanwhile, a number of tiny wells, which can be over 10 million, are created by acid etching on the substrate of a glass slide. Then the oligonucleotide-coupled beads are randomly distributed into the etched wells. The identification and quality of each bead type on the array is ensured by the manufacturer using a decoding algorithm. This algorithm employs sequential hybridizations of dye-labeled oligonucleotides or decoders, complementary to bead identifier sequences, to create a combinatorial decoding scheme for the array. In addition to the high flexibility and sensitivity of the features of the BeadArray technology, each bead type with a single oligonucleotide sequence on an array is represented an average of 30-fold, which dramatically increase the statistical power for data analysis.

3 Microarray Applications

Microarray technology allows high-throughput analysis of tens or even hundreds of thousands of genes in a single experiment, and has been applied to many different research and development areas. Originally developed as a technology for genome-wide gene-expression profiling experiments, it has now been employed in many other areas by adaptation of its basic complementary hybridization principle and in combination with other techniques. We will discuss mature applications in gene-expression profiling and genotyping in this section followed by a discussion of concerns for microarray technologies and applications and selected future trends.

3.1 Expression Profiling

A major application of the microarray technology is genome-wide gene-expression profiling [11]. Microarray technology typically employs one-color or two-color approaches, referring to the number of fluorescent dyes used. In a typical two-color expression-profiling experiment (Fig. 2), RNA from the reference sample is labeled through reverse transcription and coupled with one type of fluorescent dye, such as Cy5. RNA from the experimental sample is then labeled with a second fluorescent dye, such as Cy3. After hybridizing the samples to an array containing tens of thousands of gene-specific probes, the array is scanned with an imaging system to collect fluorescence intensities from both samples for each probe on the array. In some cases, a dye-swap is performed to minimize the fluorescent dye effects, such as dye labeling efficiency and microenvironment effects. In the one-color approach, RNA samples from the reference and experiment samples are separately labeled with the same fluorescent dye and then are hybridized to two different arrays. The difference in transcript levels for each probe in a single experiment is then calculated based on the difference in fluorescent intensities between two samples. Biological information is extracted by bioinformatic analysis of a set of microarray data, which provides a genome-wide view of gene expression in a biological process or response.

Expression profiling with microarray technology has been broadly used to investigate many basic biological processes, including embryonic development, aging [20], cancers [21] and stem cell development [22]; and has allowed the identification of a number of genes associated with these biological processes. Analyses of gene functions and pathways that they are involved in have provided rich biological information. As an example of expression-profiling experiments, several labs (including ours) have measured gene-expression profiles of aging by comparing the transcript levels of genes in older fruit flies to that in young flies using both cDNA arrays and Affymetrix chips [23–26]. Bioinformatic analysis has detected significant alteration of transcript levels in more than 3% of the genes in the fly genome in these two populations, presumably involved in a number of biological processes in aging. We have further compared the expression profile of flies to that of the nematode *C. elegans*, which is evolutionarily distant from the flies. Interestingly, this comparison has revealed conserved expression patterns between these two species. Functional studies have shown that some genes with conserved expression patterns in aging can extend lifespan in the nematode when mutated, suggesting an important role for these genes in aging. However, despite the fact that a number of expression-profiling experiments have been performed for various model organisms including humans, and thousands of aging-associated genes have been found to be differential expression between young and old individuals, functions of the aging-associated genes in the aging process are largely unknown [20]. Biomarkers for aging are still elusive, in part, due to variation in the microarray technologies used in different studies, such as different array types, variation in sample processing, and different imaging systems. This and many other microarray experiments have suggested that gene-expression profiling is valuable but is only a first step in understanding the molecular mechanisms of a biological process.

Clinical potential for gene-expression profiling has been best demonstrated in the cancer field, including the successful classification of tumor types based on gene-expression profiling. The first such classification has been attempted by Alizadeh et al. to identify different types of diffuse large B-cell lymphoma (DLBCL) [27]. DLBCL is a clinically heterogenous, non-Hodgkin's lymphoma and is difficult to separate into subgroups on the basis of morphology in order to provide the patient and physician with appropriate therapy decisions and prognosis. Researchers have profiled gene-expression patterns from lymphocytes obtained from DLBCL patients and normal humans using microarray technology. By clustering the expression data, they have identified two molecularly distinct forms of DLBCL that correlate well with two groups of DLBCL patients responding differently to therapy. Since then, a number of methods have been developed and applied to identify tumor-gene expression signatures, which are valuable in cancer diagnosis, drug discovery, and development [28, 29]. Once the reliability of the microarray technology is improved and inherent variation of biological samples can be minimized, it is foreseeable in the near future that the classification of tumors by gene-expression profiling will become a routine diagnostic method in medicine. Recent progress on gene-expression profiling of a single cell is an exciting advance in this promising field [30].

3.2 Genotyping

SNPs are the most common form of genetic variant in the human genome, occurring on average of one per ~300 bps [31]. It is estimated that more than 10 million common SNPs exist in the human genome, of which approximately 2 million have been confirmed. Some SNPs have already been shown to play an important role in gene function and response to drugs [32, 33]. SNP genotyping is critical for us to understand fully the genetic architecture of common traits underlying diseases, as well as drug or diet response. It is estimated that several hundred thousand SNP analyses are required to map a disease gene in an individual. To efficiently identify such a large number of SNPs from a single or a few of samples, several microarry-based methods have been developed for scoring SNPs. In a typical genotyping experiment using genome-wide SNPs, genomic DNA is amplified by direct PCR or by a restriction-enzyme-based-adapter-ligation PCR (Fig. 4). Amplified gDNA is hybridized to commercially available short (25-mer SNP chips from Affymetrix) or long (50-mers) oligonucleotide arrays [34–36]. SNPs can then be scored by direct allele-specific hybridization or by an allele-specific primer extension assay with labeled dNTPs. Due to the sequence complexity of genomes, most of the array-based genotyping technologies allow researchers to genotype up to 10^5 to 10^6 SNPs per sample. Illumina Inc. has recently developed an array-based whole-genome genotyping assay by combining the BeadArray technology and the enzymatic SNP scoring assay [36]. This technology has the potential to enable researchers to analyze more than a million SNPs from a single sample, and in principle has the capacity for unlimited multiplexing that is limited only by how many probes each array can hold.

Array-based genotyping has also been used to obtain a genetic map of the human genome, yielding DNA blocks (haplotypes) that can help identify disease genes associated with phenotypic traits [37]. In addition, microarrays have been used to identify viruses or microbial pathogens in diagnostics, ensuring the safety of food and medicine production; as well as to address bio-defense concerns, such as virus identification [38–41]. DNA microarray technology is a powerful tool and is gradually becoming a mature technology for a wide range of research and development applications. However, many challenges that limit its applications as a diagnostic and detection tool remain unsolved. To name a few, it is still technically challenging to measure

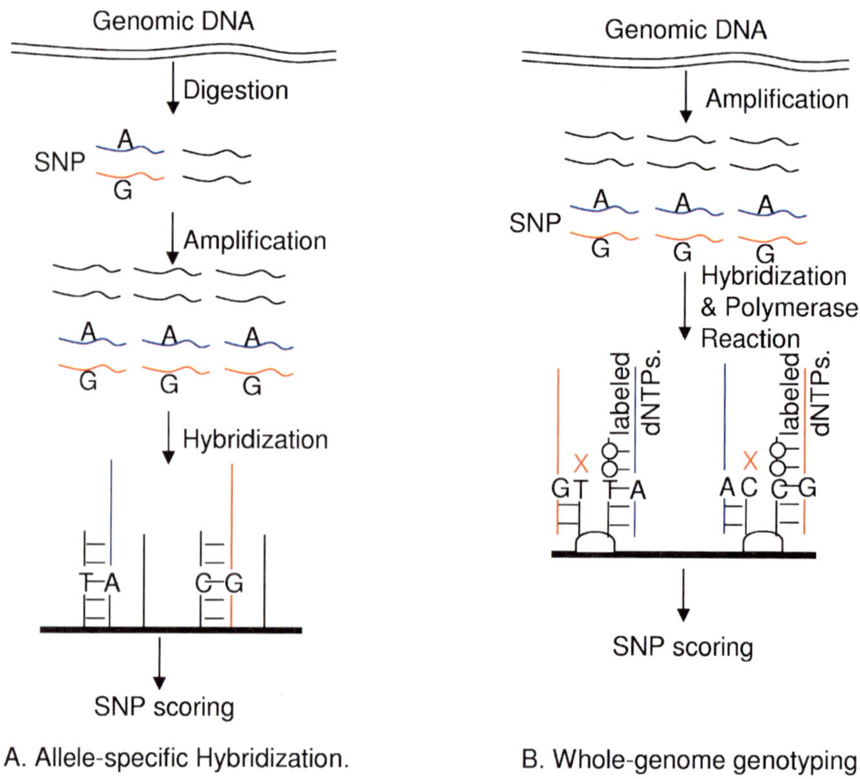

Fig. 4 Array-based SNP genotyping. **A** Allele-specific hybridization. Genomic DNA is amplified to enrich the DNA fragments with SNPs. Then the amplified gDNA is hybridized to oligo microarrays to score SNPs based on hybridization intensities. **B** Whole-genome genotyping. The entire genomic DNA is amplified and hybridized to oligo arrays with SNP-specific probes. Polymerase reaction is performed with labeled dNTPs and SNP is scored based on difference in amplification

accurately low levels of pathogens and distinguish multiple strains in a mixed sample. This is related to the sensitivity of microarray technology and will be addressed in the last section of this chapter.

4 Future Trends

4.1 Universal Microarray Platform

Generally, a microarray chip is designed to have probes matching sequences of a specific organism and to address a specific set of questions for that organism. There are two drawbacks in the current microarray platform; (1) a specific microarray is needed to address each specific application, and (2) the hybridizations occur less efficiently on an array surface than in solution. A universal microarray, also called the "zip-code" array, has been developed. This array format may ultimately address these two issues by separating the actual microarray design from the hybridization process [42, 43]. "Zip-code" arrays contain sequences that are a set of unique and distinct oligonucleotides attached covalently at known locations as the "addresses" on a microarray. The target molecules are composed of assay-specific sequences, such as gene-specific primers that are linked to the "zip-code" complementary sequences. The transcription/amplification reactions involving target molecules take place in solution, and the products are then separated and made available for scoring via hybridization to the "zip-code" microarray. In this manner, a single microarray design is employed for various array applications. Obviously, the design of the probe sequence is most critical in the universal microarray platform to cover a wide arrange of genomic sequences.

4.2 New Application Trends

Combination of DNA Microarrays and Protein Arrays for Diagnosis, Prognosis, and Treatment of Diseases

Gene-expression profiling allows for simultaneously establishing expression patterns for thousands of genes from a single specimen. It can determine if certain genes are high-expressed or low-expressed in a specific tissue type or tumor. Clinical information regarding the patient can then be correlated with gene expression. However, it is not the genes, but the proteins that are responsible for carrying out most of the cellular functions. The categorization and quantification of proteins may be necessary for characterization of cell types in health and disease states. It is therefore believed that the protein levels and activities are the best measurement of cellular functions. To address these questions, researchers have combined both genomic (DNA microarrays) and proteomic (protein arrays) means for better understanding the likely cause of diseases and hence, better diagnosis, prognosis and treatment of disease [44–50].

Early detection and correct diagnosis of cancer are highly important factors for the choice of anti-cancer treatment as well as for patient survival. Recent cDNA and oligonucleotide microarray data have documented changes in the transcriptional profile in comparison to histological normal tissues [51]. However, identification of gene-expression signatures cannot always be linked directly to the profile of expressed proteins. Detection of a protein pattern might be more reliable, especially for interpreting signaling pathways and other cellular processes that contribute to cancer development and metastasis and for interpreting treatment response. Although the application of antibody (Ab) arrays is limited by the amount of available antibodies, this technology can be a potential tool supporting DNA microarray data for cancer detection and for better understanding of cancer biology. As an example, the researchers have used the BD Clontech Ab Microarray 380 (Palo Alto, CA) to study tumor samples from 12 patients with squamous cell lung carcinoma [48]. Comparison with oligonucleotide microarray data revealed that 31% of the differentially expressed proteins correlate well with altered mRNA expression in squamous cell lung carcinomas. A histone deacetylase variant, HDAC3, was confirmed to be up-regulated by both antibody and DNA microarrays. The blockade of HDAC activity by specific inhibitors can mediate growth arrest, differentiation, and apoptosis of cancer cells. Currently, some of the specific inhibitors are in clinical trials for anti-cancer therapy due, in part, to the finding of this study.

Nuclear factor-kappa B (NF-κB) is a ubiquitously expressed, DNA sequence-specific transcription factor that has attracted widespread attention in view of its involvement in the regulation of the expression of a large number of genes involved in immune and inflammatory responses [52]. Polyphenol mangiferin (MA) has been shown to have various effects on macrophage function, including inhibition of phagocytic activity and free-radical production. To characterize further the immunomodulatory activity of MA on diverse genes related to the NF-κB signaling pathway, researchers investigated its effects on expression in activated mouse macrophages using a DNA micro-array and a cytokine protein array [46]. A good correlation was observed between the effects of MA on the mRNA levels of NF-κB-responsive genes and protein synthesis by the regulatory immune mediators. Particularly, MA decreased the protein expression of the pro-inflammatory chemokine RANTES that is also regulated by NF-κB. These results indicate that MA modulates the expression of a large number of genes that are critical for the regulation of apoptosis, viral replication, tumorigenesis, inflammation, and various auto-immune diseases, and raise the possibility that MA can be of value in the treatment of inflammatory disease and/or cancer.

Although differential protein expression correlated well with differential mRNA expression in some cases, it was not consistently observed for many signaling molecules and transcription factors. A number of studies [44–50] have raised two issues: (1) changes at the protein level may occur in a manner independent of changes at the mRNA level and (2) changes at either the mRNA or protein levels may not linked to a specific phenotypic trait.

DNA–Protein Interaction

With the novel method of on-chip chromatin immunoprecipitation (ChIP-on-Chip) [53], a protein/DNA complex is first stabilized through cross linking by formaldehyde, then fractionated through sonication, and finally purified by using an antibody raised against the protein of interest. The bound DNA within that complex is released through heat treatment and further purified from chromatins via proteinase K treatment and phenol/chloroform extraction. The purified DNA is then labeled and hybridized to arrays for sequence identification. In this fashion, the binding site of the protein in the genome can be identified. The affinity of the protein–DNA interaction and the specificity of the antibody for the recognition of the given protein are critical in the use of this method.

The "ChIP-on-Chip" method has the potential to identify all bound DNA at the genomic level even in previously unknown target regulatory regions. However, it has two significant limitations; it requires knowledge of the involvement of a specific transcription factor (TF) for a given response and provides little information about the functional consequences of the TF upon the particular genes. Recently, Fan and co-workers [54] reported a novel strategy that employs both "ChIP-on-Chip" and nuclear run-on (NRO) assay approaches to identify transcription factors that administer adaptive gene expressions and manipulate cellular responses. Nuclear run-on assays are transcription assays designed to monitor genes being transcribed in a cell nucleus at a specific time. In the near future, it is likely that more and more ChIP-on-Chips will be used to study protein–RNA and ssDNA–protein interactions and complexes involving miRNA as regulators in RNA processing [11].

Other Applications

Genomic Sequencing. Nucleic acid sequencing is a fundamental technology where some of its developers were awarded the 1980 Nobel Prize in chemistry (Paul Berg, Walter Gilbert, and Frederick Sanger). Although several technical improvements have been introduced, including ultra-thin gel technology and capillary electrophoresis, the throughput of these technologies is still limited, especially for large-scale genomic sequencing. Microarrays are therefore applied to sequencing analysis, which can be regarded as an extension of SNP analysis. The method involves immobilization of the target to be sequenced on a microarray, hybridization of this target with a very large set of labeled short probes, examination of the hybridization pattern, and computation of the DNA sequence. Vice versa, thousands of short probes can be immobilized on a microarray and hybridized to the target of interest for sequence determination. Massively parallel signature sequencing (MPSS) is therefore, a highly parallel method for DNA sequencing [55–57].

Identification of Splicing Variants. The molecular interpretations of encoded messages in a genome are responsible for variations between organisms. The transcriptional representation of exons can be analyzed by placing at least one representative of each onto an expression microarray. The ability for carrying out such combined analysis of differences in splicing and transcript levels usually depends on the quality of the genome sequence annotation. Even in the rather simply structured genome of *Saccharomyces cerevisiae* several hundred open reading frames (ORFs) were initially missed. In *Drosophila melanogaster*, the underestimation of the number of genes was even greater. Non-coding sequences transcribed are more likely to be missed. Although the number of protein-coding genes in the human genome could be relatively small, there might be many exons that are not yet accounted for. Intron–exon junction sequences can be used as alternative probes that represent exon sequences [11]. Based on the initial study carried out by Southern and his colleagues [58], the analysis of the subtle splicing differences may make it possible to sort out the linkage between structure and functionalities.

Cell-Based Assays. The completion of full-genome sequencing for humans and other organisms has made it possible to use microarrays for whole-genome expression profiling under various biological conditions, for example, disease and normal development stages. However, gene functions remain vastly unknown. An emerging challenge is how to rapidly uncover functions of genes and to identify gene products with desired properties. Ziauddin and Sabatini [59] developed a system suitable for rapidly screening large sets of cDNA and DNA constructs for genes that encode desired functions or genes that cause cellular phenotypes of interest. Using slides printed with sets of cDNAs in expression vectors, they created microarrays of living cell clusters growing on the slides and expressing the gene products. The cell clusters can be screened for a property detectable from the slide, and the identity of the responsible cDNA can be determined from the coordinates of the cell cluster with the phenotype of interest. Since then, several methods based on their work have been developed and applied for DNA, RNA, and RNAi transfection, and functional assays for inhibitor screening [60–63]. Furthermore, tissue-based assays including comparative genomic hybridization (CGH) are also emerging and provide a promising tool for systemic biology applications [64–66].

Low-density Focus Arrays. All of the microarray applications described above employ high-density microarray platforms. For instance, gene-expression profiling of a specific disease relative to a normal state is performed on high-density arrays for identification of disease biomarkers. Once these biomarkers have been identified, their ultimate utility for diagnosis and treatment of the disease in clinical setting will be accomplished by using low-density biomarker arrays. Individualized disease diagnosis would greatly demand accurate signal quantification rather than pattern visualization in gene-expression profiling. The AmpliChip CYP450 manufactured by Roche Diagnostics (Basel, Switzerland) and powered by Affymetrix microarray technology is the first array-based diagnostic test for comprehensive analyses of two genes that monitor drug efficacy.

5 Need for Standardization

Figure 2 outlines the experimental design for a typical microarray. Due to the complexity of DNA microarray technology, proper controls and standardization are required for every step of the microarray experiments in order to realize the full application potential of the technology. Below we describe existing quality-control procedures/methods, potential reference materials, and efforts towards the standardization of the microarray technology.

5.1 Sample Preparation/Quality Control

Isolation of intact RNA is a first and essential step in a microarray experiment. A common method for assessing the integrity of total RNA is to run a denatured agarose gel stained with either ethidium bromide (EtBr) or more sensitive dyes such as SYBR® Gold and SYBR Green II. Intact total RNAs on the denatured gel will have sharp, clear 28*S* and 18*S* rRNA bands for eukaryotic samples (18*S* and 26*S* for yeast; 16*S* and 23*S* for *E. coli*, et al.). The 28*S* rRNA band should be approximately twice as intense as the 18*S* rRNA band. An alternative method is to use a micro fluidic instrument, such as the Agilent 2100 Bioanalyzer (Agilent Technologies, Inc.), which will provide RNA integrity, concentration, and purity information. RNA concentration and purity can be estimated by the absorption ratio of A260/A280.

DNA microarray applications require fluorescently labeled materials, such as cDNA and aRNA [67] for hybridization on microarrays. Fluorophore in clusion can be divided into direct and indirect incorporations. The direct incorporation is accomplished by using either Cy3-dNTP or Cy5-dNTP, and the indirect one is largely done through incorporation of amino allyl-dUTP followed by a reaction with activated dye molecules. Although there are many commercial labeling kits available, new labeling methods are being developed continuously [11, 68–70]. The labeling efficiency and the activity of the labeled materials are difficult to ascertain for the assurance of micro-array performance. At present, the empirical formulas are used to estimate labeling percentage [71]. These estimations are based on absorbance measurements. In fact, impurities and even buffer compositions will affect absorbance measurements. The calculated values can be used only for the purpose of inter-laboratory quality control. Nonetheless, methods such as inductively coupled plasma-optical emission spectroscopy (ICP-OES) [72] and liquid chromatography-isotope dilution mass spectrometry (LC-IDMS) [73] have been recently developed for quantifying 20mer oligonucleotides. It is feasible in the long run that they can be used for cDNA quantitation. This will simplify the quantification of the number of fluorophores per labeled cDNA.

Biological variation is intrinsic to all organisms; it may be influenced by genetic and environmental factors, the probes designed for the array, and whether the samples are pooled or individualized. In other words, it is a challenge to assess performance of a microarray because of the complicated nature of the technology. To define the requirements for standardizing RNA-based molecular assays, a workshop titled "Metrology and Standards Needs for Gene Expression Technologies: Universal RNA Standards" was hosted by NIST [74]. The outcome of the workshop includes an assay process reference material composed of a moderately complex pool of highly characterized RNA molecules and a universal hybridization reference material. The first RNA reference material produced synthetically by in vitro transcription serves as an external spiking control and allows the assessment of the accuracy, dynamic range, sensitivity, and specificity of a given array platform under a set of conditions. This material can be used for both RT-PCR-based and array-based applications. The universal hybridization reference material, defined as a pool of 12 synthetic sequences not expressed in any known genome, is intended for variability control in array hybridization methods. The External RNA Controls Consortium is currently testing polyadenylated transcripts that can be added to each RNA sample prior to processing to evaluate the technical performance of each assay [75]. In addition to the effort led by the External RNA Controls Consortium, a consortium led by FDA/CDER has tested two mixed tissue RNA pools with known differences in tissue-selective genes that can be used specifically as rat reference materials [76]. In the expression array system from Applied Biosystems (Foster City, CA), a 24-mer oligonucleotide control labeled with a fluorescent dye is co-hybridized along with biological samples of interest to its complementary oligonucleotide that is co-deposited at every microarray feature. This oligonulceotide pair is used to ensure optimal image auto-gridding and spot finding processes as well as hybridization performance.

In recent years, concerns have been raised regarding the reliability and consistency of microarray data obtained by using different microarray platforms; low levels of concordance have been found in cross-platform microarray investigations [77, 78]. To address these concerns, the Microarray Quality Control (MAQC) project was initiated and led by the FDA to evaluate the compatibility issue of various array platforms [79]. More specifically, the study compares seven different microarray platforms using two well-characterized, commercial available RNA samples: Universal Human Reference RNA from Stratagene (La Jolla, CA) and Human Brain Reference RNA from Ambion (Austin, TX). The most promising result from the study is that the concordant detection calls between replicates of the same sample are between 70% and ~100% across six high-density microarray platforms and thousands of hybridization reactions for 12 091 common Entrez gene lists (the result of study has recently been submitted for publication). The MAQC project complements the efforts of external RNA reference materials described above by establishing two additional human reference RNA samples and an accompanying large data set that can be used by the array community to compare their results for the purposes of quality control and method validation.

To address probe-sequence effects on microarrays, we have synthesized six oligonucleotides (60–80 mer) matching different regions of the *Kanamycin* gene as probes. Upon hybridization to Cy3 and Cy5 labeled full-length *Kanamycin* cDNA, the signal intensities varied depending on the probe sequences. The ratio values of the same probe are relatively consistent upon a serial dilution of labeled cDNA. The results suggest that the probe sequences are critical for obtaining consistent results (our unpublished data).

5.2
Pre-Arrayed Slide Quality Control

In general, there are two ways to produce arrayed slides. One is by solid-phase synthesis by applying phosphoramidite chemistry. Both Affymetrix chips and Agilent chips are produced using this method. The difference between the two types of chips is that Affymetrix activates by irradiation through the masks, whereas Agilent spots building blocks one by one into the spots. The other way of fabricating arrays is by printing, including both contact printing and non-contact printing, where the probes (oligonucleotides or cDNA) are deposited on the slides by spotting pins or piezo dispenser. Most of the home-produced arrays are made by printing (contact or non-contact printing technology). To standardize the microarray process, the quality of the pre-arrayed slides is the key to reducing variations and increasing consistency, and thereby improving the reliability and reproducibility of the hybridization results. Dye-labeled random hexamer or heptamer oligonucleotides and staining dyes have often been used to examine the quality of pre-arrayed slides. They allow visualization of array spots through fluorescence-intensity measurements that ensures both adequate probe density on each spot and spot morphology. Many pre-arrayed slide producers have adopted these methods in their internal quality control processes. While these methods seem to be useful, they also have limitations: the printed slides that were used in the QC process cannot continue to be used for hybridization. In other words, the printed slides become useless once they have gone through the QC process. Moreover, issues with sensitivity and effectiveness cause additional concerns for array QC by random hexamer or heptamer oligonucleotides. To overcome the shortcomings of the conventional QC methods, an alternative microarray QC method has been developed by the scientists at Full Moon BioSystems (Sunnyvale, CA), which is designed for users to check the quality of homemade arrays and/or commercial arrays with high sensitivity. A dye-labeled reagent allows the crosslink of dye molecules to the probes immobilized on slide surfaces. The formed chemical lineage can be removed sufficiently after the fluorescence reading of the slides. The advantage of this methodology is that the arrays can still be used in further hybridization even after they have gone through the QC process. It allows users to assess arrays' quality, such as spot morphology and consistency on attachment, and produces important information on targets' quality relevant to hybridization effectiveness.

5.3 Scanner Calibration and Validation

Microarray scanners can be characterized into two categories based on the detection systems involved: charged-coupled device (CCD) based, and photomultiplier tube (PMT) based. Usually the CCD-based scanners use a white-light source with filters to generate relatively narrow wavelength excitations, while the PMT-based scanners utilize lasers to excite the fluorophores. The most commonly used scanners, such as GenePix scanners from Molecular Devices (Sunnyvale, CA) and Agilent microarray scanner, are designed for detection of Cy3 (absorption at ca. 550 nm, emission at ca. 570 nm) and Cy5 dyes (absorption at ca. 650 nm, emission at ca. 670 nm) or Alexa Fluor® 555 (absorption at ca. 555 nm, emission at ca. 570 nm) and Alexa Fluor® 647 dyes (absorption at ca. 650 nm, emission at ca. 668 nm). The GenePix scanners are equipped with two solid-state diode lasers, a 532 nm green laser with a 575 ± 17.5 nm emission filter and a 635 nm red laser with a 670 ± 20 nm emission filter. The Agilent microarray scanner is slightly different, i.e., a 532 nm YAG laser and a 633 nm helium-neon laser, which are equipped with slightly different bandwidth emission filters relative to those for GenePix scanners. Like most fluorescence measuring systems, each scanner provides relative or arbitrary fluorescent readouts. Generally, the fluorescence signals include both undesired instrument-specific and desired sample-specific signal contributions. The undesired instrument-specific contribution can be corrected or minimized with suitable setup procedures.

A microarray-standards interest group that includes most major microarray scanner manufacturers and NIST has agreed on specifications for two uniformly coated glass standards ($<10\ \mu m$ coating on 1-mm thick glass slide) for day-to-day and instrument-to-instrument fluorescence response validation. One glass standard has a typical medium intensity, corresponding to about 500 fluorophores per μm^2, and another at a relatively low intensity, corresponding to about 0.5 fluorophores per μm^2. These standards have to be photostable with an intensity uniformity of 99% or better over a sub-area of 250 μm^2 when excited at 532 nm and 633 nm. The corrected emission spectra under these two wavelength excitations will be certified for the glass standards. NIST is currently identifying candidate materials.

The microarray scanner calibration slide from Full Moon BioSystems Inc. (Sunnyvale, CA) has been developed for quantitative performance evaluations of microarray scanners in terms of dynamic range, limit of detection, channel-to-channel cross talk, uniformity of the scanners, and laser stability [80]. Briefly, two separate blocks of array spots in dilution series of Cy3 and Cy5 fluorophores are spotted on a polymer-treated glass slide. Each block consists of 28 sets of two-fold dilutions of each fluorophore coupled with three sets of buffer blanks and one set of position marker. Each column contains 12 repeats of each fluorophore concentration expressed as fluorophores/μm^2. The highest and lowest concentrations are 1.47×10^5 fluorophores/μm^2 and 1.10×10^{-3} fluorophores/μm^2, respectively. This scanner calibration slide allows users to evaluate the PMT gain settings, a critical component of micro-array scanners. Under optimal PMT settings, users can maximize their scanner's linear dynamic range and optimize the subsequent interpretation of scanner data [81, 82].

When it is desired to determine an analyte concentration in a mixture using a spectrofluorimeter, for example, it is necessary to produce a standard working curve to establish a relationship between fluorescent readouts and analyte concentrations. Such a standard curve is generated by measuring a blank and a series of samples with known analyte concentrations serving as the standard solutions. Using the plotted standard curve in the linear region (fluorescence vs. analyte concentration, Fig. 5), it is possible to de-

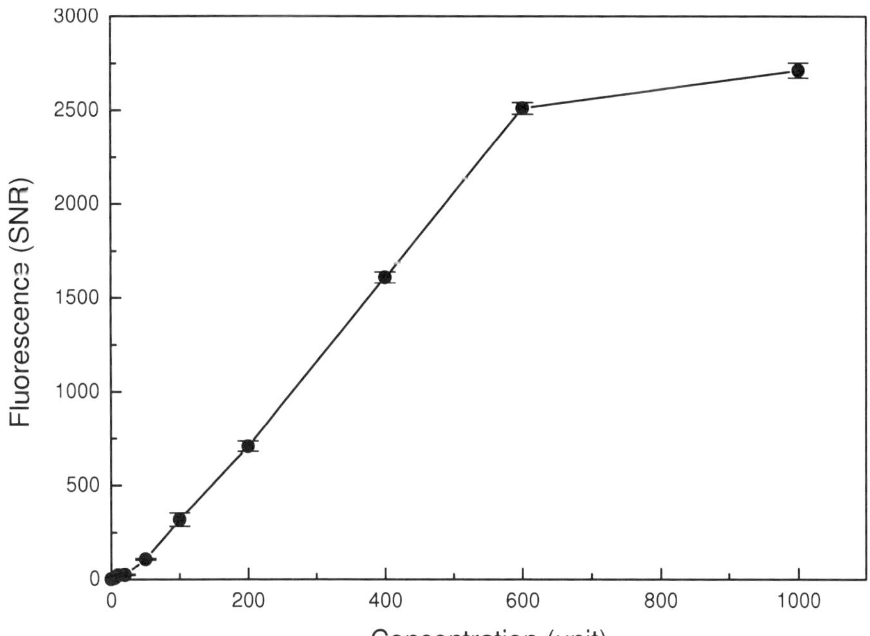

Fig. 5 A standard curve of fluorescence signal-to-noise ratio (SNR) vs. analyte concentration. SNR responds linearly at low analyte concentrations and reaches saturation at high concentrations. The *error bars* show the standard deviations

termine the analyte concentration in the mixture assuming that the analyte in the mixture experiences the same microenvironment as in the standard solutions and does not interact with other chemical substances. By allowing users to plot the relationships between the fluorescence readouts and dye concentrations, the Full Moon BioSystems' calibration slide also makes inter-system data comparison possible. Users can extrapolate the concentration of dye-labeled probes on array slides from the fluorescence readouts based on the calibration curves obtained for Cy3 and Cy5 fluorophores. The data (probe/dye concentrations) will be comparable regardless of where the data were acquired and which scanner platforms were used. Since different band-pass filters are used in different array scanners, we have further tested whether an emission spectral matching exists between the calibration slide and homemade oligonucleotide microarray slides. We found that the emission spectrum of Cy5 fluorophores on oligonucleotide array slides matches that on the calibration slide while the spectrum of Cy3 dyes is red-shifted by 5 nm with respect to that on array slides. The preliminary results also suggest that these calibration slides do photodegrade over time (a few months) with frequent usage. Users should use the slides following manufacturer's recommended procedures for usage and storage.

5.4 Reporting Data Standards

Standards such as Minimal Information About a Microarray Experiment (MIAME) [83] or MIAPE (MIAME's proteomics equivalent, MIAPE, currently in preparation) are recommended as reporting standards; however, they are not going to solve all sources of variability. Therefore, it is vital to maintain a compromise between detail and practicality in reporting data format so that compliance with the standards is not so tedious as to inhibit their adoption. A detailed description of each part and a convenient checklist are available on the MIAME Web site (http://www.mged.org/Workgroups/MIAME/miame_checklist.html).

6 Conclusions

There is no doubt that DNA microarrays will be a routine analytical technology in the near future. They have been applied broadly to address a wide variety of questions in basic and applied sciences. The applications of the microarray technology range from gene-expression profiling and genotyping to DNA–protein interactions and genome sequencing, and the list of applications keeps growing, especially when combined with other technologies such as proteomic technologies. However, many steps are involved in each microarray experiment and a number of microarray platforms exist, each with its unique features. The challenge is how to standardize the methods and materials to allow intra- and inter-comparison of microarray data collected in the same or different set of experiments. In this chapter, we have addressed the need for standardization for each step of the microarray experiments with emphasis on quality control of array fabrication and scanner calibration and verification. The proposed standards are designed for checking the quality of mRNA, fabricating slides, hybridization, and collecting, analyzing, and storing data. We believe that by providing standards for each step of the microarray process, we will help the microarray technology to realize its full potential.

References

1. Arabidopsis Genome Initiative (2000) Nature 408:796
2. The *C. elegans* sequencing consortium (1998) Science 282:2012
3. Goffeau A et al. (1997) Nature 387(suppl):5
4. Lander ES et al. (2001) Nature 409:860
5. Holt RA et al. (2002) Science 298:129
6. Waterston RH et al. (2002) Nature 420:520
7. Gibbs RA et al. (2004) Nature 428:493
8. Lindblad-Toh K et al. (2005) Nature 438:803
9. DeRisi JL, Iyer VR (1999) Curr Opin Oncol 11:76
10. Schena M, Shalon D, Davis RW, Brown PO (1995) Science 270:467
11. Hoheisel JD (2006) Nat Rev Genet 7:200
12. Marmur J, Doty P (1961) J Mol Biol 3:585
13. Lipshutz RJ, Fodor SP, Gingeras TR, Lockhart DJ (1999) Nat Genet 21:20
14. Steemers FJ, Gunderson KL (2005) Pharmacogenomics 6:777
15. Leiske DL, Karimpour-Fard A, Hume PS, Fairbanks BD, Gill RT (2006) BMC Genomics 7:72
16. Chiu SK, Hsu M, Ku WC, Tu CY, Tseng YT, Lau WK, Yan RY, Ma JT, Tzeng CM (2003) Biochem J 374:625
17. Le Berre V, Trevisiol E, Dagkessamanskaia A, Sokol S, Caminade AM, Majoral JP, Meunier B, Francois J (2003) Nucleic Acids Res 31:e88
18. Pirri G, Chiari M, Damin F, Meo A (2006) Anal Chem 78:3118
19. Hughes TR et al. (2001) Nat Biotechnol 19:342
20. Melov S, Hubbard A (2004) Sci Aging Knowledge Environ 2004(42):re7
21. Whitfield ML, George LK, Grant GD, Perou CM (2006) Nat Rev Cancer 6:99
22. Robson P (2004) Trends Biotechnol 22:609
23. Zou S, Meadows S, Sharp L, Jan LY, Jan YN (2000) Proc Natl Acad Sci USA 97:13726
24. Pletcher SD, Macdonald SJ, Marguerie R, Certa U, Stearns SC, Goldstein DB, Partridge L (2002) Curr Biol 12:712
25. Landis GN, Abdueva D, Skvortsov D, Yang J, Rabin BE, Carrick J, Tavare S, Tower J (2004) Proc Natl Acad Sci USA 101:7663
26. McCarroll SA, Murphy CT, Zou S, Pletcher SD, Chin CS, Jan YN, Kenyon C, Bargmann CI, Li H (2004) Nat Genet 36:197
27. Alizadeh AA et al. (2000) Nature 403:503

28. Ramaswamy S et al. (2001) Proc Natl Acad Sci USA 98:15149
29. Sorlie T et al. (2001) Proc Natl Acad Sci USA 98:10869
30. Tietjen I, Rihel JM, Cao Y, Koentges G, Zakhary L, Dulac C (2003) Neuron 38:161
31. Kruglyak L, Nickerson DA (2001) Nat Genet 27:234
32. Risch N, Merikangas K (1996) Science 273:1516
33. Hirschhorn JN, Daly MJ (2005) Nat Rev Genet 6:95
34. Kennedy GC (2003) Nat Biotechnol 21:1233
35. Matsuzaki H et al. (2004) Genome Res 14:414
36. Gunderson KL, Steemers FJ, Lee G, Mendoza LG, Chee MS (2005) Nat Genet 37:549
37. Gunderson KL, Kuhn KM, Steemers FJ, Ng P, Murray SS, Shen R (2006) Pharmacogenomics 7:641
38. Liu-Stratton Y, Roy S, Sen CK (2004) Toxicol Lett 150:29
39. Spielbauer B, Stahl F (2005) Mol Nutr Food Res 49:908
40. Kostrzynska M, Bachand A (2006) Can J Microbiol 52:1
41. Roy S, Sen CK (2006) Toxicology 221:128
42. Gerry NP, Witowski NE, Day J, Hammer RP, Barany G, Barany F (1999) J Mol Biol 292:251
43. Hashimoto M, Hupert ML, Murphy MC, Soper SA, Cheng YW, Barany F (2005) Anal Chem 77:3243
44. Huang R, Lin Y, Wang CC, Gano J, Lin B, Shi Q, Boynton A, Burke J, Huang RP (2002) Cancer Res 62:2806
45. Yuza Y, Agawa M, Matsuzaki M, Yamada H, Urashima M (2003) J Pediatr Hematol Oncol 25:715
46. Leiro J, Arranz JA, Yanez M, Ubeira FM, Sanmartin ML, Orallo F (2004) Int Immunopharmacol 4:763
47. Li W, Amri H, Huang H, Wu C, Papadopoulos V (2004) J Andrology 25:900
48. Bartling B, Hofmann HS, Boettger T, Hansen G, Burdach S, Silber RE, Simm A (2005) Lung Cancer 49:145
49. Pulai JI, Chen H, Im HJ, Kumar S, Hanning C, Hegde PS, Loeser RF (2005) J Immunol 174:5781
50. Tuomisto TT, Riekkinen MS, Viita H, Levonen AL, Yla-Herttuala S (2005) Atherosclerosis 180:283
51. Golub TR et al. (1999) Science 286:531
52. Baldwin AS (1996) Annu Rev Immuno 14:649
53. Radonjic M, Andrau JC, Lijnzaad P, Kemmeren P, Kockelkorn TT, Van Leenen D, Van Berkum NL, Holstege FC (2005) Mol Cell 18:171
54. Fan J, Zhan M, Shen J, Martindale JL, Yang X, Kawai T, Gorospe M (2006) Nucleic Acids Res 34:1492
55. Bains W, Smith GC (1988) J Theor Biol 135:303
56. Brenner S et al. (2000) Nat Biotechnol 18:630
57. Shendure J, Mitra RD, Varma C, Church GM (2004) Nat Rev Genet 5:335
58. Milner N, Mir KU, Southern EM (1997) Nat Biotechnol 15:537
59. Ziauddin J, Sabatini DM (2001) Nature 411:107
60. Xu CW (2002) Genome Res 12:482
61. Chang FH, Lee CH, Chen MT, Kuo CC, Chiang YL, Hang CY, Roffler S (2004) Nucleic Acids Res 32:e33
62. Narayanaswamy R, Niu WD, Scouras A, Hart GT, Davies J, Ellington AD, Iyer VR, Marcotte EM (2006) Genome Biol 7:R6
63. Neumann B, Held M, Liebel U, Erfle H, Rogers P, Pepperkok R, Ellenberg J (2006) Nat Methods 3:385

64. Kallioniemi OP, Wagner U, Kononen J, Sauter G (2001) Hum Mol Genet 10:657
65. Fedor HL, De Marzo AM (2005) Methods Mol Med 103:89
66. Abramovitz M, Leyland-Jones B (2006) Proteome Sci 4:5
67. Van Gelder RN, von Zastrow ME, Yool A, Dement WC, Barchas JD, Eberwine JH (1990) Proc Nat Acad Sci USA 87:1663
68. Zhong XB, Lizardi PM, Huang XH, Bray-Ward PL, Ward DC (2001) Proc Nat Acad Sci USA 98:3940
69. Nelson PT, Baldwin DA, Scearce LM, Oberholtzer JC, Tobias JW, Mourelatos Z (2004) Nat Methods 1:155
70. Kurimoto K, Yabuta Y, Ohinata Y, Ono Y, Uno KD, Yamada RG, Ueda HR, Saitou M (2006) Nucleic Acids Res 34:e42
71. van Bakel H, Holstege FC (2004) EMBO Rep 5:964
72. Yang I, Han M-S, Yim Y-H, Hwang E, Park S-R (2004) Anal Biochem 335:150
73. Donald CE, Stokes P, O'Conner G, Woolford AJ (2005) J Chromatogr B 817:173
74. Cronin M, Ghosh K, Sistare F, Quackenbush J, Vilker V, O'Connell C (2004) Clin Chem 50:14645
75. Baker SC et al. (2005) Nat Methods 2:731
76. Thompson KL (2005) Nucleic Acids Res 33:e187
77. Ivanova NB (2002) Science 298:601
78. Tan PK, Downey TJ, Spitznagel EL Jr, Xu P, Fu D, Dimitrov DS, Lempicki RA, Raaka BM, Cam MC (2003) Nucleic Acids Res 31:5676
79. MAQC Consortium (2006) Nat Biotechnol 24:1151
80. Zong Y, Wang Y, Zhang S, Shi Y (2003) In: Hardman G (ed) Microarrays Methods and Applications: Nuts & Bolts. DNA Press, Eagleville, p 99
81. Shi L et al. (2005) BMC Bioinformatics 6(2):S11
82. Resch-Genger U, Hoffmann K, Nietfeld W, Engel A, Neukammer J, Nitschke R, Ebert B, Macdonald R (2005) J Fluoresc 15:337
83. Brazma A et al. (2001) Nat Genet 29:365

Springer Ser Fluoresc (2008) 6: 239–263
DOI 10.1007/4243_2008_033

Published online: 4 March 2008

Comparability of Microarray Experiments from the Instrument and the Sample Site and Approaches Towards Standardization

Wilfried Nietfeld

Department Vertebrate Genomics, Max-Planck-Institute for Molecular Genetics, Ihnestr. 63–73, 14195 Berlin, Germany
Nietfeld@molgen.mpg.de

Abstract This chapter will give a short introduction into current technology related and biology related problems in the microarray technology, approaches undertaken to overcome such problems, and, finally, it will present some key features of most commercial platforms which have emerged over the past few years and may help to solve some technical problems at least related to the instrumentation side. Standard operating protocols, guidelines, as well as careful and routine checking of the instrumentation, such as spotting or hybridization devices as well as microarray reader (laser/charge-coupled device (CCD) camera), are necessary for running a microarray platform successfully. Furthermore, standardized reference material, databases, and analysis software will be necessary for generating confidential data across all platforms and laboratories and will additionally support the full power of DNA microarray technology as a research and diagnostic tool.

Keywords Bioinformatics · Fluorescence detection · Gene expression profiling · Genomics · Microarray · Molecular diagnostics · Standardization

Abbreviations

Cy3™/Cy5™	Cyanine Dye
CCD	Charge-Coupled Device
cDNA	Complementary Deoxyribonucleic Acid
DIG	Digoxigenine
DMD	Digital Micromirror Device

DNA	Deoxyribonucleic Acid
EPA	Environmental Protection Agency
ERCC	External RNA Control Consortium
EST	Expressed Sequence Tag
FDA	Food and Drug Administration
GE Healthcare	General Electric Healthcare
GO	Gene Ontology
He-Ne	Helium-Neon
Kr-Ar	Krypton-Argon
MAGE-ML	Microarray Gene Expression-Markup Language
MAQC	Microarray Quality Control Project
MFB	Measurements for Biotechnology
MGED	Microarray Gene Expression Data
μm	Micrometer
MIAME	Minimal Information About a Microarray Experiment
mRNA	Messenger Ribonucleic Acid
N	Nucleotide
NGFN	National Genome Research Network
NIA	National Institute on Aging
NIST	National Institute of Standards and Technology
nm	Nanometer
NMS	National Measurement System UK
ORF	Open Reading Frame
PCR	Polymerase Chain Reaction
PMT	Photomultiplier Tube
polyA	Polyadenylated
qRT-PCR	Quantitative Real Time Polymerase Chain Reaction
RNA	Ribonucleic Acid
SNP	Single Nucleotide Polymorphism
TIGR	The Institute for Genomic Research
™	Trademark
UK	United Kingdom
US	United States
YAG	Yttrium Aluminium Oxide

1 Introduction

For more than the last decade, the so-called biochips have become an increasingly important research tool for the biotechnology industry, molecular diagnostics, and related fields of gene expression analysis, drug screening, nucleic acid sequencing, and mutation analysis. In general, such biochips consist of a plurality of binding agents also called probes, for instance, peptides, oligonucleotides, or small molecules. Among these, DNA chips or DNA microarrays became the most important research tool. Up to ten thousands of single stranded DNA fragments representing a single gene or genome are synthesized or spotted in an array format with feature sizes in the range of 50 to 200 micrometer onto the surface of a solid support like a coated glass slide. Thus, researchers are enabled to analyze and monitor the expression levels of thousands of genes simultaneously. Applications allow (I) a genome-wide gene expression profiling by measuring relative differences in RNA transcript levels between biological samples, (II) the detection of single nucleotide polymorphisms (SNPs), and (III) sequence detection (e.g. pathogen detection) at reasonable time and costs compared to conventional techniques, e.g., northern blotting or RNase protection assays [1–9].

Since gene expression analysis using DNA microarrays is the major and the most critical application, the focus of this chapter is on this subject and general problems related to reproducibility, sensitivity, accuracy, and specificity. This chapter will give a short introduction to current technology related and biology related problems in the microarray technology, approaches that are undertaken to overcome such problems, and, finally, it will present key features of most commercial platforms which have emerged over the past few years and may help to solve some problems at least related to the instrumentation side. Standard operating protocols, guidelines, as well as careful and routine checking of all instruments, such as spotting devices, hybridization chamber, and microarray reader (laser/CCD camera), is necessary for running a microarray platform successfully. Furthermore, standardized reference material, databases, and analysis software will be necessary to generate confidential data across all platforms and laboratories and will additionally support the power of DNA microarray technology as a research and diagnostic tool.

2
General Considerations About Microarray Experiments

More than ten year ago, when DNA microarrays were first introduced, there was a huge euphoria about the power of such a research and diagnostic tool, because this technology platform enables biomedical researchers to analyze genome-wide gene expression levels in parallel by nucleic acid hybridization, it allow taking of "pictures" or "photographs" of the actual status of all gene activities in given cells or tissues, and it allows to identify trends or signatures and possible interactions in gene expression. Genome-wide alteration in gene expression can be easily measured and monitored in many biomedical applications, e.g., during the development of an organism or disease processes, etc. However, it is very difficult to obtain high quality and confident data from DNA microarray experiments from which valid biological conclusions can be drawn. The reason is the system inherent high complexity and of the physical properties of nucleic acid hybridization in complex mixtures of several thousands of different molecules with similar to complete different sequences. The labeled samples or targets to be hybridized have to match perfectly to the complementary probe (immobilized on the surface) in an overnight microarray hybridization reaction under distinct salt and temperature conditions. Thus, the high complexity of microarray experiments will have many sources of variability and errors which have to be controlled as well as possible during the whole process [10]. Therefore, important subjects and criteria for controlling are: microarray and sample quality, the labeling reactions and protocols, the efficacy of the hybridization conditions, the performance of the scanning instrument, the image acquisition and processing (grid finding and background subtraction), data normalization and analysis, quality assessment of data, and, finally, the biological interpretation of the obtained results, which will contribute to the overall uncertainty of the conclusions drawn from such experiments [11].

The comparison of DNA microarray results from apparently identical experiments, performed in different laboratories, or by different researcher, or even on different days may lead to different results. This is mainly due to a current lack of standards, standard operating protocols, and guidelines for performing and controlling the complete process of microarray experiments. The comparison to results obtained from a different microarray platform is even more challenging, because of a different probe and array design that may have been used [10–12]. The accurate measurement and quantification of absolute expression levels of all transcripts and a reliable detection of low abundance transcripts is difficult. It was shown in the past that the main problems were non-optimal design or choice of probes immobilized on the array surface. A comparison of DNA microarray results was difficult, because of technical differences that had some influences on the results: (I) the array format, e.g., incorrect probe annotations, (II) differences in the probe length (from 25–70 mers or complete cDNA fragments) and nucleotide composition, and (III) the spotting technology (pin, ink-jet spotting, or in situ synthesis). Therefore, it is recommended to validate the obtained results by a second independent method, such as quantitative real-time PCR (qRT-PCR). Due to the additional cost and time associated with independent verifications, in most studies only a limited number of transcripts have been verified, typically <20 genes [12].

The identification of regulators, which determine complex biological differentiation processes or phenotypes, e.g., in healthy or disease processes by using results from gene expression profiling, requires microarray containing probes of the whole genome together with controls for a reasonable reproducibility, accuracy, and sensitivity in the employed technology. These criteria are not that critical if microarrays are used only for screening purposes and a second independent verification platform, such as quantitative real-time PCR (qRT-PCR), is used. For diagnostic application and molecular disease classification the requirements for microarray have to be much more stringent, because regulatory agencies need solid and confident data on sensitivity, accuracy, specificity, and reproducibility of such a technique [10, 11].

The sensitivity threshold or detection limit of the current DNA microarray technology allows to detect between one and ten copies of a transcript per cell, and is probably lower for cells grown in cultures than in more heterogeneous tissues. This numbers might still be insufficient to detect alteration in gene expression of low abundant genes, such as transcription factors [13, 14]. Even an increase in gene expression by more than 100 percent may be difficult to detect by the currently used fluorescence techniques. Novel labeling technologies, such as labeling with nanoparticels or quantum dots [15], may further increase the sensitivity of the DNA microarray expression platforms. Estimating absolute cellular concentrations of transcripts requires a careful calibration of the microarray platform with known concentrations of the transcripts. DNA microarrays are being used to measure either absolute transcript abundance (i.e., signal intensities, single color) or relative transcript abundance (i.e., signal intensity ratios/dual color); For comparisons, see Fig. 1. As an example of both approaches, image files of the

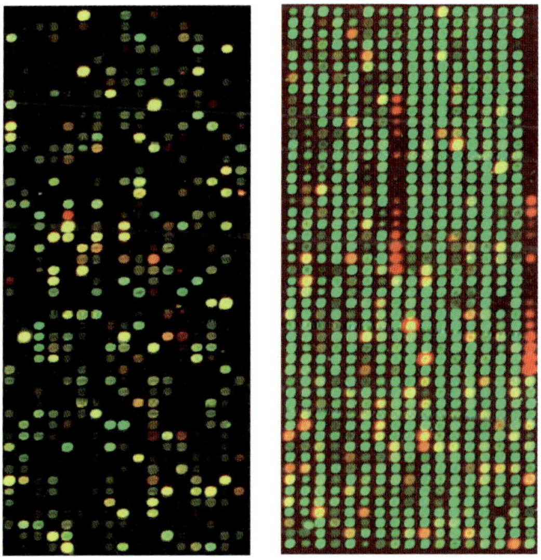

A B

Fig. 1 Image files of two different experimental designs enabling the comparison of gene expression between different samples. For direct or ratio measurement of differences in gene expression (**A**), RNA samples from two different conditions of *Arabidopsis thaliana* were labeled either with Cy3™ or Cy5™ and hybridized to an Arabidopsis microarray, respectively. For indirect or absolute measurement (**B**), one RNA sample to be analyzed was labeled with Cy5™, and the sample was co-hybridized with a Cy3™-labeled reference oligonucleotide, and measured as described before. The microarrays were scanned at excitation wavelengths of 532 and 635 nm, respectively, and read out with the respective green and red channel with a resolution of 10 μm using a Fuji FLA 8000 microarray scanner. Only sections of the corresponding images are shown two different experimental designs measuring direct (Fig. 1A) and indirect (Fig. 1B) gene expression values of the plant *Arabidopsis thaliana* during development are shown. For direct or ratio measurement, targets from two different growth stages were labeled either with Cy3™ or Cy5™ and hybridized to an *Arabidopsis* microarray. For indirect measurement of gene expression values, the target sample is labeled either with Cy5™ or Cy3™, and a conversely labeled reference oligonucleotide, which is complementary to the PCR primers used for probe generation, is spiked into the hybridization solution and cohybridized [16]. Assessing the accuracy of DNA microarray measurements requires the knowledge or availability of accurate concentrations for as many transcripts as possible. Accurate concentrations can be obtained by either spike-in transcripts or dilution experiments and measuring of transcript levels by independent techniques, such as qRT-PCR or Northern-blot analysis [17, 18]. Therefore, the estimation of ratios does not require detailed information about how the signal intensity of a given microarray probe is related to the concentration of transcript to be measured. If a probe binds specifically to its target and the produced signal intensity is proportional to the amount of transcripts hybridized, the expression ratios will reflect the relative abundance of the transcripts in the complex hybridization mixture. The reproducibility of today's DNA microarray experiments is the most readily assessable characteristic of this technology platform. Unfortunately, a technology platform can have excellent reproducibility but without necessarily producing measurements that are accurate or consistent with other platforms. This is because the reproducibility only requires that a given probe binds to the same number of labeled transcripts in repeated measurements of the same sample. Probes that are badly designed (e.g., in case of large gene family member) or where no sequence information is available (e.g., ESTs) may cross-hybridize with several other transcripts and can easily lead to highly reproducible data, but these data are useless. Therefore, the reproducibility of microarray experiments is a necessary but insufficient requirement.

For typical DNA microarray experiments with fluorescence detection (for illustration, see Chap 25; Figs. 2 and 5; Zou et al., 2008, this volume), RNA or DNA from biological samples and from control or reference material is isolated and labeled with two spectrally distinguishable dyes, such as the mainly used cyanine dyes Cy3™ (absorption at ca. 550 nm, emission at ca. 565 nm) and Cy5™ (absorption at ca. 650 nm, emission at ca. 675 nm), respectively, during a reverse transcription reaction which transcribes RNA into cDNA. The labeled molecules are then reacted or hybridized for about 16 hours with the microarray, thereby binding to their complementary sequences, which is present in the immobilized probes. After hybridization and processing, the microarrays are read out at least at two wavelengths with a microarray reader, typically a confocal laser scanner equipped with a He-Ne laser (543, 594, 612, and 632 nm) or a Kr-Ar laser (488, 514, 568, and 647 nm), or less commonly, a solid-state laser or a white light source and a PMT, or a CCD detection system with a lateral resolution of 5 to 10 micrometers, respectively. The scanning step involves excitation at two different wavelengths and subsequent recording of integral fluorescence intensities in two detection channels, typically the so-called "green" and "red" channels. The recorded signal intensities are then used for further statistical and bioinformatical analysis as well as for the biological interpretation. Usually, the scanner software is also able to combine signal intensities of both channels and to deliver a false-color image. The amount of transcript bound to a spot follows from the measured integral fluorescence intensity with abundant targets seen either in red or green and equal amounts in yellow. This enables the estimation of relative expression levels of the genes from the fluorescence intensities of the spots. For generating reliable gene expression data, several statistically analyzed and normalized repetitions and control experiments are necessary. Further bioinformatical analysis, e.g., hierarchical agglomerative clustering of the gene expression data, provides the basis for the interpretation of the alteration in gene expression of many genes in parallel [1–4].

DNA microarray experiments are traditionally performed either by ratio or direct measurement of transcript concentrations comparing two RNA samples, e.g., healthy and disease stage, or by absolute or indirect measurements comparing one RNA sample to a reference, e.g., standardized RNA or DNA, which yield either relative expression values (ratios) or absolute expression values (Fig. 1A,B, respectively). In the past, two-color (e.g., Cy3™ and Cy5™ dyes) cDNA microarrays (e.g., Agilent or self-spotted) were usually used to measure ratios, whereas single-color oligonucleotide microarrays (e.g., Affymetrix) are intended to measure absolute expression values. Two-color experiments allow a less detailed understanding of how the signal intensity of given microarray probes are related to the concentration of the measured transcript. In this case, it is important that the targets hybridize specifically to its probes, and the produced signal intensities are proportional to the concentration of transcripts.

Since reproducibility, sensitivity, accuracy, and specificity are important criteria for microarray experiments, the validation of microarray experiments by an independent technique is also crucial. Only a few published studies have performed independent validation by RT-PCR for a more comprehensive set of transcripts from about 50 to 1400 [13, 14, 18]. Real-time PCR, or so-called TaqMan Assays, provide a medium throughput technology for validation. Applied Biosystems (www.appliedbiosystems.com) has developed a microfluid card system in a 384-well format (TaqMan® Low Density Arrays), which can be individually configured, and it allows a quick and easy validation of differentially expressed genes and appropriate controls by TaqMan Assays. The Microarray Quality Control Project (MAQC) is currently producing the first comprehensive data set for human RNA, containing about 1000 genes (www.fda.gov/nctr/science/centers/toxicoinformatics/maqc/index.htm). The conclusions that can be drawn are: (I) microarray measurements reflect the existence and direction of expression changes in about 70–90% of the investigated genes, but the magnitude of microarrays tends to be different from the magnitude of other technologies, e.g., qRT-PCR, (II) microarrays tend to measure ratios more accurately than absolute expression levels, and (III) the relatively good correlation between microarray and qRT-PCR data does not necessarily mean that microarrays produce accurate estimates of gene expression ratios. Nevertheless, independent validation and spike-in results could show an empirical assessment of the accuracy of DNA microarray technology. An accurate measurement of absolute transcript levels by microarrays is currently not possible, whereas ratios can be estimated reasonably well.

If experimental data from microarray experiments were highly reproducible across various platforms and laboratories, then one could use platform-independent appropriately normalized gene expression data. This could then reduce the need to replicate experiments and would enable researchers to build universal gene-expression databases that could be used as a reference. This consideration is particularly relevant for patient material with limited amounts of mRNA. In principle, microarrays can be applied to clinical samples and have a high potency of a diagnostic tool; the application will depend on several factors, e.g., number of genes to be assayed and more importantly, whether an accurate, robust, and platform-independent quantification of the appropriate transcript level is achieved. Additionally, we have to find out whether the variation of gene expression in a certain range (higher or lower transcript turnover rates) might not have any physiological consequences and whether it is difficult to interpret. Therefore, there is a need for establishing gene expression databases to collect and store information about gene expression levels to be able to later define the range of variability in expression for every gene or transcript.

The uncertainty and reliability of the results obtained from DNA microarray experiments depend on the biological and experimental variation, as well as, to a lesser extent, on fluorescence detection techniques and data analysis. Furthermore, due to the lack of generally accepted standardized sampling protocols and biological standards to be used in every microarray experiment, problems related to biological conditions and data analysis are enhanced [5, 7]. In general, the contribution to the overall uncertainty from the biological side has the highest impact, due to genetic and environmental variation, which is difficult to measure. Fluorescence-inherent sources of error can be readily identified only if the necessary controls and standards are included in the experiment and the scanners are properly characterized. Fluorescence-based uncertainties are related to fluorophore labeling of the target [6, 19] and to the previously discussed general problems linked to measurements of fluorescence intensities, i.e., the sensitivity of the label's spectroscopic properties to chromophore microenvironment [20, 21] and to dye–dye interactions [22], as well as to instrument-specific effects. To overcome limitations in the former, e.g., different incorporation rates of the dyes or dye instability, the so-called dye swap or flour-reverse, is used with both targets being labeled with two dyes in separate reactions and hybridized to two identical microarrays with identical probe arrangements. Additional microarray scanner-specific uncertainties are related to the determination of the background signal, the method used for dye normalization, crosstalk between channels, and variations in the alignment of the optical scanner [23]. For example, for confocal reading, which is used to increase the sensitivity by rejecting light coming from other planes rather than that of the biological signals of interest, there is a strong need for accuracy of the focus and tilt adjustment of the biochip. Typically, such detection systems are less sensitive in the long wavelength region of about 670 nm as compared to a detection wavelength of about 550 nm.

3 Sources of Inconsistency in Microarray Measurements

As a first approximation, signals produced by any given microarray probe or spot can be considered as derived from sources of mainly three different types: (I) specific signals produced by the hybridized targeted fluorescently labeled transcript, (II) cross-hybridization signals produced by labeled transcripts having a significant sequence similarity, and (III) a non-specific, background signal independent of any significant sequence similarity, e.g., dust particles or contamination. Ideally, a high-specificity microarray platform would only have marginal influences on the last two sources, which in its current state is not achieved mainly due to several circumstances. In general, the probe is designed to perfectly match a complementary region of the target RNA and capture a certain number of targets proportional to its concentration. The affinity is depending on the conditions of hybridization, the sequence composition, and competition with similar sequence motifs. Nevertheless, until today, we have only had a very limited understanding of the physical properties of the microarray hybridization process, including kinetics of DNA/DNA or DNA/RNA hybridization, influences coming from sequence composition, salt or target concentration, accessibility of the probe to its target, and the diffusion of labeled transcripts in a complex hybridization solution. Signal intensities and microarray results are furthermore influenced by the presence of splice variants, which may have a high degree of homology to the probe. Since it is estimated that about 50 percent of all human genes might have splice variants, it is difficult to detect and discriminate those in a microarray experiment if the probe design is not adequately done and the complete probe sequence is not available. The overall impact of cross-hybridizations as well as a possible folding of the target transcripts is also poorly understood, but it also contributes to the variation between different probes targeting the same region of a given RNA molecule [11].

There is no doubt about the potential of DNA microarray technology to provide a powerful research and screening tool for measuring genome-wide alteration in gene expression of many genes in parallel and for detecting differentially expressed genes in physiological processes. However, some limitations of the current technology platforms, both commercial and academic, exist and have become more apparent during the past few years. The current technology status reliably allows to detect alteration in gene expression for the majority of genes. However, accurate measurements of absolute expression levels and the reliable detection of low abundance genes are still critical and strictly need the validation by different techniques, e.g., qRT-PCR. Further investigation on kinetics of nucleic acid hybridization, technological advances in hybridization and detection methods, and the standardization of the whole process of microarray experiments will certainly increase the measurement capabilities of DNA microarrays.

4 Approaches Towards Standardization

DNA microarrays definitely represent a powerful core technology in all fields of genome research. Provided that standards and quality measures are met, this technology can successfully and reliably be used in biomedical and clinical application. Currently, there is still a strong need for a better understanding of the quality of results obtained from microarray experiments. In the past, the inability to establish quality criteria has led to a relatively poor confidence in microarray results, supported by difficulties in assessing the agreement of different experiments and conflicting reports in the literature. The introduction of an international convention with standard operating protocols and guidelines will enhance research applications in life sciences, and will allow establishing gene expression profiling by DNA microarray in regulated applications, such as pharmacogenomics, toxicogenomics, and diagnostics [10, 11].

For the standardization of DNA microarray experiments, it is necessary to introduce suitable controls, which are incorporated into every experiment and are accordingly present on the array and spiked into the labeling reaction and hybridization solution, including dynamic range, ratio, negative, and contamination controls, as well as positive controls and guide-dots for orientation after scanning. Figure 2 illustrates a possible design for DNA microarrays, which allows the assessment of the quality of the experiment after hybridization and the comparison to other experiments. During fabrication of the microarrays, the ScoreCard controls (www.amershambiosciences.com), negative controls, e.g., spotting buffer or probe from different species, as well as positive and orientation controls, e.g., actin, etc., were co-spotted. Most commercial platforms and approaches already include universal reference

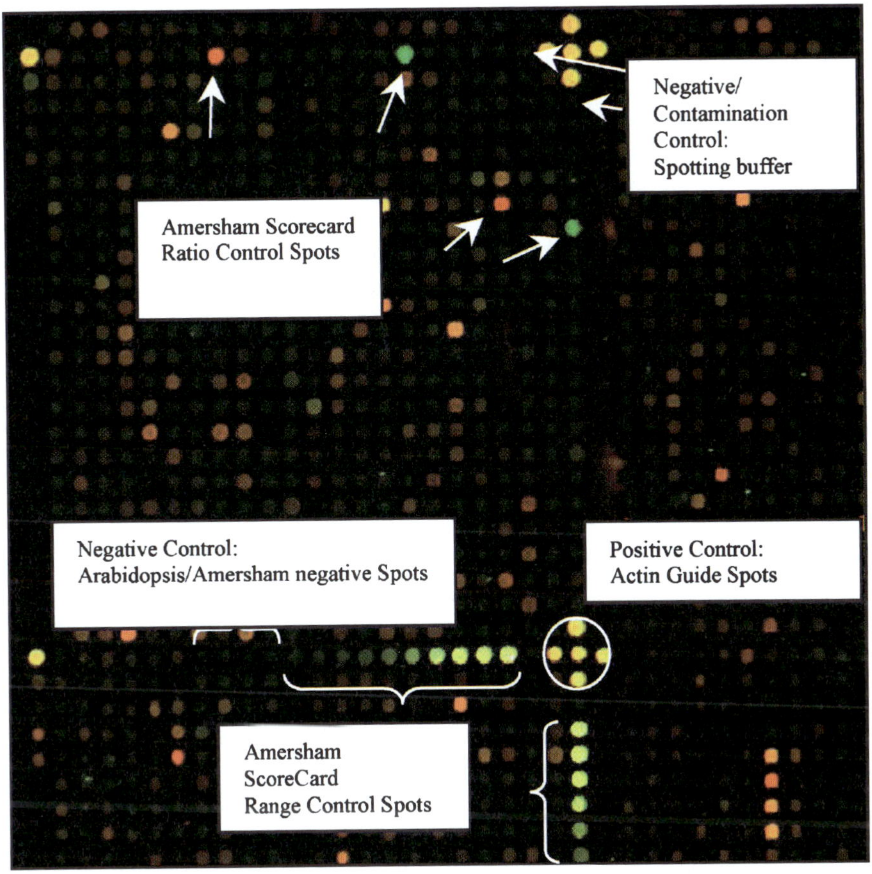

Fig. 2 DNA microarray design for gene expression profiling. Together with gene specific long oligonucleotides (60 mers) negative and contamination controls, e.g., spotting buffer alone; the ScoreCard system (www.amershambiosciences.com) providing ratio, negative and range controls, as well as actin probes for positive controls and orientation were co-spotted during fabrication of the microarrays, which allows a quality assessment after hybridization and scanning. RNAs of two different cell lines were labeled either with Cy3™ or Cy5™ and hybridized to the array. The microarrays were scanned as described in Fig. 1

samples and spike-in controls. The universal reference sample, which can be included in the hybridization solution, is intended to measure fluorescent signals across all probes immobilized on a microarray surface, which after the hybridization process leads a uniformly base level of fluorescence intensity (see also Fig. 1A,B). For the data analysis, this intensity level can be used for comparison to the co-hybridized differentially labeled transcripts. The use of universal reference materials will allow comparing data across different experiments, technology platforms, and laboratories, whereas spike-in controls are used for monitoring the experimental performance of the sample preparation in a given platform (Fig. 2).

In many cases pooled RNA samples (derived from cell lines) and reference oligonucleotides or genomic DNAs are used as the reference sample [16, 25–30]. The disadvantage of pooled RNA samples is that the relative expression of transcripts within and between cell lines may vary from batch to batch, some transcripts may be unstable, or it may be difficult to measure the absolute quantity. On the other hand, RNA reference samples can be included in the labeling reaction, allowing to measure labeling efficiency in parallel. Alternatively, the application of labeled oligonucleotides, complementary to at least part of the probe sequences, can be used for calibration [16]. This is applicable for probes which have been amplified by PCR with "universal" primer prior to immobilization or if random-mers are used (e.g., N15–N20 mers). Oligonucleotides are stable, easy to quantify, and provide uniform signals across the microarrays. The advantage of universal DNA reference samples derived from genomic DNA is that genomic DNA is readily available, relatively stable, and also leads to reasonable and uniform signals across different sequences. The main application of spike-in controls, which are normally exogenous RNA transcripts, is to monitor the labeling efficiency when added to the RNA samples prior to reverse transcription and labeling (cDNA synthesis). At the moment, the application of such universal reference materials and spike-in controls offers the possibility for limited standardization, because they are not universally applicable and cannot be traced back to any primary standard. To help realize the full potential of microarrays, development of reference materials, analytical guidelines, and standardized approaches to experimental design are required [31–33].

A series of national and international working groups and initiatives have been founded during the last years in order to develop universal reference materials, analytical guidelines, and standardized operating protocols or guidelines for the standardization of performing microarray experiments.

The working group "Microarrays", within the quality management of the German National Genome Research Network (NGFN) (www.ngfn.de), provides quality guidelines and protocols to their partners, which includes standardized protocols for the sample preparation and analysis, the production and control of spotted microarrays, as well as the data analysis and storage.

The Gene Expression Units Working Group at the Measurements for Biotechnology (MFB) (www.mfbprog.org.uk), a program supported by the National Measurement System UK (NMS) was established as part of an initiative to develop a standard unit for measuring gene expression. The objective of the working group is to recommend approaches for better standardization of procedures used in gene expression measurements through discussions and consultations. Practical assessments of recommended approaches and the development of associated methodologies, standards, and reference materials are among the aims of this work.

The External RNA Control Consortium (ERCC) is a working group that was initiated in December 2003 and is supported by about 70 partners from different private, public, academic, and regulatory organizations (www.cstl.nist.gov/biotech/Cell&TissueMeasurements/GeneExpression/ERCC.htm). The major goal is to develop a set of universal reference materials, containing external RNAs or control transcripts, which can be applied to assess the technical performance of DNA microarrays in gene expression profiling experiments. The rationale is to add such a control to a sample RNA after isolation and prior to the cDNA synthesis, which enables the researcher to evaluate whether the results are consistent with the defined performance parameter or criteria, or not. Over the past years, the ERCC has refined specifications, generated and collected control sequences, evaluated optimal polyadenylated (polyA) tail length, and identified a path forward for access and distribution of the controls. A testing phase has been initiated by the ERCC and the reference RNA controls will be evaluated in one-color and two-color microarray experiments as well as qRT-PCR. The working group also developed a specification document that was discussed and refined in a public workshop at the National Institute of Standards and Technology (NIST). Protocols for the use of external RNA controls in clinical applications have been included in the Molecular Methods 16-P document from the Clinical and Laboratory Standards Institute, and were developed in a formal, accredited, open consensus forum, which included several ERCC members. In June 2004, in a public workshop at NIST, the analysis approach was developed and is based upon the measurement of pooled transcripts at known concentrations.

The National Institute of Standards and Technology (NIST) developed a program and guidelines for standardizing instrumentation and other technical infrastructure, which is required for gene expression profiling experiments with DNA microarrays (www.cstl.nist.gov/biotech/Cell&TissueMeasurements/GeneExpression.htm). The main goal of this program is to enable measurements (science, standards, data, and models) of known quality for microarray gene expression results. This program also includes the development of approaches and standards for experimental validation and measuring microarray performance (including microarray reader and data analysis) within and between different laboratories and different platform technologies.

Recently, the US Food and Drug Administration (FDA) introduced the so-called MicroArray Quality Control (MAQC) project with the aim to provide quality control tools to the microarray community, in order to avoid procedural failures and to develop guidelines for microarray data analysis by providing the public with large reference datasets along with readily accessible reference RNA samples (www.fda.gov/nctr/science/centers/toxicoinformatics/maqc/index.htm). The MAQC project involves six FDA centers, major providers of microarray technology platforms and RNA samples, EPA, NIST, academic laboratories, and others. The project aims to establish quality control (QC) metrics and thresholds for assessing the performance achievable by various microarray technology platforms and evaluating the advantages and disadvantages of various data analysis methods. It has been decided that two RNA samples will be selected for three species, human, rat, and mouse, and differential gene expression levels between the two samples will be calibrated with microarrays and other technologies, such as qRT-PCR. Then the resulting experimental datasets will be used for assessing the precision and cross-platform/laboratory comparability of DNA microarrays, and the qRT-PCR datasets will allow the researcher to evaluate the nature and magnitude of any systematic biases which may exist between different microarray platforms and qRT-PCR. The MAQC project will support researchers to improve the microarray technology platform and foster its proper applications in discovery, development, and review of FDA regulated products. The MAQC main study of data generation was completed in October 2005, and it is expected to provide guidance on microarray quality control and data analysis by the end of 2007.

Other initiatives and working groups that have been initiated over the past years to develop and share common standards and guidelines for gene expression analysis, focusing on the bioinformatics side and computational analysis of microarray results [34, 35]. Including the Microarray Gene Expression Data (MGED) working group, which has produced a guideline for the minimal information about a microarray experiment (MIAME) to help improve comparability of microarray data and associated experimental information [36] and the MAGE-ML mark-up language [37]. Adherence to these guidelines, together with submission of data to public repositories or gene expression databases, such as ArrayExpress [38], will undoubtedly help drive forward the process of standardization in the field of microarrays. For storage of the microarray data, several databases have been established following the MIAME guidelines [39].

To make microarray data comparable worldwide, there is also a strong need for instrument standardization, e.g., for spotting, hybridization, and scanning. For the improvement of the reliability of microarray experiments from the detection side, at present, there exist only very few instrument-type standards. To improve this situation, better suited calibration slides, which can be used for every microarray scanner or calibration spots immobilized on every microarray, are desired to help adjust the scanner settings and to determine the instrument's image/spatial resolution, as well as its day-to-day performance and long-term stability (e.g., of the laser excitation source – ideally in combination with guidelines for their use). Furthermore, similarly to other fluorescence techniques, tested procedures in combination with recommended materials/systems are needed for the determination of the range of linearity of the array scanners and their sensitivity. The final goal is to improve the comparability of data generated by different instruments/laboratories and with different labels. Easy-to-use fluorescence standards designed for the comparability of fluorescence readings across instruments and laboratories should, for instance, either provide a stable output at application-relevant wavelengths and photon fluxes, or need to be easily reproduced, should have minimum local and global non-uniformities in emission, and should generate signals comparable to those observed for typical samples. In addition, the influence of spectral deviations between the fluorescence spectra of the standard and common labels needs to be examined. This renders the availability of absorption and corrected emission spectra of both the standard and the labels for application-relevant conditions, i.e., microenvironments, important. For testing of the instrument's spatial resolution, which should be in the 5-micrometer range, uniformly patterned standards are desired. Such standards must be applicable for the majority of typically used microarray readers and labels, e.g., suited for typical excitation and emission wavelengths and scanner formats. Furthermore, they should have been tested with a broad variety of instruments in a Round Robin test. With respect to the development of purpose-fit fluorescence standards, the input, not only from instrument manufacturers, but also from the community of users of the microarray technology, is required. Additionally, future technological advances of microarray readers should be considered, which would most likely provide an enhanced spatial resolution, precision, and sensitivity, as well as a higher dynamic range. Furthermore, within the community of users of the microarray technology and instrument manufacturers, it needs to be decided on whether stand-alone fluorescence standards, such as calibration slides, or the integration of fluorescence standards into slides used for the performance of microarray experiments are to be favored. The former is, for instance, sufficient to characterize the day-to-day and long-term instrument performance, whereas the latter approach – though more costly – may be tempting for an improved and more feasible comparability of the recorded fluorescence signals.

5
Comparability of Microarray Experiments

In the beginning of DNA microarray technology, the probe sequence information was very often not available or even wrong, due to the fact that complete high quality genome sequences were lacking and the use of cDNA or EST clones representing genes may have had only similar sequences. In the case of EST clones, the complete sequence of a particular clone was generally unknown and – what is more important – was not selected to yield optimal hybridization results specific to a single gene. Furthermore, ESTs only represented about half the genes identified in model species genomes. Nowadays, the situation seems to have changed, since complete genome sequences are becoming available, thus, allowing a better gene prediction and annotation of genes, and open reading frames (ORFs). Computational analysis and sequence comparison of complete genome enable a much better probe design, although discrepancies between a predicted and actually synthesized probe may not be absolutely correct and are, therefore, batch specific. Alteration in probe length and composition will have an influence on the hybridization signals after the experiment. This has to be taken into account for DNA microarrays were the synthesis is performed directly on the array surface (Agilent, Affymetrix, CombiMatrix, Illumina, NimblGen, etc.). However, probes synthesized and purified prior to spotting contain an almost homogeneous sequence population, but due to the printing, some spotting artifacts, such as inconsistent feature shape and size, may occur (Agilent (cDNA), GE Healthcare, Operon, etc.). Still, the main concern is the biological variation, i.e., growth conditions, sampling, etc. of biological sample may be different in Europe, United States, or any other place in the world. Therefore, it is also necessary to measure cross-platform as well as cross-laboratory variation not only on the microarray technology platform.

5.1 Cross-Platform Comparison

Since the introduction of the DNA microarray technology as a research tool for genome-wide gene expression profiling, there have been many concerns about the accuracy, precision, specificity, and reproducibility of the obtained experimental results. Indeed, in earlier DNA microarray studies, which were performed in different laboratories and have applied different microarray technology platforms in order to study the same or similar biological questions, the reproducibility of the obtained results was very poor and inconsistent. Kuo and coworker [40] analyzed DNA microarray data from gene expression profiling experiments on NCI60 cell lines, which were performed in their lab using the Affymetrix GeneChip platform and a spotted cDNA microarray platform used by Ross and colleagues [41]. By comparing both, the ratios and the spot intensities from the cDNA to the Affymetrix platform, they found a very poor correlation between the obtained results on differentially expressed genes. Their suggestion and conclusion was that probe-specific factors most probably influenced the results in the two platforms differently.

By using three different commercial microarray platforms (Affymetrix GeneChips, Agilent cDNA arrays, and GE Healthcare CodeLink), Tan and coworker [42] investigated the reproducibility of microarray data by hybridizing identical RNA preparations to this commercially available microarrays. By a careful analysis of the results, they found only little overlap between the lists of genes that showed significant alteration in gene expression across the used microarray platforms. Their results were in contrast to an even broader study published by Yauk and colleagues, who used six different microarray platform including two cDNA, three short, and one long oligonucleotide microarray types [43]. They compared gene expression profiles from mouse lung tissue to an immortalized lung cell line all against a universal reference RNA standard. The aim of this study was to evaluate the contribution of technical and biological impacts as well as their contribution to the quality, reproducibility, and sensitivity to the measured results by each of the platforms, and, finally, determine the correlation among these microarray technologies. Their results depict a much more positive image on microarray technologies. They show a low level of technical variability, which increased the ability to detect differentially expressed genes. They show that these differences are much more dependent on biological variation rather then on technical strings [43].

In May 2005, three additional separate studies evaluating DNA microarray technology platforms were published. They showed that the reproducibility across microarray platforms is much better than previously shown. Here, the researcher implemented carefully designed and controlled experiments using standardized protocols and data analyses [44–46]. Larkin and colleagues reported the results of a carefully designed investigation to examine gene expression changes in mouse heart in response to treatment with angiotensin II – a model for hypertension, using Affymetrix GeneChips and spotted cDNA arrays [44]. Employing standardized protocols and procedures, including the data analysis, they were able to show that for 5853 genes that were measured by both microarray platforms, in 88% of the genes the microarray platform used had no significant effect on the gene expression levels observed. This study also demonstrates that microarray measurement of gene expression can be a robust technique. Again a careful attention to the experimental protocols, data collection, and data analysis is required, but also a consistent annotation of the genes analyzed as well as the assessment of experimental quality criteria in needed. If such careful attention is paid to these criteria, the authors have shown that, for the majority of genes, the expression is independent of the platform used and biological effects have a much greater impact than platform effects [44].

Two studies which go one step further were published by Irizarry and colleagues [45] and Bammler and colleagues [46]; a comparison not only across microarray platforms but also across different laboratories was accomplished. The study of Irizarry and coworker established a consortium of ten laboratories from the area of Washington, DC-Baltimore (US), comparing data generated from three widely used platforms – Affymetrix GeneChips, spotted cDNA arrays and spotted long oligonucleotide arrays – using identical RNA samples. They determined that there were sometimes large differences between laboratories, even with the same platform, but that data from the best-performing labs (where data are reproducible within the lab) agree with each other rather well. In general, they discovered that the lab had a larger effect than the platform and that the results from the best-performing labs agreed rather well, e.g., on precision. These results suggest that it is not an inherent problem of the microarray platform per se, but rather with the application of the technology, such that data can be reproducible across both labs and platforms when good techniques are employed. Their conclusion was that improved quality assessment standards are needed and that precision assessments based on comparisons of technical replicates have to be standard operating procedures in gene expression profiling experiments [45].

This subject is illustrated even further by Bammler and colleagues, representing the Toxicogenomics Research Consortium (http://www.niehs.nih.gov/dert/trc/home.htm), who compared two standard RNA samples by using in total 12 microarray platforms in seven laboratories, and at least two standard microarray types (spotted or commercial) were used by all laboratories [46]. The consortium run comparisons using commercial mouse oligonucleotide microarrays from Affymetrix, Agilent, and Amersham, as well as spotted oligonucleotide microarrays from Compugen and Operon, and spotted cDNA microarrays from two other sources (TIGR: www.tigr.org/ and NIA: www.nia.nih.gov/). In this study, they examined the cross-platform reproducibility and the bias, which is introduced when different laboratories used the same platform by analyzing aliquots from the same RNA sample. Their initial results from RNA expression experiments showed that, although reproducibility for a platform within a single laboratory was typically good, reproducibility between platforms and across laboratories was generally poor. For about 500 genes that were present on all chosen microarray platforms the cross-platform consistency variation was between 0.11 for CodeLink vs. spotted cDNA microarrays and 0.76 for two different spotted cDNA microarray platforms. When two different laboratories used the same platform, the correlation increased and the Affymetrix platform produced by far the greatest correlation (0.91) across laboratories. However, the reproducibility of microarray results between laboratories increased when standardized protocols were implemented through the whole process: for RNA labeling, hybridization, microarray processing, data acquisition, and data normalization. The very good comparability was obtained when technical variables were standardized as much as possible. The authors also present that comparable biological themes emerged from data across different platforms and laboratories when Gene Ontology (GO) nodes are used for analyzing groups or cluster of genes representing biological themes. Important conclusions that can be drawn from this experiments are: (I) a careful assessment of the technology platform and experimental (standardized) protocols is necessary; (II) the identification of differentially expressed genes and biological pathways needs an external validation; (III) the generalizability of gene expression studies can be limited between independent laboratories and across platforms; (IV) although common gene ontology (GO) categories across laboratories and platforms were found, differences are also found and may lead to an over interpretation of the obtained results, and (V) the creation of gene expression databases incorporating results from multiple laboratories that have applied developed experimental standards and data filtering will be very helpful.

Taken together, this recently published manuscripts [44–46] show that it is possible to perform DNA microarray experiments, which are reproducible between different laboratories and technology platforms, provided that standardized protocols and methodologies are applied during the whole process of microarray experiments. Discrepancies in gene expression of the same gene which has been measured may be due to the different microarray types, which measure different splice variants of the same gene. To analyze this effect, it will be necessary to get access to the probe sequences, which, for most commercial microarrays containing short or long oligonucleotides, is no problem. In case of cDNA microarrays, this is more difficult, because in most cases only the end sequences of the cloned cDNA are available (ESTs). The resulting complete cDNA sequence is often generated from clusters of ESTs representing the same gene product. Therefore, some uncertainty exist on which intervening exonic sequences are present in the clone immobilized on the microarray. From these studies, it also became apparent that in future we will gain a better understanding of how genomes are expressed during biological processes. It seems that short or long oligonucleotides provide the more specific probes than cDNAs. For array manufacturers, it has become clear that they have to provide the full probe information and researchers have to accurately perform and record their experiments. For the data analysis, it is necessary to provide the full MIAME-compliant annotation [36] and the complete raw data sets which have been generated, including scanned image files of the arrays.

6 Commercial DNA Microarray Platforms

Nowadays, most commercial DNA microarray systems comprise DNA microarrays and reagent kits for RNA sample preparation, amplification, labeling, as well as DNA microarray processing hardware, such as hybridization chambers and station, microarray scanner, and data analysis software. In the following paragraphs, a few key features of available microarray technologies or systems that emerged during the past years are described.

A so-called photolithographic manufacturing process is used by Affymetrix to produce GeneChip arrays, containing millions of probes on 5-inch by 5-inch quartz wafer chip surfaces. Phosphoramidites of adenine, thymine, cytosine, or guanine, which contain photosensitive protecting groups, are then stepwise flooded onto chip surfaces. The so-called photolithographic produced masks help to either block or transmit light onto specific areas of the array surface during the synthesis process. The coupling of the next nucleotides occurs only in those areas, which have been de-protected through illumination. By repeating this process, usually up to 25 nucleotides are synthesized at a density of about 1.3 million unique features on the surface of the arrays. Routinely 16–22 probes per single gene are used for expression measurement and 40 for genotyping. For detecting and eliminating false or contaminating fluorescence signals next to each probe, which is fully complementary to its target sequence, the array also contains a paired "mismatch" probe, where the mismatch nucleotide is located directly in the middle of the probe. The mismatch probe serves as an internal control for its perfect match partner. The system is available with hybridization and wash station as well as a chip reader; meanwhile, Affymetrix also provides an automated sample preparation platform (www.affymetrix.com).

A similar photo-mediated synthesis for the production of microarrays is used by NimbleGen. Instead of a wafer surface and photolithographic masks, they are using glass surfaces and a digital micromirror device (DMD), allowing conventional oligonucleotide synthesis chemistry and the synthesis of up to 786 000 probes on a single microarray in a length of up to 60 nucleotides (www.nimblegen.com).

A further integrated microarray system is provided by Applied Biosystems. A key feature of the 1700 Chemiluminescent Microarray Analyzer is the special microarray surface, with three-dimensional structures allowing chemiluminescent detection chemistry in combination with fluorescent detection of internal controls, e.g., for spot morphology. During manufacturing, a unique probe (60-mer) and an internal control probe (a 24-mer) is co-spotted on the microarray surface. At the hybridization step, a complementary oligonucleotide, pre-labeled with a fluorescent dye (LIZ® dye), is included in the hybridization mixture, which is not interfering with the chemiluminescence signals. The internal control probes monitor microarray manufacturing quality, providing normalization of feature-to-feature and array-to-array variations as well as spot finding and auto gridding of the array. Controls for monitoring enzyme activity and DIG-label incorporation efficiency are synthetic bacterial control genes (*Dap, Lys, Phe, BioB, BioC*, and *BioD*), which are provided with the labeling kits. The Applied Biosystems 1700 Chemiluminescent Analyzer is a CCD camera based reader; a motorized chassis moves the microarray into the optical path to detect chemiluminescent and fluorescent signals. During loading of the arrays, the temperature is increased to 35 °C for the enzymatic reaction and multiple images are then taken to bring the microarray into focus and to measure fluorescent and chemiluminescent signals. Short (5 seconds) and long (25 seconds) read times are used in order to extend the linear dynamic range of the chemiluminescent signals (www.appliedbiosystems.com).

The Agilent SurePrint microarray technology system (www.agilent.com) is based on a non-contact ink-jet spotting technology developed by Rosetta Inpharmatics [47]. The key features are that 60-mer long oligonucleotides are either spotted or in-situ synthesized on the surface of a coated microscope slide, which produces uniformly spots onto specially-prepared glass slides, without introducing spot anomalies, resulting in consistent spot uniformity and traceability. Up to 22 000 oligonucleotides per microarray can be synthesized and multiple microarrays can be printed in parallel on a glass surface in precise quantities. Agilent has developed an own reader system for high-throughput scanning. This scanner contains a dynamic auto focus system as well as a laser power and PMT control system, which should increase the sensitivity and reduce the background. The auto focus feature is continuously adjusting the focal plane for correcting for common glass surface abnormalities and gradients within the scanner. A laser power control is integrated in the green YAG (532 nm) and red He-Ne (633 nm) lasers to continuously compensate for short-term laser fluctuations – the PMT control leads to a reduction in background signals (www.chem.agilent.com).

The company Illumina has developed a bead-based array platform for SNP genotyping and RNA profiling, which exist in two different types, the so-called Sentrix Array Matrix (96 arrays) and the Sentrix BeadChip (8 arrays) [48]. The key feature of the BeadChips is that each bead contains more than 105 immobilized probe oligonucleotides (each includes a 23- to 25-base address and a 50-base target-specific sequence), which self-assemble randomly into etched microwells. Thousands to hundreds of thousands of 3-micron features are assembled into the microwells of each array. The probes are synthesized on chemically etched fiber optic strands containing approximately 50 000 beads, which are bundled together, and two 50mer probes for each gene are synthesized as well as a set of housekeeping genes and controls. On average, each bead type is represented over 30 times on each array, providing high signal redundancy for each probe sequence. The Illumina system includes an automated confocal scanner, which reads all 96 arrays of a Sentrix Array Matrix. The features of BeadChip can be detected on other commercially available 5-micron scanners. Illumina incorporates a number of control beads, designed to routinely monitor sample quality, labeling efficiency, hybridization stringency, signal generation, and experimental noise. The labeling of samples involves the incorporation of biotin for subsequent array visualization with streptavidin-Cy3™ (www.illumina.com)

The CombiMatrix array technology, which has been recently introduced, is a modified semiconductor system adapted for biological applications [49]. Integrated circuits contain arrays of microelectrodes (94 μm in diameter), which are individually addressable using embedded logic circuitry on the chip, thus, facilitating the in situ synthesis of DNA oligonucleotides. The chip rapidly synthesizes over a thousand different molecules in parallel, each above a distinct electrode. The parallel process drastically reduces the cost and time of synthesizing hundreds or thousands of different molecules on an array surface and it is used in a variety of microarray applications. The array format can be used by conventional fluorescence labeling and detection technologies. Alternatively, CombiMatrix also provides a microarray reader that electrochemically measures hybridization events ("ElectraSense" system), which offers an alternative detection technique to conventional fluorescence detection. Common enzymes and substrates can replace relatively expensive fluorescent dyes, and relatively expensive optical systems (e.g., complex and sophisticated lasers) are replaced by less expensive electronic devices (www.combimatrix.com).

7 Conclusion

Microarray technology provides a powerful and very popular research and screening tool for analyzing differentially expressed genes in all areas of biomedical and biotechnological research. Their ability to genome-wide measure the expression of all genes simultaneously at reasonable time and costs has contributed to a proven record of success. However, certain limitations of this technology still exist and have become apparent in the past few years, when researcher from different laboratories tried to compare their results or integrate those results into public databases. For the majority of genes expressed in an appropriate range, the existence and alteration can be reliably detected and measured. However, accurate measurements of absolute levels and the reliable detection of low abundant genes are currently beyond the reach of microarray and will need a second independent technique for validation, for example, quantitative real-time PCR. Technological advances in microarray design and fabrication, detection methods and hybridization conditions, as well as global concerted efforts to further develop standard operating protocols, reference materials, and analytical and technical guidelines will certainly increase the measurement capabilities of DNA microarrays. In conjunction with clinical diagnostics, suitable microarrays will be able to contribute enormously to a personalized medicine in helping the clinicians, e.g., to analyze genotypes before therapeutic intervention or monitor disease/therapy progression.

Acknowledgements W.N. would like to thank Hans Lehrach and the Max-Planck-Society as well as Stefan Schreiber and the University of Kiel for continuous support of the research. W.N. also gratefully acknowledges financial support from the German Ministry of Education and Research (BMBF) and European Union (FP5).

References

1. Schena M, Shalon D, Davis RW, Brown PO (1995) Quantitative monitoring of gene expression patterns with a complementary DNA microarray. Science 270(5235):467–470
2. Lockhart DJ, Dong H, Byrne MC, Follettie MT, Gallo MV, Chee MS, Mittmann M, Wang C, Kobayashi M, Horton H, Brown EL (1996) Expression monitoring by hybridization to high-density oligonucleotide arrays. Nat Biotechnol 14:1675–1680
3. Schena M, Shalon D, Heller R, Chai A, Brown PO, Davis RW (1996) Parallel human genome analysis: microarray-based expression monitoring of 1000 genes. Proc Natl Acad Sci USA 93:10614–10619

4. Eisen MB, Spellman PT, Brown PO, Botstein D (1998) Cluster analysis and display of genome-wide expression patterns. Proc Natl Acad Sci USA 95(25):14863–14868
5. Eisen MB, Brown PO (1999) DNA arrays for analysis of gene expression. Meth Enzymol 303:179–205
6. Wang DG, Fan JB, Siao CJ, Berno A, Young P, Sapolsky R, Ghandour G, Perkins N, Winchester E, Spencer J, Kruglyak L, Stein L, Hsie L, Topaloglou T, Hubbell E, Robinson E, Mittmann M, Morris MS, Shen N, Kilburn D, Rioux J, Nusbaum C, Rozen S, Hudson TJ, Lander ES (1998) Large-scale identification, mapping, genotyping of single-nucleotide polymorphisms in the human genome. Science 280:1077–1082
7. Taniguchi M, Miura K, Iwao H, Yamanaka S (2001) Quantitative assessment of DNA microarrays-comparison with Northern Blot analyses. Genomics 71:34–39
8. Butte A (2002) The use and analysis of microarray data. Nat Rev Drug Discovery 1:951–960
9. Hoheisel JD (2006) Microarray technology: beyond transcript profiling and genotype analysis. Nat Rev Genet 7:200–210
10. Anderson MT, Foy CA (2005) The development of microarray standards. Anal Bioanal Chem 381:87–97
11. Draghici S, Khatri P, Eklund AC, Szallasi Z (2005) Reliability and reproducibility issues in DNA microarray measurements. Trends Genetics 22:101–1099
12. Gold D, Coombes K, Medhane D, Ramaswamy A, Ju Z, Strong L, Koo JS, Kapoor M (2004) A comparative analysis of data generated using two different target preparation methods for hybridization to high-density oligonucleotide microarrays. BMC Genomics 5:2
13. Holland MJ (2002) Transcript abundance in yeast varies over six orders of magnitude. J Biol Chem 277:14363–14366
14. Czechowski T, Bari RP, Stitt M, Scheible WR, Udvardi MK (2004) Real-time RT-PCR profiling of over 1400 Arabidopsis transcription factors: unprecedented sensitivity reveals novel root- and shoot-specific genes. Plant J 38:366–379
15. Liang RQ, Li W, Li Y, Tan CY, Li JX, Jin YX, Ruan KC (2005) An oligonucleotide microarray for microRNA expression analysis based on labeling RNA with quantum dot and nanogold probe. Nucleic Acids Res 33:e17
16. www.psb.rug.ac.be/CAGE/
17. Hubbell E, Liu WM, Mei R (2002) Robust estimators for expression analysis. Bioinformatics 18:1585–1592
18. Yuen T, Wurmbach E, Pfeffer RL, Ebersole BJ, Sealfon SC (2002) Accuracy and calibration of commercial oligonucleotide and custom cDNA microarrays. Nucleic Acids Res 30:e48
19. Bartosiewicz M, Trounstine M, Barker D, Jonston R, Buckpitt A (2000) Development of a toxicological gene array and quantitative assessment of this technology. Arch Biochem Biophys 376:66–73
20. Winzeler EA, Schena M, Davis RW (1999) Fluorescence-based expression monitoring using microarrays. Meth Enzymol 306:3–18
21. Guo Z, Guilfoylem RA, Thiel AJ, Wang R, Smith LM (1994) Direct fluorescence analysis of genetic polymorphism by hybridization with oligonucleotide arrays on glass supports. Nucleic Acids Res 22:5456–65
22. Randolph JB, Waggoner AS (1997) Stability, specifity and fluorescence brightness of multiply-labeled fluorescent DNA probes. Nucleic Acids Res 25:2923–2929
23. Perraut F, Lagrange A, Pouteau P, Peyssonneaux O, Puget P, McGall G, Menou L, Gonzalez R, Labeye P, Ginot F (2002) A new generation of scanners for DNA chips. Biosens Bioelectron 17:803–813

24. Novoradovskaya N, Whitfield ML, Basehore LS, Novoradovsky A, Pesich R, Usary J, Karaca M, Wong WK, Aprelikova O, Fero M, Perou CM, Botstein D, Braman J (2004) Universal Reference RNA as a standard for microarray experiments. BMC Genomics 5:20
25. Yang IV, Chen E, Hasseman JP, Liang W, Frank BC, Wang S, Sharov V, Saeed AI, White J, Li J, Lee NH, Yeatman TJ, Quackenbush J (2002) Within the fold: assessing differential expression measures and reproducibility in microarray assays. Genome Biol 3:research0062.1-0062.12
26. Sterrenburg E, Turk R, Boer JM, van Ommen GB, den Dunnen JT (2002) A common reference for cDNA microarray hybridizations. Nucleic Acids Res 30:e116
27. Wei Y, Lee JM, Richmond C, Blattner FR, Rafalski JA, LaRossa RA (2001) High-density microarray-mediated gene expression profiling of Escherichia coli. J Bacteriol 183:545–556
28. Williams BA, Gwirtz RM, Wold BJ (2004) Genomic DNA as a cohybridization standard for mammalian microarray measurements. Nucleic Acids Res 32:e81
29. Dudley AM, Aach J, Steffen MA, Church GM (2002) Measuring absolute expression with microarrays with a calibrated reference sample and an extended signal intensity range. Proc Nat Acad Sci USA 99(11):7554–7559
30. Weil MR, Macatee T, Garner HR (2002) Toward a universal standard: comparing two methods for standardizing spotted microarray data. Biotechniques 32:1310–1314
31. Cronin M, Ghosh K, Sistare F, Quackenbush J, Vilker V, O'Connell C (2004) Universal RNA reference materials for gene expression. Clin Chem 50:1464–1471
32. Forster T, Roy D, Ghazal P (2003) Experiments using microarray technology: limitations and standard operating procedures. J Endocrinol 178(2):195–204
33. Tumor Analysis Best Practices Working Group (2004) Expression profiling – best practices for data generation and interpretation in clinical trials. Nat Rev Genet 5:229–237
34. Leung YF, Cavalieri D (2003) Fundamentals of cDNA microarray data analysis. Trends Genet 19:649–59
35. Weeraratna AT, Nagel JE, de Mello-Coelho V, Taub DD (2004) Gene expression profiling: from microarrays to medicine. J Clin Immunol 24(3):213–224
36. Brazma A, Hingamp P, Quackenbush J, Sherlock G, Spellman P, Stoeckert C, Aach J, Ansorge W, Ball CA, Causton HC, Gaasterland T, Glenisson P, Holstege FC, Kim IF, Markowitz V, Matese JC, Parkinson H, Robinson A, Sarkans U, Schulze-Kremer S, Stewart J, Taylor R, Vilo J, Vingron M (2001) Minimum information about a microarray experiment (MIAME)-toward standards for microarray data. Nat Genet 29:365–371
37. Spellman PT, Miller M, Stewart J, Troup C, Sarkans U, Chervitz S, Bernhart D, Sherlock G, Ball C, Lepage M, Swiatek M, Marks WL, Goncalves J, Markel S, Iordan D, Shojatalab M, Pizarro A, White J, Hubley R, Deutsch E, Senger M, Aronow BJ, Robinson A, Bassett D, Stoeckert CJ Jr, Brazma A (2002) Design and implementation of microarray gene expression markup language (MAGE-ML). Genome Biol 3:RESEARCH0046
38. Brazma A, Parkinson H, Sarkans U, Shojatalab M, Vilo J, Abeygunawardena N, Holloway E, Kapushesky M, Kemmeren P, Lara GG, Oezcimen A, Rocca-Serra P, Sansone SA (2003) ArrayExpress – a public repository for microarray gene expression data at the EBI. Nucl Acids Res 31(1):68–71
39. Gollub J, Ball CA, Binkley G, Demeter J, Finkelstein DB, Hebert JM, Hernandez-Boussard T, Jin H, Kaloper M, Matese JC, Schroeder M, Brown PO, Botstein D, Sherlock G (2003) The Stanford Microarray Database: data access and quality assessment tools. Nucl Acids Res 31:94–96
40. Kuo WP, Jenssen TK, Butte AJ, Ohno-Machado L, Kohane IS (2002) Analysis of matched mRNA measurements from two different microarray technologies. Bioinformatics 18:405–412

41. Ross DT, Scherf U, Eisen MB, Perou CM, Rees C, Spellman P, Iyer V, Jeffrey SS, Van de Rijn M, Waltham M, Pergamenschikov A, Lee JC, Lashkari D, Shalon D, Myers TG, Weinstein JN, Botstein D, Brown PO (2000) Systematic variation in gene expression patterns in human cancer cell lines. Nat Genet 24:227–235
42. Tan PK, Downey TJ, Spitznagel EL Jr, Xu P, Fu D, Dimitrov DS, Lempicki RA, Raaka BM, Cam MC (2003) Evaluation of gene expression measurements from commercial microarray platforms. Nucleic Acids Res 31:5676–5684
43. Yauk CL, Berndt ML, Williams A, Douglas GR (2004) Comprehensive comparison of six microarray technologies. Nucleic Acids Res 32:e124
44. Larkin JE, Frank BC, Gavras H, Sultana R, Quackenbush J (2005) Independence and reproducibility across microarray platforms. Nat Methods 2:337–344
45. Irizarry RA, Warren D, Spencer F, Kim IF, Biswal S, Frank BC, Gabrielson E, Garcia JG, Geoghegan J, Germino G, Griffin C, Hilmer SC, Hoffman E, Jedlicka AE, Kawasaki E, Martinez-Murillo F, Morsberger L, Lee H, Petersen D, Quackenbush J, Scott A, Wilson M, Yang Y, Ye SQ, Yu W (2005) Multiple-laboratory comparison of microarray platforms. Nat Methods 2:345–349
46. Bammler T, Beyer RP, Bhattacharya S, Boorman GA, Boyles A, Bradford BU, Bumgarner RE, Bushel PR, Chaturvedi K, Choi D, Cunningham ML, Deng S, Dressman HK, Fannin RD, Farin FM, Freedman JH, Fry RC, Harper A, Humble MC, Hurban P, Kavanagh TJ, Kaufmann WK, Kerr KF, Jing L, Lapidus JA, Lasarev MR, Li J, Li YJ, Lobenhofer EK, Lu X, Malek RL, Milton S, Nagalla SR, O'Malley JP, Palmer VS, Pattee P, Paules RS, Perou CM, Phillips K, Qin LX, Qiu Y, Quigley SD, Rodland M, Rusyn I, Samson LD, Schwartz DA, Shi Y, Shin JL, Sieber SO, Slifer S, Speer MC, Spencer PS, Sproles DI, Swenberg JA, Suk WA, Sullivan RC, Tian R, Tennant RW, Todd SA, Tucker CJ, Van Houten B, Weis BK, Xuan S, Zarbl H, Members of the Toxicogenomics Research Consortium (2005) Standardizing global gene expression analysis between laboratories and across platforms. Nat Methods 2:351–356
47. Hughes TR, Mao M, Jones AR, Burchard J, Marton MJ, Shannon KW, Lefkowitz SM, Ziman M, Schelter JM, Meyer MR, Kobayashi S, Davis C, Dai H, He YD, Stephaniants SB, Cavet G, Walker WL, West A, Coffey E, Shoemaker DD, Stoughton R, Blanchard AP, Friend SH, Linsley PS (2001) Expression profiling using microarrays fabricated by an ink-jet oligonucleotide synthesizer. Nat Biotechnol 19:342–347
48. Dickinson T, Craumer W (2003) RNA profiling with bead-based array platforms (Tutorial): Sentrix arrays for focused gene expression studies across large sample sets. Gen Engin News 23:15
49. Liu RH, Nguyen T, Schwarzkopf K, Fuji HS, Petrova A, Siuda T, Peyvan K, Bizak M, Danley D, McShea A (2006) Fully integrated miniature device for automated gene expression DNA microarray processing. Anal Chem 78:1980–1986

Springer Ser Fluoresc (2008) 6: 265–282
DOI 10.1007/4243_2008_035

Published online: 5 March 2008

Microarray Technology: Unresolved Issues and Future Challenges from a Regulatory Perspective

Leming Shi[1] (✉) · Federico M. Goodsaid[2] · Felix W. Frueh[2] · Weida Tong[1]

[1]National Center for Toxicological Research, U.S. Food and Drug Administration, 3900 NCTR Road, Jefferson, Arkansas, 72079, USA
Leming.Shi@fda.hhs.gov

[2]Center for Drug Evaluation and Research, U.S. Food and Drug Administration, 10903 New Hampshire Avenue, Silver Spring, Maryland, 20903, USA

The views presented in this article do not necessarily reflect those of the U.S. Food and Drug Administration.

Abstract Despite their growing popularity in the research community, DNA microarray-based gene expression assays have not yet been approved or licensed by the U.S. Food and Drug Administration for clinical or regulatory decision-making. While recent publications demonstrated the technical reliability of microarray technology, there are several challenges that still need to be addressed so that microarray-based gene expression data can be accepted for routine use in clinical and regulatory environments. First, appropriate quality-control metrics and thresholds are needed for objectively assessing the quality of microarray data from individual laboratories. Secondly, consensus on the analysis of microarray data is needed. Thirdly, adequate evaluation and validation of microarray results is needed so that non-reproducible chance correlations can be avoided in handling high-dimensional microarray data.

Keywords Data analysis and interpretation · Microarray · Quality control metrics and thresholds · Regulatory decision-making · Reproducibility

Abbreviations

CV	Coefficient of variation
DNA	Deoxyribonucleic acid
ERCC	External RNA Controls Consortium
FC	Fold change

FDA	U.S. Food and Drug Administration
GAPDH	Glyceraldehyde-3-phosphate dehydrogenase
GO	Gene Ontology
HCA	Hierarchical cluster analysis
iHOP	Information hyperlinked over proteins
MAQC	MicroArray Quality Control project
NT	Nucleotide
PCA	Principal component analysis
PLIER	Probe logarithmic intensity error
QC	Quality control
qPCR	Quantitative polymerase chain reaction
RIN	RNA integrity number
RNA	Ribonucleic acid
SD	Standard deviation
SNR	Signal-to-noise ratio
VGDS	Voluntary Genomic Data Submission
WG	Working group

1 Introduction: The Apparent Lack of Reproducibility

DNA microarrays represent a core technology in pharmacogenomics that was identified by the U.S. Food and Drug Administration's (FDA) Critical Path Initiative as a key opportunity for advancing medical product development and personalized medicine (http://www.fda.gov/oc/initiatives/criticalpath/). The FDA issued the "*Guidance for Industry: Pharmacogenomic Data Submissions*" to facilitate scientific progress and the use of pharmacogenomic data in drug development and medical diagnostics (http://www.fda.gov/cder/genomics/). Although most microarray data currently submitted to the U.S. FDA are based on the Voluntary Genomic Data Submission (VGDS) mechanism [1], it is anticipated that such kind of data will be used routinely by sponsors to support their new drug or medical device applications to the U.S. FDA.

DNA microarray technology is a measurement tool that detects the differential gene expression in a high-throughput format in terms of the number of genes that can be simultaneously measured. Ensuring adequate data quality and appropriate analysis of the resulting data are the two fundamental challenges regarding the appropriate applications of the technology in clinical and regulatory settings.

A fundamental step in most microarray experiments is to determine one or more lists of differentially expressed genes that distinguish biological conditions, such as disease from health. Challenges regarding the reliability of microarray results have largely been founded on the inability of researchers to replicate differentially expressed gene lists across the same or highly similar experimental designs. For example, Tan et al. [2] found only four common genes using an identical set of RNA samples across three popular commercial platforms. Independent studies by the groups of Ramalho-Santos [3] and Ivanova [4] of stem cell-specific genes using the same Affymetrix platform and similar study design found a disappointing six common genes among about 200 identified in each study [5]. A comparative neurotoxicological study by Miller et al. [6] using the same set of RNA samples found only 11 common genes among 138 and 425, respectively, from Affymetrix and CodeLink platforms. All of these studies ranked genes by *P*-value from simple *t*-tests.

Criticism of and concerns about microarrays continue to appear in some of the most prestigious scientific journals [7–11], leading to a growing negative perception regarding microarray reproducibility, and hence reliability. However, in reanalyzing the data set of Tan et al. [2], Shi et al. [12] found that cross-platform concordance was markedly improved when simple fold change (FC) was used to rank order genes before determining the lists of differentially expressed genes.

Reproducibility is a fundamental requirement in scientific experiments and clinical contexts. Several plausible explanations and solutions have been proposed to interpret and address the apparent lack of reproducibility and stability of lists of differentially expressed genes from microarray studies. Larger sample sizes [13]; novel, microarray-specific statistical methods [14]; more accurate array annotation information by mapping probe sequences across platforms [2, 15]; eliminating absent call genes from data analysis [12, 16, 17]; improving probe design to minimize cross-hybridization [15]; standardizing manufacturing processes [2]; and improving data quality by fully standardizing sample preparation and hybridization procedures [18] are among the suggestions for improvement.

2 Microarray Quality-Control Metrics and Thresholds: Assessing Data Quality from Various Perspectives

The literature is populated with unreliable microarray data due to variant laboratory performance [19]. Such data should not be used to draw any conclusion regarding the reliability of microarrays or the impact of data-analysis methods. It is worth reiterating that a critical question being asked in the analysis of VGDS data is: "How does the FDA know that the DNA microarray experiment is of reasonable quality to proceed with the analysis of the generated data?"

To assess the performance of a technology, it is imperative to define a set of parameters (or metrics) that evaluate the quality from different perspectives. What metrics should be used to judge microarray data quality? How good is good enough in terms of microarray data quality? Is a data set worth the time to be analyzed? These are some of the critical questions regarding the acceptance of microarray data in a regulatory setting.

In this section, we intend to list a set of metrics that have been widely used and generally accepted by the microarray community in one way or another to assess microarray data quality. It should be noted that some of these QC metrics are platform-dependent, and may not apply to all microarray platforms. In addition, it may not be feasible for each study to evaluate a data set using all the QC metrics mentioned in this section. It is also expected that a study may use additional QC metrics for the evaluation of microarray data quality [20, 21].

There are different levels of QC metrics (Fig. 1): (1) QC metrics based on a single array (before and after hybridization); (2) QC metrics based on replicate arrays; and (3) QC metrics based on the overall quality of a study.

Single-array-based QC metrics: These metrics can be further divided into two subcategories of pre-hybridization and post-hybridization. Pre-hybridization single-array QC metrics reflect the quality of the individual steps in the microarray analysis process before hybridization:

1. RNA A260/A280 ratio (purity);
2. RNA concentration (ug/uL);
3. RNA 28S/18S ratio (integrity);

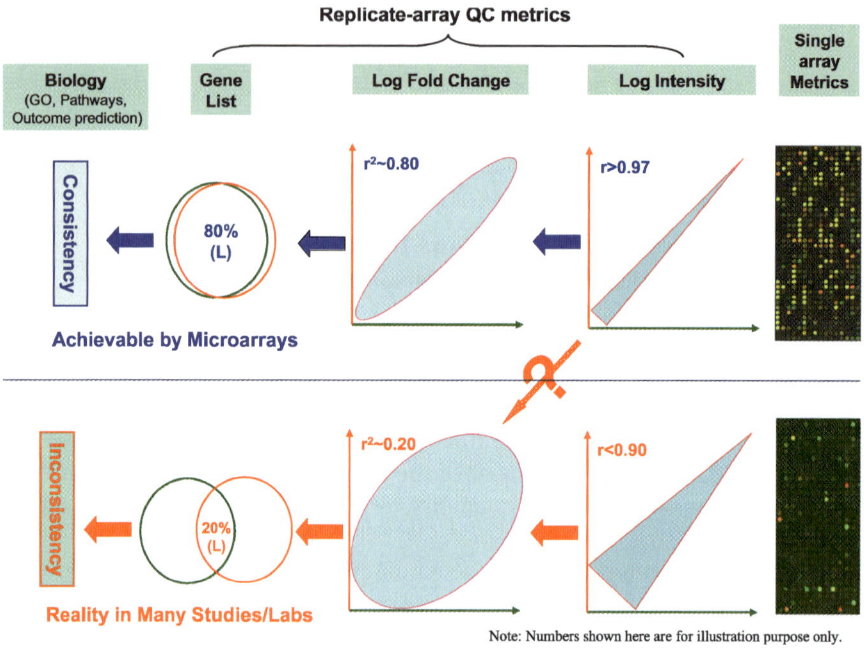

Fig. 1 Different levels of quality-control metrics: single-array, replicate-array, and whole study. The consistency of replicate arrays can be assessed by comparing intensity, fold change, or lists of differentially expressed genes

4. RIN (RNA integrity number; integrity);
5. cDNA or cRNA yield (ug);
6. cDNA or cRNA median fragment size (number of nucleotides or NTs);
7. cDNA or cRNA labeling efficiency (dye molecule/NTs);

where metrics 1–4 refer to the quality of the total RNA sample (and do not depend on the performance of a microarray platform) and metrics 5–7 refer to the quality of target preparation.

Post-hybridization, single-array QC metrics reflect the quality of a hybridized microarray chip:

1. Image defects (qualitative, visual inspection);
2. Median signal-to-noise ratio (SNR);
3. Median feature intensity;
4. SD of feature pixel intensity;
5. SD of intensity for replicate features;
6. SD of log2 ratio of replicate features for two-color arrays;
7. Spot morphology;
8. Number of unreliable features (saturation, not-found, etc.);
9. Performance of external RNA controls (spike-ins [22];
10. $3'/5'$ GAPDH ratio;
11. Global $3'/5'$ ratio;
12. % of present calls;
13. Scaling factor.

Replicate-array-based QC metrics: The use of replicate arrays increases the power of identifying differentially expressed genes. Technical replicate arrays (e.g., from a proficiency testing program [23]) allow for the assessment of platform precision at the researcher's laboratory. However, in a biological study, it is uncommon to have technical replicates for a given biological sample, but biological replicates are common that incorporate both technical and biological variations in a study. The following metrics measure the consistency in (biological) replicate arrays:

1. Consistency of % present calls;
2. Consistency of genes being called present;
3. (log) intensity correlation or coefficient of variation (CV);
4. False-positive rate (comparing replicates from the same group of samples);
5. (log) ratio (i.e., fold-change) correlation or CV;
6. Percentage of overlapping genes between two lists of differentially expressed genes;

where metrics 1–3 refer to the consistency of intensity measurement for replicate arrays within the same RNA group and metrics 5–6 refer to the consistency of ratio (fold-change) measurement for replicate array pairs in which the expression differences between two different RNA samples are measured. External RNA controls [24] have been demonstrated to behave independently of the sample and therefore can be viewed as "technical replicates" across biological replicate samples [22]. The use of external RNA controls is highly desirable for quality-control purposes.

Quality of a study: The overall quality of a microarray study, which usually attempts to identify genes differentially expressed between two groups of RNA samples (e.g., control versus treatment) or to predict clinical outcomes (e.g., good or poor drug responses) based on a patient's gene expression profile, depends not only on the within-group intensity data consistency but also on the inherent between-group differences. When the between-group differences (e.g., the true biological differences between treated vs. control animals) are large, the detected fold-change becomes more reliable. Higher within-group consistency (precision) also leads to more reliable detection of differential expression. If there is no (or little) difference between the two groups of samples being compared, then there is not much that we can do much with microarrays. It is also important to be able to interpret Gene Ontology terms, pathways and class predictions in these studies. Good laboratory practices should be followed in the design and implementation of the microarray study. For example, confounding factors should be avoided (blocked) in the experiment.

Quality-control thresholds – How good is good enough: Defining a set of QC metrics is important for microarray quality control and quality assurance. However, a QC metric without an associated acceptable threshold (cutoff) is not very useful. Therefore, a critical component of the evaluation of microarray data quality depends on the establishment of an acceptable threshold for each QC metric from the large data sets generated by community-wide standardization efforts, proficiency testing programs [23], and the VGDS archives so that the sponsors and reviewers can know whether the performance of a laboratory or operator is acceptable. Much work is needed in this area. Manufacturers should share as much QC information as possible to their users to ensure that data of unacceptable quality will not be trusted.

3 Consensus on Data-Analysis Methods: Adequate Evaluation Is Needed

Data analysis remains one of the biggest challenges in microarray-based studies due to the availability of numerous methods and the lack of consensus in the community regarding the capabilities and limitations of each method. Data analysis includes:

1. Data preprocessing and normalization;
2. Quality-control metrics and thresholds for quality assessment;
3. Handling flags and low-intensity transcripts;
4. Identification of "outlying" hybridizations;

5. Identification of lists of differentially expressed genes;
6. Classification and prediction of (clinical) outcomes;
7. Biological interpretation of microarray results.

Data preprocessing and normalization: The intensity data for each microarray feature (spot) is usually quantified from the image output of a fluorescence reader (scanner) and specialized software. It is advisable to follow the manufacturer's preferred procedures for image quantification and background subtraction and provide detailed information on software settings in reporting experimental results.

A microarray experiment is based on the analysis and comparison of multiple arrays (hybridizations). Cross-array (hybridization) reproducibility is an important criterion for judging the quality of a microarray experiment (Fig. 1). There are many variations in technical factors that can occur between arrays, leading to inconsistent microarray data. Therefore, normalization methods have been used to (partially) correct systematic variations in microarray data, e.g., the differences in the overall intensity levels of arrays. The purpose of normalization is to make truly interesting biologic variations in gene-expression levels stand out from variations due to pure technical factors. Although many options for microarray data normalization are in use, there is no consensus on the choice of one method over another. Each microarray manufacturer has its preferred normalization method for analyzing microarray data. For example, numerous normalization (or probeset-level summary) methods are available for handling Affymetrix. CEL files [25], and Affymetrix's currently preferred method includes a probe-level quantile normalization step followed by probeset-level summarization with PLIER (Probe Logarithmic Intensity ERror) plus an offset value of 16 for variance stabilization. Applied Biosystems and Illumina prefer quantile normalization, whereas Agilent (one-color) and GE Healthcare prefer median-scaling normalization [26, 27].

Handling flags and low-intensity transcripts – Increase confidence in analysis results: Each manufacturer has developed their own ways of handling low-intensity spots and assigning a flag status (e.g., **P**resent, **A**bsent, or **M**arginal) to each spot to indicate the reliability of the measured intensity value. The impact of noise filtering on data reproducibility is well documented [12, 16, 17]. If we assume that genes tiled on a microarray platform represent a random sampling of all the genes coded by a genome and that only a (small) portion of the genes coded by the genome are expected to be expressed in a single cell type or tissue under any given biological condition, it then appears reasonable to apply a noise-filtering procedure to exclude a certain percentage of spots with the lowest intensity from further analysis. The data-filtering procedure proposed by Barczak et al. [16] excludes 50% of the spots with the lowest average intensity across all arrays in a study; this is a simple approach to deal with less reliable measurements. Other approaches for dealing with flagged spots require the consideration of manufacturer flags, e.g., excluding spots that are Absent in the majority of arrays in a study. Noise filtering is important for obtaining reliable measurements of differential gene expression. Note that some of the QC metrics can be calculated before and after noise-filtering.

Identification of outlier hybridizations: It is expected that a (small) portion of the replicate arrays in a microarray study will fail and become obvious outliers. It is also reasonable to exclude such outlier arrays from further data analysis. Exploratory analysis techniques such as principal component analysis (PCA) and hierarchical cluster analysis (HCA) based on the complete gene-expression profiling data can be used to identify outlier arrays. Examination of the behavior of external RNA controls provides an alternative way of identifying outlier hybridizations [22]. However, there is no consensus on the threshold for determining outlier arrays without introducing biases. To avoid any potential bias, the exclusion of outlier hybridizations should not be based on whether an array clusters with other members of the same group.

Identification of lists of differentially expressed genes: A fundamental step in most microarray experiments is the determination of one or more lists of differentially expressed genes that distinguish biological conditions such as disease from health or treatment from control. New statistical methods for the identification of differentially expressed genes continue to appear in the scientific literature. These methods are typically promoted in terms of improved sensitivity (power) under various assumptions or conditions while retaining nominal rates of specificity. Reproducibility is a fundamental requirement in scientific experiments and clinical contexts, but is seldom emphasized in microarray literature. It is equally (if not more important) than sensitivity and specificity in certain experimental and clinical contexts. Until recently, reproducibility has not adequately been used as an essential criterion for evaluating the pros and cons of statistical methods for identifying differentially expressed genes [26, 28].

Despite the availability of numerous statistical methods for the identification of differentially expressed genes, the simple t-statistic (and slight variations) is arguably still the most widely used test statistic, and many of the various methods that exist to create lists of differentially expressed genes primarily improve upon the inference from this basic test statistic. This includes the simple unmodified two-sample t-test, Bonferroni and step-up/step-down procedures applied to the t-test, and others. Statistical significance (P) derived from the simple two-group t-test has historically been widely used as the only criterion to identify differentially expressed genes, with disappointing results reported related to reproducibility [2, 12]. It has been demonstrated [26, 28] that (1) such discordance may stem from ranking and selecting differentially expressed genes solely by a statistical significance measure (P) derived from widely used simple t-tests; (2) when fold-change is used as the ranking criterion, the lists become much more reproducible, especially when fewer genes with larger fold changes are selected; and (3) the instability of the lists of differentially expressed genes based on P cutoffs is an expected mathematical consequence of the high variability of the t-values. These observations are consistent in inter-laboratory, cross-platform, and between microarrays and TaqMan® comparisons [12, 26, 28, 29]. In addition, widely used analysis methods such as SAM (Significance Analysis of Microarrays [30] did not yield more reproducible lists of differentially expressed genes compared to fold-change ranking [26, 28].

Classification and prediction of (clinical) outcomes: Similar to the factor that many options have been proposed for identifying lists of differentially expressed genes, there are numerous methods for classification and prediction purposes in microarray data analysis. However, the merits and limitations of each method are under continuous debate [31] and the choice of a particular classification or prediction method over the others often depends on the experience of the investigator and personal preference or limitation of accessible software packages.

Earlier microarray studies claiming the superb power of microarray technology and "*de novo*" classification methods for differentiating various types of (cancer) samples or prognoses were based on an incomplete or biased gene-selection method, where the gene-selection step was incorrectly performed prior to the cross-validation process [32–34]. Such a "selection bias" has been shown to produce almost "perfect" classification results even if the data were randomly generated [34–36]. Such a bias can lead to spuriously good statistical estimates about the performance of a classification or prediction model. In fact, much of the hype related to microarray technology could be attributed to the unrealistic expectation of prediction accuracy due to such a mistaken validation procedure. This highlights another way in which the results of microarray experiments can be either over- or misinterpreted. Relating gene-expression profiles to drug efficacy or toxicity indications is a complicated task, and a large number of training-set cases are needed for improving predictive capabilities of classification models. When additional data sets are unavailable as independent test cases, the biomarker signatures may be validated using a cross-validation process. However, care should be taken not to introduce gene-selection bias in the cross-validation process [32–34]. The report on classification or prediction of (clinical) outcomes such as diagnostic or prognostic results should provide a detailed description regarding the data set, feature selection, modelling construction, and performance validation.

Biological interpretation of microarray results: The output from microarray data analysis is usually represented in the form of a set of genes that are identified as differentially expressed between groups of samples (e.g., control versus treatment). A natural next step is to identify the biologic meaning of the genes. This process demands in-depth biologic knowledge of the investigator and a comprehensive analysis of relevant information available in public or private repositories.

The biological interpretation of a list of differentially expressed genes can be conducted through the identification of Gene Ontology (GO) terms or pathways that are most significantly impacted or enriched by such a list. The GO provides structured, controlled vocabularies and classifications about the roles of genes and proteins in terms of molecular function, biologic process, and cellular component [37]. Pathways are important vehicles for interpreting the biologic functions of genes. In pathway analysis, a commonly asked question is whether genes in a particular pathway are significantly overrepresented in the list of differentially expressed genes. It is desirable to know whether the mechanism of action related to efficacy or toxicity can be inferred from the functions of pathways.

A variety of software packages for GO and/or pathway analysis are available, either free on the Web or via purchase of a commercial license [38]. Our confidence in a particular biological interpretation increases when two or more different databases or software packages lead to the same conclusion. However, this is not always the case, due to many factors such as differences in database contents or the choice of statistical approaches.

Another common approach of biological interpretation is by text mining. Medline is a rich source of biomedical information for knowledge mining such as the relationships between genes and their products. MedMiner [39] is an Internet-based hypertext program that filters and organizes large amounts of textual and structured information returned from public search engines such as PubMed and GeneCards. Information Hyperlinked over Proteins (iHOP) is an information system that converts the information in PubMed into a navigable resource where genes and proteins are hyperlinked between sentences and abstracts [40]. iHOP focuses on those sentences in which genes (proteins) co-occur. The sentences and gene models (networks) in iHOP provide a useful and concise way of summarizing gene-network information.

4
The MicroArray Quality Control (MAQC) Project

An FDA-led community-wide effort with reference RNA samples and reference data sets: The MicroArray Quality Control (MAQC) project was initiated by the U.S. FDA's National Center for Toxicological Research, Jefferson, Arkansas, on February 11, 2005 in order to address reliability concerns as well as other performance, quality, and data-analysis issues pertaining to the use of DNA microarrays for gene-expression measurements (Fig. 2, http://edkb.fda.gov/MAQC/). The first phase of the MAQC project (from February 11, 2005 to September 8, 2006) involved 137 scientists from 51 organizations including government agencies (the U.S. FDA, the U.S. Environmental Protection Agency, the National Institutes of Health, and the National Institute of Standards and Technology), manufacturers of microarray plat-

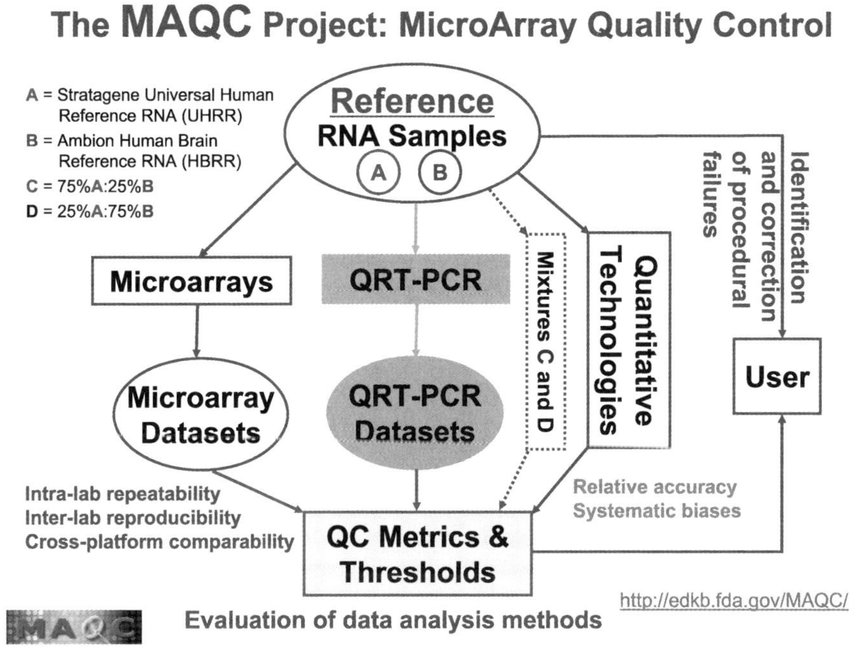

Fig. 2 An overview of the Phase I of the MicroArray Quality Control (MAQC) project: Assessment of the achievable technical performance

forms and RNA samples, microarray service providers, academic laboratories, and other stakeholders [26].

Gene-expression data on four titration pools from two distinct, commercially available reference RNA samples were generated at multiple test sites using a variety of microarray-based and alternative technology platforms, resulting in a rich data set with over 1300 microarray hybridizations and additional measurements for over 1000 genes with alternative technologies such as qPCR. The MAQC project observed intraplatform reproducibility across test sites as well as high interplatform concordance in terms of genes identified as differentially expressed. One major result was that the platforms with divergent approaches often generated comparable results of differential gene expression. In other words, the differential gene-expression patterns generated were reflective of biology regardless of the differences in technology platforms. Similar results were observed from a rat toxicogenomics data set [28], validating the major findings from data generated on reference RNA samples. Findings of the MAQC project were published in six research papers in *Nature Biotechnology*, September 8, 2006; all papers are freely available at http://www.nature.com/nbt/focus/maqc/index.html. Data are available through GEO (series accession number: GSE5350), ArrayExpress (accession number: E-TABM-132), ArrayTrack (http://www.fda.gov/nctr/science/centers/

toxicoinformatics/ArrayTrack/), and the MAQC Web site (http://www.fda.gov/nctr/science/centers/toxicoinformatics/maqc/). One important goal of the MAQC study is to assess the best performance that can be achieved by microarray technology under consistent experimental conditions so that end users will be able to judge whether the quality of their microarrays is comparable to the achievable performance of the platform. In doing so, procedural failures of a laboratory or operator may be identified and corrected before precious study samples are profiled. The commercial availability of the two reference RNA samples (A and B) coupled with the large reference data sets would also allow for the objective evaluation of new array products, reagents, or protocols.

Several unique features set the MAQC project apart from previous cross-platform comparison studies: (1) the enthusiastic participation of the microarray community in an extraordinary team effort; (2) the scale of the MAQC data set with over 1300 microarrays from more than 40 test sites and 20 microarray platforms; (3) the large number of additional gene-expression measurements with alternative technology platforms; (4) the commercial availability of the same batches of the two reference RNA samples used in the MAQC study for subsequent quality control, performance evaluations, and proficiency testing by the community; (5) the extensive sequence-based mapping of probes across platforms; and (6) last but not least, the identification of statistical explanations for some misconceptions on the comparability of microarray results.

A major challenge to the microarray user is the existence of numerous options for analyzing the same data set, which lack adequate scientific vetting of their capabilities, implications, and limitations [41]. There is a pressing need to critically evaluate currently available methods with relevant and objective criteria. For example, reproducibility has seldom been (but in the future should be) used as a critical criterion to judge the performance of data-analysis procedures. In addition, several differential gene-expression profiling studies have demonstrated that the relative expression measures (i.e., difference in transcript abundance between sample types) are more consistent than the absolute gene-expression levels. The MAQC data set is expected to be widely utilized by the community in order to reach and promote consensus on the appropriate analysis of microarray data.

Lists of genes selected solely by a statistical significance measure are inherently irreproducible: The MAQC analyses demonstrated [26, 28] that the apparent lack of reproducibility reported in previous studies using microarray assays was likely caused (at least in part) by the common practice of ranking genes solely by a statistical significance measure, for example *P* values derived from simple *t*-tests, and selecting differentially expressed genes with a stringent significance threshold (Fig. 3). The gene lists in the MAQC study were much more concordant when fold change was used as the ranking criterion. This approach also greatly reduced the impact of different normalization

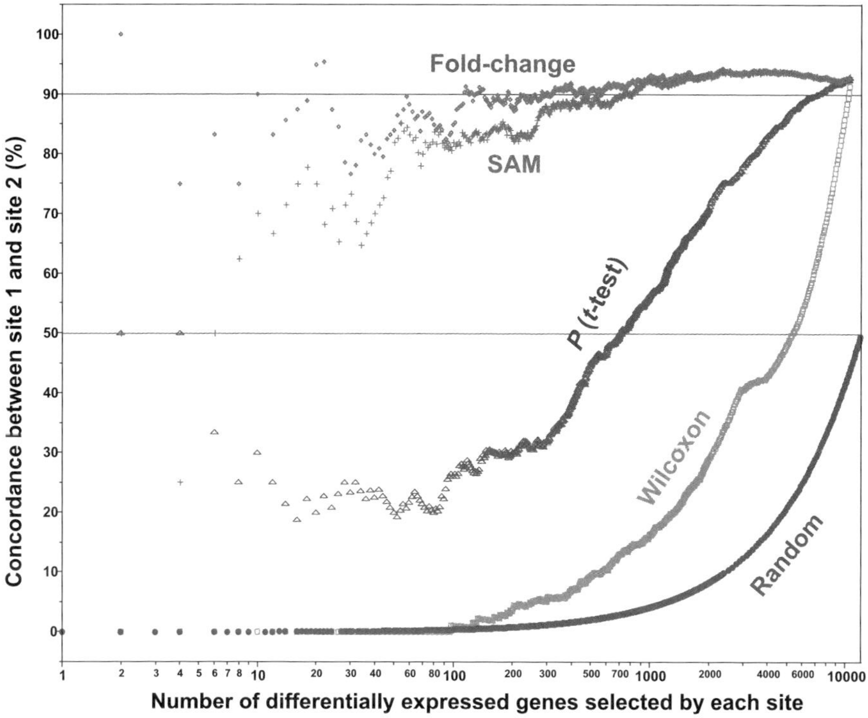

Fig. 3 The reproducibility of microarray results in terms of lists of differentially expressed genes depends on the choices of gene-selection methods and cutoff values. The fold-change-based ranking method generates more reproducible gene lists. Reproduced with permission from [26]

methods. Widely used statistical methods did not appear to improve interlaboratory or interplatform reproducibility compared to fold-change ranking. Importantly, non-reproducible gene lists led to inconsistent biological interpretations in terms of enriched GO terms and pathways [28]. Fold-change ranking plus a non-stringent *P*-value cutoff can be used as a baseline practice for generating more reproducible signature gene lists [26, 28].

The impact of data normalization is much less profound when fold change is used for gene selection: Data normalization was identified as a major factor for differences when comparing results and data interpretations performed by VGDS sponsors and FDA reviewers [1]. It should be noted that although there are many options for normalizing microarray data, when lists of differentially expressed genes are identified by the ranking of fold-change, the result is much less susceptible to the impact of normalization methods (Fig. 4). In fact, global scaling methods (e.g., median- or mean-scaling) do not change the relative rank-order of genes based on fold-change; they do, however, significantly impact gene ranking by *P*-value [26, 28, 42]. The MAQC results

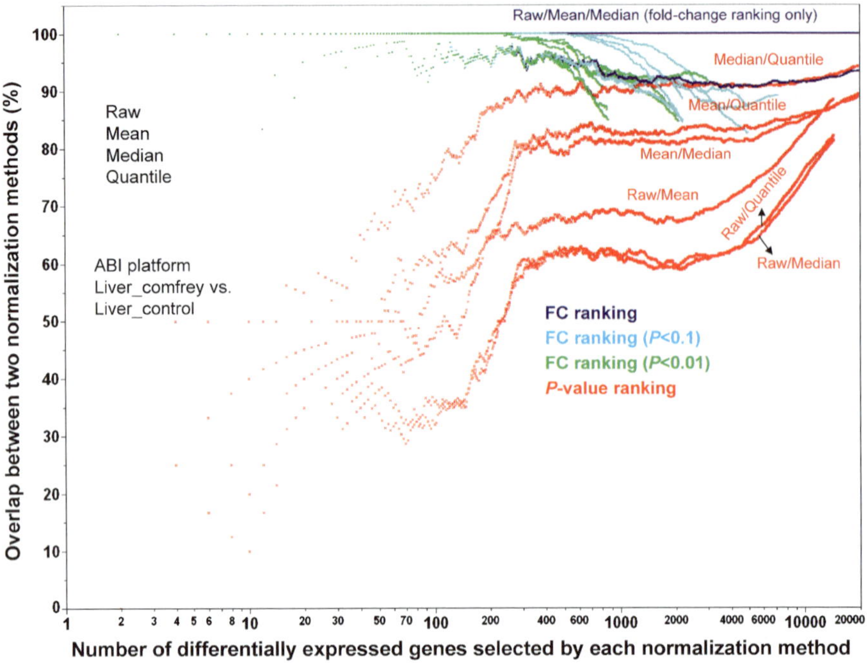

Fig. 4 The impact of normalization methods on microarray results in terms of lists of differentially expressed genes. When *P*-value is used for ranking genes, the differences between two normalization methods is dramatic (*red-colored lines*); whereas when fold change is used for ranking genes, the impact of normalization methods on the overlap of gene lists becomes much less profound. Reproduced with permission from [28]

suggest that microarray data analysis for the identification of reproducible lists of differentially expressed genes need not be as complicated and confusing as it has been practiced, and consensus on data analysis appears to be possible.

The second phase of the MAQC project – Development and validation of predictive signatures and classifiers: The MAQC Phase I (MAQC-I) has demonstrated the technical reliability of microarray technology in detecting differential gene expression. However, questions remain regarding the reliability of the technology in clinical applications such as disease diagnostics or prognostics, and tailored treatment based on gene-expression profiles [8, 9, 13, 31]. To investigate the capabilities and limitation of microarray technology in real-life applications, we have launched the MAQC Phase II (MAQC-II) efforts aimed at addressing technical and scientific issues regarding the development and validation of predictive signatures and classifiers (Fig. 5). Multiple data sets will be collected and distributed to participating organizations for independent analyses with available algorithms. The resulting classifiers will be evaluated at three different levels: within a single data set via cross-validation,

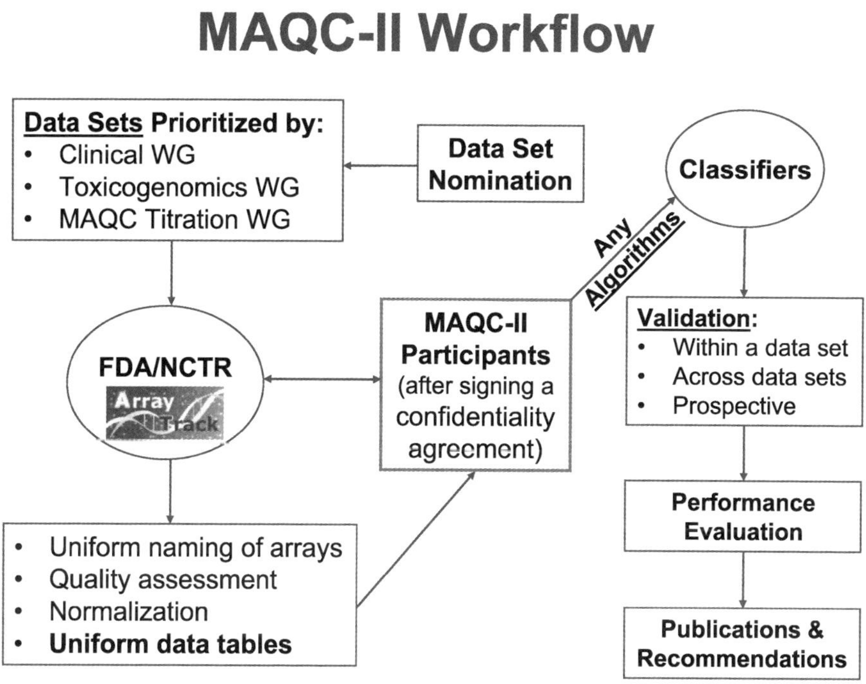

Fig. 5 An overview of the Phase II of the MicroArray Quality Control (MAQC) project: Development and validation of predictive signatures and classifiers for clinical and toxicogenomic applications. The three working groups (WGs) are working in parallel

validation across data sets from studies with the same study objectives, and prospective validation with new samples. We expect that the MAQC project, with the community's active participation, will help develop "best practices" for the generation, analysis, and application of microarray data in the discovery, development, and review of FDA regulated products.

5 Conclusions

Demonstrating reproducible technical performance is critical to the acceptance of microarray-based gene-expression data in clinical and regulatory environments. The microarray technology itself has been historically criticized whenever data were found irreproducible. Like any other analytical instrument, ensuring the technical proficiency of the laboratory that performs the microarray studies is critical for generating reproducible microarray data. As stated in the FDA's concept paper "*Recommendations for the Generation and Submission of Genomic Data*", many steps are involved in a microarray study and are critical to the overall quality of a study (http://www.fda.gov/cder/genomics/conceptpaper_20061107.pdf). Investigators should take proper measures to ensure that each step meets the proper quality-control and quality-assurance criteria. Although many metrics have been proposed for microarray quality control, there is a lack of consensus on the relative importance of these metrics. More problematically, there is little information regarding the acceptable thresholds for each quality-control metric. The establishment of quality-control thresholds will depend on the availability of large collections of historical microarray data and those from international standardization efforts such as the ERCC (External RNA Controls Consortium [24] and the MAQC [26, 28]. The community's willingness to share data is critical. Ensuring good-quality data is the essential first step toward reproducible and reliable microarray results in terms of the lists of differentially expressed genes, the biological interpretations, and the prediction of clinical outcomes.

The lack of consensus on the analysis of microarray data represents a challenge in generating reproducible results from microarray-based gene-expression studies. A common practice of identifying differentially expressed genes is by ranking all genes on the array by a statistical significance measure, e.g., the t-test statistic or its corresponding P-value [43]. However, recent publications [12, 26, 28] demonstrated that such a practice predestines non-reproducible gene lists irrespective of the platforms, sample pairs, or normalization methods. The MAQC study recommended the use of fold-change ranking plus a non-stringent P cutoff as a baseline practice in order to generate more reproducible lists of differentially expressed genes. The fold-change criterion enhances reproducibility while the P criterion balances sensitivity and specificity. It should be noted that the concurrent use of fold-change ranking combined with a P threshold, denoted as FC(P), may not necessarily be the ultimate and best way of identifying differentially expressed genes in all circumstances, but it appears to be a reasonable, straightforward baseline analysis procedure that can be used to enhance the reproducibility of lists of differentially expressed genes. Importantly, as expected, non-reproducible lists of differentially expressed genes have been shown to result in disparate enriched GO terms or pathways, hence the differing biological interpretations [28]. Therefore, to ensure reproducible biological interpretations of microarray results, the reproducibility of gene lists is required. The concurrent use of fold-change ranking with a non-stringent P cutoff is recommended as baseline practice in the analysis of microarray data for the identification of differentially expressed genes.

Acknowledgements We thank participants of the MicroArray Quality Control (MAQC) project for their dedication in generating and analyzing the large data sets. Many MAQC participants contributed to the sometimes-heated discussions on the choice of methods for identifying differentially expressed genes. The common conclusions and recommendations evolved from this extended discourse. We invite the microarray community (in particular statisticians) to participate in the debate in order to reach consensus on microarray data analysis.

References

1. Frueh FW (2006) Nat Biotechnol 24:1105
2. Tan PK, Downey TJ, Spitznagel EL Jr., Xu P, Fu D, Dimitrov DS, Lempicki RA, Raaka BM, Cam MC (2003) Nucleic Acids Res 31:5676
3. Ramalho-Santos M, Yoon S, Matsuzaki Y, Mulligan RC, Melton DA (2002) Science 298:597
4. Ivanova NB, Dimos JT, Schaniel C, Hackney JA, Moore KA, Lemischka IR (2002) Science 298:601
5. Fortunel NO, Otu HH, Ng HH, Chen J, Mu X, Chevassut T, Li X, Joseph M, Bailey C, Hatzfeld JA, Hatzfeld A, Usta F, Vega VB, Long PM, Libermann TA, Lim B (2003) Science 302:393; author reply 393
6. Miller RM, Callahan LM, Casaceli C, Chen L, Kiser GL, Chui B, Kaysser-Kranich TM, Sendera TJ, Palaniappan C, Federoff HJ (2004) J Neurosci 24:7445
7. Miklos GL, Maleszka R (2004) Nat Biotechnol 22:615
8. Michiels S, Koscielny S, Hill C (2005) Lancet 365:488
9. Ioannidis JP (2005) Lancet 365:454
10. Frantz S (2005) Nat Rev Drug Discov 4:362
11. Marshall E (2004) Science 306:630
12. Shi L, Tong W, Fang H, Scherf U, Han J, Puri RK, Frueh FW, Goodsaid FM, Guo L, Su Z, Han T, Fuscoe JC, Xu ZA, Patterson TA, Hong H, Xie Q, Perkins RG, Chen JJ, Casciano DA (2005) BMC Bioinformatics 6 Suppl 2:S12
13. Ein-Dor L, Zuk O, Domany E (2006) Proc Nat Acad Sci USA 103:5923
14. Allison DB, Cui X, Page GP, Sabripour M (2006) Nat Rev Genet 7:55
15. Mecham BH, Klus GT, Strovel J, Augustus M, Byrne D, Bozso P, Wetmore DZ, Mariani TJ, Kohane IS, Szallasi Z (2004) Nucleic Acids Res 32:e74
16. Barczak A, Rodriguez MW, Hanspers K, Koth LL, Tai YC, Bolstad BM, Speed TP, Erle DJ (2003) Genome Res 13:1775
17. Shippy R, Sendera TJ, Lockner R, Palaniappan C, Kaysser-Kranich T, Watts G, Alsobrook J (2004) BMC Genomics 5:61
18. Hoffman EP (2004) Nat Rev Genet 5:229
19. Shi L, Tong W, Goodsaid F, Frueh FW, Fang H, Han T, Fuscoe JC, Casciano DA (2004) Exp Rev Mod Diagn 4:761
20. Wilson CL, Miller CJ (2005) Bioinformatics 21:3683
21. Jones L, Goldstein DR, Hughes G, Strand AD, Collin F, Dunnett SB, Kooperberg C, Aragaki A, Olson JM, Augood SJ, Faull RL, Luthi-Carter R, Moskvina V, Hodges AK (2006) BMC Bioinformatics 7:211
22. Tong W, Lucas AB, Shippy R, Fan X, Fang H, Hong H, Orr M, Chu T-M, Guo X, Collins J, Sun YA, Wang S-J, Bao W, Wolfinger RD, Shchegrova S, Guo L, Warrington JA, Shi L (2006) Nat Biotechnol 24:1132
23. Reid LH (2005) Pharm Discov 5:20
24. Baker SC, Bauer SR, Beyer RP, Brenton JD, Bromley B, Burrill J, Causton H, Conley MP, Elespuru R, Fero M, Foy C, Fuscoe J, Gao X, Gerhold DL, Gilles P, Goodsaid F, Guo X, Hackett J, Hockett RD, Ikonomi P, Irizarry RA, Kawasaki ES, Kaysser-Kranich T, Kerr K, Kiser G, Koch WH, Lee KY, Liu C, Liu ZL, Lucas A, Manohar CF, Miyada G, Modrusan Z, Parkes H, Puri RK, Reid L, Ryder TB, Salit M, Samaha RR, Scherf U,

Sendera TJ, Setterquist RA, Shi L, Shippy R, Soriano JV, Wagar EA, Warrington JA, Williams M, Wilmer F, Wilson M, Wolber PK, Wu X, Zadro R (2005) Nat Methods 2:731
25. Irizarry RA, Wu Z, Jaffee HA (2006) Bioinformatics 22:789
26. Shi L, Reid LH, Jones WD, Shippy R, Warrington JA, Baker SC, Collins J, Longueville Fd, Kawasaki ES, Lee KY, Luo Y, Sun Y, Willey JC, Setterquist RA, Fischer GM, Dix DJ, Dragan YP, Goodsaid FM, Herman D, Jensen RV, Johnson CD, Lobenhofer EK, Puri RK, Scherf U, Thierry-Mieg J, Tong W, Wang C, Wilson M, Zhang L, Amur S, Bao W, Barbacioru CC, Bergstrom-Lucas A, Bertholet V, Boysen C, Bromley B, Brown D, Brunner A, Canales R, Cao XM, Cebula TA, Chen JJ, Cheng J, Chu T-M, Chudin E, Corson J, Corton JC, Croner LJ, Davison TS, Delenstarr G, Deng X, Dorris D, Eklund AC, Fan X-h, Fang H, Frueh FW, Fulmer-Smentek S, Fuscoe JC, Gallagher K, Ge W, Guo L, Guo X, Hager J, Haje PK, Han J, Han T, Harbottle HC, Harris SC, Hatchwell E, Hauser CA, Hester S, Hong H, Hurban P, Jackson SA, Ji H, Knight CR, Kuo WP, LeClerc JE, Levy S, Li Q-Z, Liggett W, Liu C, Liu Y, Lombardi MJ, Ma Y, Magnuson SR, Maqsodi B, McDaniel T, Mei N, Myklebost O, Ning B, Novoradovskaya N, Orr MS, Osborn TW, Papallo A, Patterson TA, Perkins R, Peters EH, Peterson R, Philips KL, et al. (2006) Nat Biotechnol 24:1151
27. Shippy R, Wolber P, Jones WD, Jensen RV, Fulmer-Smentek S, Johnson CD, Pine PS, Boysen C, Guo X, Chudin E, Sun YA, Willey JC, Thierry-Mieg J, Thierry-Mieg D, Setterquist RA, Wilson M, Lucas AB, Novoradovskaya N, Turpaz Y, Baker SC, Warrington JA, Shi L, Herman D (2006) Nat Biotechnol 24:1123
28. Guo L, Lobenhofer EK, Wang C, Shippy R, Harris SC, Zhang L, Mei N, Chen T, Herman D, Goodsaid FM, Hurban P, Philips KL, Xu J, Deng X, Sun YA, Tong W, Dragan YP, Shi L (2006) Nat Biotechnol 24:1162
29. Wang Y, Barbacioru C, Hyland F, Xiao W, Hunkapiller KL, Blake J, Chan F, Gonzalez C, Zhang L, Samaha RR (2006) BMC Genomics 7:59
30. Tusher VG, Tibshirani R, Chu G (2001) Proc Nat Acad Sci USA 98:5116
31. Simon R (2006) J Nat Cancer Inst 98:1169
32. Ambroise C, McLachlan GJ (2002) Proc Nat Acad Sci USA 99:6562
33. Ntzani EE, Ioannidis JP (2003) Lancet 362:1439
34. Simon R, Radmacher MD, Dobbin K, McShane LM (2003) J Nat Cancer Inst 95:14
35. Tan Y, Shi L, Tong W, Gene Hwang GT, Wang C (2004) Comput Biol Chem 28:235
36. Tan Y, Shi L, Tong W, Wang C (2005) Nucleic Acids Res 33:56
37. Ashburner M, Ball CA, Blake JA, Botstein D, Butler H, Cherry JM, Davis AP, Dolinski K, Dwight SS, Eppig JT, Harris MA, Hill DP, Issel-Tarver L, Kasarskis A, Lewis S, Matese JC, Richardson JE, Ringwald M, Rubin GM, Sherlock G (2000) Nat Genet 25:25
38. Tong W, Harris S, Cao X, Fang H, Shi L, Sun H, Fuscoe J, Harris A, Hong H, Xie Q, Perkins R, Casciano D (2004) Mutat Res 549:241
39. Tanabe L, Scherf U, Smith LH, Lee JK, Hunter L, Weinstein JN (1999) Biotechniques 27:1210
40. Hoffmann R, Valencia A (2004) Nat Genet 36:664
41. Eisenstein M (2006) Nature 442:1067
42. Patterson TA, Lobenhofer EK, Fulmer-Smentek SB, Collins J, Chu T-M, Bao W, Fang H, Kawasaki ES, Hager J, Tikhonova I, Walker SJ, Zhang L, Hurban P, Longueville Fd, Fuscoe JC, Tong W, Shi L, Wolfinger RD (2006) Nat Biotechnol 24:1140
43. Chen JJ, Wang SJ, Tsai CA, Lin CJ (2007) Pharmacogenomics J 7:212

Springer Ser Fluoresc (2008) 6: 283–304
DOI 10.1007/4243_2008_040

Published online: 15 March 2008

Protein Arrays and Fluorescence Detection: Applications and Limitations

Christian Maercker[1,2]

[1]University of Applied Sciences, Department of Biotechnology,
Paul Wittsack Strasse 10, 68163 Mannheim, Germany
c.maercker@hs-mannheim.de

[2]German Cancer Research Center, Genomics and Proteomics Core Facilities,
Im Neuenheimer Feld 515, 69120 Heidelberg, Germany

Abstract Within genomics, spotted microarrays have become an increasingly important research tool for the biotechnology industry, molecular diagnostics, drug screening, nucleic acid sequencing, and mutation analysis. Working with proteins on arrays is especially difficult, because the native state of proteins is extremely sensitive to pH, ionic strength, temperature, and protein concentration. Since optimized conditions differ from protein to protein, the development of in vitro protein-based assays always presents a big challenge. In this chapter, the design and applications of peptide and protein arrays will be addressed. Different labeling chemistries for the analytes and signal detection procedures, with special emphasis dedicated to quality assurance related issues, are discussed. Possible applications of protein arrays, such as epitope mapping of autoantibodies, binding assays, analysis of protein expression, and protein modification, as well as monitoring of enzymatic activity, shed light on the high versatility of protein arrays. However, much more research and development will be necessary, until they meet the requirements of laboratories involved in high-throughput screenings and diagnostics.

Keywords Antibody microarray · Fluorescence detection · Functional genomics · Molecular diagnostics · Peptide microarray · Protein microarray · Quality assurance

Abbreviations

AP	Alkaline Phosphatase
APTES	Aminopropyltrimethoxysilane
BP	Bullous Pemphigoid, an immune-mediated blistering disease
CCD	Charge-Coupled Device
cDNA	Complementary Desoxyribonucleic Acid, copy of mRNA
Cy	Cyanine Dye
DCM	Dilated Cardiomyopathy
DMSO	Dimethylsulfoxidedimethylsulfoxide
DNA	Desoxyribonucleic Acid
ECF	Enhanced Chemofluorescence
Fc	Heavy chain constant (Fc) region of immunoglobulin
FHA	Forkhaed-Associated
Fmoc	9-Fluorenylmethyl carbamate
GFP	Green Fluorescent Protein
GST	Glutathione *S*-transferase
His	Histidine
IDA	Iminodiacetic acid
IgG	Immunoglobulin G
MIST	Multiple spotting technique
NEK6	Never In mitosis, gene A-related kinase-6
NIMA	Never In mitosis, gene A
ORF	Open Reading Frame
PEG	Poly Ethylene Glycol
mPEG	Multi-arm PEG
PH	Pleckstrin Homology
PMT	Photomultiplier Tube
RCA	Rolling Circle Amplification
RNA	Ribonucleic Acid
mRNA	Messenger Ribonucleic Acid
SAT	Bead-Based Suspension Array Technology
SPR	Surface Plasmon Resonance
ST	Sulfotransferase
ST	Sulfotransferase
TIRFM	Internal reflection fluorescence microscopy

1 Introduction

Functional genome research strongly relies on the application of highly parallelized high-throughput assays and bioinformatics. Both disciplines have evolved rapidly since the early 1990s. Within genomics, biochips have become an increasingly important research tool for the biotechnology industry, molecular diagnostics, drug screening, nucleic acid sequencing, and mutation analysis [1–6]. One classical chip format is the macroarray (22 cm×22 cm nitrocellulose or PVDF membranes) with spot sizes of 400 μm in a spotting distance of 900 μm (25 344 spots per membrane. www.rzpd.de). Today, the state-of-the art tool is the microarray in the format of microscope slides or even smaller. Spotted microarrays consist of a multitude of binding agents or so-called probes, for instance, DNA, RNA, sugars, peptides, proteins, glycoproteins, oligonucleotides, or up to tens of thousands of single stranded DNA fragments, representing the genes of a genome. The probes are deposited onto the surface of a solid support, for instance, a coated glass slide, in the form of an array with spot sizes in the 50 to 200 μm range. Even spotted cells are used for high-content screening assays in pharmaceutical research.

With these arrays, RNA and protein expression, tissue-specific splice variants, gene mutations, protein binding to proteins, DNA and RNA, enzymatic activities, localization of proteins in cellular compartments, and cell differentiation processes can be measured on a whole-genome level. In comparison to other binders, working with proteins on arrays is especially difficult, because the native state of proteins is extremely sensitive to pH, ionic strength, temperature, and protein concentration. Since optimized conditions differ from protein to protein, the development of in vitro protein-based assays always presents a big challenge. This is also the reason why applications of protein arrays for high-throughput screenings and routine diagnostics are limited so far. In this chapter, the design and applications of peptide and protein arrays will be addressed with special emphasis dedicated to quality assurance related issues.

2 Peptide and Protein Arrays: Composition, Production, and Processing

2.1 Peptide Arrays

Peptide arrays are a perfect choice for various approaches in functional genome research. DNA sequence information in databases facilitates the design of peptide sequences relevant for various assays. The spot synthesis of large numbers of peptides on membrane sheets (SPOT), using Fmoc (9-fluorenylmethyl carbamate) as a protective group, promotes parallel synthesis of thousands of peptides on planar supports, such as modified cellulose or polypropylene membranes [7, 8]. Step by step, activated amino acids are pipetted onto defined positions and built up to peptides. After each pipetting step, a washing procedure removes excess amino acids and chemicals to ensure the purity of the peptide library. The spotted molecules have a length of up to 38 amino acids with a concentration between 1 and 5 nMol. To achieve a higher spot density, cellulose-bound peptides can be eluted, precipitated,

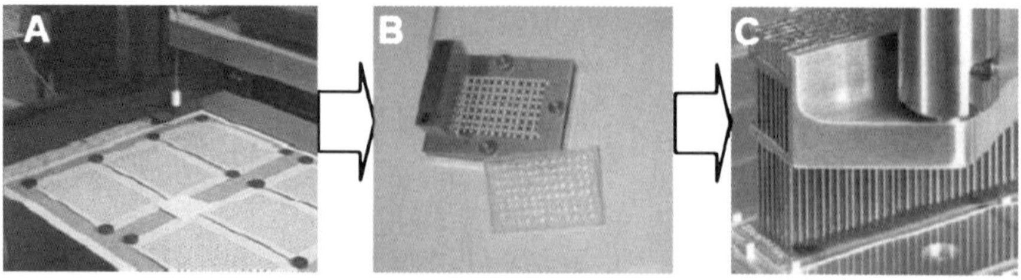

Fig. 1 Example of a peptide fabrication process. **A** Synthesis of peptides on a cellulose solid support by SPOT synthesis. **B** Separation of the assembled solid supported peptides and distribution into microtiter plates; deprotection and dissolving of cellulose in TFA followed by precipitation of the cellulose-compound conjugates in ether; dissoving the cellulose-compound in DMSO. **C** Printing of multiple copies from the DMSO solutions onto glass slides. Kindly provided by Dr. Ronald Frank, GBF Braunschweig

dissolved in dimethylsulfoxidedimethylsulfoxide (DMSO), and spotted onto glass slides (Fig. 1). Spot distances of 300 to 400 μm or 120 000 dots per standard microscope slide are available, depending on different immobilization chemistries for the peptides and slide surfaces [7, 8]. One possibility is the modification of glass slides with a multifunctional polyethylene glycol (PEG)-based polymer. The novel support material allows a versatile modification of the amino group surface density up to 40 $nmol/cm^2$ for the linkage of probe molecules [9].

High-density peptide arrays on membranes can be manufactured with the help of a modified laser printer. The colored toner particles are substituted by amino acid-toner particles, which consist of a "solid" solvent (diphenylformamide) and the amino acid. After a temperature increase, the amino acids are specifically released and coupled to the membrane or to the already synthesized peptide. This technique enables us to build up peptides with lengths varying between 15 and 20 amino acids. The low spreading caused by adjusting of the solvent by selective heating and high speed of the printer allows complexities of millions of different peptides on a single chip with a resolution as high as 1200 dpi [10].

Novel chips are covered with pixel electrodes, which facilitate an accurate positioning of molecules. Particles containing derivatives of modified amino acids are transferred to these chips with a specialized surface chemistry [11].

2.2 Protein Arrays

Although more complex and labor intensive, the interest in immobilizing whole proteins is very high, because these arrays can be used to address the functions of individual proteins. Whole proteins cannot be produced chemically. Instead, they have to be synthesized in vivo or in vitro. Well-known expression systems are bacteria (*E. coli*), yeast, insect cells, or mammalian cells. A standard method is the expression of N-terminally tagged proteins (e.g., glutathione *S*-transferase (GST) or histidine tag (6xHis)). These fusion proteins bind to specific affinity resins (e.g., glutathione or Ni-NTA agarose) so that they can easily be purified and spotted onto glass slides. For certain applications, even whole cells can be spotted without further protein purification [12]. By using high-throughput procedures, collections of several thousands of full open reading frame (ORF) proteins have already been produced from various organisms and tissues [13] (www.invitrogen.com, www.genecopoeia.com, www.rzpd.de).

Antibody arrays are the most popular protein biochips. First of all, antibodies against different antigens are relatively similar in structure, thus allowing standardized purification and spotting conditions. Secondly, spotted antibodies very often have the same binding specificities as in solution. Thirdly, proteins in solution can be bound to spotted antibodies to display their native activity, which makes them a flexible tool for different applications (see Sect. 3) [14].

Cell-free expression of proteins by in vitro transcription and translation from DNA anchored to glass surfaces with lysates from mammalian reticulocytes or wheat germ is a promising tool used to investigate the protein in its native state. The DNA template as well as a capture antibody for the expressed protein are immobilized onto the glass slide (Fig. 2), or, based on the multiple spotting technique (MIST), are captured on aminopropyltrimethoxysilane (APTES) or Ni-chelate surfaces [15, 16]. Alternatively, a bead-based suspension array technology (SAT) with biotinylated PCR products binding to streptavidin-coated polystyrene micrometer-sized beads can be used [17]. With these native proteins being immobilized in situ in large numbers, they became accessible for functional studies without the need of protein purification.

Since the accessibility of the protein on the array is an important issue for signal detection, adequate methods to immobilize spotted proteins on surfaces have been developed. For example, the heavy chain constant (Fc) region of immunoglobulin G (IgG) binds specifically to protein A or protein G. To increase the self-adhering ability on hydrophobic surfaces, a Fc binding protein has been designed and it consists of a Fc binding domain of protein G and a hydrophobic domain of elastin [18]. Other proteins can be coupled to agarose microbeads in silicon microcavities. Alexafluor-488-linked antibodies binding to these proteins can generate signals which are measured with a charge-coupled device (CCD) camera with a detection limit of 1ng/ml [19]. A special solution for polyhistidine-tagged protein molecules are high-density PEG-coated silicon surfaces, which are obtained from the reaction of a multi-arm PEG (mPEG) molecule with a chlorine terminated Si(111) surface. Four out of the eight arms on each immobilized mPEG molecule are accessible for linking to the chelating iminodiacetic acid (IDA) groups for the binding of Cu(2+) ions. The resulting Cu(2+)-

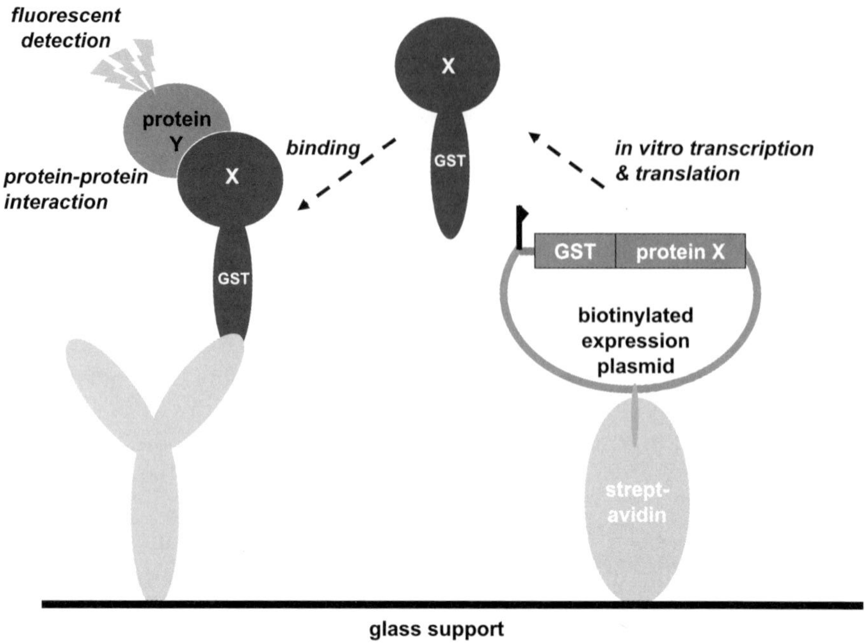

Fig. 2 Protein binding assay on a self-assembling protein microarray. Biotinylated DNA, encoding a fusion protein between GST and a protein X, binds to the streptavidin coated glass slide. After overlay with in vitro transcription and translation extract, the protein is expressed and binds to the anti-GST antibody on the slide. If protein Y, which was fluorescently labeled before, binds to protein X, the spot on the glass slide emits a fluorescent signal, which can be detected by a laser scanning (see Sect. 3). Figure kindly provided by Dr. Manfred Koegl, DKFZ Heidelberg

IDA-mPEG-Si(111) surface is shown to specifically bind 6xHis-tagged protein molecules, including green fluorescent protein (GFP) and sulfotransferase (ST), but otherwise retains its inertness towards nonspecific protein adsorption. Therefore, proteins are immobilized without the need of prepurification [20].

Immobilization of probes within three-dimensional hydrogels offers many advantages over two-dimensional surface immobilization. In particular, the increase in immobilization capacity as compared to surface microchips, aqueous surrounding of immobilized compounds, absence of contacts with hydrophobic surface, and the stabilizing effect of the gel appear to be especially important for proteins. Application of fluorescence Cy5-labeled mouse IgG and fluorescence quantification by scanning laser confocal microscopy showed fluorescence intensities with attomole level sensitivity for the 3D films measured in direct capture immunoassays. This is 3- to 10-fold higher compared to analogous 2D surfaces. However, 2D surfaces reported equal or greater sensitivity on a per-molecule basis [21].

2.3
Labeling Chemistry

Besides surface chemistry and presentation of the protein on the array, the label of the analyte is critical for signal intensity and specificity. Most common procedures employ antibodies which are directly linked to fluorescent dyes or enzymes, such as alkaline phosphatase, which convert a substrate into a fluorescent signal. With secondary antibodies, multiple binding of the secondary to the first antibody and, therefore, signal amplification is possible. Still, methods to amplify signal intensity on small spots on the array and to limit concentration of spotted proteins and binders are being developed. One example of an interesting solution is the "rolling circle signal amplification". Antibodies bound to the target are linked to a DNA molecule which is extended by DNA polymerase and labeled afterwards by hybridization of

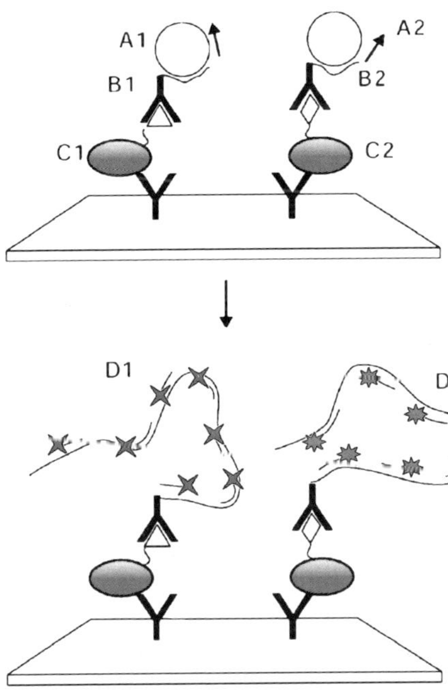

Fig. 3 Schematic representation of two-color rolling circle amplification on antibody microarrays. Two pools of proteins are respectively labeled with digoxigenin (*C1*, digoxigenin-labeled protein, digoxigenin represented by the triangle) and biotin (*C2*, biotin-labeled protein, biotin represented by the diamond). DNA oligonucleotide-primers conjugated to anti-digoxigenin (*B1*) and anti-biotin (*B2*) bind to captured proteins, followed by hybridization of DNA circle *A1* and *A2*. Polymerase extends the primers using the circles as templates. Cy5-labeled oligonucleotides (*D1*) and Cy3-labeled oligonucleotides (*D2*) are hybridized to the extended DNA strands, producing signal amplification in *two colors* [22]

the antibody-linked DNA to fluorescence dye-labeled oligonucleotides [22] (Figs. 3 and 4).

For typical protein microarray experiments with fluorescence detection, proteins from biological samples and from controls are isolated and labeled with two spectrally distinguishable dyes, such as the cyanine dyes, Cy3 (absorption at ca. 550 nm, emission at ca. 565 nm) and Cy5 (absorption at ca. 650 nm, emission at ca. 675 nm), by NHS-ester linkage, for instance. The labeled biomolecules are then reacted with the microarray, thereby binding to antibodies present in the spotted probes. Alternatively, antibody arrays are overlayed with unlabeled proteins, which are detected again by specific antibodies against the different proteins bound to the array. This leads to more specific and reproducible results. However, the assay is also more complex

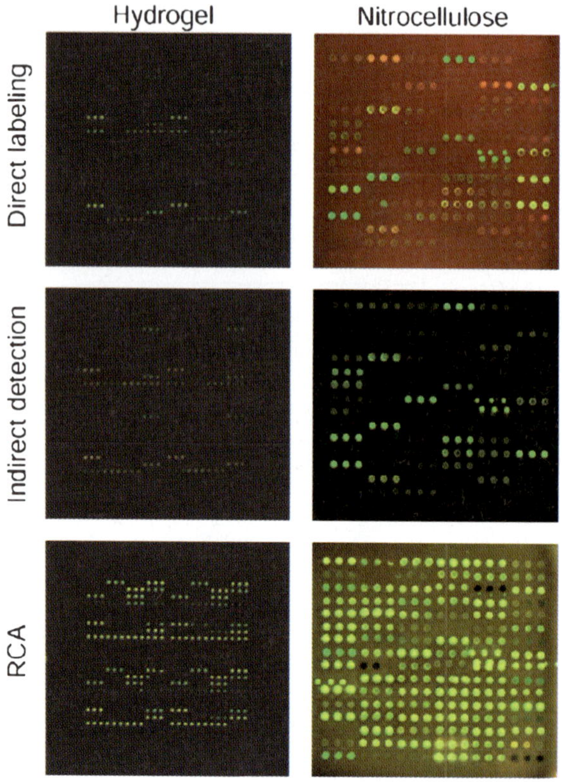

Fig. 4 Representative images of antibody microarrays with different surfaces and after different labeling procedures. A serum sample was incubated on antibody microarrays prepared on hydrogels (*left*) and nitrocellulose (*right*) and detected with direct labeling (*top*), indirect detection (*middle*), and rolling circle amplification (RCA, *bottom*). The same serum sample was used in *each color* channel for each experiment. Scanner settings were identical within microarrays performed on the same substrate [22]

and more cost intensive. If the signal is amplified by a labeled secondary antibody against the antibodies recognizing the target, two different antibodies from two different hosts (e.g., mouse and rabbit) are needed for each protein of interest.

The first antibody also can be bound to a secondary antibody (e.g., an anti-mouse antibody), which is fluorescently labeled or linked to an enzyme (e.g., alkaline phosphatase) to convert a nonfluorescent substrate into a fluorescent one that is subsequently detected (e.g., ECF, enhanced chemifluorescence. www6.gelifesciences.com).

Reverse-phase arrays contain unlabeled proteins which are probed with fluorescently labeled antibodies (see also Sects. 3.5 and 4).

These labeling methods are relatively easy to handle and labeling kits are also available to standardize the procedures as well as possible. However, the direct labeling of proteins sometimes leads to problems. Firstly, the specific labeling efficiency may differ between experiments. This is a strong disadvantage if results have to be quantitated or experiments within a series have to be compared. Secondly, the label might be responsible for changes in protein conformation, which might influence the binding activity or enzymatic capabilities of the analyte. Therefore, many groups propagate label-free signal detection systems as the state-of-the-art techniques (see Sect. 2.4).

2.4
Signal Detection

The final readout of an array experiment is very much dependent on high-end instruments more than any other step within the whole procedure. These machines not only have to have the capability to produce exact, sensitive, and reproducible measurements, but they also should be equipped with a flexible software which is easy to handle and which delivers data that can be understood and exchanged by all experts in the field.

Typically, spot intensities on a microarray are measured with an array reader, a confocal laser scanner equipped with a He–Ne laser (543.5, 594, 612, and 632.8 nm) or a Kr–Ar laser (488, 514, 568, and 647 nm) or less commonly, a solid-state laser or a white light source and a photomultiplier tube (PMT) or a CCD detection system with a lateral resolution of 5 to 10 μm, respectively. This step involves excitation at two different wavelengths and subsequent recording of integral fluorescence intensities in two detection channels, the so-called green and red channels [23]. Usually, the scanner software is able to combine signal intensities of both channels and to deliver a false-color image. The amount of analyte bound to a spot follows from the measured integral fluorescence intensity with abundant targets seen either in red or green and equal amounts in yellow. This allows for the estimation of relative gene expression levels from the fluorescence intensities of the spots. For the generation of reliable data, several repetitions and controls are necessary with the obtained expression values being statistically analyzed and normalized. Further bioinformatic analysis, e.g., hierarchical agglomerative clustering of the gene expression data, provides the basis for the interpretation of the alteration in gene expression of many genes in parallel [24–26]. Microarray experiments are performed either by ratio or direct measurements comparing two protein samples, for example, healthy and disease stage, or by absolute or indirect measurements comparing one sample to a reference, for instance, standardized labeled antibody. Therefore, this yields either relative values (ratios) or absolute values. To improve the reliability of the data, suitable controls are introduced that are incorporated into every experiment and are accordingly present on the array and spiked into the binding reaction or the enzymatic assay. This includes range, ratio, negative and contamination controls, as well as a positive control and a guide-dot for orientation after scanning.

While fluorescence detection of labeled proteins is a standard methodology, facilitated by excellent scanners and fluorescent dyes, it still produces artifacts in some cases, when the sterical integrity of the protein is disturbed by the incorporated dye or when label reproducibility is not satisfying. This is the reason why label-free methods are currently being discussed and may yield certain advantages over fluorometric detection. An example are surface plasmon resonance (SPR) biosensors for the identification of protein–protein interactions [27]. Several companies offer SPR systems to analyze up to 500 samples simultanously.

Another interesting label-free application is the direct quantitative measurement of protease activity on a glass array. The probe is nonfluorescent in the absence of the protease and the fluorescence intensity increases depending on the protease activity, due to the cleavage of the quencher-containing moiety on the glass slide [28]. This method works without labeling of proteins, however, at the same time, standard dyes and fluorescence detection systems are applicable for this approach, which should make this approach extremely interesting for future developments.

2.5
Variables Influencing Array Experiments

The reliability of the results obtained from microarray experiments depends on the biological and experimental variation as well as – to a smaller extent – on fluorescence detection and data analysis. Furthermore, due to the lack of generally accepted standardized sampling protocols and biological standards to be used in every microarray experiment, problems related to biology and data analysis are amplified [29]. In general, the contribution to the overall uncertainty from the biological side has the highest impact due to genetic and environmental variations, which are difficult to measure. Fluorescence-inherent sources of error can be more easily identified only if the necessary controls and standards are included in the experiment and if the scanners are properly characterized. Fluorescence-based uncertainties are related to fluorophore labeling of the target [30] and to problems linked to measurements of fluorescence intensities, i.e., the sensitivity of the label's spectroscopic properties to chromophore microenvironment [31, 32], to dye–dye interactions [33], and to instrument-specific effects. To overcome limitations in the former, e.g., different incorporation rates of the dyes or dye instability, the so-called dye swap or fluor-reverse is used with both targets being labeled with two dyes in separate reactions with identical arrays and identical probe arrangements. In addition to the instrument-related effects already discussed, microarray scanner-specific uncertainties are linked to the determination of the background signal, the method used for dye normalization, crosstalk between channels, and variations in the alignment of the optical scanner [34]. To increase the sensitivity by rejecting light coming from other planes than that of the biological signals of interest, confocal reacting is used. However, this method needs accuracy of the focus and tilt adjustment of the biochip. Typically, detection systems are less sensitive in the long wavelength region of ca. 675 nm as compared to a detection wavelength of ca. 565 nm. However, long wavelengths, as long as near infrared, are getting more and more popular. New dyes with long excitation wavelengths together with compatible detection systems are on the market. A strong advantage of this system is that background fluorescence is very low in most cases. Therefore, specialized spotting surfaces which otherwise would not be applicable in protein array experiments, like nylon, nitrocellulose, or activated glass surfaces, can be used [35].

3 Applications of Protein Arrays

3.1 Antibody Epitope Mapping

Antibodies are a versatile tool for many applications in research, diagnostics, and therapy of diseases. Immunizing mice, rabbits, rats, or other animals with the appropriate antigen or screening phage display libraries are commonly used methods to isolate these molecules. Additionally, several groups are trying to generate specific binders against all human proteins [36].

The quality of an antibody, which is decisive for any assay, is determined by specificity for a certain epitope and its affinity towards a recognized antigen. Therefore, characterization of capture molecules is a major challenge for protein array based systems. Recombinant antibody technologies, such as human combinatorial antibody library (HuCAL®, MorphoSys, Munich, Germany), allow the fast generation of highly specific binders to nearly any given

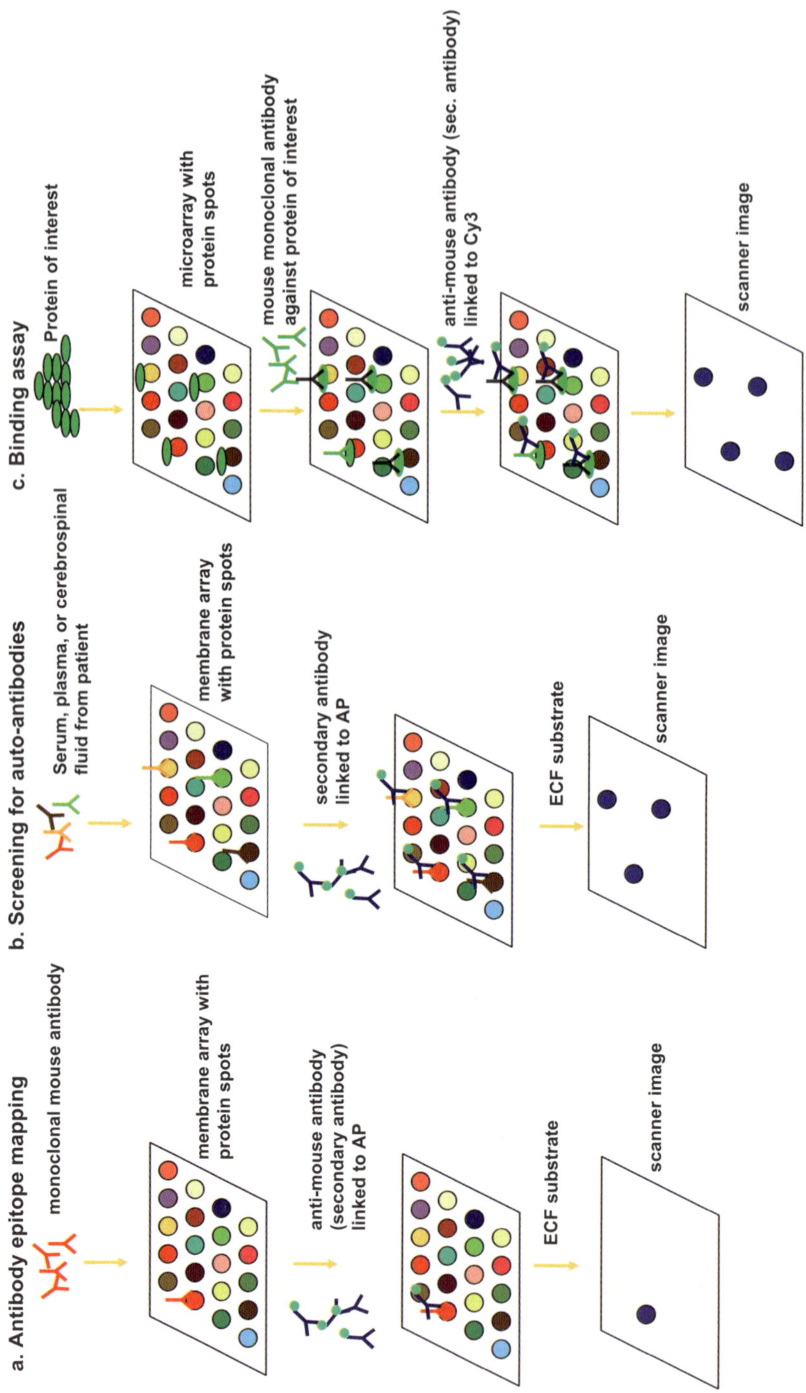

Fig. 5 a–c Different applications of protein arrays. For details, see text

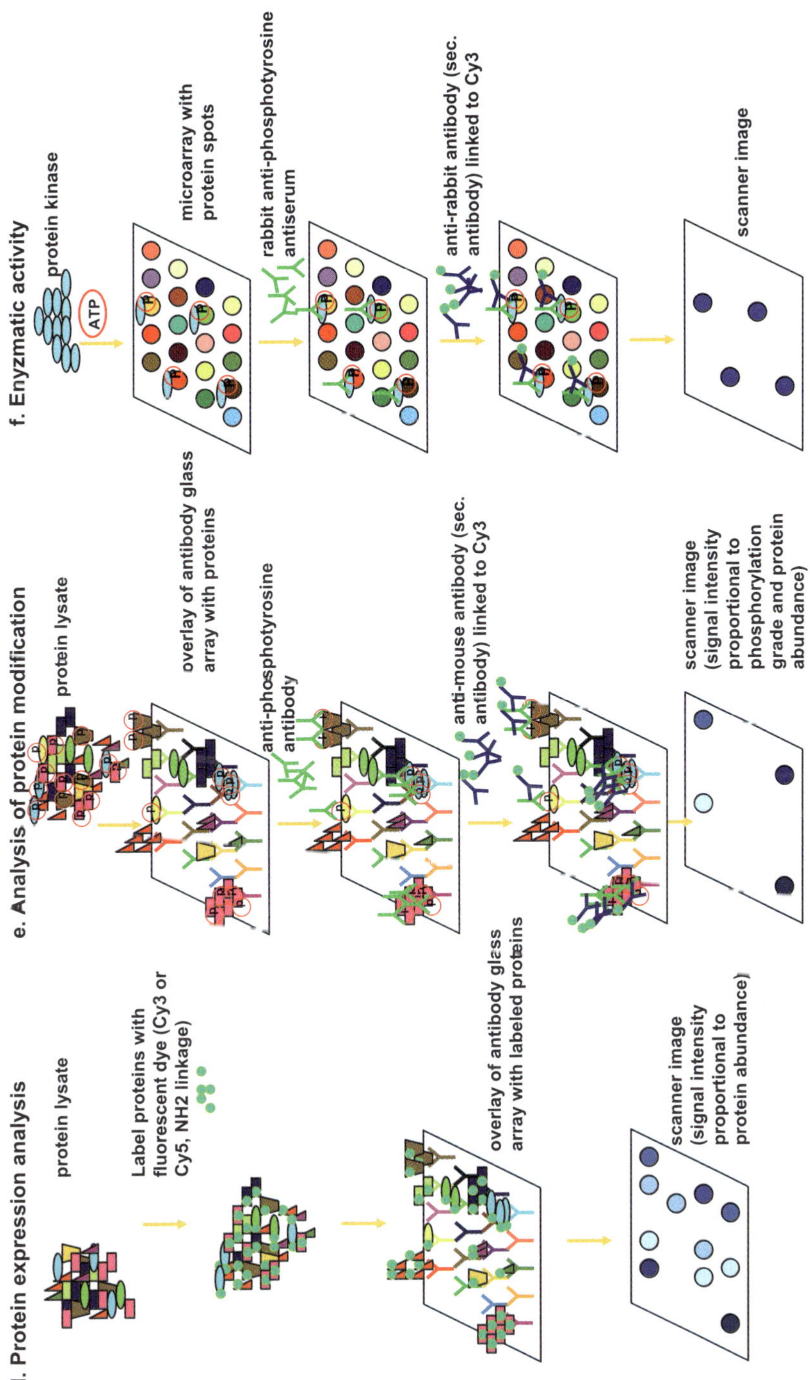

Fig. 5 d–f Different applications of protein arrays. For details, see text target molecule. Although antibody libraries comprise billions of members, it is not the selection process but the detailed characterization of the preselected monoclonal antibodies that presents the bottleneck for the production of high numbers of specific binders.

The size of the epitope recognized by the antibody is five to eight amino acids, which is normally much smaller than the antigen used for immunization. Therefore, for most assays, it is of great interest to find out which epitope exactly is recognized by the antibody. A very attractive method to address this problem is to incubate the antibody of interest with peptide arrays, protein arrays, or even spotted protein expression libraries. The antibody (e.g., monoclonal mouse antibody) binds to epitopes represented on the array. The first antibody can be detected by a secondary antibody carrying a fluorescent label (see Sect. 2.3). Arrays with short peptide sequences representing amino acid sequences of a certain antigen are a common tool for this application (Fig. 5a) [37]. An example for proof of principle is the epitope fine mapping of antibodies raised against the extracellular domain of the CD11b protein [38].

3.2 Screening for Autoantibodies

Many complex diseases are caused by autoimmune responses. Examples are diabetes mellitus type I, rheumatoid arthritis, sarcoidosis, and multiple sclerosis. Even in diseases like cancer, specific autoantibodies are involved [39]. Therefore, the targets of these antibodies are important prognostic and diagnostics markers. Serum screening on protein arrays allows us to identify binders in blood serum, blood plasma, or other body fluids. For instance, Epstein–Barr virus proteins in cerebrospinal fluid were found to be putative targets of the immune response in multiple sclerosis (Fig. 5b) [40]. Multiplexing of immunoassays – as it is necessary for clinical applications – will enable us to investigate multiple antigens in a single experiment, for example, virus antibodies in human serum [41]. For the straightforward detection of autoimmune diseases, such as bullous pemphigoid (BP), an immune-mediated blistering disease, and dilated cadiomyopathy (DCM), it is also important to discriminate between immunoglobulin subclasses like IgG1, IgG2, IgG3, IgG4, and IgM, which can be achieved by, for instance, IgG3-specific secondary antibodies [42, 43].

3.3 Binding Assays

The identification of binding partners is an important step in describing the function of a protein. Therefore, numerous microarray-based assays have been developed to show binding of proteins, nucleic acids, and other compounds to proteins immobilized on slides or membranes. Possible probes are short peptide motifs that serve as recognition modules for the assembly of multiprotein complexes. Examples are SH3 and SH2 (Src homolgy 2 and 3), PDZ (a protein domain originally identified in proteins PSD-95, DLG, and ZO-1), pleckstrin homology (PH), and forkhead-associated (FHA). The peptides were fused to glutathione-S-transferase for preparation of the protein-domain microarray [44]. A global protein interaction study was performed with a yeast proteome chip containing 5800 open reading frame polypeptides, which were probed with biotinylated calmodulin. In addition to known calmodulin kinases and caclineurins, 33 new potential binding partners of calmodulin were identified. Arrays of this kind also can be used to study protein–small molecule or protein–lipid interactions [13, 45]. To gain insights into transcription and DNA replication, the analysis of protein–DNA interactions and the determination of their sequence requirements are of central importance. To follow this issue, different domains of bacterial replication initiator DnaA, spotted on FAST slides (glass slides covered with a nitrocellulose membrane, Whatman Schleicher and Schuell), were probed with Cy5-labeled, double-stranded, 21-mer oligonucleotides [46]. Protein–carbohydrate interactions are essential for many biological processes including normal tissue growth and repair, cell–cell adhesion, inflammation, cell growth, fertilization, viral replications, parasitic infection, as well as tumor–cell motility and progression. Alterations of gycosylation events are also involved in a number of diseases.

An important step in functional proteome analysis is the identification of binding partners of individual polypeptides. To approach this problem, it is possible to overlay an array with the protein of interest. The protein will bind to any binding partner on the array, and the resulting protein–protein interaction can be detected by a sandwich of a first antibody recognizing the binding partner and a labeled secondary antibody (Fig. 5c). It is also possible to label the potential binding partner directly before incubation with a peptide or protein array and scanning [47]. The more sophisticated immobilization of proteins on beads in a density of more than 100 microbeads per μm^2 in combination with total internal reflection fluorescent microscopy (TIRFM) permits monitoring of protein–protein interactions, for example, between antigen and antibody (compare Sects. 3.1 and 3.2). DNA–protein interactions and interactions with other compounds in extremely small volumes of target solutions, such as extracts from single cells, can be measured in a very high throughput [48].

3.4
Protein Expression Analysis

The most common way to detect gene expression is to label mRNA or its DNA copy (cDNA) and to hybridize DNA microarrays containing probes representing all genes of an organism. This high-throughput approach is easy and efficient, however has some disadvantages: The expression rate of RNA, the turnover rate, and the timepoint of expression are not necessarily sufficient to give information about localization, abundance, half-life of the proteins, or post-translational modifications. These disadvantages can be compensated by performing expression analyses on the protein level. Combining both approaches enables us to correlate the mRNA information with the protein information. Antibodies on the array are directed against the proteins to be analyzed. Proteins from cells or tissues are either directly labeled with a fluorescent dye (e.g., Cy3, Cy5) or after prefractionation steps. Labeling by NHS–ester linkage is normally used to be able to detect the proteins proportional to protein abundance in a tissue or a certain cell type, respectively (Fig. 5d) [49]. Applying this method, regulation of certain transcription factors was found to be potentially involved in spinal muscular atrophy [50]. Antibodies or single-chain Fv antibodies generated via phage display libraries [51] can also be used for protein binding studies (see Sect. 3.3).

Compared to Western blot analysis, these assays not only possess a higher degree of sensitivity, but they also can be scaled up for throughput screening of hundreds of proteins in complex biofluids, such as blood [52, 53]. Other diagnostical applications are quantitative immunoassays for plant and bacterial toxins [54, 55] or biomarkers in complex diseases, like cancer and rheumatoid arthritis [56–58].

3.5 Analysis of Protein Modification

Specific phosphorylation of single amino acids within a protein is a very important posttranslational modification, which is relevant for many signaling processes in the cell. Detection of tyrosine phosphorylation is possible on antibody arrays by utilizing antibodies specifically directed against this modified amino acid (see also Sect. 3.6). Again, antibody arrays are overlayed with protein lysates. The antiphosphotyrosine antibody recognizes all tyrosine phosphorylated proteins attached to the specific antibodies on the array. A secondary antibody, linked to a fluorescent dye (e.g., Cy3), leads to detection of signals with intensities proportional to protein concentration and phosphorylation grade (Fig. 5e) [59, 60]. Today, even serine phosphorylations can be detected by the same method. The more specific approach is the use of different antibodies specific for nonphosphorylated and phosphorylated conformation of each individual protein. An example is the analysis of signal transduction pathways in stimulated T lymphocytes on reverse-phase microarrays [61]. However, due to complexity and costs of biological material, this method is not frequently applied in array format. The same is true for other protein modifications, such as sumoylation, ubiquitination, and acetylation, which are also detectable by antibodies.

3.6
Measurement of Enzymatic Activity

The detection of enzymatic activities with immobilized proteins is still limited. Only a very small number of studies describing this approach have been published so far. Major challenges are to keep the proteins on the array in the native state and also to bind cleavage products to the respective spots, for instance. One reason for this is the unavailability of appropriate substrates and products for the biochemical reactions of interest. Typically, reactions are performed in solution. Therefore, many array-based approaches work with microwells or nanowells containing the reaction partners.

On solid protein microarrays, it is possible, for example, to perform kinase assays. Proteins which are spotted on chemically derivatized glass slides can be phosphorylated by enzymes to identify substrates of protein kinases (Fig. 5f) [62]. A peptide chip was used successfully in an activity assay of the nonreceptor tyrosine kinase c-Src [63]. Arrays spotted by the SPOT technique permitted to determine the substrate specificity of the NIMA (never in mitosis, gene A)-related kinase-6 (NEK6) [64].

A label-free method has been proven to be applicable for the verification of protease activity. Caspase-3, a key player in apoptosis, was incubated with a glass array together with fluorphore and quencher. The probe was nonfluorescent in the absence of caspase-3 and the fluorescence intensity increased depending on the caspase-3 activity due to the cleavage of the quencher-containing moiety on the glass slide [28].

The development of enzymatic assays in array format stands at its beginning. However, highly parallelized whole-genome assays are a must for the progress in functional genomics research, and in pharmaceutical development. Therefore, there will be many new assays available in the near future which will fulfill high demands regarding robustness, velocity, and compatibility with recognized detection systems.

4
Quality Assurance of Array Data

Absolute quantification of proteins with spotted arrays is difficult. Normally, both samples (e.g., control and disease) are compared for relative quantification of protein abundance. Concerning the biological material to be investigated, high linearity and excellent sensitivity without the need for labeling of sample proteins are possible with reverse-phase protein microarrays (see also Sects. 2.3 and 3.5). For instance, lysates are prepared from cultured cells of microdissected tissues by laser capture microdissection and assembled in a microarray format. The whole repertoire of sample proteins that represent the state of individual tissue cell populations undergoing disease transitions are immobilized on the array. To monitor cancer progression, for example, the microarrays are screened with specific antibodies for the presence of defined target proteins [65].

The quality of assays on protein arrays also very much depends on the surface of the array and the specific signal intensity. Sputtering multilayers of thin metal films, metal oxides, and metal nitrides onto glass wafers are possibilities to improve the signal-to-background ratio [66]. If gold surfaces or aminated glass slides are coated with calixcrown derivatives with a bifunctional coupling property, direct or sandwich fluorescence immunoassays have a sensitivity of as low as $1-10 \times 10^{-18}$ g/l analyte protein. For example, protein–protein interactions (e.g., between integrin and extracelular matrix proteins) can be measured with these kinds of chips [67].

At present, only very few standards are available to improve the reliability of microarray experiments from the detection side. These are intensity standards designed, for example, for control of the performance of array scanners, channel crosstalk, and effect of PMT voltage, as well as for comparison of performance of different instruments (e.g., Clondiag Chip Technologies GmbH; Full Moon Biosystems Inc.). One of them offers some means for control of the spatial resolution [68]. Some microarray reader manufacturers have integrated a self-test using an internal light source, which runs at the start of the instruments and before every scan (e.g., Fuji Photo Film Co.) to check PMT settings, whereas others (e.g., Axon Instruments Inc.) provide specific calibration slides for an evaluation of the instrument performance.

Better suited calibration slides or calibration spots on every microarray are desired to help adjust the scanner settings and to determine the instrument's image/spatial resolution and its day-to-day performance and long-term stability – ideally, in combination with guidelines for their use. Furthermore, tested procedures in combination with recommended materials/systems are needed for the determination of the range of linearity of the array scanners, their dynamic range, and sensitivity. The final goal is to improve the comparability of data generated by different instruments/laboratories and with different labels. Easy-to-use fluorescence standards that meet these requirements should, for instance, either provide a stable output at application-relevant wavelengths and photon fluxes or need to be easily reproduced, should have minimum local and global nonuniformities in emission and should generate signals comparable to those observed for typical samples. In this respect, also the influence of spectral deviations between the fluorescence spectra of the standard and common labels needs to be examined. This renders the important availability of absorption and corrected emission spectra of both the standard and the labels for application-relevant conditions important, i.e., microenvironment(s). For testing of the instrument's spatial resolution, which should be at least in the 5 μm-range, uniformly patterned standards are desired. Such standards must be applicable for the majority of typically used microarray readers and labels, for example, suited for typical excitation wavelengths and emission wavelength intervals as well as scanner formats. Moreover, they should have been tested with a broad variety of instruments in a Round Robin test. For the development of such purpose-fit fluorescence standards, input not only from instrument manufacturers, but also from the community of users of the microarray technology is required. Additionally, future technological advances of microarray readers which will most likely provide an enhanced spatial resolution, precision, sensitivity, and a higher dynamic range should be considered. Furthermore, it needs to be decided on within the community of users of the microarray technology and instrument manufacturers, whether standalone fluorescence standards, like calibration slides, or the integration of fluorescence standards into slides used for the performance of microarray experiments are to be favored. The former is, for instance, sufficient to characterize the day-to-day and long-term instrument performance, whereas the latter approach – though more costly – may be tempting for an improved and more feasible comparability of the recorded fluorescence signals.

To make protein microarray data comparable worldwide, there is a strong need for instrument standards, application of biological reference materials, and standardized procedures/protocols for the performance and evaluation of such experiments. For the reduction of biology- and data evaluation-related uncertainties, the microarray research community has been discussing the introduction of quality controls (QCs) and SOPs as well as biological reference materials for a long time. Standards are based on genomic DNA, a reference RNA pool, or a reference oligonucleotide for measuring more precisely absolute or indirect gene expression values [69, 70] (see Chapter 26 by W. Nietfeld). The introduction of similar standards for protein microarrays would be very desirable.

In conclusion, protein arrays are a versatile tool for various applications. Despite the need for much more research and development, much progress has been made in publicly and privately funded initiatives during the last years. Due to the great impact of protein microarrays on any aspect of functional genomics, they will find their way into more and more laboratories involved in high-throughput screenings and diagnostics in the near future.

Acknowledgements Crucial contributions to this chapter by Ute Resch-Genger are greatly acknowledged. I also would like to thank Jocelyn Faberman for improving the language of this chapter, also Harald Seitz and Wilfried Nietfeld (both from Max Planck Institute for Molecular Genetics, Berlin) for critical reading of the manuscript and helpful discussions. The author's knowledge of protein arrays is very much based on his experiences as a member of the RZPD German Resource Center for Genome Research between 2000 and 2006. His work was funded by the German Federal Ministry of Education and Research (BMBF) within the National Genome Research Network (NGFN2).

References

1. Pirrung MC (2002) How to make a DNA chip. Angew Chem Int Ed 41:1276
2. Epstein JR, Biran I, Walt DR (2002) Anal Chim Acta 469:3
3. Lemieux B, Asharoni A, Schena M (1998) Mol Breeding 4:277
4. Hedge P, Qi R, Abernathy K, Gay C, Dharap S, Gaspard R, Hughes JE, Snesrud E, Lee N, Quackenbush J (2000) BioTechniques 29:548
5. Christensen CBV (2002) Arrays in biological and chemical analysis. Talanta 56:289
6. Biochips (Market survey) (2003) New Drugs, p 26
7. Frank R (2002) Comb Chem High Throughput Screen 5:429
8. Frank R (2002) J Immunol Methods 267:13
9. Beyer M, Felgenhauer T, Bischoff RF, Breitling F, Stadler V (2006) Biomaterials 27:3505
10. Bischoff FR, Stadler V, Breitling F (2002) Biospektrum 5:654–657
11. Beyer M, Nesterov A, Block I, König K, Felgenhauer T, Fernandez S, Leibe K, Torralba G, Hausmann M, Trunk U, Lindenstruth V, Bischoff FR, Stadler V, Breitling F (2007) Science 318:1888
12. Cahill DJ, Nordhoff E (2003) Adv Biochem Eng Biotechnol 83:177
13. Zhu H, Bilgin M, Bangham R, Hall D, Casamayor A, Bertone P, Lan N, Jansen R, Bidlingmaier S, Houfek T, Mitchell T, Miller P, Dean RA, Gerstein M, Snyder M (2001) Science 293:2101
14. Sanchez-Carbayo M (2006) Clin Chem 52:1651
15. Ramachandran N, Hainsworth E, Bhullar B, Eisenstein S, Rosen B, Lau AY, Walter JC, LaBaer J (2004) Science 305:86
16. Angenendt P, Kreutzberger J, Glokler J, Hoheisel JD (2006) Mol Cell Proteomics 5:1658
17. Nord O, Uhlen M, Nygren PA (2003) J Biotechnol 106:1
18. Tanaka G, Funabashi H, Mie M, Kobatake E (2006) Anal Biochem 350:298
19. Li S, Floriano PN, Christodoulides N, Fozdar DY, Shao D, Ali MF, Dharshan P, Mohanty S, Neikirk D, McDevitt JT, Chen S (2005) Biosens Bioelectron 21:574
20. Cha T, Guo A, Jun Y, Pei D, Zhu XY (2004) Proteomics 4:1965
21. Charles PT, Goldman ER, Rangasammy JG, Schauer CL, Chen MS, Taitt CR (2004) Biosens Bioelectron 20:753
22. Zhou H, Bouwman K, Schotanus M, Verweij C, Marrero JA, Dillon D, Costa J, Lizardi P, Haab BB (2004) Genome Biol 5:R28
23. Schena M, Shalon D, Davis RW, Brown PO (1995) Science 270:467
24. Eisen MB, Spellman PT, Brown PO, Botstein D (1998) Proc Natl Acad Sci USA 95:14863
25. Granjeaud S, Bertucci F, Jordan BR (1999) Bioassays 21:781
26. Eisen MB, Brown PO (1999) Method Enzymol 303:179
27. Malmqvist M (1999) Biochem Soc Trans 27:335
28. Han A, Sonoda T, Kang JH, Murata M, Niiidome T, Katayam Y (2006) Comb Chem High Throughput Screen 9:21
29. Butte A (2002) Nat Rev Drug Discovery 1:951
30. Cregger M, Berger AJ, Rimm DL (2006) Arch Pathol Lab Med 130:1026
31. Winzeler EA, Schena M, Davis RW (1999) Method Enzymol 306:3
32. Guo Z, Guilfoyle RA, Thiel AJ, Wang R, Smith LM (1994) Nucleic Acids Res 22:5456
33. Randolph JB, Waggoner AS (1997) Nucleic Acids Res 25:2923
34. Perraut F, Lagrange A, Pouteau P, Peyssonneaux O, Puget P, McGall G, Menou L, Gonzalez R, Labeye P, Ginot F (2002) Biosens Bioelectron 17:803

35. Loebke C, Sueltmann H, Schmidt C, Henjes F, Wiemann S, Poustka A, Korf U (2007) Proteomics 7:558
36. Persson A, Hober S, Uhlen MA (2006) Curr Opin Mol Ther 8:185
37. Reineke U, Kramer A, Schneider-Mergener J (1999) Curr Top Microbiol Immunol 243:23
38. Poetz O, Ostendorp R, Brocks B, Schwenk JM, Stoll D, Joos TO, Templin MF (2005) Proteomics 5:2402
39. Storr SJ, Chakrabarti J, Barnes A, Murray A, Chapman CJ, Robertson JF (2006) Expert Rev Anticancer Ther 6:1215
40. Cepok S, Zhou D, Srivastava R, Nessler S, Stei S, Bussow K, Sommer N, Hemmer B (2005) J Clin Invest 115:1352
41. Zhang W, Huang J, Zhou MF, Chen LY, Ding YP, Cao HJ, Geng YY, Wang SQ (2005) Mol Diagn 9:81
42. Buschman KE, Seraly M, Thong HY, Deng JS, Draviam RP, Abernethy JL (2002) J Cutan Pathol 29:82
43. Horn S, Lueking A, Murphy D, Staudt A, Gutjahr C, Schutte K, König A, Landsberger M, Lehrach H, Felix SB, Cahill DJ (2006) Proteomics 6:605
44. Espejo A, Cote J, Bednarek A, Richard S, Bedford MT (2002) Biochem J 367:697
45. Schweitzer B, Predki P, Snyder M (2003) Proteomics 3:2190
46. Kersten B, Possling A, Blaesing F, Mirgorodskaya E, Gobom J, Seitz H (2004) Anal Biochem 331:303
47. Ziechner U, Schonherr R, Born AK, Gavrilova-Ruch O, Glaser RW, Malesevic M, Kullertz G, Heinemann SH (2006) FEBS J 273:1074
48. Sasuga Y, Tani T, Hayashi M, Yamakawa H, Ohara O, Harada Y (2006) Genome Res 16:132
49. Hultschig C, Kreutzberger J, Seitz H, Konthur Z, Bussow K, Lehrach H (2006) Curr Opin Chem Biol 10:4
50. Anderson K, Potter A, Baban D, Davies KE (2003) Brain 126:2052
51. Wingren C, Steinhauer C, Ingvarsson J, Persson E, Larsson K, Borrebaeck CA (2005) Proteomics 5:1281
52. Olle EW, Sreekumar A, Warner RL, McClintock SD, Chinnaiyan AM, Bleavins MR, Anderson TD, Johnson KJ (2005) Mol Cell Proteomics 4:1664
53. Hamelinck D, Zhou H, Li L, Verweij C, Dillon D, Feng Z, Costa J, Haab BB (2005) Mol Cell Proteomics 4:773
54. Rubina AY, Dyukova VI, Dementieva EI, Stomakhin AA, Nesmeyanov VA, Grishin EV, Zasedatelev AS (2005) Anal Biochem 340:317
55. Rucker VC, Havenstrite KL, Herr AE (2005) Anal Biochem 339:262
56. Yeretssian G, Lecocq M, Lebon G, Hurst HC, Sakanyan V (2005) Mol Cell Proteomics 4:605
57. Urbanowska T, Mangialaio S, Hartmann C, Legay F (2003) Cell Biol Toxicol 19:189
58. Ellmark P, Belov L, Huang P, Lee CS, Solomon MJ, Morgan DK, Christopherson RI (2006) Proteomics 6:1791
59. Abdollahi A, Hahnfeldt P, Maercker C, Grone HJ, Debus J, Ansorge W, Folkman J, Hlatky L, Huber PE (2004) Mol Cell 13:649
60. Maercker C (2005) Biosci Rep 25:57
61. Chan SM, Ermann J, Su L, Fathman CG, Utz PJ (2004) Nat Med 10:1390
62. MacBeath G, Schreiber SL (2000) Science 289:1760
63. Houseman BT, Huh JH, Kron SJ, Mrksich M (2002) Nat Biotechnol 20:270
64. Lizcano JM, Deak M, Morrice N, Kieloch A, Hastie CJ, Dong L, Schutkowski M, Reimer U, Alessi DR (2002) J Biol Chem 277:27839

65. Paweletz CP, Charboneau L, Bichsel VE, Simone NL, Chen T, Gillespie JW, Emmert-Buck MR, Roth MJ, Petricoin EF III, Liotta LA (2001) Oncogene 20:1981
66. Matyushin V, Gandhum A, Stich N, Schalkhammer T, Hagen WR, Mayer C (2004) J Nanosci Nanotechnol 4:98
67. Lee Y, Lee EK, Cho YW, Matsui T, Kang IC, Kim TS, Han MH (2003) Proteomics 3:2289
68. Adelheim K, Emantraut E, Kaiser T, Tuchscheerer J (2002) New Drugs, pp 22–23
69. Dudley AM, Aach J, Steffen MA, Church GM (2002) Proc Nat Acad Sci USA 99:7554
70. Weil MR, Macatee T, Garner HR (2002) Biotechniques 32:1310

Part IV
Flow Cytometry

Springer Ser Fluoresc (2008) 6: 307–342
DOI 10.1007/4243_2008_037

Published online: 11 March 2008

Flow Cytometry: Instrumentation, Applications, Future Trends and Limitations

Robert A. Hoffman

BD Biosciences, 2350 Qume Drive, San Jose, CA 95131, USA
bob_hoffman@bd.com

Abstract Flow cytometry is the measurement of cells or other particles flowing past one or more sensing zones. Most commonly the sensing zone is a focused beam of light, and particles are characterized by the scattered and fluorescent light pulses that are created as the particle passes through the light beam. Unlike microscopic image analysis, flow cytometry generally measures the total signal from a particle and does not provide morphological detail on the shape or size of the sources of signal within the particle. The strength of flow cytometry is rapidly providing measurements from multiple fluorochrome stains along with several intrinsic measures of particle characteristics. In some flow cytometers particles with selected measurement properties can be physically sorted and collected. This chapter provides an overview of the technology and application of flow cytometry. All major aspects of the technology are introduced, and references to original literature are provided for readers interested in additional details. Special attention is given the use of fluorescence measurements.

Keywords Cell sorter · Flow cytometry · Fluorescence · Immunofluorescence · Laser · Light scatter

Abbreviations

NA	Numerical aperture
ECV	Electronic cell volume
ADC	Analog to digital converter
Log	Logarithmic
Log amp	Logarithmic amplifier
RF	Radio frequency
DC	Direct current
LED	Light emitting diode
CW	Continuous power
APD	Avalanche photodiode
PMT	Photomultiplier
CCD	Charge coupled device
CV	Coefficient of variation
FSC	Forward light scatter
SSC	Side light scatter
FITC	Fluorescein isothiocyanate
PE	Phycoerythrin
APC	Allophycocyanin
Re	Reynolds number
MESF	Molecules of equivalent soluble fluorophore

1 Introduction

This chapter provides an overview of the technology and application of flow cytometry. All major aspects of the technology are introduced, and references to original literature are provided for readers interested in additional details. Special attention is given the use of fluorescence measurements. For more information, recent books by Givan [1], Ormerod [2], and Diamond and DiMaggio [3] give a good overview of flow cytometry from a user perspective with a focus on applications. Watson's introductory book gives more emphasis to instrumentation [4]. *Practical Flow Cytometry* by Shapiro is a comprehensive book on flow cytometry covering in detail both instrumentation and applications [5].

2
General Concept of Flow Cytometry with Brief History

2.1
Brief History

Flow cytometry is the measurement of cells or other particles flowing past one or more sensing zones. Most commonly the sensing zone is a focused beam of light, and particles are characterized by the scattered and fluorescent light pulses that are created as the particle passes through the light beam. Unlike microscopic image analysis, flow cytometry generally measures the total signal from a particle and does not provide morphological detail on the shape or size of the sources of signal within the particle. The strength of flow cytometry is rapidly providing measurements from multiple fluorochrome stains along with several intrinsic measures of particle characteristics. In some flow cytometers particles with selected measurement properties can be physically sorted and collected.

A second type of sensing zone passes electrical current through a small orifice or channel to measure the resistance or electrical impedance of particles. In this case the passage of a particle causes a momentary change in the electrical current, which is detected as a voltage pulse by the data acquisition electronics. When live cells and other non-conductive particles interrupt a constant or low frequency current, the resulting voltage pulse is proportional to the volume of the particle and also proportional to a shape factor [6]. This measurement is often called the electronic cell volume (ECV).

In broad scope, the technology of flow cytometry began in the 1940s, was first commercialized in the 1950s, widened its technology base in the 1960s, grew rapidly in the 1970s and has been steadily maturing for the last 25 years. Although concepts of optical flow cytometers were known and had limited use prior to the 1950s, the first flow cytometer that had wide use was an electrical impedance sensing instrument invented by Wallace Coulter. Based on the principal that cells behave as electrical insulators and displace conductive saline, an electrical signal is generated due to the increased electrical impedance (resistance) in a small fluid-filled channel (orifice) when there is an electrical current in the orifice and a cell is present. By flowing cells suspended in saline through a small orifice between two saline-filled chambers, it was possible to count the number of cells, determine their concentration and measure the relative volume or size of the cells. The impedance sensing method was soon adapted to routine measurements of red blood cells in clinical hematology and then more widely for other blood cells and general particle counting and sizing.

Injection of a particle suspension into a secondary sheath flow provides control over the location and cross-section size of the particle stream within the sheath flow This concept of hydrodynamic focusing of particle suspensions was developed for blood cell counting by Crosland–Taylor in 1953 and has been used in nearly all optical flow cytometers since that time [7, 8].

The fundamental technologies for fluorescence flow cytometry, cell sorting and multiparameter analysis were demonstrated in the 1960s and refined and applied to a wide variety of applications in the 1970s. Flow cytometers with optical sensing were developed in the 1960s independently by Dittrich and Göhde in Germany and Kamentsky in the US. Göhde's instrument was based on an epi-fluorescence microscope design optimized for high precision measurements of DNA content [9]. Kamentsky's first instrument used UV absorption to measure DNA content of cells [10]. An early Kamentsky design also included a fluidic switch to sort selected cells [11]. The first cell sorter using the now standard technique of charging selected particle-containing droplets was developed by Fulwyler at Los Alamos National Laboratory in 1965 [12]. Fulwyler combined a method for ink jet printing developed by Richard Sweet at Stanford [13] with electronic impedance sensing. This early instrument was followed by sorters using fluorescence to identify cells to be selected. In the early 1970s the first laser-based sorter was developed at Stanford by Herzenberg's group [14].

Computers for data acquisition and analysis became common in the 1980s and greatly simplified the use of multiparameter analysis and sorting. Flow cytometry became widely used in both research and clinical applications in the 1980s, with several companies offering analysis-only instruments that could be easily used by most laboratory personnel. Flow sorting, which was also available from several companies by the 1980s, became an essential tool for studying cells in complex mixtures and biological systems. The invention of monoclonal antibodies targeting specific subsets of immune system cells in blood, bone marrow and other tissues combined with the analytical and sorting capability of flow cytometry was a powerful tool in the fields of immunology, hematology, cancer research and developmental biology.

2.2 Analysis only Instruments

A block diagram of a typical flow cytometer is shown in Fig. 1.

Sensing zones with their associated excitation sources are at distinct locations along the sample stream. Multiple optical measurements can be made in

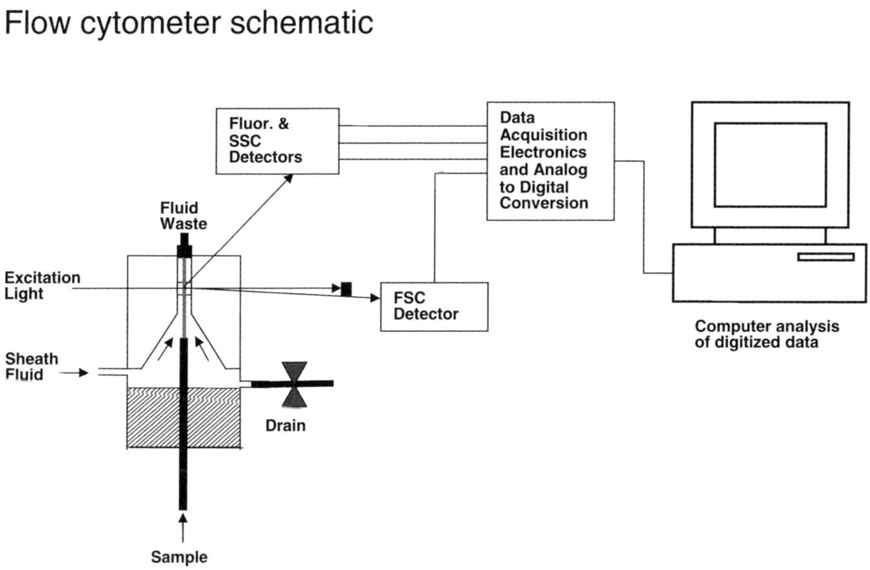

Fig. 1 Schematic block diagram of a typical flow cytometer. The flow cell shows sheath flow surrounding the sample injected from the tube in the center, which provides hydrodynamic focusing to position and control the size of the sample stream. A focused excitation light beam, typically from a laser, intercepts the sample stream and produces scattered and fluorescent light from particles as they flow past the focused excitation light. Forward scattered light (FSC) is detected at low angles. Sideward scattered light (SSC) and fluorescence are typically detected in a range centered around 90° relative to both the direction of flow and direction of excitation light. For each particle detected, output pulses from the scatter and fluorescence detectors are amplified and processed to determine parameters such as pulse peak or integrated signal per pulse. Digital values for the set of parameters acquired for each particle is sent to a computer. Data may be analyzed in real time as it is accumulated or off line using the data saved as list of parameter measurements for each particle

a sensing zone. For example, several fluorescence detectors and two or more light scatter detectors could make measurements from one laser beam focused on the sample stream. In some instruments multiple excitation sources are applied at a single sensing zone location. Two or more laser beams could be focused at the same point on the sample stream, or ECV and optical measurements could be made simultaneously in one sensing zone.

Each optical or electrical impedance detector is processed through a data acquisition channel consisting of an amplifier, a pulse processing section, and a digitizing section. In the last stage of data acquisition, digitized data from all the channels is combined into a single, multiparameter "event" that describes the particle. Digitization of the data causes a range of data values to be assigned to a single "bin" of data. If data are binned on a linear scale, each bin has an equal range of values. If data are binned on a logarithmic scale, successively higher bins include an increasingly wide range of values. An artifact in the display of logarithmically binned data at the low end of the scale is discussed in later sections.

An additional very useful parameter is the relative time at which an event occurred. A comparison of a measured parameter vs. time shows any variation in the parameter during the data acquisition. This may be used to monitor a specific time varying particle characteristic. If the particle measurement is not expected to vary with time, then monitoring the measured parameter vs. time indicates the stability of the instrument during the data acquisition.

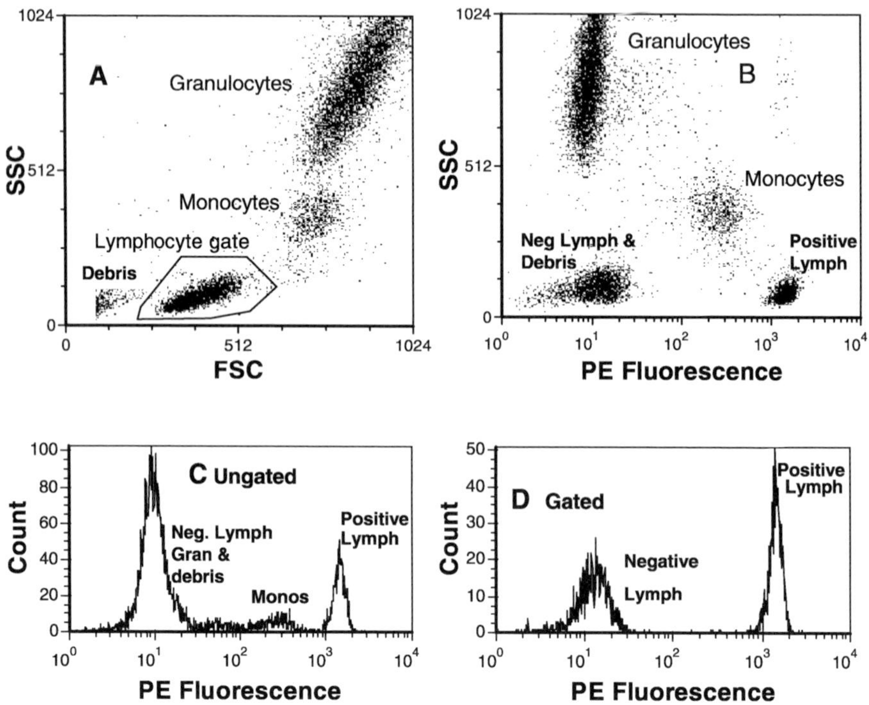

Fig. 2 Examples of correlated 2-parameter plots (or dot plots) and histogram data displays. Data are for a human blood sample stained with CD4 antibody conjugated to PE followed by lysis of red blood cells and a buffer wash. Panel **A** is a typical light scatter plot which distinguishes the major classes of white blood cells – lymphocytes, monocytes, and granulocytes. Events due to unidentified particles with low scatter signals are labeled "debris". The events corresponding to lymphocytes are shown bounded by a region of interest that can be used to gate additional analysis based on fluorescence. Panel **B** is a plot of SSC vs. PE fluorescence of all the events shown in panel **A**. CD4-lymphocytes stain brightly and are designated "positive". Monocytes have higher SSC than lymphocytes and stain less brightly with CD4. Granulocytes and some lymphocytes do not stain with CD4, and the population is designated "negative". Panel **C** is a histogram of the number of events vs. PE fluorescence for all cells (ungated). Panel **D** is a histogram of number of events vs. PE fluorescence for only those cells falling within the lymphocyte region or gate in panel **A**

If there are multiple sensing zones, the known time for particle travel between sensing zones is used to correctly correlate data from each sensing zone to a particle. The data from multiple events are structured in a list format, which conceptually can be described as a spreadsheet, where each column represents a measured parameter and each row contains data from a single particle.

The data are analyzed most simply as single parameter histograms as in Fig. 2c and d or as correlated plots of two parameters (usually referred to as dot plots) as in Fig. 2a and b. Analysis of multiparameter data typically consists of setting regions of interest in histograms or dot plots. The process of analyzing only events that are within selected regions of interest is called "gating", and the regions of interest are referred to as "gates". Events within the gate can be selectively displayed in other histograms or dot plots as illustrated in Fig. 2a and d. Data analysis is usually much more complex than these simple illustrations. Logical combinations of gates are routinely used to define and count subpopulations of events. Multidimensional gates of three or more parameters are also possible with some software, which use automated clustering algorithms to define regions in the multidimensional data space.

2.3 Measuring Particle Concentrations

The concentration of particles in a sample is determined in two different ways. The most straightforward approach is to measure the volume of sample that was used to produce the particle events that were detected. Various types of positive displacement pumps (commonly syringe pumps) can be used to deliver known volumes or known sample flow rates. Some instruments use sensors to detect displacement of a fixed amount of fluid from the sample container.

A second approach common in flow cytometry is the addition of a known number of reference particles to the sample. During sample acquisition events from both reference and sample particles are measured and can be counted separately. The ratio of sample particle count to reference particle count is used to calculate the concentration of the sample particle [15].

2.4 Cell Sorting

In addition to measuring characteristics of particles, selected particles can be physically separated from a mixture of different types of particles. The gate or gate combination that defines the subset of particles can initiate a process that moves the selected particle to a separate path and allows them to be collected either as isolated single particles or collected in bulk as a purified population. The two fundamentally different ways used to separate the particles are droplet sorting and enclosed fluidic sorting. Droplet sorting can select particles at rates up to 25 000/s. Enclosed fluidic sorters are much slower-selecting cells at rates in the range of 300/s to 1000/s.

The rate at which particles can be sorted is determined by coincidence statistics in a sorted drop [16, 17]. The particle arrival rate should be much less than the number of drops generated per second. Various modes of sorting can optimize maximum sort throughput (enrichment of selected cells with lower purity), high purity (with fewer sorted particles/s), sorting of one or a predetermined number of particles in a container, and many additional modes with intermediate characteristics of the purity of the sorted population. Drop generation rates of 100 000/s allow good sort purity and yield at particle rates up to 25 000/s.

Droplet sorting is based on an early method of ink-jet printing developed by Richard Sweet in which a fluid jet from a nozzle is vibrated under conditions that cause the stream to break up into uniform droplets [13]. Measurements made on particles in a region of fluid flow that is smooth and laminar establishes criteria for deciding whether to sort a particle. As a drop is breaking off from the fluid jet, an electrical charge can be momentarily applied to the stream, and drops breaking off while charge is applied will carry a portion of the charge. The drops pass through a strong electric field, and charged drops containing selected particles are deflected into a collection container. For droplet sorting, the time that it takes for a particle to travel from the sensing zone to a drop breaking off is accurately and precisely known. Only the drop (or drops immediately preceding and following) most likely to contain the selected particle is charged. For some applications, it is advantageous to charge all drops except those likely to contain a particle of interest and collect the uncharged, undeflected drops. Figure 3, a composite of stroboscopic photographs of the sorting process, shows the concept.

The first droplet sorter was developed by Fulwyler in 1965 using ECV as the particle measurement parameter. Droplet sorting using optical measurements was demonstrated by Herzenberg's group in 1969 and is the approach that has been widely used and developed for commercial cell sorters [12, 18, 19]. A laser (or lasers) focused on the fluid stream is used for fluorescence and light scattering measurements. In many droplet sorters the laser is focused on the fluid jet. This approach, commonly called the "stream in air" method, has the advantage of simplicity, a short time between analysis and drop break-off, and a constant particle velocity across the jet. Particles at the center and near the edge of the jet travel at the same velocity. Thus the time from when a particle is measured to when it is in a droplet breaking off is not critically dependent on the location of the particle within the jet. There are optical disadvantages using the stream in air approach, however. The jet acts as a cylindrical lens, and any oscillations in the jet at the point of laser intercept cause variation in the fluorescence or light scatter signals. The jet also strongly reflects and refracts the laser light- creating significant background

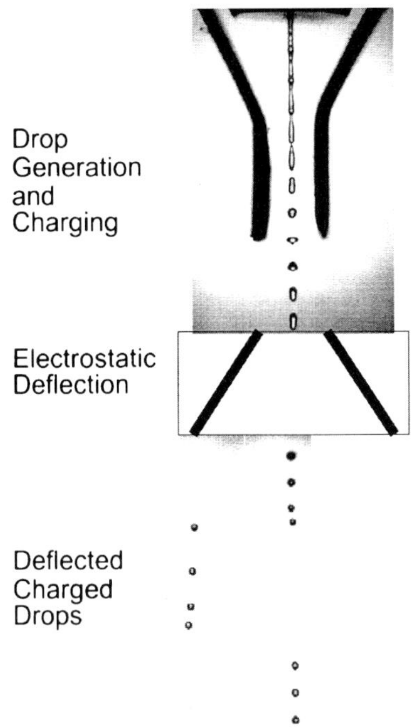

Fig. 3 Stroboscopic photos of drop generation and deflection illustrate the droplet sorting process. Acoustic energy applied to a fluid stream emerging from a small orifice can cause the stream to break into droplets at a well defined distance from the orifice. Stable, periodic drop formation occurs over a limited range of frequency, fluid velocity and other physical parameters. Since the time at which a drop is about to break off the jet is well defined, a pulse of electrical charge can be applied to the stream just as the drop is breaking off. Charged drops are deflected by an intense electric field and take a sideward trajectory. Four successive drops were charged in this illustration. For particle sorting, only drops expected to have wanted particles can be charged. Separate stroboscopic photos of the drop generation and deflection regions were combined in this composite figure

that must be blocked out from fluorescence and light scatter detectors. Blocking the laser beam usually requires also blocking a portion of the fluorescent or scattered light signal.

An alternative approach to droplet sorter design is to make the optical measurements in a cuvette (typically with internal dimensions of 100–500 μm) with optically flat walls and to create the sort jet with an orifice at the output of the cuvette. This design has several optical advantages. Optical background from the laser intercepting the cuvette walls can be mostly eliminated by using imaging optical designs and appropriately placed apertures. High numerical aperture (NA) light collecting lenses can be used to gather more fluorescent light – e.g., coupled designs with NA of 1.2 or greater can be used. Also, the particle velocity is usually lower in the cuvette than in the droplet forming jet. So the particles are illuminated for a longer time by the laser and generate a larger fluorescence signal. The disadvantage of using a cuvette with a separate jet forming orifice is that particle velocity is less well defined inside the cuvette. A parabolic flow profile develops inside the cuvette and only particles near the center of the cuvette travel at the same velocity. In practice this is not a significant limitation since the particles must be well centered in the cuvette for precise optical measurements.

Enclosed fluidic sorting was first described by Kamentsky [11], who used a stepper motor driven syringe to move selected cells to an alternate flow path in an optical flow chamber. Other versions of fluidic sorting use the same general principle. After the cells pass through the sensing zone, there is a delay before they reach a point where alternative fluid paths (waste and sorted) are possible. In most implementations of fluidic sorting, the waste and sorted fluid paths are fixed, and a fluidic disturbance causes selected particles to move from the waste path to the sorted path. This approach is being used in microfluidic flow cytometers [20].

A different approach is used in the BD Biosciences "catcher tube" fluidic sorter design. In this design the sort channel is a movable tube whose entrance is normally positioned in sheath flow but outside the sample core stream. When a selected particle is to be sorted, the "catcher tube" momentarily moves into the sample core stream, capturing the selected particle and allowing it to flow to a collection container.

3 Instrumentation Details

Most flow cytometers have been designed to make both light scatter and fluorescence measurements. Some instruments also incorporate an orifice for impedance sensing. The great variety of biological applications that flow cytometry can address is due to availability of fluorescence probes for specific cellular targets. This section will focus on design and performance details that affect fluorescence measurements and give brief overviews of other instrumentation features.

3.1 Fluidics

The critical fluid path in a flow cytometer always is a small channel. For impedance sensing and droplet sorting the channel is a short orifice. For most other flow cytometers that have optical sensing, the channel is a glass capillary. When fluid initially enters a closed channel, all elements of the fluid move at the same velocity, but viscous drag at the walls of the channel slow fluid at or near the wall. Fluid at the center of the channel moves faster than near the walls. At sufficient distance into the channel, the velocity distribution across the channel is parabolic. Flow from this point on down the channel remains parabolic with no flow at the channel walls and the velocity at the center of the channel is twice the average velocity. Flow of particles in a closed channel or flow cell closely follows the laminar flow lines of the fluid. But particles near the wall tend to be pushed toward the center of the channel. Shear forces on the particles tend to orient asymmetric particles with their long axis in the direction of flow. Specially designed flow cells have been used to enhance and maintain particle orientation in the direction of flow.

The simplest approach to fluidics is to simply push or pull the sample suspension through the flow cell. This sheath-free approach is an advantage for field use or where transportation of sheath fluid is expensive or inconvenient. In most situations where this approach is used, the sample is introduced with a volumetric pump such as a syringe. Data acquisition electronics must account for particles traveling at different velocities depending on how far they are from the center of the channel. A disadvantage of this approach for optical sensing is that the flow cell walls become coated with material from the sample, and frequent cleaning is necessary. Also, the channel must be relatively small in cross-section in order to have the particles concentrated near the center of the channel. For mammalian cell work, channel cross-sections are in the range of 70–100 μm. The small cross-sections make the channels more susceptible to clogging.

Most flow cytometers use sheath flow to constrain the sample core stream to a well defined region in the centre of flow. This approach, called hydrodynamic focusing, provides flexibility in the volume rate of sample delivery while maintaining essentially constant particle velocity. Typically less than 1% to no more than 5% of total fluid volume is from the sample. The most common situation would have the ratio of sample to sheath flow no more than 0.02. With sample flow constrained to a very small region near the centre of a channel, there is no significant difference in particle velocity at the centre or edge of the sample stream even when the total flow velocity profile is parabolic. Furthermore, changes in sample flow rate have little effect on particle velocity. Since the particle velocity is constant, the cross-sectional area of the sample core stream is proportional to the volume rate of sample flow.

By varying the sample core stream cross-section the particle event rate can be controlled. This allows a wide range of sample concentrations to be analyzed without event rates becoming too high for the data acquisition electronics to process. An additional advantage of hydrodynamic focusing is the ability to control uniformity of illumination of the particles when the excitation beam has a Gaussian distribution, as is standard with laser excitation. Figure 4 shows a Gaussian distribution with an indication of a typical sample stream width used for a high sample flow rate (e.g., 1 μl/s). A high sample flow rate can be used for applications such as immunofluorescence, for which the biological variation contributes CV's of 20% or greater and uniform illumination is less important. At a low sample flow rate (e.g., 0.1 μl/s), the small sam-

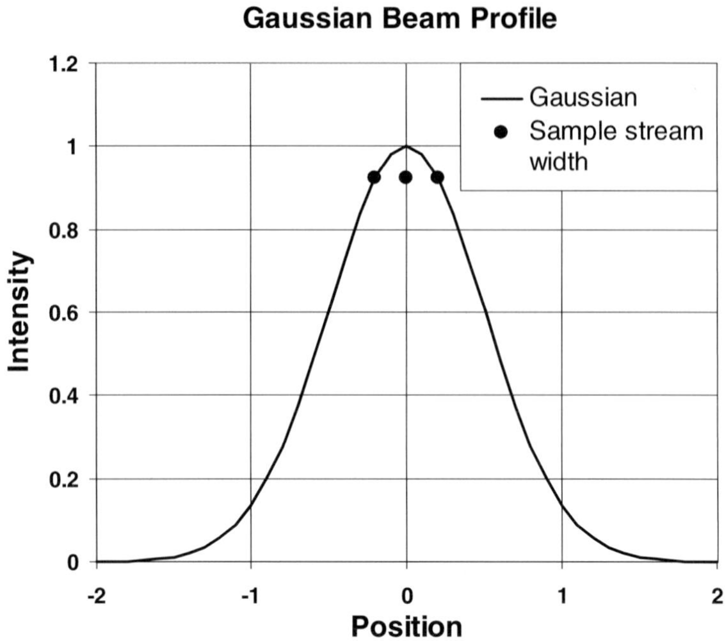

Fig. 4 Illustration of the Gaussian intensity profile typical of laser illumination of a sample stream. The *dotted region* at the *top of the profile* defines the position within a sample stream for which the excitation intensity on a particle would vary by 5% or less. Most of the excitation energy is not applied to the sample stream – only the brightest and most uniform region of the focused beam. The relative position is in units of the $1/e^2$ dimension of the distribution, which is typically in the range of 20–60 μm

ple core stream diameter allows more uniform illumination of particles and lower variance of fluorescence measurements for applications such as DNA content of cells. Intermediate sample flow rate and core stream diameter may also be used. In any case the sample core can be adjusted so that illumination non-uniformity does not add significantly to the total measured variance.

It is important to keep the flow cell clean and free of bubbles which could disturb the position of the sample stream. In typical systems the sample stream position must be maintained within 5 μm or less in order to have good alignment with the optical system.

Many different approaches to delivering sheath and sample flows are used. The simplest approach is to apply vacuum to the exit side of the flow cell and pull both sheath and sample into the sensing area. The ratio of sample to sheath flow is determined by the relative fluidic resistance in each path. A majority of flow cytometers use a pressurized container to deliver sheath flow. Depending on the flow cell dimensions and application, the sheath pressure varies from about 20 kPa to over 400 kPa. Other approaches to delivering sheath fluid that have been used in commercial instruments include syringe pumps, gear pumps, and peristaltic pumps. For slow flow instruments, gravity feed can be used by placing the sheath container a meter or more above the flow cell [21, 22].

Sample flow is commonly produced by pressurizing the sample container at a slightly higher pressure than the sheath. For delivering sample at known rates or delivering known volumes, syringe pumps or peristaltic pumps have been used. Very rapid sampling from microwell trays has been accomplished by running a peristaltic pump continuously, briefly aspirating a slug of sample with a probe attached to the pump, and moving the probe to the next well while aspirating an air bubble, which separates the sample slugs [23].

3.2 Electronics

The signal pulse or pulses from the sensing zones are first amplified to raise the signals to appropriate levels for pulse analysis and digitizing with analog to digital converters (ADCs). In most systems, the amplifiers also remove any constant background signal so that only the transient pulse due to particle passing through the sensing zone is measured. This is called baseline restoration. Only pulses whose amplitude on a selected measurement channel (or channels) is greater than a preset threshold (or trigger level) are analyzed further by the electronics. The amplified, baseline restored pulses are processed in one of two ways, referred to as analog or digital.

For analog signal processing, an analog circuit, such as a peak detector or integrator, measures a characteristic of the pulse that corresponds to the total light emitted from the particle. Other circuits could measure additional pulse features such as the pulse width. The pulses may be amplified by a logarithmic amplifier (log amp) prior to peak detection. Or the pulse peak or pulse integral signals could be log amplified. Log amps are used because for many applications, e.g., analysis of immune function cells, the signal range is three or more decades. Typically wide dynamic range data are displayed on a four-decade logarithmic scale. This wide dynamic range is difficult to achieve with high accuracy using log amps, and accuracy of the log response is usually compromised at the very low and very high parts of the scale.

Two approaches to dealing with the imperfect logarithmic response of log amps are used. One is to measure an individual log amp's response to input voltage over the entire four-decade range, and use the known response to calculate a corrected result. A lookup table of true response can be used to convert the measured response to true logarithmic results. The second approach is to not use log amps, but instead use ADCs with at least four decades of dynamic range. Depending on the resolution desired in the lowest decade, an ADC with 16 or more bit resolution must be used, and 18 or more bits of resolution are needed to avoid coarse resolution in the lowest decade. An alternative to such a high resolution ADC is to use automatic range switch-

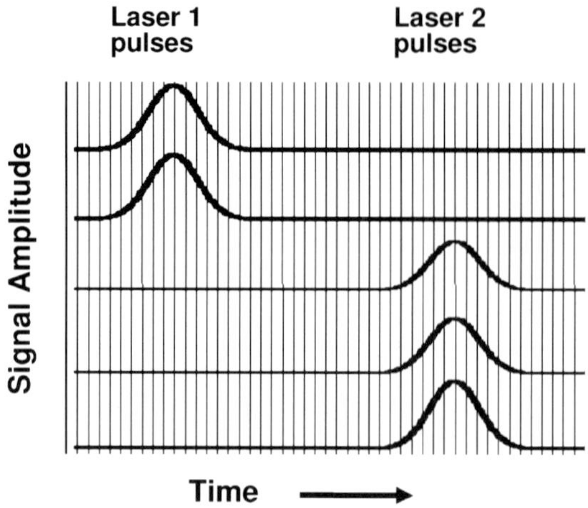

Fig. 5 Digital pulse processing using continuous digitization. A continuously sampling ADC digitizes the pulses periodically as indicated by the *vertical time slice lines*. The five curves represent five different measurements channels during the transit of a single particle through two successive laser beams. The greatest digitized value represents the pulse peak. The sum of all the values sampled during the duration of the pulse represents the pulse integral or total emitted signal. Since the time between pulses excited by different lasers is constant, the digital electronic processor accounts for the time delay and identifies pulses appropriately as belonging to the same particle

ing electronics in which the signal is amplified in separate high and low gain paths, and the high or low gain path is selected for digitization depending on the signal level [5].

The digital approach to signal processing uses continuous high speed digitizing (also called sampling) of the detector output. The time between samples of the detector output must be much less than the duration of a signal pulse. Sampling the pulse many times allows the pulse to be reconstructed and measured using computation on the digital information. Pulse peak, pulse area, pulse width or other pulse characteristics are computed from individual samples of the pulse. For example, pulse peak is the highest sampled value, and pulse area is the sum of all the sampled values. Figure 5 illustrates digital signal sampling.

3.3 Impedance Sensing

Impedance sensing or ECV was described in historical context above. The ECV signal is proportional to $1/d^4$, where d is the cross-section diameter of the orifice. Electronic noise limits the practical maximum diameter of impedance orifices to about 100 μm for work with typical cells and particles such as blood cells. Some commercial and lab-built flow cytometers have combined electrical impedance and optical measurements. With the assumption that cells are spheres, ECV can be used to measure cell surface area and fluorescence per unit area of cell surface [24–26].

Impedance sensing is not limited to using constant electrical currents (DC). When a radio frequency (RF) current (typically a few MHz) is used, a cell membrane no longer insulates the interior of the cell, and the electronic signal is affected by both the cell size and internal conductivity [27, 28]. The ratio of the RF/DC impedance signals, called opacity, is a measure of the electrical property of the cell interior. Leif found the buoyant density of erythrocytes correlated with opacity. Combined with ECV and light scatter measurements, the RF measurement (termed conductance in the reference) has been used in automated hematology analyzers to discriminate different types of white blood cells [29].

3.4 Optics

The optical design of a flow cytometer is the most critical factor defining performance. The excitation and emission designs must combine to give adequate fluorescence sensitivity and alignment stability. Spatial resolution is not very important, but performance is improved if the emission optics can reject most light (such as scatter or reflection from the capillary) that does not come from the sample stream. Most optical designs create an image of the sensing zone and use an aperture in the image plane to pass light from the sample stream and reject background light.

3.4.1 Excitation

Both lasers and incoherent sources are used for excitation. Lasers are by far the most common and will be considered first. Arc lamps and light emitting diodes (LEDs) are incoherent sources useful for fluorescence work.

3.4.1.1 Lasers

Nearly all lasers used for flow cytometry are single mode with a Gaussian intensity distribution, which allows predictable beam shaping and focusing. Multimode lasers are occasionally used, but produce less predictable focused illumination on the sample stream. Beam shaping to produce a more flat top intensity distribution is possible but not yet common. Laser wavelengths from 325 nm to 650 nm are in routine use. There are no particular restrictions on laser wavelength, which is chosen to match appropriate fluorochromes intended for use in an instrument.

All types of lasers, including gas, ion, dye, diode pumped solid state, and diode lasers, are used. The most common lasers used are 488 nm argon ion or solid state lasers, 633 nm helium neon, violet and red diode lasers, 532 nm solid state, and solid state or ion UV lasers. Most lasers used operate in a continuous power (CW) mode, but a mode-locked solid state UV (355 nm) laser with a 100 MHz pulses repetition rate is used in flow cytometers from several manufacturers. The high repetition rate of mode locked lasers is considered quasi-cw for flow cytometers, where the particles are illuminated for at least 1 μs and are excited by 100 or more mode-locked excitation pulses. Pulsed lasers are potentially useful if they can be triggered asynchronously to emit only when a particle is in position to be illuminated. Q-switched lasers operating at a fixed, periodic repetition rate are not very useful since there is no way to synchronize their emission when a particle is present.

Lasers are usually focused to an elliptical spot on the sample stream. The focused spot size (measured at the $1/e^2$ or 13.5% intensity level) in the direction perpendicular to flow is chosen to provide adequately uniform illumination across the maximum sample stream core size and possibly some tolerance for the sample stream to move. The depth of focus of the laser spot must be large enough to encompass the largest sample core size and position variation in the direction of laser propagation. Typical laser spot size dimensions across the sample stream are 60–100 μm. The depth of focus (for 15% increase in spot size) for a laser focused to a spot size of at least 60 μm is more than 1 mm, which is larger than the internal dimension of most flow cells. The spot size in the direction of flow is narrower to minimize the size of the sensing zone and the possibility of particle coincidence. Typical dimensions of the spot size in the flow direction are 5–30 μm, which give depths of focus of 17–600 μm respectively for a worst case red laser beam.

3.4.1.2
Incoherent Sources (Lamps, LEDs)

Although less intense sources than lasers, arc lamps and LEDs provide useful levels of excitation in both the UV and visible range. Arc lamps have been used in several commercially available flow cytometers either as the only light source or as an additional source in a laser-based instrument. Peters has compared arc lamp and laser-based flow cytometers for DNA content analysis [30]. The author has shown LEDs to be useful light sources for some applications with relatively bright levels of fluorescence, but LEDs are not used in any commercially available instruments at this time. The intensity of LED excitation under optimal conditions is about 1% as great as with a laser of the same total power. Arc lamps or LEDs are used in an epi-fluorescent configuration as in standard fluorescence microscopes. The same lens is used to illuminate the sample and collect the fluorescent light. In a typical arrangement a dichroic mirror reflects light from the lamp into the lens and the same mirror passes the fluorescent light to the emission optics and detectors. Critical illumination can be used to image the arc or LED emitter onto the sample stream for maximum intensity. Or Köhler illumination can be used to provide more uniform illumination [30].

3.4.2
Efficiency of Fluorescence Collection

Most flow cytometers designed for high fluorescence sensitivity use immersion lenses to collect fluorescence. Immersion lenses are much more efficient than air spaced lenses. The highest possible NA, 0.9, for an air spaced lens gives only 13% light collection efficiency compared to 28% for a 1.2 NA immersion lens [31]. When laser excitation is used, the fluorescence collection lens also collects side scatter light. Since the fluorescence collection lens is not used for high resolution imaging, the effective depth of "focus" is much larger than if the lens were used as in a microscope. A very blurred image is acceptable as long as the emitted light passes through any limiting aperture in the optical path.

Stream-in-air droplet sorters cannot use immersion lenses, and working distance limits the practical NA in this case to about 0.6 or 5% collection efficiency. Stream-in-air sorters use higher power lasers to partially compensate for the lower light collection efficiency.

The numerical aperture of the lens used for epi-illumination with an incoherent light source is a critical factor in producing the optical signal. For example a 1.2 NA lens collects 28% of light emitted isotropically from water and a 0.6 NA lens collects only 5.4% [31]. In an epi-illumination system the total efficiency is the square of the collection efficiency, so a 1.2 NA lens is 28 times more efficient and produces 28 times the signal compared to a 0.6 NA lens.

3.4.3
Filters and Dichroic Mirrors

Although some special purpose and lab-built flow cytometers use a grating or prism to disperse collected light into selected wavelength bands, all commercially available flow cytometers currently use arrays of dichroic mirrors and color selection filters. To minimize the effect of the angle of light on the performance of a mirror or filter, emission is either collimated or constrained to a low angle (typically $< 6^\circ$) in any part of the emission optics that contains mirrors or filters. Because fluorescence is much weaker than scattered light, the combination of filters and mirrors should reject excitation light by a factor of 10^6 or more at the fluorescence detectors.

Many different geometrical arrangements of the dichroic mirrors are used. Since reflection from a dichroic mirror is generally more efficient that transmission, designs that minimize the number of times light must pass through a dichroic mirror are more efficient.

3.4.4 Polarization Effects

Transmission and reflection characteristics of an interference filter or dichroic mirror can depend on whether the incident light is polarized parallel or perpendicular to the plane of incidence. When the incident light is perpendicular to the optical surface, the angle of incidence is zero and there is generally no difference in transmission or reflection for the two polarization conditions. The greater the angle of incidence deviates from zero, the greater is the difference in reflection and transmission for the two polarization conditions. If fluorescence is excited by a polarized laser, which is a typical situation, the fluorescence is likely to be partially polarized [32, 33]. Thus different configurations of dichroic mirrors or filters can produce different measured intensities from the same fluorescence signal. If fluorescence is transmitted through depolarizing optical elements such as a multimode optical fiber, the polarization dependence of dichroic mirrors and filters is reduced or eliminated.

3.4.5 Detectors

The pulses from flow cytometers typically have durations of 1–20 μs depending on the particle velocity and illumination spot size. Photodiodes, avalanche photodiodes (APDs) and photomultiplier tubes (PMTs) are used as detectors. Nearly all commercially available instruments use individual detectors. But some specialized or lab-built instruments use detector arrays. Beisker has demonstrated use of a charge coupled device (CCD) for non-imaging flow cytometry measurements [34].

Photodiodes are limited by electronic noise since they have no internal gain. Photodiodes are used to detect forward light scatter or extinction, typically the brightest optical signals. They are sometimes used for side scatter or very bright fluorescence measurements.

APDs are used in linear gain mode in some commercial instruments. With internal gain of 100 or more, APDs provide better sensitivity than photodiodes and can be used for all but the most sensitive fluorescence measurements in the visible range (400–700 nm). For fluorescence above 700 nm, APDs have much higher quantum efficiency than PMTs and can be comparable to PMTs in resolving dimly fluorescent from unstained particles. The Geiger, or avalanche, mode of APD operation is not generally useful for flow cytometry since the maximum photon counting rate is about 100 MHz. In a 10 μs pulse this allows a maximum count of only 1000 photons, which limits detection to only very weak signals. For even the brightest signals in a 10 μs pulse, Poisson statistics would cause the CV of the data to be more than 3%.

The signal/noise in an APD or PMT can be expressed in terms of the of number of photoelectrons (for a PMT) or electron–hole pairs (for an APD) and an excess noise factor as

$$S/N = \frac{N_S}{\sqrt{E(N_S + N_B)}}, \tag{1}$$

where E is an excess noise factor from the amplification process, and N_S and N_B are the number of signal and background photoelectrons or electron–hole pairs respectively.

APDs have high quantum efficiency, but signal to noise is limited by excess noise generated in the amplification process. Excess noise increases with gain in an APD, so an optimal signal to noise condition is usually defined as the gain at which APD noise is equal to the electronic noise of the preamplifier. A useful compromise in the author's experience produced an excess noise factor of 4. APD gain is temperature dependent, so for stable operation the detector should either be cooled or include temperature compensation circuitry.

PMTs are the most common detector for fluorescence measurements. A PMT combines good quantum efficiency (up to 40% in the visible range and 5–10% at 800 nm), high gain to greater than 10^7 and low excess noise factor (usually less than 1.4). Unlike APDs the excess noise factor of a PMT decreases with increasing gain, so it is best not to operate a PMT at too low voltage. The excess noise factor of a PMT increases significantly if the photoelectrons emitted by the photocathode are not collected efficiently at the first dynode [35]. Poor photoelectron collection occurs if the voltage between the photocathode and first dynode is too low, but above a certain voltage essentially reaches an optimum. Photoelectron collection at the first dynode affects both gain and excess noise factor. Excess noise factor is difficult to measure, but it is easy to estimate the voltage where photoelectron collection is near optimal by measuring signal as a function of PMT voltage. In most of the range, there is a linear relationship between the log of the PMT output signal and the log voltage across the PMT. Below a certain voltage, the signal is less than predicted by this relationship because photoelectron collection is not optimal. For best signal to noise, the PMT should not be operated below this voltage.

For accurate measurements it is also important that the anode current not exceed a maximum peak or average value. If the current is too high, the signal response will deviate from linearity. Electrostatic forces in the PMT and limitations in current from the high voltage supply will cause PMT output pulses to drop below the expected value. Deviation from a linear response due to limited PMT current can be caused by background light causing constant current draw, high average pulse repetition rate and high average current per pulse.

4 Optical Measurements

4.1 Light Scatter – Total and Polarization Components

Light scatter measurements are common feature of most flow cytometers. The typical configuration is shown schematically in Fig. 1. Scatter angles are measured relative to the direction of the light beam. Scatter at small angles is commonly called forward scatter (usually abbreviated to FSC), and scatter at angles around 90° is commonly called side scatter (usually abbreviated to SSC). Most commercial flow cytometers use laser excitation for light scatter, but incoherent light sources are also used in some instruments. Light scattering from particles is a complex phenomenon. A complete theoretical analysis is available for homogeneous spheres (Mie scatter) and some other simple shapes [36, 37]. Figure 6 shows an example of forward scatter predicted by Mie theory vs. particle diameter for polystyrene spheres, where scatter is collected over an angular range of 0.4–3.75°.

Forward scatter at angles near 1° can be used under some conditions to estimate the size of cells, but light scatter depends on factors other than size. There is usually a range of particle size and refractive index over which forward light scatter decreases or is constant as particle size increases [38].

Axial light loss (also called extinction), which measures the loss of light from the illuminating beam, is an alternative to measuring the scattered light

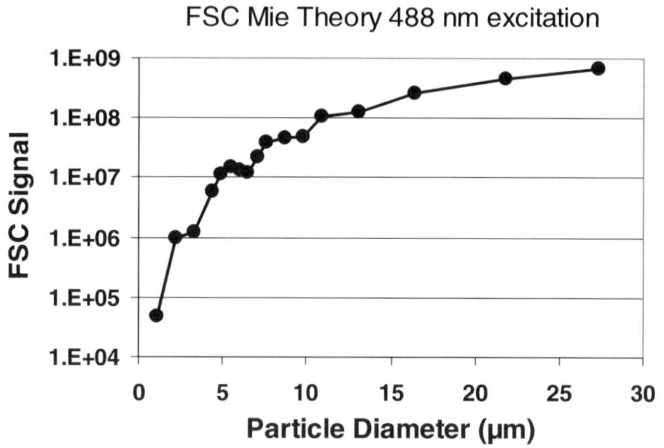

Fig. 6 Theoretical forward light scatter signal for polystyrene spheres with 488 nm excitation. For each particle diameter indicated by a *black circle*, the theoretical Mie theory light scatter intensity was calculated for angles from 0.4 to 3.75° in 0.05° increments. Total signal was defined as the sum of intensities over the range 0.4–3.75°. Computations were performed using the software program MiePlot available from www.philiplaven.com

for estimating particle size [39–41]. For most particles larger than a few microns, the most intense scattering is near 0°. If the axial light loss sensor is set to measure only light lost from angles very near 0° (< 1°), the signals are nearly the same magnitude (but inverted) compared to light scatter. Under this condition, axial light loss can be an effective way to measure light scatter in an angular range that cannot be directly measured with the scattered light.

Like forward scatter, side scatter is a complex function of size, refractive index and the angular range covered by the detector. Side scatter is also dependent on the microstructure in a cell, commonly called granularity. Side scatter is routinely used to help identify the three main classes of white blood cells- lymphocytes, monocytes and granulocytes. A typical plots of SSC vs. FSC for blood samples prepared by lysing red cells is shown in Fig. 2a. Since side scatter is orthogonal to the direction of the excitation beam, it is effectively a dark field measurement with little background light. In most flow cytometers, side scatter is more sensitive than forward scatter for detection of submicron cells and particles. The practical limitation for detecting submicron particles by light scatter is the concentration of particles in sheath and buffer liquids that scatter similarly to particles in the sample. Liquids can be filtered to remove most particles above 0.1 μm, which is approximately the practical minimum size particle that can be reliably detected. Steen has used a special setup for recirculating sheath to continuously reduce the submicron particle concentration in order to detect viruses by light scatter [22].

Polarized light scatter is most useful in the side direction. Depolarized side scatter has been used to distinguish eosinophils from other granulocytes in blood [42]. Using an automated hematology analyzer that incorporates a measure of depolarized side scatter, Fawzi et al. found that red blood cells infected with malaria parasites produced a distinctive and diagnostic depolarization pattern [43]. Olson found depolarized forward scatter useful in discriminating certain marine phytoplankton [44].

4.2
Pulse Width Sizing

The duration of a light pulse (fluorescence or scatter) as a particle passes through a light beam can be used to measure the diameter of the particle. The duration of the light pulse is determined by both the size of the focused light beam in the direction of particle flow and the size of the particle. To a good approximation, if the component due to height of the light beam is subtracted from the total pulse duration, the remaining time is proportional to the particle diameter (or length if the particle is not spherical). Two methods to measure pulse width have been used in commercial instruments. Both methods are independent of the amplitude of the pulse. One method measures the time that a pulse is above a defined constant fraction of the pulse peak [45]. The second method uses the ratio of pulse area divided by pulse peak as measure of pulse width [46]. Both methods give a linear relationship between pulse width and particle diameter for particles whose diameter is at least half as large as the $1/e^2$ height of the laser beam. Theoretical relationships between focused laser beam height, particle diameter, and the different measurements of pulse width were derived by Sharpless et al. [46]. A set of bead size standards can be used to calibrate the pulse width measurement. The particle diameters determined by pulse width can be used to estimate particle volume and surface area, and these can be used to normalize particle size dependence of fluorescence measurements.

4.3 Fluorescence

4.3.1 Accurate Measurement of Fluorescence

The primary measurement of fluorescence in flow cytometry is proportional to the amount of fluorophore on the particle, and fluorescence is integrated over the entire particle. The integrated fluorescence signal in some instruments is estimated by using a focused excitation beam that is larger than the particle. If all parts of the particle are uniformly illuminated, the peak amplitude, called pulse height, of the fluorescence signal is proportional to the total fluorescence emitted.

If the particles are not uniformly illuminated, e.g., if they are illuminated by a narrowly focused laser beam, the peak of the fluorescence pulse is not an accurate measure of the amount of fluorophore. Lasers, which are the most common light source for flow cytometry, normally have a Gaussian intensity distribution, shown in Fig. 4. The size of a laser beam is usually given in terms of the point at which the beam intensity has decreased to $1/e^2$ (13.5%) of maximum. Since a laser beam has a Gaussian distribution, only particles much smaller than the $1/e^2$ width will be uniformly illuminated. Larger particles centered in a Gaussian laser beam are illuminated less intensely at the edge than at the center. To a reasonable approximation (less than 15% error), pulse height is proportional to amount of fluorophore and independent of particle size if the particle dimension is less than half of the $1/e^2$ size of the laser beam. If pulse height accuracy of less than 2% is required, then the particle should be less than one fifth the $1/e^2$ size of the laser beam. The laser beam is often focused to an elliptical spot on the sample stream with the smaller beam dimension in the direction of particle flow. If the focused laser beam has an elliptical shape at the sample stream, then the beam size limitations on measurement accuracy refer to the smaller dimension of the focused beam.

If the width of the laser beam across the sample stream is much larger than the particle size and the integral (or pulse area) of the fluorescence pulse is determined, the measurement is proportional to the amount of fluorophore on the particle and independent of the particle size. The integral collects and sums the fluorescence as the particle moves through the excitation beam, so it is proportional to the total fluorescence emitted by the particle.

In many applications that stain particles with multiple fluorophores, the fluorescence emission from two or more fluorophores can overlap in the region where one has peak emission. A simple situation is shown in Fig. 7 for fluorescein and phycoerythrin. To measure the amount of signal due to a single fluorophore, e.g., the amount of PE fluorescence at 580 nm, the spectral overlap of interfering fluorophores can be computed and removed. Figure 8 shows examples of FITC and PE data compensated for spectral overlap. The data are corrected so the median value in the "color" where spectral overlap occurs is equal to the median for the stained population. For example the median "PE" value for CD45-FITC stained lymphocytes is the same as the median value for unstained lymphocytes. The CD45-FITC stained cells in Fig. 8c have the same median PE value as the unstained cells. The apparently higher median is an artifact of displaying the data on a log scale. Most of the events in the compensated CD45-FITC population are actually off scale in the lowest displayed "PE" value bin, less than or equal to 10. This display artifact is discussed in detail using the examples in Fig. 9.

Another effect demonstrated in Fig. 8 is the additional spread in data values due to added optical background from spectral overlap. The PE stained cells in Fig. 8d show a wider range of FITC values than unstained cells, And the FITC stained cells in Fig. 8c show a much greater spread in PE values than unstained cells. Optimizing the antibody-fluorochrome combinations to minimize the impact of spectral overlap contributions to optical background on the cells is important in multi-color experiments. Maecker et al. describe the problem in detail and suggest approaches to select optimal fluorochrome-antibody combinations [47].

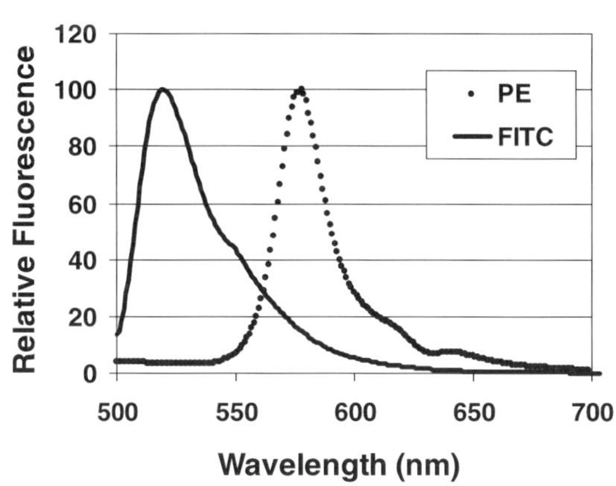

Fig. 7 Fluorescence emission spectra of FITC and PE conjugated antibody solutions. Excitation in the spectrofluorometer was at 488 nm

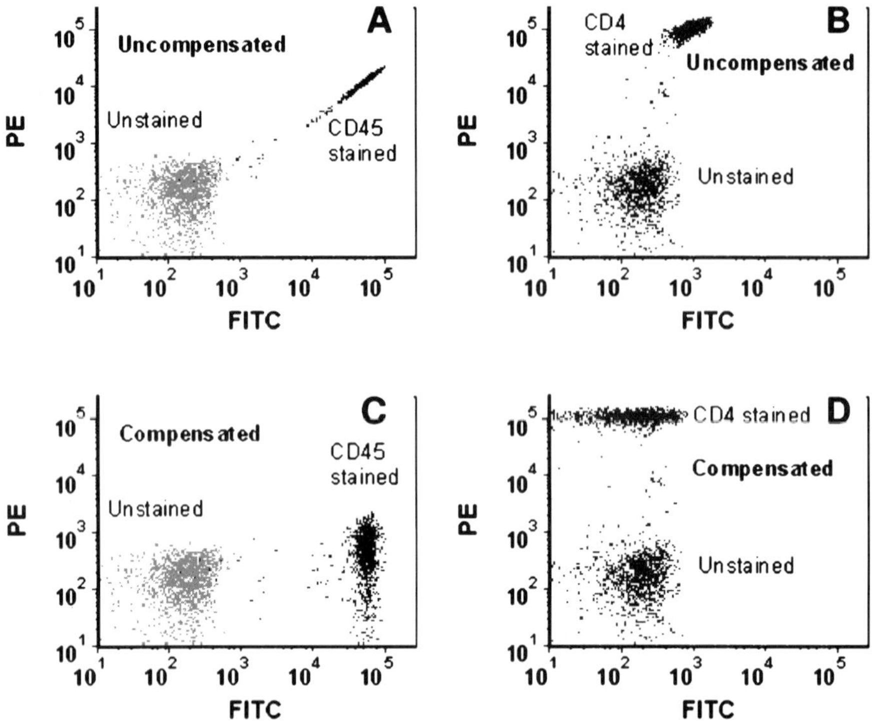

Fig. 8 Illustration of spectral overlap and compensation. Panels **A** and **C** are overlays of data from unstained lymphocytes and lymphocytes stained with CD45 antibody conjugated to FITC. Panel **A** is uncompensated data showing significant FITC signal the detector channel intended to measure PE fluorescence. Panel **C** shows the same data compensated for the spectral overlap of FITC into the PE detector, where a constant 22% fraction of the FITC signal is subtracted from the PE signal for each event. Panels **B** and **D** show uncompensated and compensated data for lymphocytes stained with antibody to CD4 conjugated to PE. In this case the spectral overlap of PE into the FITC detector is much smaller, and a constant 0.8% fraction of the PE signal is subtracted from the FITC signal for each event

If a sample is stained with only two different fluorophores, the correction for spectral overlap is a simple solution of two linear equations, and spectral overlap can be compensated using analog electronics. When at least some pairs in a set of three or four fluorochromes have no significant spectral overlap, analog electronic compensation is still possible. But for more complex situations involving multiple fluorophores, a computer is used to determine the corrections based on the measured amount each fluorophore generates in each detector channel [48–50]. The underlying physics and mathematics results in solving a set of linear equations, but matrix mathematics is used and the correction terms are not easily interpreted in an intuitive way.

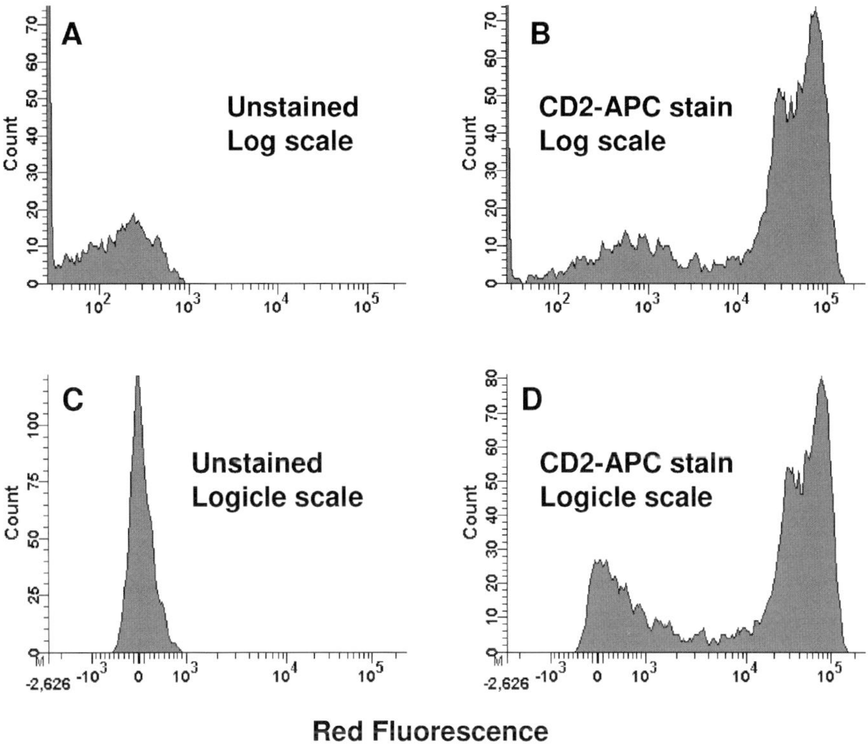

Fig. 9 Illustration of a histogram binning artifact for logarithmic scales. Panel **A** is a histogram of unstained lymphocytes for which the median red fluorescence value is 35 and the range of values is large. Naive interpretation of the histogram would suggest two populations, one with median value near zero and a second population with a median near 350. This apparent second, higher population is an artifact of binning data on a log scale. Panel **C** shows data on a logicle scale that is linear at lower values (approximately − 1000 to 1000) and transitions to logarithmic scaling at higher values. On a linear scale the data are observed to be a nearly symmetric population with median value 35. Panels **B** and **C** compare log and logicle scales to display histograms of lymphocytes stained with antibody to CD2 conjugated to APC. The log scale histogram in panel B seems to show three populations with median values near zero, 600 and 40 000. Panel **D** shows the low value samples displayed on an essentially linear scale to be a single asymmetric population. Data were obtained with a digital acquisition system that measures variation both above and below the nominal "zero" baseline

4.3.2 Fluorescence Lifetime

Fluorescence lifetimes of stained cells have been measured by some investigators. To make these measurements an acousto-optic or electro-optic modulator is placed in the excitation laser beam, typically causing a sinusoidal modulation of laser power at a frequency of 10 to 100 MHz [51–56]. The fluorescence lifetime (or lifetimes) of the emitted fluorescence is determined from the phase shift and/or change in depth of modulation of the fluorescence compared to a reference signal. Steinkamp and colleagues at Los Alamos National Laboratory have published a fairly extensive range of application examples [57]. But as of this writing there are no commercially available flow cytometers that include the fluorescence lifetime measurement.

4.3.3 Fluorescence Polarization

Fluorescence polarization or anisotropy has been measured in flow cytometers. Since an excited fluorescent molecule can move and rotate, the polarization of emitted fluorescence tends to be random unless the environment inhibits rotation of the molecule. A standard expression of the degree of fluorescence polarization is *P*, defined as

$$P = \frac{a - b}{a + b}, \tag{2}$$

where a is the fluorescence intensity with polarization in the direction of excitation source and b is the fluorescence intensity with polarization orthogonal to that of the excitation. *P* can vary from 0 (completely unpolarized emission) to 0.5 (completely polarized).

Investigators have built or adapted commercial flow cytometers to measure fluorescence polarization [58–60]. In the early 1980s the application of flow cytometric fluorescence anisotropy focused on changes in the apparent rigidity of cytoplasm and in membrane fluidity [61–63]. Under some conditions, anisotropy is a useful measure of energy transfer between donor and acceptor fluorochromes and can be used to detect clustering of receptors [64, 65]. Van den Engh and colleagues have used flow cytometry to measure polarization anisotropy of fluorochromes used to stain DNA and as tags to antibodies for surface antigens on cells. Emission from PE-labeled antibody was nearly isotropic (no emission polarization), while fluorescein-tagged antibody and all DNA staining showed polarization values in the range of 0.1 to 0.3 [32]. They also found that the polarization of DNA stains varied with the type of stain and the staining concentration [66]. Optical components, particularly dichroic mirrors, in flow cytometers can be polarization sensitive in their transmission or reflection. Since fluorescence from stained particles usually is partially polarized, fluorescence intensity measurement in instruments with different optical components and configurations can vary. The authors caution that the calibration of fluorescent intensity with standard particles, which may have a different degree of polarization than stained cells, should take polarization into account when establishing the conditions under which they can be used.

4.3.3.1
Data Acquisition and Software

Data acquisition software is integrated with the hardware on commercially available flow cytometers. The data are either acquired or can be exported in an industry standard format, FCS, currently at version 3.0 [67]. Flexible, general purpose analysis software for multiparameter data are available from instrument manufacturers and third party companies. Automated acquisition and/or analysis software for a specific test is often used for clinical applications.

For multicolor data analysis, automated compensation for spectral overlap is often critical. Display of compensated data on a log scale can sometimes lead to misleading interpretation of what appear to be distinct populations, but are in fact an artifact of the binning of events displayed on a log scale [68]. The added noise for spectral overlap also causes broadening of fluorescence data on cells stained with multiple fluorochromes, and the log display makes comparison with unstained or dim control populations difficult. Alternatives to log display are being implemented in analysis to aid in data interpretation. The alternative displays use a transformation of the linear data to a function that is essentially linear for low values and smoothly changes to a log function at high values [69, 70]. Figure 9 compares fluorescence histograms displayed on logarithmic and Logicle (also called bi-exponential) scales. The data of unstained cells displayed on log scale, Fig. 9a, appear to have a bimodal distribution with many events in the lowest channels and a second, broad population in higher channels. This apparent bimodal distribution is an artifact of the logarithmic display, in which successively higher bins of data include an increasing wide range of data values [68]. A similar artifactual bimodal distribution is seen with the logarithmic scale for unstained population in Fig. 9b. When data are displayed using the Logicle scale (Fig. 9c and d), which is essentially linear for the lowest data values, the unstained cell population is correctly displayed as being a single nearly symmetrical population with median near zero.

4.3.3.2
Automation

Automated loading of samples onto the flow cytometer is efficient and is sometimes used to acquire data for later off-line analysis. Automated software analysis of the data for each sample is often combined with sample loading and generally improves reproducibility of the results compared to manual analysis of each sample. Automation of sample preparation saves operator time and can increase the reproducibility of results. Most flow cytometers used for multiple purposes have samples prepared on a separate device and then loaded as a rack or carousel of test tubes or in a multiwell microplate.

Specialized flow cytometers, such as automated hematology analyzers, often combine sample preparation, sample loading and automated analysis in one instrument.

Automated data analysis usually requires data that meets pre-defined criteria or expected ranges of the data. While instrument standardization and performance qualification are always important, they are even more critical when acquisition and analysis are automated. Standardized instrument setup is important to ensure data are in the range expected by the automated analysis software. Equally important is a minimum level of data quality that allows populations of cells or particles to be reliably distinguished and counted.

5 Limitations

5.1 Fundamental Physical limitations

5.1.1 Particle Velocity

Reynolds number, Re, is a dimensionless quantity that characterizes fluid flow in a tube.

$$\mathrm{Re} = \frac{dv}{\nu}, \tag{3}$$

where d is the tube diameter, v is the average velocity, and ν is the kinematic viscosity.

Laminar flow is maintained when Reynolds number is less than 2300. Thus the maximum velocity that maintains laminar flow is inversely proportional to tube diameter. For water at 20 °C, the kinematic viscosity is 0.01 cm^2/s. The equivalent tube diameter for a representative flow cell is 0.025 cm, for which case the maximum average laminar flow velocity is 920 cm/s or 9.2 m/s. If flow is parabolic, the velocity at the center is twice the average, so the maximum fluid and particle velocity is 18.4 m/s for this case.

5.1.2 Background Light

The unavoidable sources of background light are intrinsic autofluorescence or Raman scatter of the particles and Raman scatter from water. Cellular autofluorescence is often from flavins, which absorb blue light strongly and emit in the green and yellow range [71, 72]. For most cells there is a significant reduction in autofluorescence if green excitation is used instead of blue. Cells stained with phycoerythrin-tagged antibody show three to five times greater signal to autofluorescence ratio when excited with green (e.g., 532 nm) excitation compared to 488 nm excitation. Raman scatter from water has peak emission at 580 nm when excited with 488 nm light and is the major source of background for fluorescence in the yellow to orange range. Using green excitation at 532 nm shifts the Raman scatter to 643 nm and eliminates this source of background for phycoerythrin and other yellow-emitting fluorophores.

5.1.3 Fluorescence Yield

Fluorescence from a fluorochrome cannot be increased indefinitely by increasing the excitation power. One fundamental limitation is the finite fluorescence lifetime. The maximum rate at which a fluorescent molecule can be excited is limited by the time, typically 1–100 ns, that it spends in an excited state before emitting a photon. The result is that as excitation power increases beyond a certain amount, the fluorescence begins to approach a limiting intensity. Some fluorochromes can enter long lived triplet states, during which fluorescence and excitation are not possible. Destruction of the fluorophore through photodestruction can also occur [73]. Doornbos et al. have reported on theoretical and experimental studies of the maximum fluorescent signal that could be generated for several fluorophores commonly used in flow cytometry [74].

5.2 Practical Limitations for Fluorescence Measurements and Their Determination

5.2.1 Linearity

Good linearity of fluorescence measurements is fundamental to quantitative results. If logarithmic amplifiers are used, their response must be known in order to use the data quantitatively. A simple method to determine how accurately and over what range a fluorescence measurement is linear has been described by Bagwell et al. [75]. Two particles with different fluorescence intensity, e.g., differing by a factor of 2, are analyzed together. The PMT voltage is varied in increments and the mean fluorescence of each particle is measured at each step. The expected relationship is that the ratio of the measured fluorescence from the two particles is constant, since the input to the PMT does not change. There are several ways to analyze the data to show by how much and at what part of the range the response deviates from linearity. The simplest analysis is to plot the measured ratio or deviation of the ratio from the expected value vs. the measured mean of one of the particles. This ratio approach can also be used to measure the response of logarithmic amplifiers [76, 77].

5.2.2
Measurement Precision at High Signal Levels

Measurement precision or CV at high signal levels is primarily limited by the alignment of the sample stream to the excitation light. In some instruments, particularly stream in air sorters, the user can adjust the alignment. In most currently manufactured bench top analyzers, there is no user adjustment to the optical alignment, and the solution to misalignment is typically to clean the flow cell of bubbles or debris. Brightly fluorescent beads with known low fluorescence CV are used to check alignment.

5.2.3
Measurement Precision at Low Signal Levels (Sensitivity)

Fluorescence sensitivity characterizes the ability of an instrument to resolve a dimly fluorescent population from noise or an unstained population. Sensitivity is limited by amplifier noise in detection electronics, photoelectron Poisson statistics, and background light. (see Eq. 1) Each of these contributions can be measured using fluorescent particles. For the user of a flow cytometer it is important to know if sensitivity is outside the bounds expected for their particular instrument. If fluorescence sensitivity is less than expected, knowing which of several possible factors is outside specification allows quick resolution of the problem.

Electronic noise from amplifiers and other components in the data acquisition system can add to the overall variance of fluorescence measurements at the low end of the range. And electronic noise limits the useful dynamic range of the measurement. When PMTs are used, it is possible to amplify the signals sufficiently to make electronic noise insignificant, but the dynamic range is limited. The effect of electronic noise can be visualized and in many systems quantitatively measured. The general approach is to run a sample of beads or other particles with relatively bright fluorescence and low CV when the particles are measured at a PMT gain that puts the data in the upper decade of a logarithmic histogram display. The CV of the fluorescence distribution will increase when the PMT voltage is lowered sufficiently so that electronic noise is a significant contribution to total variance. If the data can be accurately measured in the lowest range (which may be a problem with logarithmic amplifiers), a plot of standard deviation vs. mean intensity will tend toward an asymptotic value. This asymptotic value is a measure of electronic noise in histogram channel units and is a constant contribution to measurement variance.

Since resolution of two dimly fluorescent populations depends on multiple factors, simple measurements of signal to noise do not predict performance. The most basic situation is that both background light and photoelectron statistics affect sensitivity [78]. Background, B, and detection efficiency, Q, have been defined in terms of molecules of equivalent soluble fluorophore (MESF) and photoelectrons generated per MESF respectively [79]. *Q* and *B* can be measured using LED light flashes [80]. And by running samples of properly characterized fluorescent beads, *Q* and *B* can be measured on any flow cytometer [79]. The measurement of *Q* in the most fundamental units of photoelectrons per MESF, provides a way to compare instruments or predict the performance of an instrument for resolution of dimly fluorescent particles [68, 79].

In many applications using multiple fluorochromes, the largest contribution to background fluorescence can be from spectral overlap. For example, a lymphocyte stained with CD45-FITC has yellow fluorescence in the region used to measure PE equivalent to that produced by 8000 PE molecules. Thus measurement of PE on the same cell is done in the presence of a background equivalent to 8000 PE molecules and the ability to detect less than 8000 PE molecules on the cell is compromised. In multicolor immunofluorescence applications it is important to choose antibody-fluorochrome combinations carefully to minimize the spectral overlap contribution to interfering background [47].

6 Applications and Future Trends

Many applications have been mentioned above in describing instrumentation or flow cytometry measurements. The range of applications in flow cytometry are too broad to be discussed in any detail here, but some general areas will be reviewed.

Immunofluorescence analysis of blood cells is a common application in clinical research and diagnostics. Routine applications use 2–4 different fluorochrome-tagged antibodies to identify and count specific subsets of white blood cells. For example, a mixture of four antibodies (CD45, CD3, CD4 and CD8) each tagged by a different fluorochrome allows identification and counting of all white blood cells, all lymphocytes, all T-lymphocytes and the two main subsets of T-lymphocytes. The concentration of CD4 cells in blood is used to diagnose AIDS and to monitor the response to therapy in AIDS patients. Intracellular staining for specific proteins such as cytokines requires cell fixation and permeabilization [81, 82]. Measurement of cytokine expression also requires blocking the protein secretion pathway [83]. Detailed information about immune system function can be obtained using these techniques to measure the production of cytokines in subsets of CD4 or CD8 T-lymphocytes in response to exposure to a virus or other antigen [84, 85]. This immune function assay is complex and standardization is necessary in order to get reproducible cell counts [86]. Flow cytometry is also an important tool for investigation and isolation of stem cells [87].

Instruments capable of detecting 16 different colors are available, and 8 to 10 color flow cytometry is possible with many instruments. Multicolor (sometimes called polychromatic) flow cytometry makes use of eight or more different antibodies or other fluorochrome probes to obtain detailed information about the composition and function of immune system cells [88, 89]. Twelve-color flow cytometry has been used in vaccine development to rapidly assess immune response to a formulation [90]. Multicolor flow cytometry is also used to characterize the subsets of blood or bone marrow cells in patients with leukemia or lymphoma [91].

The use of bead mixtures rather than planar arrays for multiplexed assays is a growing flow cytometry application [92–95]. For example, a sandwich immunoassay uses antibodies on the bead to capture antigen, and a second, fluorescently labeled antibody binds to another epitope on the antigen to provide a quantitative readout. By using different sizes of beads or, more commonly, beads that are dyed with varied amounts of one or more fluorescent dyes, distinct subsets of beads in a mixture can be identified in a flow cytometer. Each type of bead can perform a separate assay. Mixtures of beads can be added to a single sample to perform many assays simultaneously. Nolan and Mandy reviewed the state of this technology and discuss new multiplexing applications and strategies [96].

Flow cytometry is used to identify and count microorganisms in research, environmental, and industrial applications [97–99]. General purpose flow cytometers are often used for this purpose in laboratories. An automated analyzer for microorganisms (Advanced Analytical Technologies) and a field portable instrument (BioDETECT) are also available.

In contrast to microorganisms, flow cytometers are also used to analyze and sort large particles including small animals. Most stream in air droplet sorters can be configured to sort particle up to about 200 μm, and specialized sorters are used for even larger particles. The COPAS sorter from Union Biometrica, uses a jet of air to deflect the sample stream except when a desired particle is to be collected. The air momentarily stops blowing on the sample stream, which allows the particle to be deposited into a container. COPAS has been used to sort a variety of large particles including nematodes and cell clusters of pancreatic islets [100]. A laboratory built fluidic switching sorter has been built to sort genotype specific Drosophila embryos using a fluorescent protein marker [101].

Microfluidic flow cytometers have been built in many laboratories. Huh et al. provide an excellent review of the technology [102]. Microfluidics allows simplified construction of the flow cell and often the associated fluidics. Design tradeoffs for other subsystems such as optics, detectors and data acquisition electronics is the same as other flow cytometers. For microfluidic flow cytometers the trade-off usually emphasizes simplicity rather than high performance. But several groups are working on using many parallel analysis and sorting channels in microfluidic devices for high throughput cell sorting [17].

Useful performance for several applications has been demonstrated. Two-color fluorescence analyses including cell viability, protein expression and immunofluorescence were demonstrated on an Agilent microfluidic instrument [103, 104]. Immunofluorescence analysis is being developed on a microfluidic sorting device from Micronics [105]. A microfluidic sorter has been demonstrated by a group at the Technical University of Denmark [106]. A microfabricated flow cytometer measuring DC and RF impedance opacity was developed by Cheung et al. and used to characterize red blood cells [107].

Flow cytometry is used or integrated into special purpose instruments for a wide variety of applications. The most widely used flow cytometers are automated hematology analyzers, which use one or more of the basic sensing methods – impedance, light scatter or fluorescence. FACSCount™, a simple 2-color analyzer manufactured by BD Biosciences, is dedicated to counting CD4 cells for diagnostics and therapy monitoring of HIV/AIDS patients in resource poor countries such as Africa [108, 109]. In the dairy industry, flow cytometers are incorporated into automated analyzers made by Bentley or Foss that measure somatic cell and bacterial contamination in raw milk [110, 111]. In the livestock industry, special purpose flow cytometers have been used to measure the viability and concentration of bull and boar sperm used for in vitro fertilization [112, 113]. A commercial company, XY, Inc., uses fluorescence cell sorting to separate X and Y bearing sperm for use in sex-determined in vitro fertilization in the livestock industry and other non-human sex selection applications [114, 115]. For environmental study of phytoplankton, a specialized and technically advanced flow cytometer, the Cytobuoy, has been developed. The Cytobuoy can be configured as a laboratory flow cytometer, moored in position in the ocean or fresh water with remote transmission of data, or placed in a pressurized vessel for incorporation into a remote controlled submarine, the Cytosub [116, 117].

References

1. Givan AL (2001) Flow Cytometry: First Principles. Wiley, New York
2. Ormerod MG (2000) Flow Cytometry: A Practical Approach. Oxford University Press, Oxford
3. Diamond RA, Demaggio S (eds.) (2000) In Living Color: Protocols in Flow Cytometry and Cell Sorting. Springer, Berlin
4. Watson JV (2004) Introduction to Flow Cytometry. Cambridge University Press, Cambridge
5. Shapiro HM (2003) Practical Flow Cytometry, 4th edn. Wiley, Hoboken
6. Kachel V (1976) J Histochem Cytochem 24:211
7. Crosland-Taylor PJ (1953) Nature 171:37
8. Crosland-Taylor PJ, Stewart JW, Haggis G (1958) Blood 13:398
9. Dittrich W, Gohde W (1969) Z Naturforsch B 24:360
10. Kamentsky LA, Melamed MR, Derman H (1965) Science 150:630
11. Kamentsky LA, Melamed MR (1967) Science 156:1364

12. Fulwyler MJ (1965) Science 150:910
13. Sweet RG (1965) Rev Sci Instrum 36:131
14. Hulett HR, Bonner WA, Barrett J et al. (1969) Science 166:747
15. Stewart CC, Steinkamp JA (1982) Cytometry 2:238
16. Keij JF, van Rotterdam A, Groenewegen AC et al. (1991) Cytometry 12:398
17. Leary JF (2005) Cytometry A 67:76
18. Bonner WA, Hulett HR, Sweet RG et al. (1972) Rev Sci Instrum 43:404
19. Hulett HR, Bonner WA, Sweet RG et al. (1973) Clin Chem 19:813
20. Fu AY, Spence C, Scherer A et al. (1999) Nat Biotechnol 17:1109
21. Habbersett RC, Jett JH (2004) Cytometry A 60:125
22. Steen HB (2004) Cytometry A 57:94
23. Kuckuck FW, Edwards BS, Sklar LA (2001) Cytometry 44:83
24. Schwartz A, Sugg H, Ritter TW et al. (1983) Cytometry 3:456
25. Steinkamp JA, Hiebert RD (1982) Cytometry 2:232
26. Steinkamp JA, Kraemer PM (1974) J Cell Physiol 84:197
27. Hoffman RA, Britt WB (1979) J Histochem Cytochem 27:234
28. Hoffman RA, Johnson TS, Britt WB (1981) Cytometry 1:377
29. Warner BA, Reardon DM (1991) Am J Clin Pathol 95:207
30. Peters DC (1979) J Histochem Cytochem 27:241
31. Goodwin PM, Ambrose WP, Martin JC et al. (1995) Cytometry 21:133
32. Asbury CL, Uy JL, van den Engh G (2000) Cytometry 40:88
33. Beisker W, Eisert WG (1981) Anal Quant Cytol 3:315
34. Beisker W (1997) Proc SPIE 2982:420
35. Hakamata T (2006) Photomultiplier Tubes – Basics and Applications. Hamamatsu Photonics K.K. Electron Tube Division, Hamamatsu City
36. Kerker M (1983) Cytometry 4:1
37. Salzman GC, Wilder ME, Jett JH (1979) J Histochem Cytochem 27:264
38. Mullaney PF, Crowell JM, Salzman GC et al. (1976) J Histochem Cytochem 24:298
39. Sharpless TK, Bartholdi M, Melamed MR (1977) J Histochem Cytochem 25:845
40. Steinkamp JA (1983) Cytometry 4:83
41. Stewart CC, Stewart SJ, Habbersett RC (1989) Cytometry 10:426
42. de Grooth BG, Terstappen LW, Puppels GJ et al. (1987) Cytometry 8:539
43. Fawzi ZO, Fakhro NA, Nabhan RA et al. (2003) Trans R Soc Trop Med Hyg 97:71
44. Olson RJ, Zettler ER, Anderson OK (1989) Cytometry 10:636
45. Leary JF, Todd P, Wood JC et al. (1979) J Histochem Cytochem 27:315
46. Sharpless TK, Melamed MR (1976) J Histochem Cytochem 24:257
47. Maecker HT, Frey T, Nomura LE et al. (2004) Cytometry A 62:169
48. Bagwell CB, Adams EG (1993) Ann NY Acad Sci 677:167
49. Roederer M (2001) Cytometry 46:357
50. Roederer M (2001) Cytometry 45:194
51. Steinkamp JA, Crissman HA (1993) Cytometry 14:210
52. Cui HH, Valdez JG, Steinkamp JA et al. (2003) Cytometry A 52:46
53. Deka C, Cram LS, Habbersett R et al. (1995) Cytometry 21:318
54. Deka C, Sklar LA, Steinkamp JA (1994) Cytometry 17:94
55. Pinsky BG, Ladasky JJ, Lakowicz JR et al. (1993) Cytometry 14:123
56. Steinkamp JA (1994) Methods Cell Biol 42 Pt B:627
57. Crissman HA, Steinkamp JA (2001) Methods Cell Biol 63:131
58. Arndt-Jovin DJ, Jovin TM (1976) Prog Clin Biol Res 9:123
59. Keene JP, Hodgson BW (1980) Cytometry 1:118
60. Schaap GH, de Josselin de Jong JE, Jongkind JF (1984) Cytometry 5:188

61. Bock G, Huber LA, Wick G et al. (1989) J Histochem Cytochem 37:1653
62. Epstein M, Norman A, Pinkel D et al. (1977) J Histochem Cytochem 25:821
63. Fox MH, Delohery TM (1987) Cytometry 8:20
64. Bene L, Fulwyler MJ, Damjanovich S (2000) Cytometry 40:292
65. Chan SS, Arndt-Jovin DJ, Jovin TM (1979) J Histochem Cytochem 27:56
66. Uy JL, Asbury CL, Petersen TW et al. (2004) Cytometry A 61:18
67. Seamer LC, Bagwell CB, Barden L et al. (1997) Cytometry 28:118
68. Wood JC (1998) Cytometry 33:260
69. Bagwell CB (2005) Cytometry A 64:34
70. Parks DR, Roederer M, Moore WA (2006) Cytometry A 69:541
71. Aubin JE (1979) J Histochem Cytochem 27:36
72. Benson RC, Meyer RA, Zaruba ME et al. (1979) J Histochem Cytochem 27:44
73. White JC, Stryer L (1987) Anal Biochem 161:442
74. Doornbos RM, de Grooth BG, Greve J (1997) Cytometry 29:204
75. Bagwell CB, Baker D, Whetstone S et al. (1989) Cytometry 10:689
76. Gandler W, Shapiro H (1990) Cytometry 11:447
77. Schmid I, Schmid P, Giorgi JV (1988) Cytometry 9:533
78. Wood JC, Hoffman RA (1998) Cytometry 33:256
79. Chase ES, Hoffman RA (1998) Cytometry 33:267
80. Steen HB (1992) Cytometry 13:822
81. Jung T, Schauer U, Heusser C et al. (1993) J Immunol Methods 159:197
82. Prussin C, Metcalfe DD (1995) J Immunol Methods 188:117
83. Nylander S, Kalies I (1999) J Immunol Methods 224:69
84. Maino VC, Maecker HT (2004) Clin Immunol 110:222
85. Suni MA, Maino VC, Maecker HT (2005) Curr Opin Immunol 17:434
86. Maecker HT, Rinfret A, D'Souza P et al. (2005) BMC Immunol 6:13
87. Gratama JW, Orfao A, Barnett D et al. (1998) Cytometry 34:128
88. Baumgarth N, Roederer M (2000) J Immunol Methods 243:77
89. Roederer M, De Rosa S, Gerstein R et al. (1997) Cytometry 29:328
90. Roederer M, Brenchley JM, Betts MR et al. (2004) Clin Immunol 110:199
91. Wood B (2006) Arch Pathol Lab Med 130:680
92. Lisi PJ, Huang CW, Hoffman RA et al. (1982) Clin Chim Acta 120:171
93. McHugh TM, Stites DP, Casavant CH et al. (1986) J Immunol Methods 95:57
94. Morgan E, Varro R, Sepulveda H et al. (2004) Clin Immunol 110:252
95. Nolan JP, Sklar LA (2002) Trends Biotechnol 20:9
96. Nolan JP, Mandy F (2006) Cytometry A 69:318
97. Davey HM (2002) Methods Cell Sci 24:91
98. Shapiro HM (2000) J Microbiol Methods 42:3
99. Winson MK, Davey HM (2000) Methods 21:231
100. Fernandez LA, Hatch EW, Armann B et al. (2005) Transplantation 80:729
101. Furlong EE, Profitt D, Scott MP (2001) Nat Biotechnol 19:153
102. Huh D, Gu W, Kamotani Y et al. (2005) Physiol Meas 26:R73
103. Chan SD, Luedke G, Valer M et al. (2003) Cytometry A 55:119
104. Palkova Z, Vachova L, Valer M et al. (2004) Cytometry A 59:246
105. Lancaster C, Kokoris M, Nabavi M et al. (2005) Methods 37:120
106. Wolff A, Perch-Nielsen IR, Larsen UD et al. (2003) Lab Chip 3:22
107. Cheung K, Gawad S, Renaud P (2005) Cytometry A 65:124
108. Levin A, Brubaker G, Shao JS et al. (1996) Int J STD AIDS 7:288
109. Young NL, Ponglertnapakorn P, Shaffer N et al. (1997) Clin Diagn Lab Immunol 4:783

110. Gonzalo C, Boixo JC, Carriedo JA et al. (2004) J Dairy Sci 87:3623
111. Gunasekera TS, Veal DA, Attfield PV (2003) Int J Food Microbiol 85:269
112. Christensen P, Knudsen DB, Wachmann H et al. (2004) Theriogenology 62:1218
113. Christensen P, Stenvang JP, Godfrey WL (2004) J Androl 25:255
114. Garner DL (2001) J Androl 22:519
115. Johnson LA (2000) Anim Reprod Sci 60-61:93
116. Dubelaar GB, Geerders PJ, Jonker RR (2004) J Environ Monit 6:946
117. Dubelaar GB, Gerritzen PL, Beeker AE et al. (1999) Cytometry 37:247

Springer Ser Fluoresc (2008) 6: 343–370
DOI 10.1007/4243_2008_047

Published online: 7 May 2008

Flow Cytometry Quality Assurance

Robert M. Zucker

U.S. Environmental Protection Agency, Office of research and Development,
National Health and Environmental Effects Research Laboratory,
Reproductive Toxicology Division (MD-67), Research Triangle Park, NC 27711, USA
zucker.robert@epa.gov

Abstract A flow cytometer needs to be evaluated prior to its operation to insure that it is aligned with good sensitivity and the fluidics are functioning properly without any restrictions or blockage. Two simple performance tests have been described to determine if a flow cytometer is functional with good sensitivity. The first test determines if the system is properly aligned with a clean flow cell that contains no fluidic blockage. Using uniform single-intensity beads, the coefficients of variation (CVs), peak channels, and histogram distributions are measured to assess the system for alignment and functionality. The second test monitors the sensitivity of the system by using a series of four to six multi-intensity beads. The test measures dye contamination, the cleanliness of the system and the amount of background light scatter that may be contained in the system. By comparing the means and standard deviations of the first two peaks, a value can be determined which relates to the sensitivity of the system. Failure to obtain good CV and sensitivity values in these two tests will compromise the ability to detect dim fluorescence from the background and will ultimately affect the precision of the system. These multi-intensity beads can also be used to determine the linearity of the system. It is important to run these tests at the flow rate at which the samples will ultimately be measured, as the values will change as the flow rate increases.

Keywords Alignment beads · Flow cytometry · Lasers · Multi-intensity beads · Quality assurance · Quantification · Spectroscopy

1 Introduction

Prior to measurement of molecules of equivalent soluble flurochrome (MESF) and quantification of a flow cytometer, the machine should be aligned and all fluorescence and scatter parameters should be optimized for best resolution. In the alignment procedure, one tries to get the cluster of beads to have the highest intensity with the tightest histogram in one dimension or a small cluster in two dimensions. These features usually relate to a low coefficient of variation (CV) measurement. The alignment is normally done by a service engineer, but it can and should be checked daily by the investigator prior to starting an experiment. Usually the investigator can observe that the beads or reference particles are located in the same reference channel at the same Photomultiplier tube (PMT) voltage setting each day with a similar CV value. Measurement of the CV values relates to the daily reproducibility of the machine. These two variables, CV and reference channels, are two factors that can be measured by investigators to insure the machine is equivalent to the previous day's setup. These tests should be done daily for equipment reproducibility. A number of different particles have been used as reference particles, including biological particles (trout red blood cells, chicken red blood cells, thymus cells) or uniform fluorescence beads (hard dyed beads). Generally, hard dyed beads are more reproducible than a biological standard.

Prior to the accurate assessment of fluorescence from microscopy or flow cytometry fluorescence equipment, the investigator must ensure that the equipment is functioning correctly. Unlike confocal microscopy Quality assurance (QA) procedures, in which many tests have to be made prior to making quantitative measurements, flow cytometers are much simpler to evaluate [1–3]. Two simple tests can be used to assess the fluidics, sensitivity and alignment of the flow cytometry system. Figure 1A–C shows three different conditions of flow cytometry (slow flow rate, A; fast flow rate, B; and blockage, C). The alignment of the system can be checked by using beads of uniform size and fluorescence emission. These beads have been designated alignment beads [4–8]. The acquired data should be measured on a linear scale and not a logarithmic scale (Figs. 2 and 3) The second test uses a mixture of identically sized beads that have different intensities to determine the cleanliness and sensitivity of the system. This cocktail of beads usually contains between four and eight different intensity beads including a bead with no fluorescence, a bead with dim fluorescence and one to six other populations of beads that have various levels of brighter fluorescence [4–10] (Figs. 4, 5, and 6). It is essential to test the

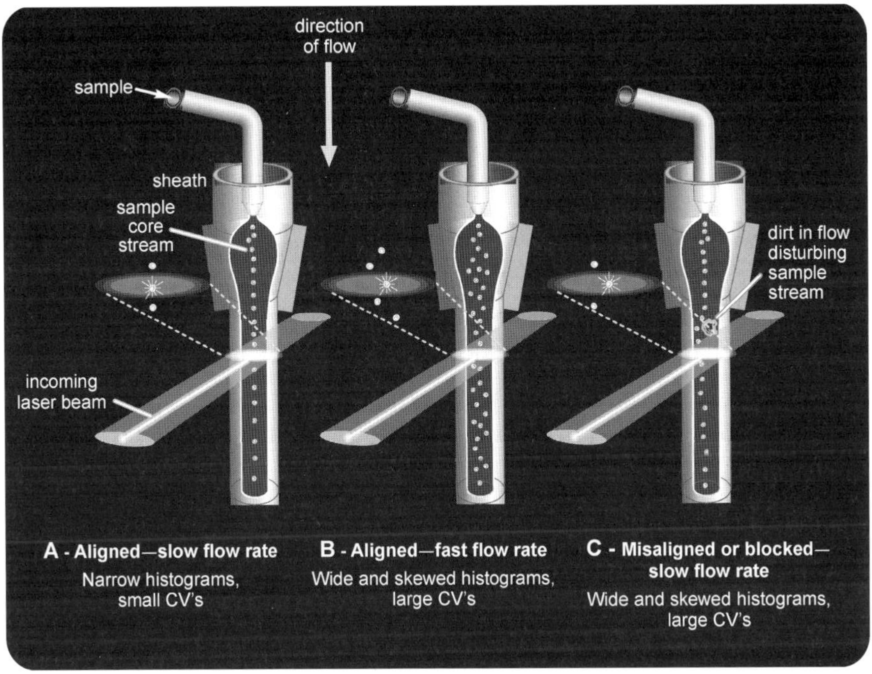

Fig. 1 Diagram of beads in relatively single file in a flow cell yielding a reliably narrow coefficient of variation (*CV*). **A** Increasing the flow rate will spread the region of detection relative to the laser beam increasing the CV and about a 1% and broadening the histogram. **B** Blockage of the flow cell **C** shifts the stream and broadens the histogram in a nonsymmetrical manner. If the alignment is not correct as shown in Tables 1–3, there is usually a large skewness in the histogram distribution and the CV will be increased greatly at higher speeds

alignment and sensitivity of flow cytometers to ensure that they are working correctly prior to making MESF measurements, DNA analysis and other fluorescence measurements described in other chapters of the book that involve flow cytometers [5–7, 11–17].

In order to quantify fluorescence, the system has to be calibrated using a standard. Abe Schwartz originally devised the MESF concept, in which the absolute amount of fluorescence on cells can be determined if a reference regression line is established [6, 9, 16]. Usually this reference regression line is determined by running a series of six ro eight populations of beads of known fluorescence intensity. The test can then measure the following: linearity, detection threshold, dynamic range and windows of analysis. If done correctly, each channel can be designated a specific value and cells that occupy a specific region can be identified by the amount of fluorescence located on the surface. It is suggested that the reader refer to other chapters in this book for the details and to Abe Schwartz reprints on the subject [10, 17].

Sensitivity is an important parameter in detecting not only dim fluorescence from background, but in the ability to resolve particles in a flow cytometer. Although the multi-intensity beads of MESF have been used for this endpoint, the data are not well established.

The performance of flow cytometers has been described in terms of precision, sensitivity and accuracy by Howard Shapiro [5]. Precision describes the extent to which identical values are obtained from measurements of identical particles. Precision can be described as the CV, while sensitivity is the ability to detect particles above background. Accuracy is the degree in which the measurement performs compared to true values. In this chapter, the values derived from precision and sensitivity measurements are used to determine proper performance and daily QA of a flow cytometer. The CV is expressed as the standard deviation (SD) divided by the mean. The CV can also be derived by measuring the full width at half maximum (FWHM) and dividing by the mean channel. In this chapter, the work of Ortyn et al. and Jim Wood [18, 19] is extended. They proposed that the distribution resolution metric, Rd, or the relative sensitivity of a flow cytometer equals the mean of the dim population minus the mean of the null population divided by the sum of the SD of these two lower intensity populations. Ortyn's equation is $Rd = M2 - M1/(SD1 + SD2)$. Jim Wood called it the "resolution parameter" and it is described in [19].

This chapter will help define the CV test used to check flow cell blockage and alignment and the sensitivity test will be used to evaluate the machine contamination and ability to measure low-level fluorescence. A machine that is set up correctly will pass the CV test and the sensitivity test and these values should be measured daily.

2 Precision

Precision of a flow cytometer can be measured with a set of identical brightly stained beads that has a low CV. These beads are often referred to as alignment beads as they can also be used for that endpoint [5–7]. The precision of a machine may be decreased by a number of parameters, including laser light fluctuations, failing laser, poor alignment of the optics, and disturbances of the fluid flow. Generally, the primarily suspect for poor precision in flow cytometers is the system fluidics. Blockage and dirty tubing will also decrease the system precision (Figs. 1, 5, and 6). If and when blockage occurs the system should be flushed, cleaned, and then reevaluated. The evaluation process consists of simply running a bead population that has a small internal CV at a constant flow rate. The shapes of the resulting scatter and fluorescence histograms are observed and their resultant CV is then measured (Figs. 2 and 3). This reference bead population should be run at a reproducible flow rate and

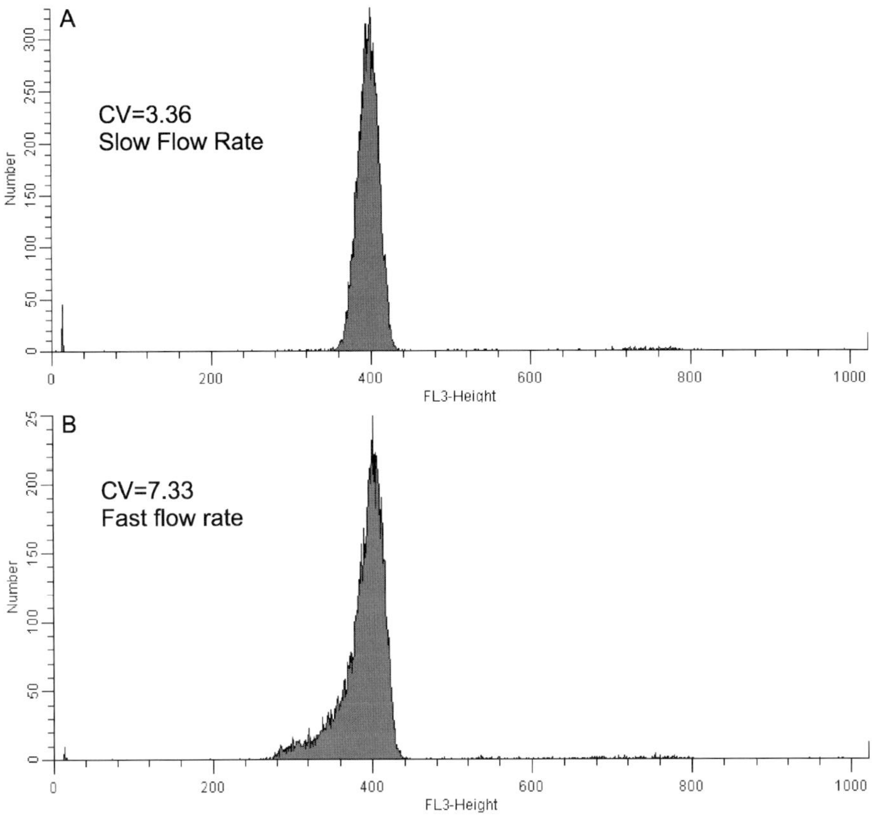

Fig. 2 Evaluation of red fluorescence (FL3) at slow (**A**) and fast (**B**) flow rates. The skewness at higher speeds indicates a slight misalignment of the system. The increase of the CV correlates with the increased skewness of the distribution. Large 10 μm Coulter beads show this type of skewness better than smaller 2–3 mum beads (Duke 3 μm or MP 2.5 um)

placed into the same approximate channel each day. Alterations in the CV or mean channel position indicate that the machine has changed its acquisitions parameters from the previous day's operation and the issue should be further investigated.

2.1 Factors

There are a number of variables that can affect the CV measurements. The relationship between the core sample size and the laser beam shape determines the CV and the histogram shape. Increasing the sample flow rate while holding the sheath flow rate constant will widen the sample stream in an aligned system. In a misaligned system, the histogram distribution will be dramati-

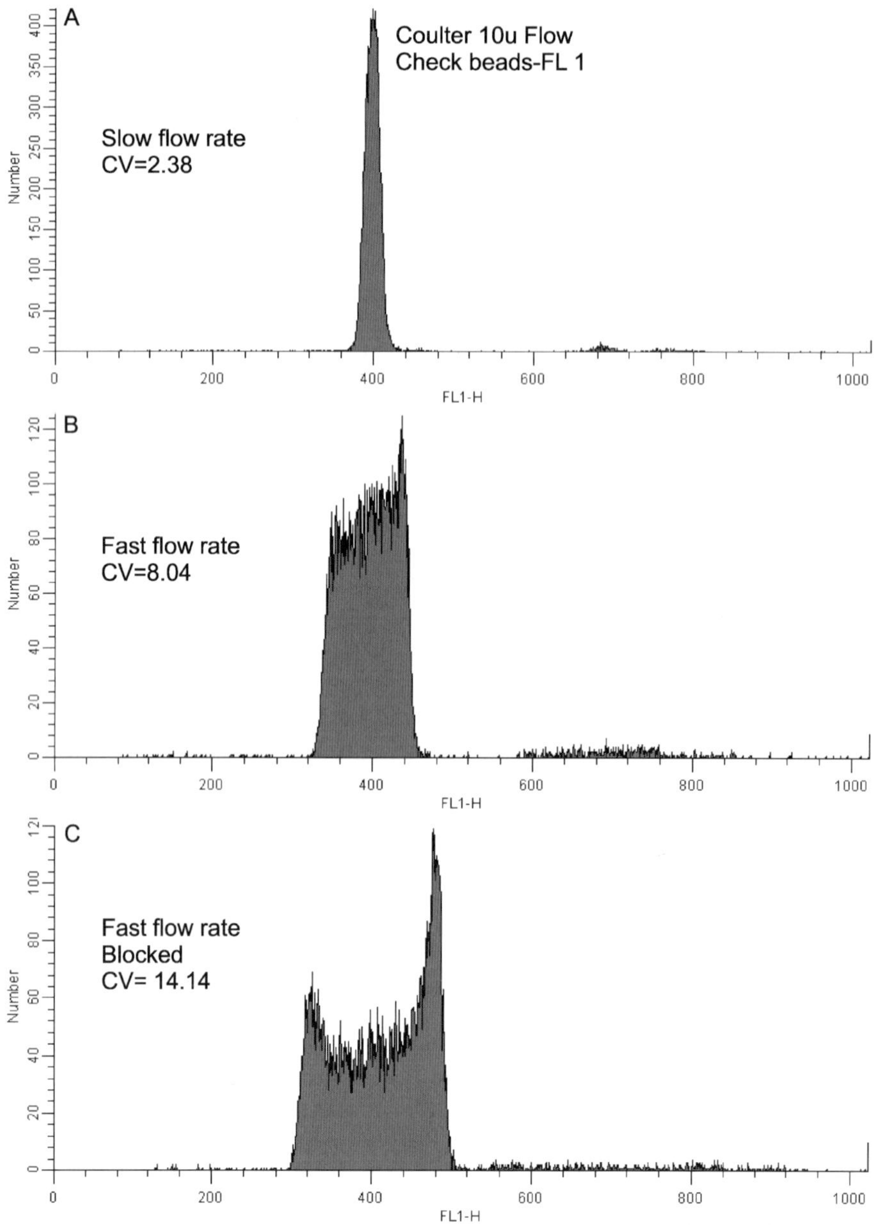

Fig. 3 Evaluation of FL1 at slow **A** and **B** fast flow rates. The skewness at higher speeds indicates a misalignment of the system. If blockage occurs in a misaligned system**C**, the CV will increase and the histogram distribution will widen. The CV values are shown in Table 2b

cally changed by increasing the sample flow rates. Laser fluctuations, failing or dying lasers, PMT voltages and system electronics are other factors that can affect this CV measurement. More photons detected due to higher numerical aperature (NA) objectives or higher laser power will decrease the CV value. The higher NA objectives used in collection gather more light, but will be more susceptible to these alignment issues. Higher powered lasers may bleach the sample and these lasers are costly to install and maintain. However, the majority of variations in the CV measurement will be the result of problems with the system's fluidics.

If the system is properly aligned and functioning properly, the stream occurs at the center of the flow cell, in the region where there is maximum laser light. If there is an obstruction in the flow cell, the stream may be moved lateral to the optical axis of the laser beam and away from the peak intensity position at the center of the beam (Fig. 1C). In this case, the histogram distribution will show skewness to lower values and a larger CV. This is due to the beads intersecting the laser beam in a region that has lower laser intensity. Normally the laser light will illuminate a spot whose intensity is maximum and uniform across the sample stream as shown in Fig. 4 of Chapter 34 of this book [12]. If for some reason this intersection is off-center then the laser profile will not be uniform across the core stream. This can occur if the optics is not aligned in the center of the measurement field. This will generate larger CVs as beads located within the core pass through the different portions of the laser beam (Fig. 1). An example of this larger CV is shown in Figs. 2 and 3. If one observes the cluster of beads in a two-dimensional cytogram of side scatter versus forward scatter, the cluster will widen as the sample flow rate increases, which generally suggests an aligned system. However, if the cluster moves greatly to the left and results in lower intensity values, it suggests that the alignment of the system is not ideal and the sample stream is not in the center of the flow cell (Figs. 2 and 3). This test is especially important if one uses the equipment at medium or fast flow rates. In our experience, we have observed that the service personnel have specifications for QA at low flow rates but will sometimes check the alignment using beads at faster flow rates. The CV can also widen if the laser tube is reaching the end of its lifetime (dying) and changes the size of the illuminating spot in the flow cell. The PMTs of a FACSCalibur are defined in the following way: 488 laser excitation yields FL1 (green fluorescence), FL2 (orange fluorescence), FL3 red fluorescence and 633 excitation of FL4 (far-red fluorescence). Alignment beads of 2.5 μm (A 7302, Molecular Probes) yielded approximately 1% CV in FL2 and FL3 on a system with a good laser and nearly 3% on a system with a dying laser. An increase in the operating voltages of the laser is another indicator that the laser is reaching the end of its lifetime. Lasers can introduce noise into the system if they are operating subpar.

After proper alignment of our flow cytometer [2], it yielded consistently low CVs that did not change significantly when the flow rate increased from slow (12 mul/s) to faster speeds (60 mul/s). In our flow cytometers, the CVs of the fluorescent beads were between 1–2% for Coulter 10u beads and MP 2.5u beads.

2.2 Beads

Ideally it would be useful to have a large number of monodisperse beads with a tight CV (Coulter 10u) for alignment. This yields a better representation of the fluidic pattern and it is easier to observe potential problems using them, as they are similar to cells in size. However, these Coulter beads are relatively expensive compared to alignment beads from other companies (i.e., Molecular Probes (MP), Bangs, Dukc, and Polysciences) that have sizes between 2 and 3 mum and have been used routinely for many years by numerous investigators and Becton Dickenson (BD) service personnel. In this study we have used Coulter 10 mum (Flow Check fluorospheres), Duke (3 mum (XPP-1251, alignment beads) and Molecular Probes 2.5 mum (A7302 align flow) for alignment of the flow cytometers. The CVs of the Duke 3 mumare about 1/2 % greater than the 2.5 mum MP and about 1% greater than the 10 mum Coulter FlowCheck beads. These values are relative and may change due to different lots in the future. As long as a standard value can be determined, any of these beads appear to be acceptable for alignment endpoints. Usage of these alignment beads in a flow cytometer consists of just adding a few drops (usually one to three drops) depending on their concentration to a few milliliters of solution (water or saline) and then running the sample at a constant rate of approximately 500 counts/s. Higher concentrations and flow rates can affect the core dynamics and thus the CV. The sample is run on a linear scale and the histogram peak of the distribution is placed into approximately 40% of full scale (channel 400 on a scale of 1024). This enables one to see doublets at approximately 80% of full scale (channel 800 of 1024 scale) and have enough channels to observe the shape of the entire histogram.

The data that one obtains from an alignment test are the following: peak channel at a specific PMT voltage, mean channel, histogram shape, and CV. These data should be obtained from all the scatter and fluorescent channels. It is important to have a bead that is fluorescent with a broad emission spectrum that emits into all three PMT channels. The forward scatter signal has a different light path than the right-angle scatter signal and could be affected very differently in any alignment procedure. Generally, all the bead manufacturing companies produce a bead that has a dye that can be excited in the 488 nm laser and will emit fluorescence with a broad spectrum that emits into the green, orange and red detectors. It is important to standardize the channel of detection (i.e., 400), as the location of the histogram will affect the CV. Beads placed in lower channels will increase the CV as there are fewer channels in the histogram. The PMT has an optimum voltage for operation below which the signal to noise is low. The use of the lower channels may also show an increased CV due to the operation of the PMTs at suboptimal levels. PMTs will yield a high CV when they are operated at low PMT voltages. This observation has been described in recent BD technical notes and presentations to optimize flow cytometers. However, as shown previously by our laboratory in both confocal microscopy and flow cytometry techniques, higher PMT voltages, as expected, reveal more noise resulting in a slightly larger CV [1, 2, 27]. In systems that have a filter in the sheath line, there will be a general slowing of the sheath flow rate over time as the filter becomes clogged. This will result in a higher peak fluorescence channel for the same sample (slow—medium–fast) flow rate [4] as the flow rate slows and the particle remains in the laser excitation region for a longer period of time, thus yielding higher fluorescence. It is important to record the channel number and CV for a specific flow rate daily, as any changes from this reference value represent changes in the actual fluidic system that need to be addressed with either cleaning or a service call. The change in CVs can be used as a first indicator to monitor the condition and stability of the machine. Generally, with the newer machines the user does not have the capacity to align the equipment. Thus, the precision of the system can be monitored by the widening of the CV. A change in the temperature of the saline can affect the reactive index and flow rate of the stream and thus the CV measurements.

2.3 Histogram Shape

The histogram should be a symmetrical distribution with a low CV (Figs. 2 and 3) when measured at a slow flow rate. An increase in the CV suggests there is a problem with the way the sample stream intersects the optical laser light. If the CV is large then it will be observed as skewness to lower intensity values. Increasing the sample flow rate relative to the sheath flow rate will increase the CV as the stream widens. If the system is in alignment there will be just a widening of the bead cluster, but if it is out of alignment the cluster will move from its original position and widen. Since each PMT has its own alignment optical path, this change in cluster dynamics with flow rate can affect the individual PMTs differently (Table 1).

In most of the BD FACSCalibur flow cytometers, the flow rate is described as slow, medium, or fast which is equivalent to approximately 12 mul/s, 30 mul/sand 2u1m l/s respectively. These values are approximate and can be adjusted either higher or lower by turning an internal knob. A slower rate will result in lower CVs. The flow cytometers tested at the EPA showed that the PMTs had acceptable CVs for 2.5 mum beads (MP) at slow flow rates (usually below 3% for fluorescence channels and below 4% for scatter parameters). In one machine, increasing the flow rates to faster speeds yielded skewness to lower values and a larger CV with FL3, while just a general broadening of

Table 1 Alignment data. Flow cytometer #1 (BD FACSCalibur): CV of 2.5 μm MP beads (A-7302 alignment beads) and 10 μm Coulter beads (Flow Check fluorospheres) were run at slow (12 mul/s) and fast speeds (60 mul/s). The increase in coefficient of variation (CV) is expected at faster speeds. The large increase in red fluorescence (*FL3*) suggests an alignment problem resulting form the detector being out of alignment with the fluid stream excitation point that is only detected at fast flow rates. The histogram distribution usually widens in a nonsymmetrical manner with increased intensities in lower intensity channels (Fig. 2). *FL1* Green fluorescence, *FL2* red fluoresence, *FS* forward scatter, *SS* side scatter

Flow #1 Parameter	MP 2.5 Slow	MP 2.5 Fast	Coulter 10 Slow	Coulter 10 Fast
FS	3.77	4.06	2.93	3.91
SS	2.54	4.67	3.52	4.23
FL1	2.6	3.73	2.53	3.13
FL2	2.44	3.47	2.7	3.33
FL3	3.09	8.46	2.94	7.64

the distribution in the other PMTs. The CVs on this machine increase about 1% between slow and fast flow rates in an aligned system using 2.5 mum MP beads. A number of examples are shown in Tables 1–4 in which the increase was greater than 1%, illustrating either a misaligned flow cytometer or a dirty fluidic system in the flow cytometer.

2.4 Bead Size

Monodisperse beads ranging in sizes between 2 mum and 10 mum have been used to check the alignment of flow cytometers. We have found that with 2 mum beads (Polysciences #18604 Fluorobrite calibration grade) there is skewness to lower and higher values with about 5% of the population on the EPS flow cytometers. It is better to use either 2.5 mum beads (MP) or 3 mum beads (Duke) to assess the alignment of flow cytometers. The CVs of the 2.5 mum beads from MP are about 1/2 % lower than the 3 mum beads from Duke. Although it is best to use the population of beads that has the smallest CV it is also preferable to use the larger bead to get a better representation of the stream dynamics. The 90° scatter signal can be very broad, depending on the particle used. It may be related to MIE scatter and the surface of the bead [12]. The 5–6 μm particles will yield very broad distribution while 2–3 μm and 10 mumyield better side scatter distributions. MIE scatter may affect 2 mum beads also. Polysciences 2 mum beads yield a cytogram distribution in which the majority of the beads are in a cluster with about 5% being either greater or less than the cluster. Naturally this will affect the measuring and the visualization of the major bead cluster and the

Table 2 a Alignment data. FACSCalibur #2: CV of 2.5 μm MP beads and 10 μm Coulter beads at slow (12 mul/s) and fast speeds (60 mul/s). Machine was inactive for a long period of time and was relatively dirty. The alignment of the system was made on this relative dirty machine. Values were excellent; 1–2% CVs for the fluorescent channels and abut 2.5% for the scatter channels. After 3 days it was found that forward scatter was way out of alignment. The CVs increased greatly with both beads as the flow rate was increased. The CVs of Coulter beads greatly increased in this misaligned machine at faster flow rates. Cleaning with Clorox and Contron 70 improved the system performance (Table 2b). Another alignment by a factory representative is necessary on the clean fluidic machine

Flow #2 Misaligned/Dirty	MP 2.5 Misaligned		Coulter 10 Misaligned	
Parameter	Slow	Fast	Slow	Fast
FS	28.7	40.32	14	24
SS	6.11	8.34	3.18	11.3
FL1	5.65	10.52	3.76	10.6
FL2	6.01	11.28	3.61	10.76
FL3	5.72	9.86	3.66	10.9

Table 2 **b Alignment data FACSCalibur #2:** CV of 2.5 μm MP beads (A-7302 align flow), Duke Cytocal multifluor alignment beads (Duke XPP1251), and 10 μm Coulter beads (Flow Check) at slow (12 mul/s) and fast speeds (60 mul/s). All the lines in the machine were cleaned with 10% Clorox followed by water and 10% Contron 70. There was improved performance compared to the dirty machine shown in Table 2a. The CVs of all beads increased at fast flow rates in this misaligned machine with the greatest increase being shown by the Coulter beads. Upon blockage, the histogram distributions widen and the CVs of all parameters increased (Fig. 3)

Flow #2 Parameter	MP 2.5 Slow	MP 2.5 Fast	Duke 3 Slow	Duke 3 Fast	Coulter 10 Slow	Coulter 10 Fast	Blocked Fast
FS	8	6.96	4.85	7.08	1.62	7.04	10.54
SS	3	7	6.19	9.65	3.38	10.87	16.54
FL1	2.17	5.71	2.38	8.55	2.38	8.04	14.14
FL2	1.82	5.92	2.79	7.4	2.2	8.12	13.97
FL3	1.95	5.74	3.82	7.47	2.19	8.27	14.47

CVs in the histograms. We prefer to use 2.5 mum (MP), 3 mum (Duke) or 10 μ (Coulter) beads for alignment purposes. Generally if the systems are aligned with larger beads it will also be aligned with smaller beads but the reciprocal situation of being aligned with smaller beads does not translate to larger beads.

The 2.5 mum (MP) and 3 mumbeads (Duke) showed a broad forward scatter signal (about 4–5%) and a narrow side scatter signal (2–3%). However, using 10 mum fluorospheres (Coulter), the side scatter showed a larger signal

Table 3 Alignment data. Sample and sheath lines of two flow cytometers were cleaned with Clorox followed by Contron 70. Flow #1 has large CVs as the laser is dying and the laser spot has an increased size. Flow #2 is a newly acquired machine that needs to be aligned again as the dirty conditions due to inactivity have altered the fluidic pattern. Cleaning a machine with Clorox water, Contron 70 and water will change the fluidic patterns and it is suggested that this be thoroughly done prior to professional alignment during a service call

Parameter	Flow #1 MP 2.5 Slow	Flow #1 MP 2.5 Fast	Flow #2 MP 2.5 Slow	Flow #2 MP 2.5 Fast
FS	3.19	4.06	5.81	7.16
SS	2.54	4.67	2.98	7.88
FL1	2.6	3.73	2.47	8.25
FL2	2.44	3.43	2.17	7.74
FL3	3.09	8.46	2.3	7.13

Table 4 **Alignment data. Flow cytometer #2 (BD FACSCalibur):** CV of 2.5 μm MP beads (A-7302 alignment beads) and 10 μm Coulter beads (Flow Check fluorospheres) were run at slow (12 mul/s) and fast speeds (60 mul/s). The increase in CV is expected at faster speeds. This flow cytometer correctly aligned yields fluorescence data between 1 and 2% compared to flow cytometer #1 shown in Table 1. The forward scatter of 2.5 beads is large while the side scatter of 10u beads is large, indicating that both beads are measuring something different either from alignment or the bead scatter characteristics

Flow #2 Parameter	MP 2.5 Slow	MP 2.5 Fast	Coulter 10 Slow	Coulter 10 Fast
FS	4.75	5.65	1.17	1.57
SS	1.95	2.41	3.56	3.40
FL1	1.54	1.85	1.21	1.83
FL2	1.14	1.53	1.06	1.47
FL3	1.28	1.68	1.25	1.51

(3–4%) than the forward scatter (2%). It has been reported that larger beads represent cells better and therefore are a better particle to check alignment of a flow cytometer. In our hands, both particles are acceptable but we would not emphasize the values of CVs of scatter signals, as they appear to change with the size and type of particle. It can be said that if a machine is aligned using a large particle, it will generally be aligned for a smaller particle; however, the reverse is not always true. Both particles work, but flow systems that use small particles (2–3 u) for alignment, followed by measuring cells that have sizes of 8–15 μm cells may not have the optimum alignment in their scatter signals. This point has been debated and the user should be aware that there is some disagreement among the experts and companies on the acceptable size of particle that should be used for alignment.

2.5 Summary

The CV of the bead population is a measure of the alignment of the system and the overall functionality of the system. A low CV indicates a good alignment (Table 4) while a high CV indicates either an alignment problem or a fluidic problem (Tables 1–3). Increasing the flow rate should increase the CV, but the histogram distribution should stay relatively symmetrical. Deviations from a symmetrical pattern as shown in Fig. 2 indicate a system that is not aligned optimally.

The alignment bead test is very simple to perform and is the foundation for assessing flow cytometer alignment and performance. The test is performed in the following way: use a single intensity bead with a very small CV. The bead is placed in channel 400 on a scale of 1024. The PMT voltages and CVs are measured for each fluorescence and scatter parameter. The beads should be run at slow and fast settings. The faster settings widen the sample core which ultimately will increase the CV. An increase of approximately 1% usually occurs when the sample is run at fast speeds. Manufacturer specifications are not readily available, so we can not state what the manufacture believes the increase in bead CV should be. However, if much higher CV values are obtained it indicates the sample stream is not at the center of the core. We have illustrated representative CV data in Tables 1–3 and Figs. 2 and 3. It is clear that when the CV measurement changes, the conditions of analysis have changed. A change in flow rate or a blockage of the system will change the CV (Fig. 2, Tables 1–3).

3 Sensitivity

The fluorescence sensitivity of an instrument is the ability of a machine to distinguish a dim fluorescence signal from the other interfering background signals [5, 26]. If sensitivity evaluation is to become part of the standard testing on a flow cytometer, then a method that is both simple to understand and easy to perform will be necessary for the flow cytometry community.

Scientists involved in flow cytometry use multi-intensity beads to evaluate the functionality of their machines. However, the interpretation of the data derived from these beads have not been consistent among different investigators [7–17]. Normally these multi-intensity beads contain between six and eight different intensity bead populations. There is one blank bead and one very dim bead and four to six other beads of various intensities. These

Fig. 4 Comparison of populations of multi-intensity 7 beads and 4 beads. Manufacturers make bead cocktails of six-to-eight multi-intensity beads to evaluate flow cytometers. Duke Scientific manufactured a bead cocktail that removed three-intensity bead populations from their standard multi-intensity bead cocktail, allowing all of the remaining four beads to be observed on an analog log scale. Populations #3 and #5 and #7 were removed from the seven-bead population leaving #1, #2, #4 and #6. A comparison of the seven-beads formulation (Duke FCMC, **A** and **B**) with a new four-peak beads (XP4-1637, **C**) formulation are shown. Ashows that in order to get all the seven beads on scale on a FACSCalibur, the blank beads population is only 1/2 on the *x*-axis scale. In order to evaluate low-level sensitivity, it is better to place the blank bead on scale and eliminate the higher intensity bead population (#7) (Fig. 4B). This will enable the calculation of Rd which is related to the differences in means between the dim and blank population divided by the sum of the SD of both populations. The four-bead population allows for the calculation of the flow cytometer detection efficacy Rd using the dim and blank beads and it is easier to observe four multi-intensity beads that cover the entire scale than the seven beads which have values that are larger than the log scale

beads are useful in acquiring a linear regression line that could be used to determine if the logarithmic amps are working properly, and if interpreted correctly they could provide information on system sensitivity. Newer systems use linear collection electronics and then convert the information into log data in the software, bypassing the need for the log amplifier technology. Transforming these bead peak intensity data into other useful information to determine the sensitivity and status of flow cytometers is not routinely done by the majority of flow cytometry operators. Some investigators use these beads as markers that fall into a window of detection channels, while others use these particles to determine the CVs in log scale (Fig. 4). Other investigators try to separate two closely spaced beads to show that resolution exists in the system. There is still some controversy and confusion that surrounds the use of the beads and the interpretation of the linear regression line derived from the six to eight different intensity beads. For this reason, these multi-intensity beads are not currently used by most flow cytometer laboratories to determine the functionality and sensitivity status of their flow cytometers.

3.1 Background Noise and Efficiency

Eric Chase and Bob Hoffman have described a concept in which the ability to resolve dimly fluorescent populations depends on the efficiency, Q, in converting fluorescent molecules located in the flow cell illumination zone into photoelectrons that could be detected in the PMT and the amount of background light reaching the PMT from interfering light sources [19–22]. These multi-intensity beads have been used to describe B and Q measurements. Q is a measure of detection efficiency, while B is a measure of background noise. B varies considerably more than Q with the operating conditions. De-

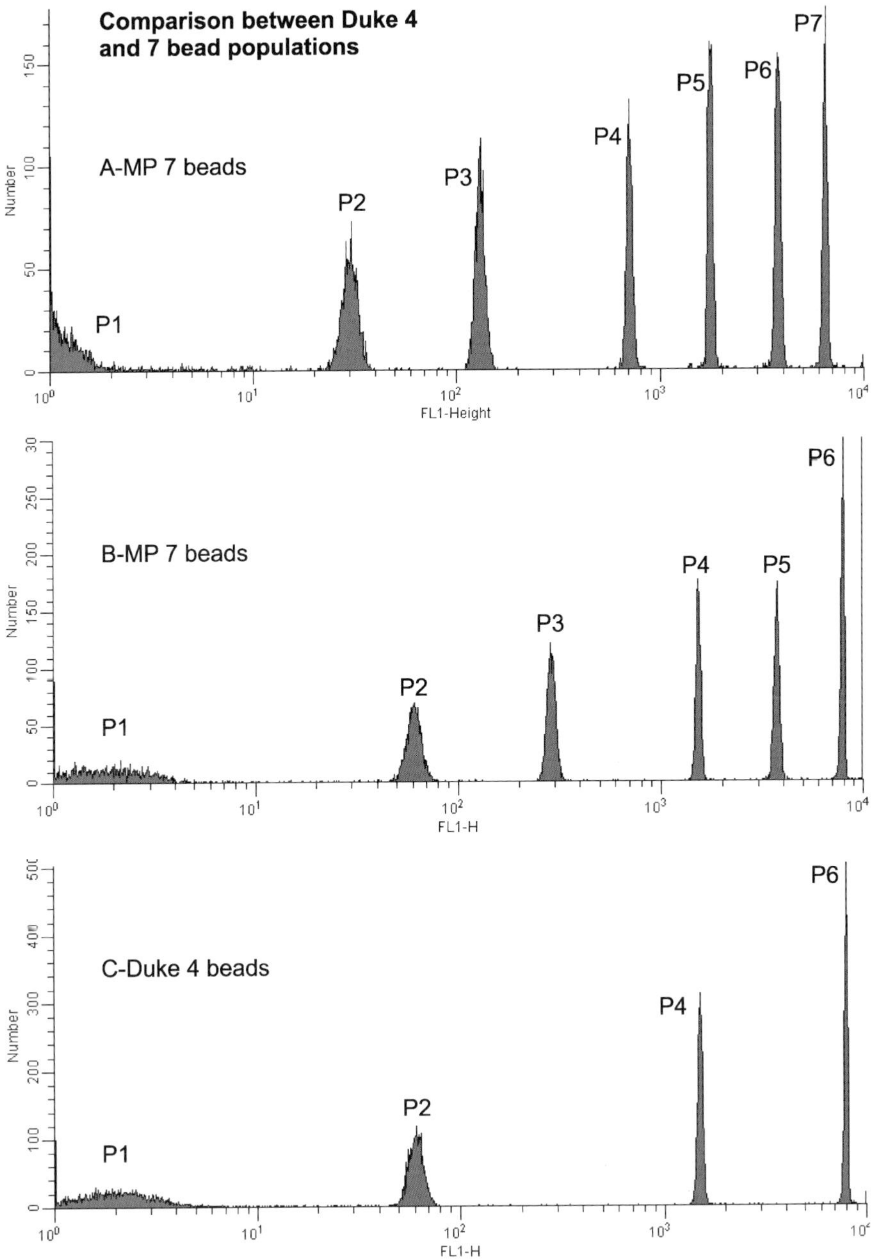

creasing the laser light will generate fewer photoelectrons, which will result in an increase in CV value of the bead histogram. Decreasing Q has the effect of broadening all of the bead distributions. The CVs are larger in the lower decades and smaller in the higher decades. Increasing B has the primary effect of broadening the dimmer peaks and pushing the distributions up the scale. Most importantly, increasing B can inhibit the machine's ability to distinguish dim objects from the background. B also includes contributions from photons emitted by fluorochome molecules not associated with the one of interest. Dye contamination of the system will increase the background and interfere with the ability to detect dimly stained cells. The foundation of this hypothesis is based on the detection of signal from noise which has been described in many other microscopic and other fluorescence systems [5, 23–26].

This concept of B and Q is not widely accepted by the flow cytometry community and therefore it is not being used in a practical way. However, it is reportedly used by the manufacturing community to test the reliability of their equipment. Recently, there is one flow cytometry service company in America (Cytek) that uses this B/Q approach to evaluate the condition of flow cytometers during their routine preventative maintenance service calls. To evaluate the machine sensitivity, a need still exists to develop methods that are easy to perform and easy to interpret.

3.2 Flow Cytometer Detection Efficacy

Dim cellular fluorescence is often measured on flow cytometers, as many probes only stain cells with a minimum number of surface molecules. In order to detect low-level fluorescence, an instrument has to provide very good sensitivity and the operator needs to perform tests to determine if the machine is operating correctly. Observation of the blank and dim beads characteristics can subjectively assess the machine performance. We are proposing to use the approach that has been presented by Orytn that measured the ability of the machine to detect low levels of fluorescence and thus the sensitivity of the machine [18]. He proposed the following equation to compare flow cytometer detection efficacy: Rd = M2 – M1/(SD1 + SD2) where M is the mean of the first two peaks and SD is the standard deviation of the first two peaks. This is equivalent to Wood's resolution factor, proposed 8 years earlier [19]. Like Ortyn, we are proposing to focus on the dim and blank bead as opposed to drawing linear regression lines. These lines appear to be useful to detect linearity of the amplifiers and the electronics of the system. When the value of Rd is large, it means there is considerable separation between the blank and dim peaks. These two populations are also observable and cleanly separated in histograms and cytograms. When Rd is small (less than 1.5), one can not determine the difference between the two bead peaks, as the lower fluorescence regions are very continuous and the SDs and CVs of the lower two populations are very large. We use a cytogram of FL1 with either FL2 or FL3 that can easily demonstrate the size of the blank and dim bead clusters. Subjectively if the blank and dim beads are on scale, we can observe a cluster of them in the lower left corner of the cytogram. A tight cluster signifies an aligned system while a broad cluster signifies a system that is not as sensitive. It can also be observed on the fluorescent histograms by the clear separation of the blank and dim peaks into two populations. Subjectively, a diffusion of the two populations into each other occurs on a flow cytometer with poor sensitivity and a separation of these two peaks is observed with a system that has a relatively higher sensitivity. We illustrate poor sensitivity with two examples: dye contamination and fluorescent light contamination. The use of Rd provides a quantitative means to characterize this typically subjective observation and also provides a quantitative quality control parameter to characterize the performance of a given flow cytometer over time.

In Jim Woods 1998 sensitivity paper, he shows a log plot that diverges at low channels from a straight line [19]. The plotting of two fluorescence parameters yields such a line. Observations show that the cluster at low channels diverges greatly when the sensitivity is bad and is more of a tight cluster when the sensitivity is good.

3.3 CVs and PMTs

Generally the CVs are higher in the lower channels for a number of reasons. The primary reason being the photon statistics associated with light detection. This process is characterized by a Poisson distribution where the SD is equal to the square root of the mean. Therefore, as the mean number of photons decreases, the SD becomes a larger fraction of the mean thereby increasing the CV [5, 18, 19, 26]. There are fewer channels for the histogram distribution, which also increases the CV. Increasing the PMT voltage also will introduce noise into the system [23–28]. Unlike microscopy, where averaging and longer exposure can reduce the noise and improve the signal to noise, flow cytometry has only one pass across the laser beam so the inherent noise of the system can affect the ability to resolve dim particles from background values.

Various manufacturers (Duke, Spherotech, Bangs, and MP) have produced a series of multi-intensity beads that cover the entire four-decade logarithmic scale of most flow cytometers. They have been used to measure the linearity of the log amplifiers and in some cases to determine the relative amount of molecules that can be detected on a cell. In using the Duke multi-intensity beads (Duke FCMC Multifluor beads) we have found that most of the 7 peaks can be resolved in the three PMT channels on a FACSCalibur. As shown in Fig. 4, either the blank bead is not on scale (Fig. 4A) or the seventh bead is not on scale (Fig. 4B). Duke has produced a four-bead population (XP4-1637 Multifluor beads) without the #3, #5 and #7 intensity beads leaving a blank, dim, and #4 and #6 intensity population of beads which can easily be displayed on a logarithmic scale (Fig. 4C). In our opinion the four-bead cocktail from Duke will facilitate the application and acceptance of these beads for sensitivity measurements.

3.4 Contamination

We found a diminished ability to resolve distributions of blank or unstained beads and dim beads when fluorescein or propidium iodide (PI) contaminated the flow cytometer fluidic lines (Figs. 5 and 6). The degradation occurred primarily in FL1 and to a lesser extent in FL2 and FL3. PI contamination exhibits the same effect primarily in FL2 and FL3 (Table 5). Ortyn has described the difference between the peak #1 (blank or unstained fluorescence) and peak #2 (dim fluorescence) to be a relative measure of the sensitivity of the system (Table 5) [18]. It is best to have the blank bead (P1) around 10 which is the first decade and away from the *y*-axis to do this test. Dirty systems that are contaminated with fluorescent dyes will affect the ability to observe P1 blank bead from P2 dim bead (Figs. 5 and 6).

Fig. 5 **A** Fluorescein contamination. A clean flow cytometer (**A, C**) was intentionally contaminated with a solution of fluorescein (**B, D**) which obscured the resolution of the blank and dim populations. Cleaning the machine with 10% Clorox followed by water and 10% Contron 70 resulted in the separation of the two populations again (**A, C**). The fluorescein contamination spilled over into the orange fluorescence (*FL2*) channel obscuring the separation of the blank and dim populations in green fluorescence (*FL1*) and FL2. Calculation of Rd showed a 2× increase in a clean machine (Table 5a) ▶

Table 5 a Fluorescein contamination. Application of Rd to a clean and contaminated system. Rd = M2 – M1/SD1 + SD2 where M is the mean of the peaks and SD1 and 2 are standard deviations of the distribution. The blank bead population (*P1*) and the dim bead population (*P2*) are compared. A clean machine was contaminated with fluorescein for 5 min and then washed with 10% Clorox for approximately 5 min followed by 10% Contron 70 for 5 min. The system was run at the start, after contamination (dirty), and after cleaning. The Rd was twice as much in a clean machine as a dirty machine. This confirms the data shown in the histograms of the bead populations is shown in Fig. 5. The dim and blank peaks were clearly separated in a clean machine

Treatment	Bead Mean Peaks (first 4 of 7) P1 Blank	P2 Dim	P3 Med	P4 High	Rd
FL 1 Clean Start	4	74	360	1863	5.19
FL 1 Dirty	15	76	355	1886	2.07
FL 1 Clean After	7	71	336	1734	5.25

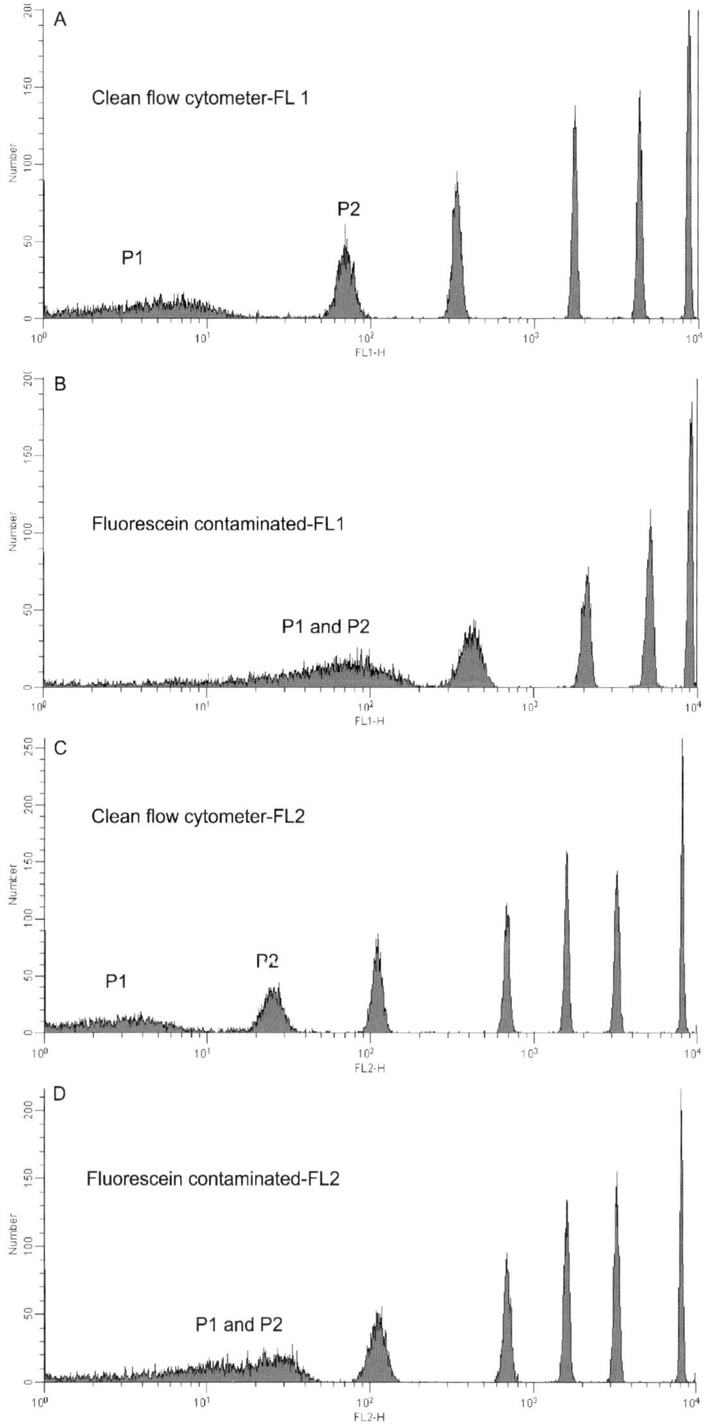
A
Clean flow cytometer-FL 1
P1
P2
Number
FL1-H
B
Fluorescein contaminated-FL1
P1 and P2
Number
FL1-H
C
Clean flow cytometer-FL2
P1
P2
Number
FL2-H
D
Fluorescein contaminated-FL2
P1 and P2
Number
FL2-H

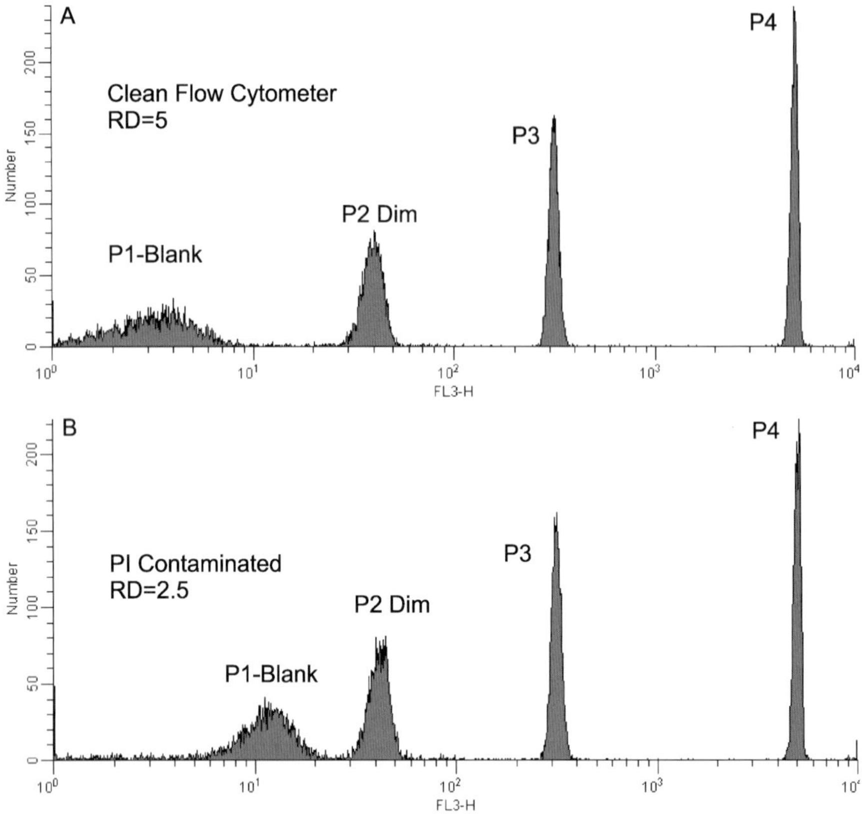

Fig. 6 Propidium iodide (*PI*) contamination. A clean flow cytometer (**A**) was contaminated by PI that was used for viability and DNA assays (**B**). This moved the blank bead population into the dim bead populations region. Cleaning the machine with Clorox followed by Contron 70 resulted in the return of separation of the two populations. Calculation of Rd showed an increased value in the clean machine. The blank and dim beads are separated by a large distance in a clean machine while they are nearly overlapping in a dye-contaminated system

These systems will have reduced sensitivity in the detection of cellular fluorescence.

To test this hypothesis, we contaminated the machine with a concentrated solution of fluorescein for FL1 and FL2 or PI for FL2 and FL3 (Fig. 5). After this dye contamination, it was difficult to distinguish the dim and blank bead populations, as the blank bead had increased in fluorescence intensity and approached the intensity values of the dim bead. The machine needs to be cleaned with 10% Clorox, followed by water and 10% Contron 70 prior to future analysis of dim fluorescence. Other detergents i.e., Tween, NP40 and SDS have also been used but have not been critically evaluated for the condition of the tubing.

Table 5 **b** Propidium iodide (PI) contamination. The system was measured with 4 multi-intensity population beads (Duke, XP4-1637). These are similar to the seven-peak beads (Duke FL3MC Cytocal) with the #3 and #5 and #7 removed leaving only the #1, #2, #4 and #6 bead populations to be counted. The system was contaminated by running PI-stained cells in the DNA cell cycle assay or PI-dead cells in viability samples. This contaminated only FL2 and FL3 and not FL1. The system was then cleaned with Clorox followed by Contron 70 for 5 min each. The Rd of the clean system was greater than the dirty system with the FL2 and FL3 channels being affected by PI fluorescence. This is shown in the histograms in Fig. 6

	Bead Mean Peaks				
Treatment	P1 Blank	P2-Dim	P3 Med	P4 High	Rd
FL1clean	3.3	64	1495.0	8007	5.2
FL2 clean	5.6	46.8	906.0	6472	5.5
FL3 clean	3.7	38.3	312.0	4976	5.0
FL 1 dirty	2.5	61.4	1504.0	7897	6.4
FL 2 dirty	8.3	49.0	905.0	6479	4.5
FL 3 dirty	14.0	42.3	321.6	4976	2.5

According to Ortyn, the relative sensitivity can be evaluated by measuring the means of the unstained or blank particle and the dim particle and their respective standard deviations M1 – M2/(SD1 + SD2) = Rd [18]. This Rd value is a relative measure of the sensitivity of the system. It should be noted that by choosing an Rd value, such as 1.5 and using a two-population bead set containing a blank bead population and a dimly labeled bead population with a known mean number of molecules on the dimly labeled beads, the Rd formulation can be used to determine an absolute measure of sensitivity. However, for QC (Quality Control) purposes, this may not be particularly useful. We have shown in Table 4a,b and Figs. 5 and 6 that a contaminated system will have difficulty distinguishing between the dim and blank beaks and the Rd value will be about half as much in a dirty system as in a clean system. The SDs and CVs of the populations would be increased in the contaminated, less sensitive system.

3.5
Light Pollution and Contamination

In order to simulate ambient light scatter, the hood of a flows cytometer was raised and the fluorescence lights left on. The machine was run with incandescent overhead lights or fluorescent lights. Fluorescent lights in the room produced photons which contaminated the light path of an opened machine and decreased the sensitivity in the FL1 and FL2 channels (Fig. 7, Table 6). Since the fluorescence lights do not emit in the far red, they did not greatly

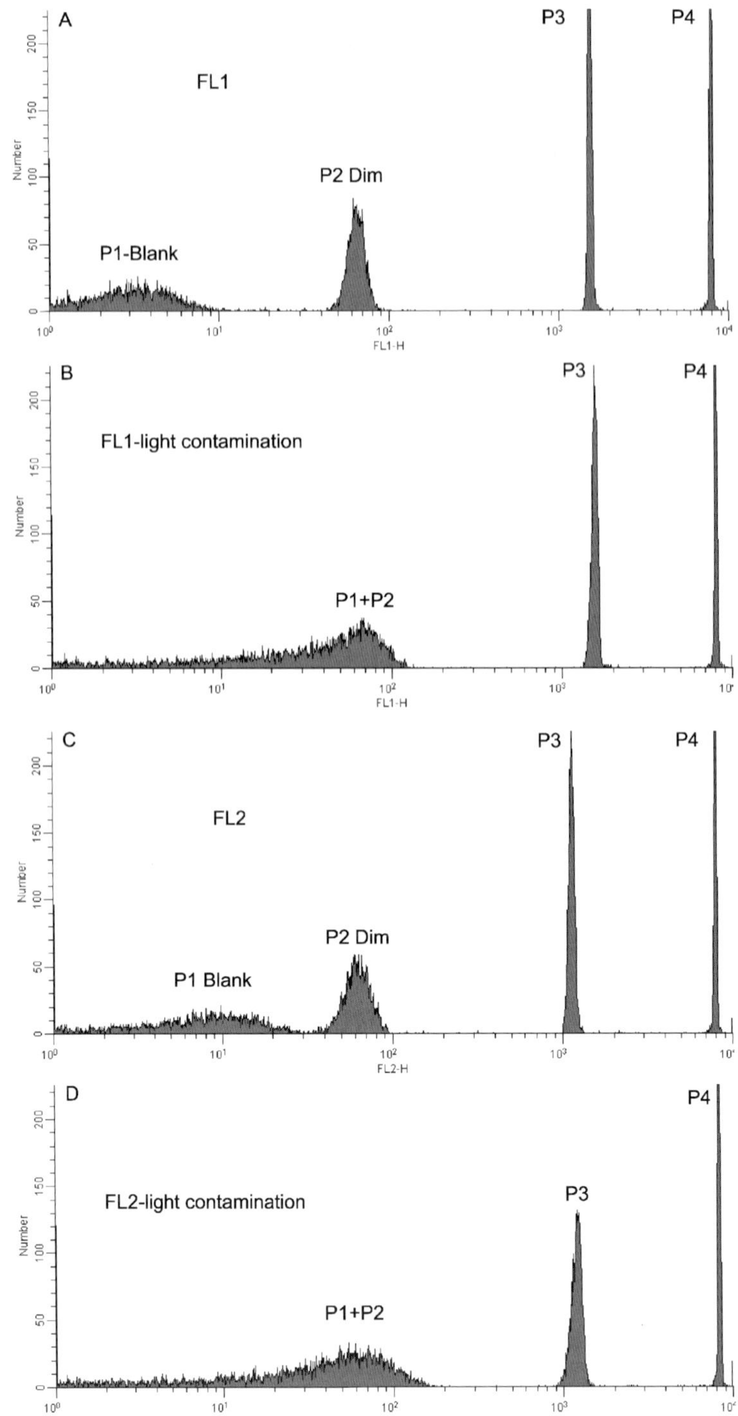
A
FL1
P1-Blank
P2 Dim
P3
P4
Number
FL1-H
B
FL1-light contamination
P1+P2
P3
P4
Number
FL1-H
C
FL2
P1 Blank
P2 Dim
P3
P4
Number
FL2-H
D
FL2-light contamination
P1+P2
P3
P4
Number
FL2-H

◀ **Fig. 7** Light contamination. Comparison of FL1 and FL2 showing apparent different sensitivities of the two PMTs (photomultiplier tube) (**A, C**). An illustration of the Rd concept is shown by the position of the blank and dim beads in FL1 and FL2. FL1 shows more sensitivity than FL2 as there is greater separation between the blank and dim peaks in FL1 compared to FL2 on this flow cytometer; by opening the flow cytometer top, fluorescent lights contaminated the blank and dim fluorescence peaks eliminating the ability of the machine to detect low-level fluorescence (**B, D**). It is suggested to run flow cytometers with incandescent light so there is no light contamination. Scatter light in the system might also decrease the ability to detect low-level fluorescence, which would be detected by these multi-intensity beads

Table 6 Light contamination. The system was evaluated using 4 multi-intensity bead populations (Duke, XP4-1637) and correlates to the histograms shown in Fig. 7. These beads are similar to the 7 peak beads (Duke FL3MC Cytocal) with the #3 and #5 and #7 removed leaving only the #1, #2, #4 and #6 bead populations to be counted. The system was contaminated by exposing the flow cell to fluorescence lights. The Rd of the closed system is higher than the contaminated system. FL2 appears to be less sensitive than the other PMTs. FL3 and FL4 are not as sensitive to fluorescence lights as FL1 and FL2

	Bead Mean Peaks				
Treatment	P1Blank	P2-Dim	P3 Med	P4 High	Rd
FL1 Align	3.2	59.3	1409	7302	6.61
FL2 Align	8.5	53.0	924	6721	3.24
FL3 Align	5.1	59.0	422	7168	4.58
FL4 Align	4.2	172.0	846	6270	6.45
FL1 Cont	14.7	61.7	1445	7551	1.88
FL2 Cont	19.2	62.3	976	7183	1.47
FL3 Cont	9.9	58.0	423	7235	2.34
FL4 Cont	10.1	173.0	844	6269	4.4

contaminate the FL4 population. Fluorescence lights affected FL3 but not to the same degree as FL1 and FL2, as shown by the Rd value in Table 6 and with the histograms in Fig. 7. By comparing the position of the FL1 and FL2 bead peaks, this ambient light test showed that the sensitivity of FL1 was greater than FL2, for unknown reasons. The sensitivity is related to the large value of the Rd and the large separation of the unstained and dimly stained beads. Light contamination or light scatter can destroy the machine's ability to detect dim fluorescence from cells or particles.

In summary, we find that multi-intensity beads can be employed to measure fluorescence dye contamination and misalignment resulting in excessive background signals. Understanding where the peaks should be located on a log scale for a given PMT setting is key to understanding how the system is functioning on a daily basis. Poor alignment and contamination can change the position of these peaks. The way to make flow cytometers more sensitive in detecting weak signals is to lower the background by using good fluidic, electronic and optical components and to align the instrument correctly. This results in a high fluorescence detection efficiency with narrow fluorescence distributions (low CV values) [19].

3.6 Errors

To perform an amplifier linearity check with a multi-intensity bead set, scientists often attempt to view all populations on the same scale. This approach may conflict with measuring sensitivity if a portion of the blank bead population cannot be seen on the scale. In order to evaluate sensitivity it is important that the entire blank population be on scale. Duke has produced a four-peak bead set that allows the investigator to evaluate the sensitivity of the flow cytometer by comparing the blank and dim bead while using one or two high-intensity beads to assess alignment of the system. However, we prefer to use a separate alignment bead for the precision and alignment measurement. As recommended by Ortyn et al. [18], the difference in means between these two beads is divided by the sum of the SD of each population. If a system is working, one can determine a value for the difference between the dim and blank bead population that relates to system sensitivity. If the system has poor sensitivity or is contaminated with fluorescent dyes as shown in Figs. 5 and 6, the Rd value will decrease and one cannot determine two separate populations in their proper relative locations for the dim and blank beads.

3.7 Peak Position

We and others have found that the instrument does not have optimal resolution and sensitivity, when the bead particles are placed in the lower channels. Becton Dickenson attributed this to low PMT voltages, which will amplify the noise of the amplifier (BD technical notes). In order to get the particles in these channels, the PMT voltage was decreased, which meant the distribution was contained with fewer channels. Fewer channels will inevitably result in a high CV of the population. Newer digital machines should be able to use a population of three to four different beads to measure daily QC alignment changes in addition to monitoring the sensitivity. These data will be helpful in monitoring stability and day-to-day consistency. It is recommended that CV and peak channel be measured at a specific PMT settings daily.

3.8 Molecules of Equivalent Soluble Fluorochromes

Currently, a method developed by Abe Schwartz uses a mixture of different intensity beads to evaluate the sensitivity of a system [10, 13]. He expressed the detection limits in terms of molecules of equivalent soluble fluorochromes (MESF). The beads have a number of different intensities, including one bead population of unstained beads. The beads have a fixed number of dye molecules and will occur in specific log channels. An equation of a least squares regression line can be calculated. The resulting equation is then used to calculate the number of MESF units corresponding to the channel number of a blank bead. Using this approach, the number of molecules of fluorescence that can be calculated is related to the sensitivity of the machine. This approach and value has been described by Hoffman, and Gaigalas and Wang in chapters [12, 13] in this book, which are critical for the determination of the number of fluorescence molecules on a cell population. Unfortunately this technique is not being applied to flow cytometers in a consistent manner. Some investigators are using the multiple-intensity beads to check the windows of detection, while other investigators are using the histogram distributions in log measure CVs on analog machines. We maintain that this approach is confusing, as there are too many peaks that intentionally span the whole log scale. It would be better to use fewer beads consisting of a null bead a dim bead and one or two other higher intensity beads such as the Duke scientific four-peak set (XP4-1637). This is easy to construct as the bead maker can add any number or type of beads into their bead cocktail. These multi-intensity beads should be used for sensitivity measurements and they should not be used for alignment and CV measurements on an analog system. In digital systems, the data is collected in linear scale and then converted into log data. Some of the higher intensity peaks may be used to calculate CV. It may be optimal to have a set of beads for alignment that is different from the set that is used for sensitivity to keep the methodology directed to the endpoint being evaluated.

In summary, the sensitivity of the machine can be evaluated by comparing the dim bead relative to the blank bead. The distribution of the blank bead population should be observed away from the ordinate, so that statistical values of the SD mean and CV can be obtained. We maintain that it is important to observe the blank bead in the first decade and the dim bead in the second decade and compare these two beads using the Rd method. Duke has prepared a population of four beads to set up the machine which helps clarify the procedure allowing all the beads to be on scale (Fig. 4). By using fewer bead populations, it becomes easier to interpret the derived bead histograms and cytograms. Having too many intensity beads in the counting solution is ultimately a distraction for the determination of this sensitivity endpoint. If a straight line across the 3–5 log decades with correlation coefficient is warranted to check the amplifiers, it can also be accomplished with four peaks instead of six to eight peaks. However six-to-eight peak beads will yield better evaluation of linearity of the machine if this is the desired endpoint.

The higher the PMT voltage, the better the Rd setting. However, if we push the fourth peak off scale the three remaining peaks are measured at full PMT voltage settings, which is not recommended. Since the Rd values are relative for a machine it is recommended to test the systems using both the third and fourth intensity beads with these voltages adjusted for the fourth peak to be located between channel 900 and 1000. This results in the location of the blank bead population on scale in lower channels (1–10 log units).

4 Summary

We have provided two simple performance criteria to determine if a flow cytometer is aligned and functional. Using uniform single-intensity beads and measuring the CVs, peak channels, and histogram distributions, the system can be assessed for alignment and functionality. The second test monitors the cleanliness of the system and the amount of background light scatter. This test is related to the alignment. Failure to obtain good values in these tests will compromise the ability to detect dim fluorescence from background and will affect the precision of the system. It is important that some of these bead intensities are similar to the intensity of the samples measured.

Government disclaimer: Although the research described in this article has been supported by the United States Environmental Protection Agency, it has not been subjected to Agency review and therefore does not necessarily reflect the views of the Agency and no official endorsement should be inferred. Mention of trade names or commercial products does not constitute endorsement or recommendation for use.

Acknowledgements Thank are extended to Jim Birk, Ernie Thomas and Bill Ortyn for their helpful comments in the discussion of the QA aspects of the paper. I wish to thank Mike Brinkly for providing us with four and seven multi-intensity beads. I wish to thank John Havel for the excellent representation of the flow cell. I wish to thank Ellen Lorang and Connie Meecham for proof reading and formatting the paper.

References

1. Zucker RM (2006) Confocal slide based system performance. Cytometry 69A:659–676
2. Zucker RM (2006) Confocal microscopy slide based systems: instability. Cytometry 69A:677–690
3. Carter D (1999) Practical considerations for collecting confocal images. Methods Mol Biol 122:35–57

4. Zucker RM, Elstein KH, Gershey EL, Massaro EJ (1990) Increasing sensitivity of the ortho analytical cytofluorograph by modifying the fluid system. Cytometry 11:848–851
5. Shapiro HM (2003) Practical Flow Cytometry, 4th edn. Wiley, Liss, New York
6. Tanke HJ (2005) Digital fluoresence microscopy. In: Robinson P (ed) Current Protocols in Cytometry unit 2.5. Wiley-Liss, New York
7. Horan PK, Loken MR (1985) Practical guide for the use of flow systems. In: van Dilla MA, Dean PN, Laerum OD, Melamed MR (eds) Practical Guide for the Use of Flow Systems in Flow Cytometry: Instrumentation and Data Analysis. Academic Press, London, England, pp 259–280
8. Hoffman RA (2001) Standardization and quantitation in flow cytometry. Methods Cell Biol 63:299–340
9. Kraan JJ, Gratama JW, Keeney M, Jean-Luc D'Hautcourt (2003) Senting up and calibration of a flow cytometers for multicolor immunophenotyping. J Biol Regulators Homeostatic Agents 17:223–233
10. Schwartz A, Marti GE, Poon R, Gratama JW, Fernández-Repollet E (1998) Standardizing flow cytometry: A classification system of fluorescence standards used for flow cytometry. Cytometry 33:106–114
11. Gratama JW, D'Hautcourt JL, Mandy F, Rothe G, Barnett D, Janossy G, Papa S, Schmitz G, Lenkei R (1998) Flow cytometric quantitation of immunofluorescence intensity: problems and perspectives. Eur Working Group Clin Cell Anal 33:166–78
12. Hoffman RA (2007) Flow cytometry: instrumentation, applications, future trends and limitations. In: Resch-Genger U (ed) Chapter 34
13. Gaigalas AK, Wang L (2007) Approaches to quantitation in flow cytometry. In: Resch-Genger U (ed) Chapter 36
14. Zenger VE, Vogt R, Mandy F, Schwartz A, Marti GE (1998) Quantitative flow cytometry: inter-laboratory variation. Cytometry 33:138–145
15. Purvis N, Stelzer G (1998) Multi-platform, multi-site instrumentation and reagent standardization. Cytometry 33:156–165
16. Powell MK, Whitfield W, Redelman D, Henderson LO, Vogt RF Jr (1998) Titration of a CD45-FITC conjugate to determine the linearity and dynamic range of fluorescence intensity measurements on lymphocytes. Cytometry 33:219–224
17. Schwartz A, Gaigalas AK, Wang L, Marti GE, Vogt RF, Fernandez-Repollet E (2004) Resolution of dimly fluorescent particles: a practical approach. Cytometry B Clin Cytom 57:1–6
18. Ortyn WE, Hall BE, George TC, Frost K, Basiji DA, Perry DJ, Zimmerman CA, Coder D, Morrissey PJ (2006) Sensitivity measurement and compensation in spectral imaging. Cytometry 69A:852–862
19. Wood JCS (1998) Fundamental flow cytometer properties governing sensitivity and resolution. Cytometry 33:260–266
20. Givan A (2001) Flow Cytometry First Principles, 2nd edn. Wiley-Liss, New York
21. Chase ES, Hoffman RA (1998) Resolution of dimly fluorescent particles: a practical measure of fluorescence sensitivity. Cytometry 33:267–279
22. Wood JCS, Hoffman RA (1998) Evaluating fluorescence sensitivity on flow cytometers: an overview. Cytometry 33:256–259
23. Art J (1995) Photon detectors for confocal microscopy. In: Pawley J (ed) Handbook of Biological Confocal Microscopy, 2nd edn. Plenum, New York, pp 183–195

24. Pawley JB (1994) Sources of noise. In: Stevens J (ed) Three- Dimensional Microscope Data Sets in Three Dimensional Confocal Microscopy: Volume Investigations Of Biological Specimens. Academic, New York, pp 47–94
25. Pawley J (2006) Fundamental limits in confocal microscopy. In: Pawley J (ed) Handbook of Biological Confocal Microscopy, 3rd edn. Plenum, New York, pp 20–41
26. Zucker RM, Price OT (2001) Statistical evaluation of confocal microscopy images. Cytometry 44:295–308
27. Zucker RM, Price OT (2001) Evaluation of confocal system performance. Cytometry 44:273–294
28. Zucker RM (2005) Evaluation of Confocal Microscopy System Performance. In: Taajets D (ed) Cell Imaging Techniques. Humana, Totowa, New Jersey, pp 77–135

Springer Ser Fluoresc (2008) 6: 371–398
DOI 10.1007/4243_2008_042

Published online: 12 March 2008

Approaches to Quantitation in Flow Cytometry

A. K. Gaigalas (✉) · Lili Wang

National Institute of Standards and Technology (NIST), 100 Bureau Drive,
Gaithersburg, MD 20899-8312, USA
adolfas.gaigalas@nist.gov

Abstract Flow cytometry is used to measure the fluorescence intensity (FI) from labels bound to antigens present on the surface of T and B cells. The T and B cells are associated with the human adaptive immune system and the amounts of surface antigens, such as CD4, CD8, and CD20, are used for diagnostic purposes. To estimate the number of a specific antigen expressed on the surface, the cells are incubated in a solution containing antibodies specific to that antigen and the antibodies are conjugated to fluorophores that provide the fluorescence signals used to detect the presence of the antibodies on the cell surface. The fluorophore on the antibody, microsphere, or cell is called a label. The antibody gives biological specificity and the fluorophore provides a mechanism for readout. Quantitation of cytometer measurements is accomplished by the comparison of the fluorescence signal of labeled cells with the fluorescence signal of labeled reference microspheres. The fluorescence signals from labeled cells and labeled microspheres are converted to a fluorescence yield (FY) where FY is defined as the product of the number of fluorophores and the fluorophore quantum yield. The comparison of fluorescence provides the basis for an estimate of the number of molecules of equivalent soluble fluorophores (MESF) associated with the labels bound on the cell surface. A procedure is outlined for assigning MESF values to microspheres followed by a demonstration of the assignment of MESF values to microspheres with immobilized R-phycoerythrin. The MESF values are used to obtain an estimate of the antibodies bound to the cell (ABCe). The final step in the quantitation process is the conversion of the ABCe values to an estimate of the number of specific antigens on a surface of a cell. With proper care, the ABCe value may be a good indicator of the total number of specific antigens present on the cell surface. Examples are given of the determination of the number of CD4 and CD20 antigens on lymphocytes. The MESF quantitation strategy has been applied to the calibration of multicolor flow cytometers.

Keywords Antibodies · Calibration · Flow cytometry · Fluorescence · MESF · Microspheres · Quantitative

Abbreviations

A	Measured absorbance
ABC	Antibody binding capacity or antigen molecules per cell
ABCe	Antibodies bound per cell
BD	Becton Dickinson Biosciences
C	Concentration in mol/L
B-CCL	B cell chronic lymphocytic leukemia
CDXX	Surface antigen on T and B cell, XX is a code for the antigen
CLSI	Clinical and Laboratory Standards Institute
CV	Coefficient of variation
DU	Digital unit, output from a analog to digital converter
FITC	Fluorescein isothiocyanate
FL1	First fluorescence channel in a flow cytometer
FI	Fluorescence intensity or fluorescence signal
FY	Fluorescence yield
HIV	Human immunodeficiency virus
I_0	Power of incident light
LED	Light emitting diode
MESF	Molecules of equivalent soluble fluorophore
N	Number density of microspheres
ND	Neutral density filter
OD	Optical density associated with a neutral density filter
P_{adj}	Ratio of the illumination power of a calibration and analyte measurement
PBS	Phosphate buffer saline
PMT	Photomultiplier tube detector
QC	Quality control
QY, ϕ	Quantum yield
R-PE	R-phycoerythrin
SRM	Standard reference material
SSC	Side scattering channel in a flow cytometer

$\varepsilon(\lambda)$ Molecular extinction coefficient at wavelength λ
Ω Efficiency of instrument for collecting fluorescence

1 Introduction

The accuracy of quantitative flow cytometry measurements is crucial for a large number of in-vivo diagnostics in the clinical setting. Examples include measurements involving blood, bone marrow, lymph nodes, leukemias, lymphomas, HIV, immune testing, stem cells, cytokine positive cell, and antigen specific T-cell detection. Complex, multicolor, multi-laser instruments are now routinely used for the in vitro selection and sorting of cells for ex vivo manipulation and subsequent clinical use. However, the incomparability of the results from different laboratories and across different instrument platforms is a serious barrier that has needed to be overcome for the advancement of this field. The paper describes procedures that may be used in clinical laboratories in the future to insure comparability.

The measurements carried out with flow cytometers are specific cases of immunoassays whose goal is to determine the number of antigens in the system under study. The procedure is to incubate the biological sample with antibodies designed for the antigen of interest, and then determine the number of antibodies that have remained bound in the biological sample. The common method for determining the number of antibodies in the biological sample is to attach fluorophore labels to the antibodies and then to use fluorescence signal as an indicator of the number of labeled antibodies.

The major concern of immunoassays is the reduction of possible errors that can skew the estimate of the number of antigens. Errors are present in all of the many steps that comprise a typical immunoassay. The step that will be the subject of this chapter is the detection and quantitation of the fluorescence signal. Quantitation will mean the conversion of the fluorescence signal into a numerical estimate of the number of antigen molecules present in the biological sample. Quantitation has been pursued by many over the last several decades [1–5] and has been recently detailed in a CLSI report I/LA 24-A [6]. The discussion below summarizes previous work to develop scientifically rigorous procedures for presenting flow cytometer measurements in terms of the number of antigens on a cell surface. The pursuit of quantitation in flow cytometry is based on a vision that a universal quantitation scheme is possible, that manufacturers of reagents are willing to implement the scheme, and that the presentation of flow cytometer results in terms of the number of antigen molecules will be a great benefit to the health providers.

The discussion of quantitation will start with a summary of the properties of fluorophores which determine their efficacy for signaling the presence of antibodies. A detailed discussion of fluorescence spectroscopy can be found in a textbook by Lakowicz [7]. Furthermore, an excellent summary of the basic concepts of fluorescence measurements is given in a recent review [8].

2 Fluorescence Signal

Most molecules absorb electromagnetic radiation over a limited range of wavelengths. The absorption leads to the population of a molecular orbital of higher energy and a concomitant disappearance of a photon (the photon is the basic energy unit of the radiation field). The energy difference between molecular orbitals determines which wavelength of the incident radiation field will be absorbed with high probability. Thus each molecule has a characteristic absorption spectrum which measures the likelihood of photon absorption at different photon wavelengths. (In practice the solvent environment of the molecule will influence the details of the absorption spectrum). The salient features of the absorption spectrum are the existence of a wavelength of maximum absorption and a large spread of wavelengths around the maximum at which absorption is significant. The absorbance at any wavelength can be converted into a molar extinction coefficient, $\varepsilon(\lambda)$, by dividing the absorbance by the known molar concentration of the absorbing molecules. The absorption probability is related to the product of the power of the incident radiation, I_0, and the molar extinction coefficient $\varepsilon(\lambda)$ at the wavelength of the incident radiation. For applications in flow cytometry, molecules are chosen which have very large extinction coefficients at the wavelength of illuminating light.

The absorption of a photon leaves the molecule in an excited molecular orbital. There is an equilibration process in the excited state during which energy is lost to the surroundings so that any emission wavelength is usually longer than the wavelength of the absorbed photon. The difference in the wavelengths of maximum absorption and maximum emission is called the Stoke's shift. The large difference in the Stoke's shift between fluorescein and R-phycoerythrin makes it possible to excite fluorescein and R-phycoerythrin with the same wavelength (usually 488 nm) and detect the emission in different wavelength regions.

The radiative decay of the excited state is but one of many relaxation paths. Internal conversion, during which the electronic energy is converted to vibrational energy and dissipated to the environment, is another relaxation path. The conversion of electronic spin from singlet to triplet results in a long lived excited state which does not radiate and is susceptible to chemical reactions. The presence of alternate relaxation pathways reduces the number of photons emitted by excited states. The relative fraction of excited states that decay radiatively is given by the parameter called the quantum yield (QY).

The competition between radiative decay and other decay pathways leads to the dependence of QY on the environment of the fluorophore. The above discussion can be summarized by a relation between the fluorescence intensity/signal (FI) and the properties of the measurement instrument and the fluorophores.

$$\mathrm{FI} = \Omega I_0 \varepsilon(\lambda_{\mathrm{ex}})\phi C\,. \tag{1}$$

Here Ω represents the efficiency of the transport of the radiation from the emitting fluorophores in the solution to the detector, I_0 is the power of the laser beam that illuminates the sample, $\varepsilon(\lambda_{\mathrm{ex}})$ is the molar extinction coefficient at the wavelength of the illuminating laser beam, ϕ is the quantum yield, and C is the concentration of the fluorophores in mol/L. Equation 1 shows that the FI depends on five independent properties associated with the fluorophores and the measurement instrument. The large number of variables is the reason for the difficulty in quantitating the fluorescence signal.

3 Conceptual Framework for Quantitation

3.1 Fluorescence Yield

Soluble fluorophores (analyte) can be quantitated using reference solutions with known concentrations of the same fluorophore in the same buffer as the analyte. In such cases the molar extinction coefficient and the quantum yield of the fluorophores are the same for the analyte and the reference solutions, and the fluorescence signal from both is proportional to the number of fluorophores (see Eq. 1). In the more common case, the analyte fluorophores and the reference fluorophores are in different environments (e.g., one is in solution and the other fluorophore is immobilized on a microsphere). In this case the quantum yields may be different for the two fluorophores, and the equality of fluorescence signal from two solutions does not imply equal numbers of fluorophores. In what follows we discuss the notion of molecules of equivalent soluble fluorophore (MESF) as a method for quantitating fluorescence signal in the case where quantum yields differ for the fluorophores in the analyte and the reference solutions.

The discussion below is based on the concept of fluorescence yield, which is defined as the product of the fluorophore concentration and their quantum yield [9]. The fluorescence yield of a number of fluorophores is a measure of the number of photons emitted if every fluorophore starts out in the excited state. For fixed solution conditions, the fluorescence yield is a well-defined property of the solution. Just as it is valid to compare the quantum yield of two fluorophores, it is also valid to compare the fluorescence yield of any two solutions of fluorophores.

3.2 Comparison of Fluorescence Yield and the Definition of MESF

Assume that the instrument factor, Ω, and the illumination, I_0, are the same for fluorescence measurements carried out on two solutions, s1 and s2. The equality of total integrated fluorescence signal measured for the two solutions implies the relation given in Eq. 2:

$$C_{s1}\varepsilon_{s1}(\lambda_x)\phi_{s1} = C_{s2}\varepsilon_{s2}(\lambda_x)\phi_{s2} \,. \qquad (2)$$

Equation 2 shows the well-known fact that for fixed instrument conditions, the fluorescence signal is determined by the product of three independent variables: the concentration of fluorophore, the probability of absorbing a photon, and the probability of emitting a photon. If the above measurements are repeated with a different illumination wavelength (change instrument conditions), the fluorescence signals may not be equal for the two solutions since the extinction coefficients for the two solutions may be different at the new wavelength. The absorption is a strong function of wavelength, and thus the resulting integrated fluorescence signal will depend on the wavelength of illumination. To eliminate the dependence on excitation wavelength, the fluorescence signal measurement needs to be normalized by the molecular extinction coefficient at the illumination wavelength. Such a normalization changes the comparison of fluorescence signal into a comparison of fluorescence yield. If a series of measurements of fluorescence yield are performed for different concentrations of one of the fluorophores then there will be a specific concentration at which the fluorescence yields from the two solutions will be equal implying the relationship in Eq. 3:

$$C_{s1}\phi_{s1} = C_{s2}\phi_{s2} \,. \qquad (3)$$

The equality of the measured fluorescence yields would be instrument independent. All instruments that measure the ratio of fluorescence signal and extinction coefficient would give the same response for solutions having the same fluorescence yields (Eq. 3). The equality would be true even if the absorption and emission spectra were very different for the two fluorophores. A practical procedure for converting a comparison of fluorescence signal into a comparison of fluorescence yields is to work with the same fluorophore in the reference solution and in the analyte solution so that the extinction coefficients are reasonably matched in the two solutions and cancel out during the comparison of fluorescence signal. The equality of fluorescence signal is then equivalent to the equality of fluorescence yield. For most of the following discussion we will assume that the analyte and the reference fluorophores are the same.

4 Practical Assignment of MESF Values

4.1 Absolute Determination of Values of MESF of Microspheres

The proposed procedure for assigning MESF values to microspheres with immobilized fluorophores involves several steps [10]. First, measure the FI from serial dilutions of a reference solution of fluorophores (such as SRM 1932 [11]) to calibrate a fluorometer. The calibration provides a relation between FI and a concentration of fluorophores in the reference solutions. Next, prepare a suspension of microspheres with immobilized fluorophores (same as the reference fluorophores) and measure their concentration using an instrument, such as the Coulter Multisizer 3. The concentration of microspheres must be kept sufficiently low ($<10^6$/ml) so that the scattering does not lower the fluorescence signal appreciably. Finally measure the fluorescence signal of the microsphere suspension using the same fluorometer as in the calibration. During the data analysis, the calibration is used to convert the fluorescence signal of the microsphere suspension into a concentration of equivalent reference fluorophores. Finally the equivalent concentration of reference fluorophores is divided by the concentration of microspheres to obtain the MESF value of a single microsphere. The various steps are summarized in the flowing table and discussed in more detail below.

Summary of steps in the MESF assignment:

Step	Description of action
1	Measure FI of serial dilutions of solutions of reference fluorophore.
2	Make a calibration curve using FI in step 1 and the known concentrations of reference fluorophore solutions. Equation 9 is an example
3	Measure the concentration of microspheres. Quantitate aggregates.
4	Measure FI from microsphere suspension
5	Use calibration in step 2 to relate the FI in step 4 to a concentration of reference fluorophores. This is the equivalent concentration of reference fluorophores. Equation 6
6	Divide equivalent concentration by microsphere concentration. Equation 5

4.2 Instrument Validation

The apparatus used for the assignment of MESF values needs to be validated. The validation provides a characterization of the response of the apparatus and an estimate of error in the values of the measured fluorescence intensity [12, 13]. Some of the steps in instrument validation are outlined below.

The wavelength calibration of the fluorometer can be checked using the argon ion laser lines, mercury lamp lines, or neon lamp lines.

The linearity of the fluorometer should be checked. A possible method is to pass the output of a light emitting diode (LED) through a variable neutral density (ND) filter and focus the output on the end of a bifurcated optical fiber. One of the arms of the optical fiber can be placed at the location of the sample in the fluorometer. The other arm can be placed in front of a photodiode which is known to be linear within 0.1% over its entire dynamic range. The output of the photo diode can be compared to the output of the fluorometer over a range of optical density (OD) values associated with the ND filter.

Check the spectral response of the fluorometer. This can be performed by the use of a calibrated lamp with a known output over a range of wavelengths, e.g., 340 nm to 800 nm [14]. The output port of the calibrated lamp can be placed at the location of the sample in the fluorometer. The intensity of the calibrated lamp should be adjusted with neutral density filters so that the fluorometer response is in the linear range. The spectral correction factor can be obtained by dividing the normalized reference lamp output by the normalized measured FI. Multiplying the measured fluorometer signal by the correction factor corrects for the variability of the fluorometer response over the wavelength range.

Measure the background. To measure the fluorescence signal from low concentrations of reference fluorophores, it is necessary to subtract the spectrum of the buffer from the spectrum of the solution containing the fluorophores. If an ion laser is used for illumination, a laser line filter is a critical component in the fluorometer since without it, the spectra may include significant contribution from laser plasma lines. Both the buffer and fluorophore spectra can be collected for a sufficient time to insure that the fluorometer is operating in the linear region. The water Raman band is usually the largest contribution to the background. If the samples are contained in capillary flow cells, Raman bands from the flow cell walls can be significant. In any case, after subtraction of the background, the fluorophore emission spectrum should be clearly recognizable. Background subtraction is not critical for the collection of spectra from solutions with higher concentrations of fluorophore. The subtracted spectrum should be corrected for the spectral response of the detector as described above. Background correction for suspensions of microspheres with immobilized fluorophores has to be performed with suspensions of microspheres without any immobilized fluorophore.

In the case where the microsphere measurements are carried out after the calibration measurement with the reference solutions, great care is needed to eliminate possible contamination. Prior to the microsphere measurements, the cuvette or the capillary flow cell should be washed for several hours. As a rule of thumb, the cleaning is sufficiently good when the fluorometer signal is within instrumental error of the FI from the buffer solution.

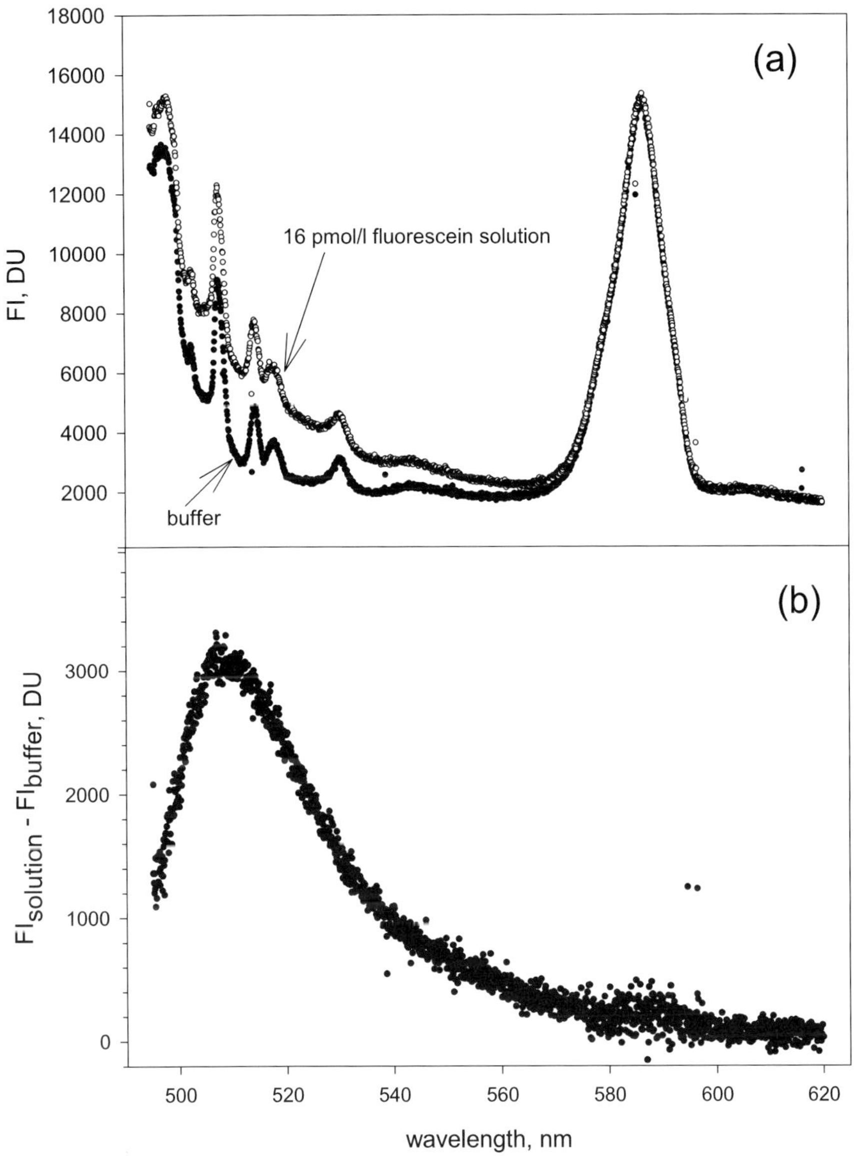

Fig. 1 **a** The *open circles* show the fluorescence spectrum from a solution with 16 pmol/l fluorescein flowing in a capillary cell. The excitation wavelength is 488 nm. The broad band at 585 nm is the water Raman line, while the peaks around 510 nm are the Raman lines from the cell walls. The *solid circles* show the spectrum taken with the buffer flowing in the capillary cell. (The Raman lines from the cell walls are absent if a cuvette is used to hold the sample). **b** The spectra shown in **a** are subtracted yielding the fluorescence spectrum of fluorescein. The spectrum is corrected for variation in the detector spectral response and integrated over all wavelengths to yield the final value of the fluorescence intensity associated with that solution

Since the measurement of the reference solutions and the microsphere suspensions can take several hours, it is necessary to monitor the power of the illumination. A glass slide was used to reflect part of the incident laser beam onto a silicon detector that was connected to a calibrated power meter.

After the fluorescence measurements are finished, the same algorithm should be used to analyze the calibration and analyte data. We used a Mathcad program to read in the spectra and apply all of the corrections. Graphical display of the changes produced by the corrections was monitored to verify that the corrections were proceeding normally. Examples of the measurement and analysis are shown in Figs. 1, 2. The open circles in Fig. 1a show the fluorescence spectrum of a solution of fluorescein with a concentration of 16 pmol/l, excited by 488 nm laser line. The full circles in Fig. 1a show the spectrum of the buffer. The dominant feature in both spectra is the water Raman line at 585 nm. Both solutions were flowing through a capillary cell so that glass Raman lines are apparent in the wavelength region 495 nm to 540 nm. Figure 1b shows the spectrum obtained by subtracting the buffer

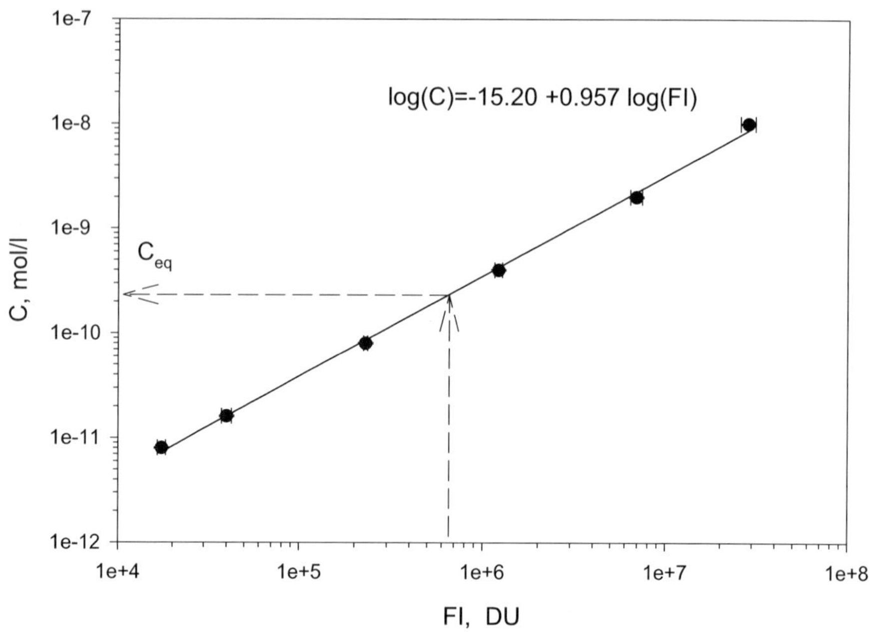

Fig. 2 The log of the fluorescence intensity (FI) associated with different dilutions of fluorescein reference solution are plotted on the x-axis. The log of the concentration of fluorescein (C) associated with each solution is plotted along the y-axis. The *solid circles* show the measured results and the *solid line* gives the best linear fit to the data. A slope in the range 0.95 to 1.05 is acceptable. The *dashed lines* give a graphical display of the procedure used to associate an effective fluorophore concentration (C_{ef}) with a fluorescence intensity measured for a microsphere suspension

spectrum from the solution spectrum. The fluorescein spectrum is apparent. The spectrum was integrated to obtain the final fluorescence intensity for the fluorescein solution.

Measurement such as shown in Fig. 1b were performed for six dilutions of the fluorescein SRM. The solid circles in Fig. 2 shows the fluorescence intensities plotted as a function of the concentration of the six dilutions of the fluorescein SRM. The solid line in Fig. 2 shows the calibration obtained by fitting the log of the concentration of each solution versus the corresponding log of the fluorescence intensity.

The fluorescence spectrum from microsphere solutions looks very similar to that shown in Fig. 1b. The fluorescence intensity obtained for a particular microsphere population provides a point on the horizontal axis on Fig. 2 such that a vertical line (dotted line in Fig. 2) drawn from this point intersects the calibration line at a specific concentration which is called C_{eq}. These lines are shown by the dotted lines in Fig. 2. The value C_{eq} together with the measured microsphere concentration gives a value of MESF for that sphere population. This procedure will be revisited below.

4.3 Measurement of Microsphere Concentration

Both the Multisizer 3 and the flow cytometer measurements show the presence of microsphere doublets, which are permanent joining of two microspheres. Some aggregation of colloidal particles is expected, and the amount depends on previous treatment of the suspension (e.g., sonication), age, and solvent. The concentration measurements were performed with the same suspension as the fluorescence measurements. Furthermore, the two measurements were performed on the same day. Figure 3 shows an example of the output obtained from the Multisizer 3. The horizontal axis gives the diameter of the microsphere and the vertical axis gives the number of events recorded during the time it took a given amount of the suspension to flow through the instrument. The larger peak at approximately 7 μm corresponds to single microspheres while the broader peak between 8.3 μm and 9.7 μm corresponds to aggregates of two microspheres. Adding the numbers in the two regions and multiplying by a dilution factor gives an estimate of the microsphere concentration in the original suspension.

The simplest assumptions are that the Multisizer 3 gives the correct concentration of doublets in the suspension used for fluorescence measurements and that the fluorescence intensity from the doublets is twice the fluorescence intensity from single microspheres. Therefore in comparing the fluorescence yield of a suspension and solution we multiply the doublet concentration by a factor of 2 relative to the singlet concentration.

$$N_{\text{Total}} = N_{\text{Single}} + 2N_{\text{Double}} \,, \tag{4}$$

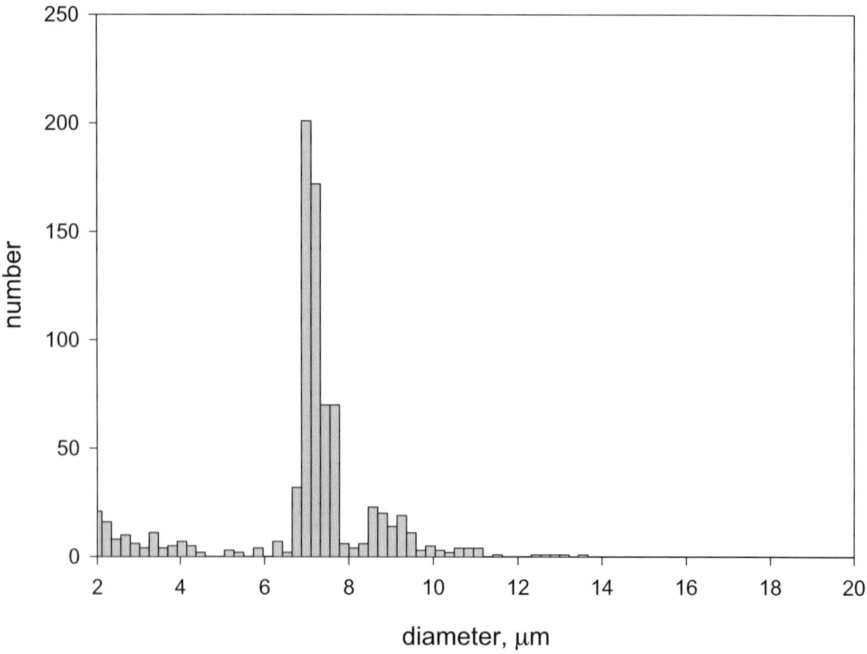

Fig. 3 The histogram shows the distribution of particle diameters measured for a suspension of microspheres. The suspension was obtained by diluting the suspension used in the measurement of fluorescence intensity. The peak around 7.4 μm corresponds to single microspheres passing through the orifice of the Coulter Multisizer 3 instrument. The broader peak at approximately 9 μm corresponds to permanent associations of two microspheres. The microsphere number concentration was obtained by summing the histogram peaks, multiplying by the dilution factor, and adjusting for the known sampled volume

where N_{Single} is the number concentration of single microspheres, N_{Double} is the number concentration of double microspheres, and N_{Total} is the total concentration used in comparing fluorescence yields. The assumption that the fluorescence intensity from a double microsphere is twice the fluorescence from a single microsphere is qualified by such consideration as distortion of the illuminating light and changes in quantum yield at the point of contact between the microspheres. Since the cytometer measures the fluorescence of single microspheres, these considerations are critical for the comparison of cytometer and fluorometer measurements.

The measured polarization anisotropy for the microsphere suspensions was about 0.08. This small value was rationalized by the large tether length of the chain of carbon molecules ($n = 7$) that immobilized the fluorophore to the microsphere surface. A long tether permits considerable rotational freedom. The systematic effects due to polarization differences between solution and microspheres were neglected.

The comparison of the fluorescence signal from solutions of fluorophore and suspension of microspheres with immobilized fluorophore requires a correction for differences in absorption probability of soluble and immobilized fluorescein. Fluorescence excitation spectra show that the absorption spectra of the soluble and immobilized fluorescein are shifted relative to each other. The fluorescence excitation spectra allow an estimate of the change in relative absorbance at 488 nm due to the spectral shift. However, at present there is no measurement of the absolute value of the molar extinction coefficient for fluorescein immobilized on microspheres. The adjustment for possible differences in the molar extinction coefficient is left for a later time.

4.4 Computation of MESF Values

Equation 5 gives the relation to calculate the MESF values for the microspheres with immobilized fluorophores.

$$\mathrm{MESF} = \frac{6.022 \times 10^{23}}{1000} \frac{C_{\mathrm{eq}}}{N_{\mathrm{spheres}}} , \tag{5}$$

N_{sphere} is the number density (cm^{-3}) of fluorophore labeled microspheres and C_{eq} is the molar concentration (mol/L) of soluble fluorophore which gives the same fluorescence signal as the suspension of microspheres. The constant in Eq. 5 is a conversion factor between molar and number concentrations i.e., Avogadro number and 1000 ml/l. The equivalent concentration of soluble fluorescein is determined using the calibration curve determined with the reference solutions of fluorophores.

$$C_{\mathrm{eq}} = 10^{-\mathrm{intercept}} (\mathrm{FI} \cdot P_{\mathrm{adj}})^{\mathrm{slope}} , \tag{6}$$

where *intercept* and *slope* are the linear fit parameters describing the relationship between the logarithm of the observed fluorescence signal and the logarithm of the concentration of fluorescein in the reference solutions. Analysis of results from repetitive calibrations gave an uncertainty of about 3% in the value of *slope* and about 2% in the value of *intercept*. FI in Eq. 6 is the measured fluorescence signal of the microsphere suspension. The ratio of the average of the power readings taken during the calibration and microsphere measurements is set equal to P_{adj}, which multiplies FI in Eq. 6 and compensates for possible differences in illumination intensity.

The values of MESF are assigned using Eqs. 4, 5 and 6 using the six measured values: *slope*, *intercept*, FI, P_{adj}, N_{singles}, and N_{doubles}. The procedures used to measure the six parameters also provide an estimate of the error associated with each measured value. Once the uncertainties for the six values are known, Eq. 5 can be used to assign an overall uncertainty of the MESF value.

4.5
Relative Determination of MESF – Cytometer Measurement

The objective of this measurement is to ensure that the MESF assignments obtained with the fluorometer are linear. The microspheres with immobilized fluorophore are passed through a flow cytometer, and the response is measured. Each population of microspheres, each with a different amount of immobilized fluorophores, produces a population of fluorescence pulses which are characterized by a mean pulse height and a standard variation. The flow cytometer response is known to be linear, therefore the mean fluorescence pulse height of the different populations of microspheres should correlate linearly with the assigned values of MESF. Thus, while the cytometer does not provide an independent measurement of the MESF values, it does place a stringent constraint on the consistency of the relative assigned values. The MESF values obtained with the fluorometer can be adjusted slightly (<5%) so that they correlate linearly with the means of the fluorescence pulse distributions measured with the flow cytometer.

4.5.1
Cytometer Characterization

The cytometer at NIST was constructed to be as simple as possible with all physical processes open to inspection. An air-cooled argon ion laser provided the source of 488 nm illumination. The laser beam was focused to a waist of approximately 20 μm by a spherical lens with a focal length of 50 mm. The focal point was located in the flow channel of a cytometer flow cell adapted from a FACSCalibur cytometer made by Becton Dickinson Biosciences. The sample was pumped by a syringe pump, and the sheath fluid was pumped by the pressure in the sheath fluid container vessel. A flow meter in the sheath flow line gave an indication of the flow rate. The laser beam passed through a glass plate whose orientation provided a sensitive adjustment of the beam position in the scattering plane. The flow cell contained the collection optics which focused the emitted light on photomultiplier detectors (PMTs). The usual arrangement of a dichroic mirror and filters selected the fluorescence and elastically scattered light components. The outputs from the two PMT were processed by digital electronics provided by Becton Dickinson Biosciences (BD FACSDiVa system). An oscilloscope provided a visual monitor of the pulses associated with the side scattering channel (SSC) and the first fluorescence channel (FL1) of the detection electronics.

4.5.2
Linearity and Dynamic Range

The linearity and dynamic range of the flow cytometer was checked using the output of a stable LED. The output of the LED was chopped and split into two parts one of which was directed toward the scattering channel PMT (SSC) and the second part was directed through a ND filter towards the fluorescence channel PMT (FL1). The chopper rotation was adjusted to give pulses that approximated the pulses from the microspheres in the flow cell. This adjustment was performed by visual inspection of the oscilloscope traces. The linearity measurements were performed by noting the geometric mean channel of the detected pulses in FL1 for a given value of OD (optical density) associated with the neutral density filter. The plot of the log of the relative pulse height versus the log of the illumination given by 10^{-OD} resulted in a straight line with an average slope of 1.027 ± 0.008.

4.5.3
Noise Characteristics

The noise characteristics of the flow cytometer are a sensitive indicator of the quality of its operation. CV is defined as the standard deviation of the fluorescence pulse amplitude distribution divided by the mean amplitude, $\mathrm{FI}_{\mathrm{mean}}$, of the pulse distribution. The CV is related to $\mathrm{FI}_{\mathrm{mean}}$ by

$$(CV)^2 = \frac{1}{K \cdot \mathrm{FI}_{\mathrm{mean}}} + \frac{\sigma_{\mathrm{E}}^2}{\mathrm{FI}_{\mathrm{mean}}^2} + \sigma_{\mathrm{S}}^2 , \tag{7}$$

where σ_{E} summarizes electronic sources of noise, K is the photon transfer constant for the PMT, and σ_{S} summarizes source of noise that depends linearly on signal. A linear relation between $(CV)^2$ and $1/FI_{\mathrm{mean}}$ implies that the noise in the pulse height is due to the statistics of photoelectron production at the cathode of the PMT, and the cytometer is responding as expected. The constant value of CV^2 at high values of $\mathrm{FI}_{\mathrm{mean}}$ is a measure of the noise component which is proportionate to the signal. A likely source of this noise is mechanical vibrations of the chopper or, in the case of flowing microspheres, fluctuations in the intensity of the illuminating light and the inherent non uniformity of the microspheres themselves.

4.5.4
Microsphere Measurements

Alignment microspheres from Spherotech Corp. were used to align the cytometer laser beam. (The alignment was sufficiently good when the FL1 pulse mean channel was between 180 000 and 200 000, and the CV % was better than 4%.) After the alignment, the five populations of the reference micro-

(a)

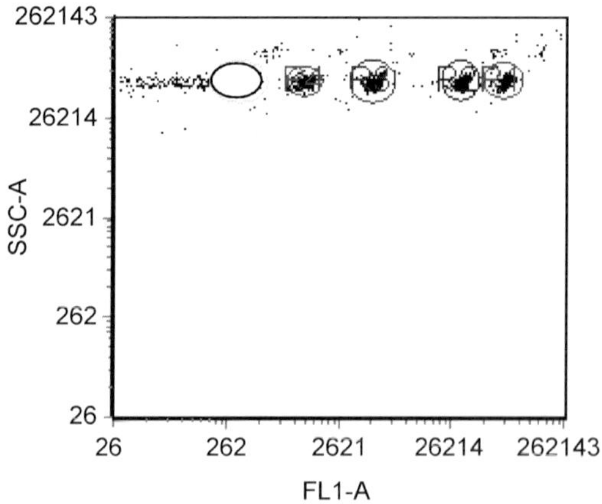

(b)

Region #	# of Events	X Geometric	Y Geometric	% of Gated Cells	% of All Cells
None	30000	20987.58	63138.92	100.0	100.0
1	4736	80167.84	61753.79	15.79	15.79
2	4609	32133.73	61723.15	15.36	15.36
3	6633	5554.8	61645.0	22.11	22.11
4	4476	1330.09	61596.18	14.92	14.92
5	4498	335.6	61863.24	14.99	14.99

(c)

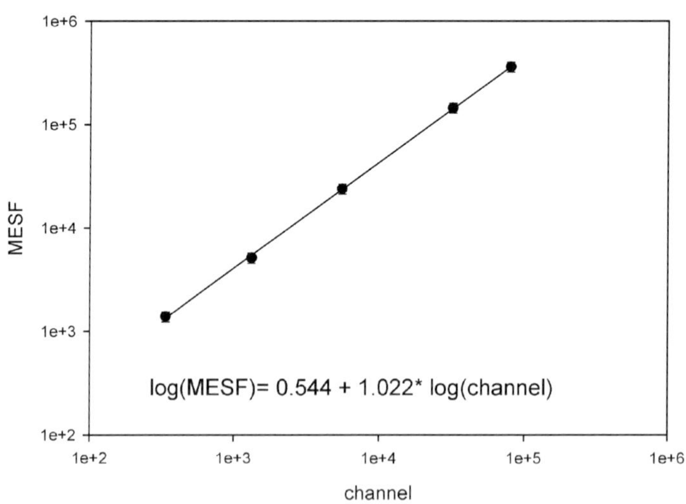

◀ **Fig. 4** Calibration of the response of a flow cytometer in terms of MESF. **a** Distribution of scattering and fluorescence light pulses from five populations of microspheres with different amounts of immobilized fluorescein. The side scattering signal (SSC-A, *Y*-axis) is relatively constant for all five populations while the average magnitude fluorescence of the fluorescence signal (FL1-A, *X*-axis) varies over two and a half magnitudes. **b** A compilation of the mean geometric values for the distribution of data points in Fig. 3a. **c** A plot of the logarithm of the assigned MESF value of each population as a function of the logarithm of the associated mean magnitude of the fluorescence pulses. A fit of the log (MESF) values to a linear function of log (FI) constitutes a calibration of the cytometer response

spheres were mixed and pumped through the cytometer and the corresponding fluorescence and scattering peaks recorded. In all cases the single and double microsphere signals were resolved. A scatter plot (Fig. 4a) was made where the *Y* axis corresponds to the scattered light pulse amplitude, while the *X* axis corresponds to the fluorescence pulse height. As shown in Fig. 4a, the scattering signal is relatively constant for the five populations, while the fluorescence signals differ substantially and form five distinct populations. The population centered at the *X* value of 336 corresponds to the microspheres with the least amount of fluorophores. This population is distinct from the blank microsphere population which forms a wide background with values of *X* less than 260.

Figure 4b tabulates the geometric means of the scattering and fluorescence signals associated with each of the five microsphere populations. The geometric mean was chosen to represent the location of the population because this parameter is less sensitive to outliers, and because it is used widely in representing data from flow cytometers.

Figure 4c shows the plot of the log of the geometric mean of fluorescence pulses associated with each microsphere population as a function of the log of the assigned MESF value for each microsphere population. As described above, the MESF assignments were performed using the fluorometer. Since the cytometer is a linear device, the measured mean pulse heights should be correlated linearly with the MESF values which are proportionate to the number of fluorophores on the microsphere and hence to the fluorescence signal. We made small adjustments (<5%) in the assigned MESF values, by requiring that they produce a linear response in the cytometer.

5
Application of MESF

The MESF calibration has been used in quantitative measurements of ligand receptor interactions [15]. A recent study used quantitative cytometry with MESF calibration to report the level of ZAP-70 expression in B cells [16]. The authors conclude that "The MESF unit provides a powerful tool to compare flow cytometry analysis of intracellular protein levels in an entire cell population, in a precise quantitative manner over extended periods of time and across instruments." This view deserves strong consideration. However, in a large portion of assays performed with flow cytometers a small extension of the MESF calibration would provide even more biologically significant information. The extension involves assigning MESF values not only to cells but also to the labeled antibodies which are used to label the cells. The objective is to estimate the number of antibodies bound to the cell (ABCe). The quantity ABCe is still a step removed from the number of total antigen molecules per cell [17] or what is also called antibody binding capacity [18]. The later quantities, which are referred to as ABC (without the small e at the end), can be inferred from ABCe if there is information on the valency of binding between the antigen on the surface and the labeled antibody specific to the antigen. It seems worthwhile to expend a small additional effort and strive to report flow cytometer results in terms of ABCe values.

5.1 Conceptual Basis for Determining ABCe

The same procedure that was described for the microspheres could be applied to the assignment of MESF values to labeled antibodies in solution. These values will be called $MESF_{Ab}$. Given the MESF values for the microspheres, it is possible to calibrate a flow cytometer response and assign a MESF value to a cell incubated with the antibodies labeled with fluorophores. The assigned value will be denoted by $MESF_{cell}$. The antibodies, the microspheres, and the cells are assumed to be labeled with the same fluorophore and in the same buffer. In this case the equality of FI provides the same equivalence between fluorophore concentrations in the three environments. Therefore it is valid to estimate of the number of antibodies bound per cell by the relation given in Eq. 8.

$$\mathrm{ABCe} = \frac{\mathrm{MESF}_{\mathrm{cell}}}{\mathrm{MESF}_{\mathrm{Ab}}} \,. \tag{8}$$

The validity of Eq. 8 is based on the assumption that a labeled antibody has the same FY on the microsphere, the cell and in solution. In that case, dividing the FY of a cell by the FY of a single antibody gives an estimate of the number of antibodies on the surface of the cell. The ABCe values obtained by Eq. 8 may have to be adjusted to reflect the actual number of antigens on the cell surface. The adjustment may be needed because not all antigens on the surface will bind an antibody. The adjustment factor could be found by measuring the ABCe value of a biological control with a known number of antigen molecules expressed on the cell surface. However, such controls may not be easy to obtain. In the following, MESF values are assigned to microspheres with immobilized R-phycoerythrin (R-PE), and the microspheres are used to obtain ABCe values for antigens called CD4 and CD20. A detailed example of the MESF assignment is presented below.

5.2
Assignment of MESF Values to Microspheres with Immobilized R-PE

A stock solution of R-PE was obtained from BD and had a nominal protein concentration of 1.47 mg/ml. Assuming a molecular mass of 220 000, the molar concentration of R-PE in the stock solution was 6.12×10^{-6} mol/L. Solutions of R-PE were prepared by dilution of the stock solution in PBS containing 0.022% (mass/mass) Tween 20 surfactant. Table 1 summarizes the data used to obtain six serial dilutions of the reference solution of R-PE. The concentrations of the diluted solutions were found by weighing the empty vial (column 2 in Table 1), the vial with the buffer (column 4), and the vial with added R-PE solution (column 6) and performing the appropriate arithmetic. The results are shown in the last column of Table 1. The weighing uncertainties (col 3, 5, and 7) were found from the standard deviation of the values from multiple weighings. The uncertainties due to weighing are small compared to the uncertainty in the concentration of the reference R-PE solution.

The solutions of R-PE were placed in a semi-micro cuvette and illuminated with 488 nm laser line from an argon ion laser. The observed spectra were normalized to 1 s collection time, corrected for spectral response, and integrated. The resulting fluorescence signals (FI) in digital units (DU) are shown in Table 2 together with the measured errors.

The errors in FI were obtained from repeated measurements. The FI and R-PE concentration values were converted to logarithms and plotted in a similar manner as shown in Fig. 2. For the data in Table 2, the linear relation (obtained using linear regression program in SigmaPlot 9) is best described by

$$\log(\mathrm{PE}) =- 16.937 + 0.979 \cdot \log(\mathrm{FI}) . \tag{9}$$

The last column in Table 2 gives the predicted values of the log of the concentration of R-PE. The errors in the fit parameters are not shown.

Table 1 A list of masses used to calculate the concentration of R-phycoerythrin in serial dilutions

Solution index	Mass of vial, g	Error g	Vial+ Buffer g	Error g	Vial + Buffer + R-PE, g	Error g	R-PE g	Buffer g	R-PE mol/l
0									6.12e-6
1	8.261	6.0e-4	18.421	1.5e-3	18.522	9.6e-4	0.100	10.160	5.99e-8
2	8.395	1.3e-3	18.558	2.6e-3	18.665	8.1e-4	0.108	10.163	6.28e-10
3	8.287	5.4e-4	16.392	8.6e-4	18.408	7.7e-4	2.015	8.105	1.25e-10
4	8.252	1.2e-3	16.365	5.3e-4	18.388	3.4e-4	2.023	8.113	2.50e-11
5	8.370	2.1e-3	16.504	1.2e-3	18.512	9.1e-4	2.008	8.134	4.94e-12
6	8.415	1.9e-3	16.556	2.0e-3	18.562	1.6e-3	2.006	8.141	9.77e-13

Table 2 A list of solutions of R-PE and the corresponding measured fluorescence signal (FI)

R-PE mol/l	Error mol/l	FI	Error	log(PE)	log(FI)	Predicted log(PE)
9.77e-13	1.0e-13	98 000	6000	−12.01	4.99	−12.05
4.94e-12	6.0e-13	639 000	50 000	−11.31	5.81	−11.26
2.50e-11	5.0e-12	3 080 000	60 000	−10.60	6.49	−10.59
1.25e-10	1.0e-11	15 500 000	600 000	−9.90	7.19	−9.90
6.28e-10	4.0e-11	76 200 000	500 000	−9.20	7.88	−9.22

The number concentration of the microspheres in the suspensions provided by BD was measured using a Coulter Multisizer 3 counter. The concentration, N_{sph}, was 20.2×10^6/mL. For fluorescence measurements, the microsphere suspension was diluted by a factor of 10 in the same buffer used for the R-PE solution calibration. The fluorescence spectra of the microspheres were measured under identical conditions that were used for the R-PE solution measurements. The spectra were treated as described previously and the resulting spectra were integrated to obtain a FI of 6.27×10^6 DU (digital units). This value of FI was used in Eq. 9 to obtain an equivalent concentration of soluble R-PE, $C_{\mathrm{PE}} = 5.14 \times 10^{-11}$ mol/L, that gave the same FI as the bead suspension. This value of C_{PE} was used in Eq. 10

$$\mathrm{MESF} = \frac{N_{\mathrm{a}}}{1000} \frac{C_{\mathrm{PE}}}{N_{\mathrm{sph}}} \tag{10}$$

to obtain a MESF $= 15500 \pm 3000$ for the microspheres. The error is an estimate obtained by propagating the errors of the quantities in Eqs. 9, 10.

The above determination of MESF requires a linear fit to the entire range of concentrations of R-PE. A simpler technique is to use a single solution of R-PE which gives a FI very close to that of the microsphere suspension and assume a linear response of the fluorometer. Suppose that FI is obtained for a R-PE solution with a concentration C and $\mathrm{FI_b}$ is obtained for a microsphere suspension with a number concentration N_{b}. Since the fluorometer is assumed to be linear, the fluorescence signal from the microsphere suspension, $\mathrm{FI_b}$ could also be obtained using a solution of R-PE with a concentration $C_{\mathrm{eq}} = (\mathrm{FI_b}/\mathrm{FI})C$. Thus the MESF value is calculated by

$$\mathrm{MESF} = \frac{N_{\mathrm{A}}}{1000} \frac{C_{\mathrm{eq}}}{N_{\mathrm{b}}} = \frac{N_{\mathrm{A}}}{1000} \frac{\mathrm{FI_b} C}{\mathrm{FI}} \frac{1}{N_{\mathrm{b}}} \,. \tag{11}$$

Using Eq. 11 with a R-PE concentration $C = 2.50 \times 10^{-11}$ mol/L, the corresponding $\mathrm{FI} = 3.0 \times 10^6$ DU, and previous values for N_{b} and $\mathrm{FI_b}$, Eq. 11 gives an MESF value of 15 700. The MESF assignment provided by the manufacturer is 22 000. The difference in assigned MESF values suggests a need to develop commonly accepted methodology for assigning MESF values. An alternate method for the assignment of MESF values was reported that utilized dissolution of the microsphere material [19]. Calibrations using these microspheres gave reasonable results [20].

5.3 CD20 Measurements

Numerous attempts have been reported to use flow cytometry to quantify receptor expression on lymphocytes as potential disease biomarkers [21–23]. CD20 expression in B-cell chronic lymphocytic leukemia (B-CLL) is one of the best examples of such a biomarker. The following is a description of measurements demonstrating the determination of ABCe values for CD20 in normal and diseased samples.

In addition to CD20, measurements were also performed on the expression of CD4 on T cells. CD4 is a good biological control since there is relatively low interpersonal variation in the expression of CD4 on T lymphocytes [24–27]. Unimolar PE antibody conjugates and PE-labeled microspheres were used to obtain ABCe values. The assumption is that each CD4 antibody has one and only one R-PE label and that the fluorescence yield is the same for R-PE conjugated to antibodies, R-PE on the microspheres, and R-PE in the reference solution. These assumptions have been well documented for the reagents and microspheres used in this study, but it may not apply to all R-PE conjugates [28, 29]. Given these assumptions, the ABCe value is given by

$$\mathrm{ABCe} = \frac{\mathrm{MESF}_{\mathrm{cell}}}{\mathrm{MESF}_{\mathrm{Ab}}} = \frac{\mathrm{MESF}_{\mathrm{cell}}}{1} = \mathrm{MESF}_{\mathrm{cell}} \,. \tag{12}$$

The mean value of ABCe determined using unimolar CD4-PE conjugate is 36 800. The measured ABCe values are somewhat dependent of sample preparation protocols, namely the fixation procedure used in the present study. The measured ABCe values for CD4-PE are relatively constant between healthy blood donors and B-CLL patients.

The measurement was repeated with standard grade anti-CD4 PE. The averaged number of ABCe determined using standard grade anti-CD4 PE and based on calibration curves generated by QuantiBRITE PE Quantitation kits is 34 400 with a CV of 5.5%. This value is lower than that found above for unimolar antibodies and the value reported by Davis and co-workers, ∼38 000 PE molecules using standard grade CD4 PE from BD Biosciences and freshly prepared blood samples without fixation step [7]. It is worth noting that quantitative measurements of ABCe values, using unimolar conjugate of CD4 PE and standard grade PE conjugates may differ. The commercially available standard grade PE conjugates contains a small fraction of unlabeled antibodies (personal communication, Robert Hoffman, BD Biosciences). Unlabeled antibodies bind to the surface epitope and reduce the number of labeled antibodies bound to the cell surface. This gives smaller MESF values for the cell resulting in lower apparent ABCe values. The standard grade PE conjugates could be used in quantitative measurements if their effective MESF values were measured. The effective MESF value takes into account possible hindrances to the binding of the labeled antibody to the receptor on the cell surface.

The same methodology described above for unimolar PE antibodies was adapted to measure CD20 expression levels on B cells, in terms of the ABCe values. The mean value of ABCe obtained for B-CLL patients (21 700) is more than six times lower than that for healthy individuals (143 500). The values for healthy blood donors are consistent with those reported by Bikoue et al. [24] and by Gratama and co-workers [1]. Considering the use of different reagents, such as the standard grade CD20 PE from BD and different quantitation methods, i.e., quantum simply cellular and MESF approaches, in [30–33], it is difficult to compare CD20 expression levels on normal B lymphocytes in terms of the ABCe values. The present measurements shows much less variation in CD20 expression among numerous B-CLL patients (42.0% CV).

Measurements were also carried out on CD4 and CD20 quantitation using monoclonal antibodies labeled with fluorescein isothiocyanate (FITC) and RM 8640 [34] as the cytometric calibration standard. The assigned values of MESF for CD4 were relatively consistent between normal blood donors and B-CLL patients with a mean MESF value of 15 200. According to the same calibration curves, the mean MESF value of CD20 FITC was calculated to be 81 700 for healthy persons. To resolve the MESF number to antibodies bound per cell (ABCe), a MESF value of FITC-labeled antibodies is needed. Given that the sample preparation protocols are the same for R-PE-labeled and FITC-labeled antibodies, the effective MESF value of the FITC-labeled antibodies is estimated by the ratio of the measured MESF value for the CD4 FITC-stained lymphocytes (15 200) and the MESF value for unimolar CD4 PE-stained cells (36 800). The ratio has a value of about 0.41 which gave $MESF_{Ab\text{-}FITC} = 0.41$. The mean MESF value obtained for CD20 FITC-labeled lymphocytes (81 700), can now be converted to ABCe value for CD20 FITC-stained B cells by the ratio of the two values, $81\,700/0.41 \approx 199\,300$. The estimated value is larger than the ABCe value determined using unimolar CD20-PE conjugate (143 500). The calculated ratios of ABCe values for CD20 FITC stained B cells from normal donors and B-CLL patients using CD4 as a biological calibrator are in general agreement with the values measured using unimolar CD20-PE conjugate. The accurate assignment of effective MESF values to antibody conjugates is under active study in our laboratory using the procedure employed to assign MESF values to microspheres with immobilized fluorescein (RM 8640). The results obtained in the present study strongly suggest that consistent ABCe values are possible with different fluorophores and preparative methods, but the reagents, calibrators, and methods must be evaluated carefully to insure such consistency [35]. Since ABCe values do not depend on the details of the fluorescence measurement, they provide a universal scale for conjugate binding and indirectly for receptor expression.

5.4 Cytotrol® Measurements

The following is a demonstration of the contents of an assay kit that may be sold by a reagent manufacturer. The two main items are the microspheres with assigned MESF values and the labeled antibody with its own value of MESF. The user would use the microspheres to calibrate the flow cytometer and measure the MESF of the analyte cells (Cytotrol® is a product of Beckman Coulter Inc.). The value of the MESF assigned to the labeled antibodies will be used to divide the value of the cell MESF to obtain an estimate of the ABCe. Assignment of MESF to Cytotrol® was performed by first calibrating a cytometer using RM 8640 calibration beads. The MESF value was found to be $41\,000 \pm 5000$. The MESF values for T4-FITC antibodies for CD4 molecules on the Cytotrol surface were found by calibrating a fluorometer with SRM 1932 and then comparing the FI of a T4-FITC solution with an equivalent solution of SRM 1932. The MESF of the T4-FITC antibody was found to be 1.16. Dividing the MESF of Cytotrol® by the MESF of the T4 antibody gives a value of $35\,000 \pm 6000$ for ABCe which corresponds to the number of labeled T4 antibodies bound to Cytotrol®. The number of CD4 antigens on Cytotrol® is expected to be about 40 000. Therefore the ABCe needs to be adjusted upwards. As mentioned previously, the assignment of MESF to labeled antibody is performed in solution and does not take into account possible problems of coupling to the antigens on the surface of a cell. Effect of a surface on the antibody-antigen binding can not be discounted. For example, both Huang et al. [36] and Buranda et al. [37] report that the forward rate constants, k_f, found for biotin binding to streptavidin at surfaces were order of magnitude lower than in solution, with a value of about $10^7\ M^{-1}\ s^{-1}$. In the case of Cytotrol®, the ABCe value of $35\,000 \pm 6000$ is reasonably close to the expected number of binding sites for CD4 (40 000), it is likely that in this case the surface of the cell does not affect significantly the binding of the antibody to the antigen.

However, consider the following hypothetical case. Suppose that the antibody is labeled and several fluorophores are attached to each antibody. The assignment of $MESF_{Ab}$ in solution gives a high value. Suppose furthermore that the labeling decreases the binding of the labeled antibody to its antigen on the surface of the cell. When the labeled antibody is incubated with the cells, only a small amount of the labeled antibody will bind to the surface antigen. A cytometer measurement of the labeled cells will give a small value of cell MESF and the division of the cell MESF by the antibody $MESF_{Ab}$ will yield a very small value for antibodies bound on cell (ABCe). This value of ABCe will not reflect the true number of antigens on the cell surface.

Consider another hypothetical case. Suppose that the fluorophore used in labeling does not change the binding properties of the antibody, however the fluorophore is sensitive to pH (e.g., fluorescein). Suppose furthermore that the pH at the interface of the cell or microsphere is different from the pH in the bulk of the solution. Assignment of MESF to antibody in solution will reflect no pH change since both the antibody and SRM are in solution. The MESF assigned to microspheres will reflect a change in pH between the surface and the bulk of the solution. The calibration of the flow cytometer with the microsphere will propagate the pH difference to the value of MESF assigned to the cell. Dividing the cell MESF by the antibody MESF will contain a bias due to the difference of pH between the surface and the bulk. The two hypothetical scenarios point to some of the checks that should be performed by the manufacturers of reagents with assigned MESF values to both the microspheres and labeled antibodies. Biological systems that can be used to check for biases have been described [38, 39].

5.5 Application to Multicolored Cytometer Measurements

The quantitation protocol described above can be generalized to assays involving many different fluorophores. The most straightforward procedure is to provide a microsphere MESF calibrant for each fluorescence channel and a MESF value for each of the labelled antibodies. The procedures for assigning MESF values to reagents and microspheres can be automated.

The manufacturer of an assay kit would need to assign MESF to each of the labeled antibodies in terms of solutions of the labeling fluorophore. The same solutions would be used to assign MESF values to the microspheres with immobilized fluorophores. Ideally the assignments could be carried out in a central laboratory accessible to all manufacturers.

With experience it may be possible to devise procedures using a single reference solution for all fluorophores. This path would require knowledge of more photophysical properties of the fluorophores in different environments. This is demonstrated by the attempt to assign MESF values to microspheres with immobilized R-PE using reference solutions of fluorescein. The main difference is that instead of comparing FI directly, the FI is divided by the appropriate extinction coefficient to obtain the corresponding FY, and then the values of FY are compared. The procedure can be summarized by Eq. 13 below that relates the ratio of the measured fluorescence signal and extinction coefficient from solutions of FITC and PE on the left to properties of the two solutions on the right.

$$\begin{aligned} \frac{FS_{FITC}}{\varepsilon_{FITC}(\lambda_{ex})} &= \frac{\Omega I_0 \varepsilon_{FITC}(\lambda_{ex}) \phi_{FITC} C_{FITC}}{\varepsilon_{FITC}(\lambda_{ex})} = \Omega I_0 \phi_{FITC} C_{FITC} \\ \frac{FS_{PE}}{\varepsilon_{PE}(\lambda_{ex})} &= \frac{\Omega I_0 \varepsilon_{PE}(\lambda_{ex}) \phi_{PE} C_{PE}}{\varepsilon_{PE}(\lambda_{ex})} = \Omega I_0 \phi_{PE} C_{PE} \, . \end{aligned} \tag{13}$$

Equation 13 shows that comparing the ratio of the FI and extinction coefficient of two solutions is the same as comparing the FY of the two solutions (it is assumed that the instrument factor, ΩI_0, is the same and that proper spectral corrections were performed prior to integrating the fluorescence spectra). The FI for FITC was divided by 82 and the FI of PE was divided by 2200. (Only the relative value of the two molar extinction coefficients is needed, therefore the values are scaled down by 1000 for computational convenience.) A calibration of the FY of FITC was performed by dividing the FI measured for each reference solution of FITC by 82 and plotting the concentration values of the reference solutions as a function of the corresponding values of FY. A line was obtained relating FY to the concentration of FITC. The measured FI from the microspheres with immobilized PE was divided by 2200, and the FITC calibration was used to obtain an equivalent concentration of FITC that gives the same FY. The resulting value of MESF for the microspheres with immobilized PE was 9550. That is, each microsphere has a FY that is equivalent to the FY of 9550 soluble FITC molecules. Measurements described above showed that each microsphere with immobilized R-PE has a FY equivalent to 15 500 molecules of soluble R-PE. Thus it appears that the FY of 9550 molecules of soluble FITC is equivalent to the FY of 15500 molecules of soluble R-PE. This result was checked by comparing the FY of the actual solutions of FITC and R-PE. The FI from FITC and R-PE solutions were converted into FY as described previously and compared. It was established that the FY of one molecule of FITC is equivalent to the FY of 1.7 molecules of R-PE. That is it takes fewer molecules of FITC to produce the same FY as R-PE. The result found using solutions of FITC and R-PE is the same as the result obtained from the comparison of microsphere FITC MESF value and microsphere R-PE MESF value. The definition of FY implies that the ratio of equivalent concentrations of FITC and R-PE (ratio = 0.6) is equal to the ratio of quantum yields of R-PE and FITC. The QY of R-PE relative to the QY of fluorescein SRM solution was obtained by measuring the FI and absorbance, A, for each of the two solutions and by taking the ratio FI/A for each of the two fluorophore solutions. (Recall that the measured absorbance is the product of the molar extinction coefficient, the molar concentration, and the path length commonly set to 1 cm). Specifically

$$\frac{QY_{PE}}{QY_{FITC}} = \frac{FS_{PE}}{A_{PE}} \frac{A_{FITC}}{FS_{FITC}} \, . \tag{14}$$

The result was $QY_{PE}/QY_{FITC} = 0.92 \pm 0.03$. This ratio is not consistent with the ratio of QY of 0.6 found from the comparison of FY. The discrepancy between the two values of relative QY is most likely due to differences in the nature of the two measurements. The measurement of relative QY requires the ratio of measured FI and the measured absorbance, A, and there is no need to know the concentration and molar extinction coefficient individually. The FY measurement requires knowledge of the extinction coefficients and concentrations separately. Thus the measurement of FY requires more information. Comparison of FY between different fluorophore requires an accurate correction of the spectral response of the fluorometer, and an accurate measure of the extinction coefficients. Neither is critical if FY is compared between solutions of the same fluorophore. This is why comparing FY between solutions containing the same fluorophore is more practical.

6 Conclusion

The quantitation of cytometer measurements in terms of the number of antibodies bound per cell (ABCe) is feasible and desirable. A central requirement is the establishment and acceptance of a methodology for assigning MESF values to microsphere calibrants and to labeled antibodies. The methodology would allow suppliers of reagents for flow cytometer based assays to assign MESF values in a uniform manner. In the case of multicolored flow cytometer assays, there would be a separate set consisting of microspheres and labeled antibody for each color. Quality control (QC) of the operation of the flow cytometer would still be necessary as would measurement of controls. However once QC is performed, the calibration and the details of the assignment of ABCe values could be incorporated into the software so that the users of flow cytometers would be presented only with a set of ABCe values and an estimate of error. The software could also include possible adjustments to ABCe values to obtain a better estimate of the actual number of antigens on the surface of a cell. The intricacies associated with the measurement of the fluorescence signal and its interpretation would be the responsibility of the instrument manufacturers and suppliers of reagents. The user would be presented with ABCe values (and the likelihood that the ABCe values are an accurate estimate of the antigen numbers on the cell). This information would be of direct biological significance. An alternate approach to quantitation in multicolor flow cytometer measurements, would use microspheres with MESF assignments to establish a linear fluorescence intensity scale for each channel. A biological standard would be used to pin the fluorescence intensity scale to the number of antigens on the surface of a cell.

Disclaimer: Certain commercial equipment, instruments, or material are identified in this paper to foster understanding. Such identification does not constitute an endorsement by NIST, nor does it imply that the materials or equipment identified are necessarily the best available for the purpose.

References

1. Gratama JW, D'hautcourt J-L, Mandy F, Rothe G, Barnett D, Janossy G, Papa S, Schmitz G, Lenkei R (1998) Cytometry 33:166
2. Lenkei R, Gratama JW, Rothe G, Schmitz G, D'hautcourt JL, Arekrans A, Mandy F, Marti G (1998) Cytometry 33:188
3. Schwartz A, Repollet EF, Vogt R, Gratama JW (1996) Cytometry 26:22
4. Serke S, Lessen A, Huhn D (1998) Cytometry 33:179
5. Zenger VE, Vogt R, Mandy F, Schwartz A, Marti GE (1998) Cytometry 33:138
6. Marti GE, Robert F Vogt J, Gaigalas AK, Hixson CS, Hoffman RA, Lenkei R, Magruder LE, Purvis NB, Schwartz A, Shapiro HM, Waggoner A (2004) CLSI, I/LA24-A, Wayne, PA
7. Lakowicz JR (1999) Principles of Fluorescence Spectroscopy. Kluwer Academic Press/ Plenum Press, New York
8. Jameson DM, Croney JC, Moens PDJ (2003) In: Marriott G, Parker I (eds) Biophotonics, Part A (Methods in Enzymolgy), vol 360. Academic Press, Amsterdam, p 1
9. Schwartz A, Wang L, Early E, Gaigalas A, Zhang Y-Z, Marti GE, Vogt RF (2002) J Res Nat Inst Stand Technol 107:83
10. Gaigalas AK, Wang L, Schwartz A, Marti GE, Robert F, Vogt J (2005) J Res Nat Inst Stand Technol 110:101
11. Certificate of Analysis, Standard Reference Material 1932, Fluorescein Solution (2004) Nat Inst Stand Technol, (http://ts.nist.gov/ts/htdocs/230/232/232.htm)
12. Resch-Genger U, Pfeifer D, Monte C, Pilz W, Hoffmann A, Spieles M, Rurack K, Hollandt J, Taubert D, Schonenberger B, Nording P (2005) J Fluorescence 15:315
13. Hollandt J, Taubert RD, Seidel J, Resch-Genger U, Gugg-Helminger A, Pfeifer D, Monte C, Pilz W (2005) J Fluorescence 15:301
14. DeRose PC, Early EA, Kramer GW (2007) Rev Sci Instrum 78:033107
15. Fay SP, Posner RG, Swann WN, Sklar LA (1991) Biochemistry 30:5066
16. Kay S, Herishanu Y, Pick M, Rogowski O, Baron S, Naparstek E, Polliack A, Deutsch VR (2006) Cytometry Part B Early View
17. Ginaldi L, Martinis MD, D'Ostilio A, Marini L, Loreto F, Modesti M, Quaglino D (2001) Am J Hematology 67:63
18. Barnett D, Storie I, Granger V, Whitby L, Reilly JT, Brough S, Garner S, Lawry J, Richards S, Bell AE, Shenton BK (2000) Clinical & Laboratory Haematology 22:89
19. Oonishi T, Uyesaka N (1985) J Immunol Method 84:143
20. Oonishi T, Sakashita K, Uyesaka N (1988) J Immunol Method 115:159
21. Hultin LE, Matud JL, Giorgi JV (1998) Cytometry 33:123
22. Denny TN, Stein D, Mui T, Scolpino A, Holland B (1996) Cytometry (Communications in Clinical Cytometry) 26:265
23. Lenkei R, Andersson B (1995) J Immunol Method 183:267
24. Bikoue A, George F, Poncelet P, Mutin M, Janossy G, Sampol J (1996) Cytometry (Communications in Clinical Cytometry) 26:137
25. Davis KA, Abrams B, Iyer SB, Hoffman RA, Bishop JE (1998) Cytometry 33:197
26. Poncelet P, Carayon P (1985) J Immunol Method 85:65
27. Poncelet P, Poinas G, Corbeau P, Devaux C, Tubiana N, Muloko N, Tamalet C, Chermann JC, Kourilsky F, Sampol J (1991) Res Immunol 142:291

28. Iyer SB, Hultin LE, Zawadzki JA, Davis KA, Giorgi JV (1998) Cytometry 33:206
29. Pannu KK, Joe ET, Iyer SB (2001) Cytometry 45:250
30. Marti GE, Faguet G, Bertin P, Agee J, Washington G, Ruiz S, Carter P, Zengler V, Vogt R, Noguchi P (1992) Ann New York Acad Sci 651:480
31. Almasri NM, Duque RE, Iturraspe J, Everett E, Braylan RC (1992) Am J Hematology 40:259
32. Cinaldi L, Martinis MD, Matutes E, Farahat N, Morilla R, Catovsky D (1998) J Clinic Pathol 51:364
33. D'Arena G, Musto P, Cascavilla N, Dell'Olio M, Renzo ND, Carotenuto M (2000) Am J Hematology 64:275
34. Certificate of Analysis, Reference Material RM 8640, Microspheres with Immobilized FITC (2005) Nat Inst Stand Technol, (http://ts.nist.gov/ts/htdocs/230/232/232.htm)
35. Purvis N, Stelzer G (1998) Cytometry, p 33
36. Huang S-C, Stump MD, Weiss R, Caldwell KD (1996) Anal Biochem 237:115
37. Buranda T, Jones GM, Nolan JP, Keij J, Lopez GP, Sklar LA (1999) J Phys Chem B 103:3399
38. Bikoue A, Janossy G, Barnett D (2002) J Immunol Method 266:19
39. Robert F Vogt J, Cross GD, Henderson LO, Philips DL (1989) Cytometry 10:294

Part V
Fluorescence Immunoassays

Springer Ser Fluoresc (2008) 6: 401–428
DOI 10.1007/4243_2008_039

Published online: 12 March 2008

Immunoassays: Basic Concepts, Physical Chemistry and Validation

Matthias Seydack

8sens.biognostic GmbH, Robert-Roessle-Str. 10, 13125 Berlin, Germany
seydack@biognostic.de

Abstract Clinical biochemistry teaches that the presence, elevation, or decrease in the concentration of certain proteins or other species in living organisms has a diagnostic function. Thus, the need for rapid, early, and specific analyses of human samples has been the major driving force for the development of appropriate analytical tools. This chapter is meant to give a succinct introduction into the basic mechanisms of immunoassays followed by more detailed information on validation and regulatory aspects. Given the scope of the book, those points that may be of special relevance to emission-based assays are highlighted whenever deemed reasonable. Due to the introductory character of this chapter and limitations in space, there is a clear focus on representative aspects and key issues.

Keywords Assay validation · ELISA · Fluorescence · Immunoassay · Label

Abbreviations

Ab	Antibody
Ag	Antigen
ELISA	Enzyme-linked immunosorbent assay
ε	Molar decadic extinction coefficient
F/P	Fluorescent molecule-to-protein ratio
FIA	Fluorescence immunoassay
FITC	Fluorescein isothiocyanate
FRET	Förster-type resonance energy transfer
Φ_{em}	Quantum yield of emission (fluorescence or luminescence)
IVD	In vitro diagnostic test
LOD	Limit of detection
M	Marker/Label
MTP	Microtiter plate
ROC	Receiver-operator characteristic
S/N	Signal-to-noise ratio

1 Introduction: Antibodies as an Analytical Tool

Clinical biochemistry teaches that the presence, elevation, or decrease in the concentration of certain proteins or other species in living organisms has a diagnostic function. Medical, ethical, and economic need for rapid, early, and specific analyses of human samples (i.e., blood, plasma, saliva, stool, and urine to name the most important examples) has been the major driving force for the development of appropriate analytical tools. By their very nature, these samples are extremely heterogeneous. Therefore, very specific "recognition units" are required in order to bind and quantify one type of protein ("analyte") against a background of many different proteinaceous and other components. The only kind of "recognition unit" that fulfils this task is the same that nature itself designed for this purpose: antibodies. Since in living organisms antibodies are produced by the immune system, tests using one or more binding reactions of antibodies are hence termed "immunoassays". Their analytical power reaches far beyond protein quantification: since appropriate antibodies can selectively bind to hormones, enzymes, or haptens as well, such biochemical structures can be analyzed, too. In contrast to biochemical recognition in nature, quantification also requires a reporting step of the binding event. Thus, there is always the need for a "marker", usually attached to one kind of antibody used. Depending on the type of marker used, its presence can be recorded, e.g., by an absorption or fluorescence measurement. After calibration, the amplitude of the obtained signal can then be converted into an analyte concentration.

This chapter is meant to give a succinct introduction into some basic aspects of immunoassays. Given the scope of the book, those points that may be of special relevance to emission-based assays are highlighted whenever deemed reasonable. The basic principles, however, are independent of the detection mechanism. Given the introductory character of this chapter and limitations in space, it should be noted that a number of important developments will only be mentioned in a very sketchy manner or will even be completely left out. The most important examples are immuno-PCR, flow injection assays, protein chips, assays based on surface plasmon resonance (SPR), and biosensors. Important though they are, these areas would justify their own chapter and are extensively reviewed elsewhere [1–6].

2 Antibodies and Recognition Reaction

2.1 Antibodies

Antibodies are glycoproteins that belong to the group of immunoglobulins (Ig) and possess antigen-binding sites. In nature, antibodies mainly occur in the blood and in other extracellular fluids but also function as receptors on the surface of B-lymphocytes within the adaptive immune response. Apart from binding to their respective antigen, they may additionally recruit other cells and molecules of the immune system that finally destroy the antigen. These two functions are spatially separated into the antigen-binding regions (Fab) and the constant (or crystallizable) region (Fc). All antibodies share the same basic structure: four polypeptide chains, where two identical pairs form a mirror-symmetrical structure. Two of the chains are "heavy chains" (H-chains) and the other two "light chains" (L-chains). They are associated with each other by disulfide bridges (Fig. 1).

There are two types of L-chains (lambda and kappa) and five different main classes of H-chains. These are structurally different and govern the allocation of the antibody to one out of five classes of immunoglobulins (IgM, IgD, IgE, IgG or IgA) and to a corresponding sub-class (e.g., IgG_1, IgG_2, IgG_3 or IgG_4). About 85% of the antibodies found in human serum belong to the IgG class at a typical concentration of 8–18 gL^{-1}, whereas dimeric IgA (0.9–4.5 gL^{-1}) and pentameric IgM (0.6–2.8 gL^{-1}) make up for most of the remaining 15%. Taken together, immunoglobulins represent around 20% of all serum proteins. The molecular weights of these biopolymers are 150 kDa (IgG and IgD), 900 kDa (IgM), 150 or 600 kDa (IgA) and 190 kDa (IgE), respectively.

The chemical composition of the accessible surface of an average protein is 55% non-polar, 25% polar, and 20% charged [7]. The composition of the recognition sites is similar to this except that the proportion of charged surface is slightly lower. Since there is a huge variety of surface groups capable of interacting with the micro-environment, changes in pH, solvent/buffer polar-

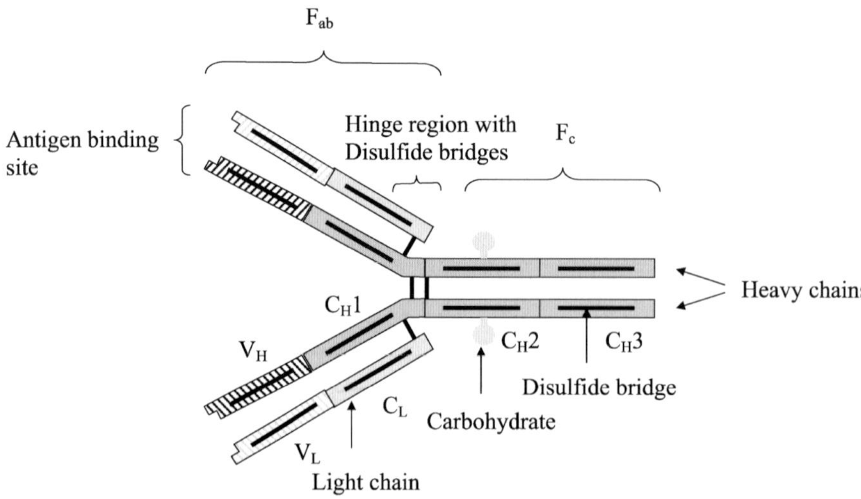

Fig. 1 Schematic representation of an antibody. V_L: variable part of the light chain, V_H: variable part of the heavy chain, C_L constant part of the light chain, C_H1, C_H2, C_H3: constant parts of the heavy chain, F_{ab}: fragment antigen binding, F_c: fragment constant/crystallyzable. Adapted from [61]

ity, ion strength, etc., may all affect colloidal stability, binding and recognition properties of the antibody or antigen (Fig. 2).

Depending on the production process, three groups of antibodies can be distinguished. (a) Polyclonal Abs ("antisera") are produced by living animals after having been vaccinated with the corresponding immunogen. Thus, an antiserum consists of antibodies produced by different cells. In effect, polyclonal Abs are not exactly identical but consist of different isotypes and recognize different epitopes on their respective antigen (cf. Sect. 2.2). (b) Monoclonal Abs, by contrast, are produced by identical hybridoma cells, consist of one isotype only, and are thus highly specific to exactly one and the same epitope of their antigen. They are preferred for most diagnostic applications. For sandwich immunoassays (cf. Sect. 3.3), workable couples of monoclonal Abs have to be found that bind to their antigen without sterical hindrance. Nevertheless, assay development usually starts with antiserum because it is often cheaper and assays can be established with less time and effort. An assay that does not work with antiserum will most probably also not work with monoclonal Abs. (c) Sometimes, only fragments of monoclonal Abs are preferable, e.g., for inhibiting unspecific binding. These fragments often consist of one (Fab) or both ($F(ab)_2$) antigen-binding sites only, the Fc part being separated. They can be obtained from the complete Ab by enzymatic cleavage. Recently, genetically produced Ab and fragments have played an increasing role. Tiny Ab fragments, consisting of only the variable region (V_H and V_L)

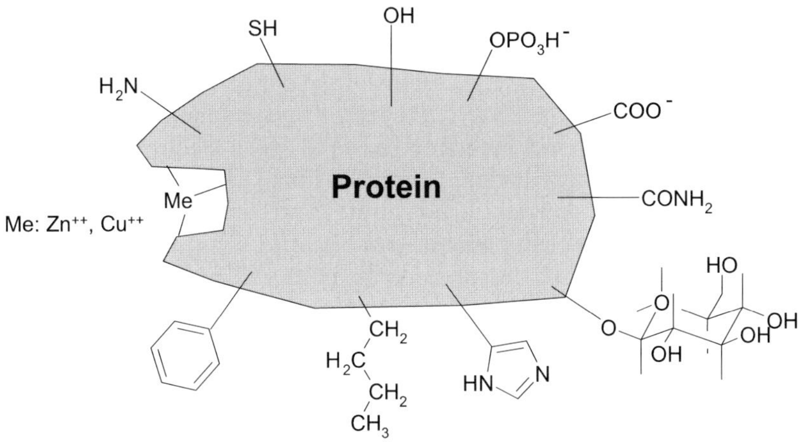

Fig. 2 Chemical groups and amino acid side chains of a protein capable of interacting with the environment chemically or physically, e.g., with coupling reagences, surfaces, particles and other proteins. Adapted from [11]

of one antigen-binding site (single-chain fragment variable, scFv) have been produced in this way.

Like other proteins, the polypeptide chains of antibodies consist of amino acids. Three of these are aromatic: tryptophan (Trp), tyrosine (Tyr) and phenylalanine (Phe). By virtue of this aromaticity, all proteins show a marked absorption peak at 280 nm, which is mainly due to tryptophan and can be used for measuring the overall protein content of a sample. As a first approximation, a protein solution of $c = 1.0\,\mathrm{mg\,mL^{-1}}$ measured in a 1-cm cuvette shows an absorption of $A_{280} = 1.0$. For a precise estimation, several empirical formulas have been suggested which, however, require the number of tryptophan, tyrosine, and disulfide groups of the protein to be known [8–10]. In comparison to colorimetric methods, such as the widely used BCA and Lowry tests or the Bradford method, where the protein reacts chemically to produce a dye that is subsequently measured absorptively, the protein absorption method is non-destructive but less sensitive by at least one order of magnitude [6, 11, 12].

More problematic for fluorescence immunoassays is the fact that the aromatic amino acids are fluorescent themselves. This autofluorescence ($\lambda_{max} \sim 348$ nm) needs to be allowed for whenever signals from the markers are weak, which is often the case, and excitation wavelengths are in the UV.

2.2 Recognition Reaction

Antibodies (Abs) are capable of recognizing and binding to matching surfaces, usually of proteins but also of viruses and haptens (e.g., anti-FITC antibodies that specifically bind to fluorescein isothiocyanate). The recognition reaction itself is based on a framework of non-covalent binding reactions between the antigen-binding F_{ab} regions of the antibody (paratopes) and the surface structure of the analyte (epitopes). The dominant contributions to the binding process are (a) hydrogen bonds between electronegative atoms, (b) electrostatic attraction between complementary charges, (c) hydrophobic interaction between nonpolar carbohydrates and, finally, (d) van-der-Waals forces due to electron-density fluctuations of neighboring atoms. The average number of intermolecular hydrogen bonds in an antibody–antigen (Ab–Ag) complex is known to be in the order of 10 [13].

Although the exact shape of the contact region of antibodies and other proteins varies greatly, the contact surface area it buries (the "hypervariable region") is of similar size with $16 \pm 3.5\,\mathrm{nm}^2$ [13]. Depending on the overall size of the antibody, these interface areas form about 5–20% of the total accessible surface area of the protein [14, 15].

Kinetic studies have been pursued on a number of Ab–Ag complexes and such investigations are typically performed either optically (by using surface plasmon resonance), thermodynamically (by isothermal calorimetry), or by equilibrium dialysis and evaluation of subsequent Scatchard plots [12]. Results for monoclonal antibodies binding to different types of antigen show that while there is only little variation in association rate constants ($k_{ass} = 10^5$–$10^7\,\mathrm{M^{-1}\,s^{-1}}$), the range of dissociation rate constants is far broader ($k_{diss} = 1\,\mathrm{s^{-1}}$ and $10^{-7}\,\mathrm{M^{-1}\,s^{-1}}$). Thus, affinity changes are predominantly due to changes in k_{diss} [16, 17]. Comparing the association rate constants with the collision rates of typical proteins in aqueous medium, which is about $10^9\,\mathrm{M^{-1}\,s^{-1}}$, one can derive that about one in 100 of the collisions leads to association for the most stable complex and one in 10^4 in the most unstable. Since interfaces cover only 5–20% of each protein surface, this means that a high proportion of collisions at the target surface leads to stable Ab–Ag complexes. Some authors discuss a two-step mechanism where, initially, only a fraction of the possible interactions leads to the formation of a loose complex which then undergoes small conformational changes to form a stable structure with correct orientation at the interface [13].

The capability of an antibody to bind to an antigen by forming a 1 : 1 complex is termed "affinity". It is usually described by means of a dissociation constant, K_D, that can approximately be derived from the law of mass action:

$$K_D = \frac{k_{ass}}{k_{diss}} = \frac{[Ag]\,[Ab]}{[AgAb]}\,, \tag{1}$$

K_D varies between 10^{-5} and $10^{-12}\,\mathrm{M^{-1}}$ [18].

3
Assay Formats – Assay Setups

3.1
Solid-Phase Assays: Surfaces and Binding

Most immunoassay formats require one kind of antibody (or analyte) to be immobilized ("coated") on a solid surface whereas the other components remain in the fluid medium, i.e., in the reaction buffer or sample matrix. Depending on the setup, the solid phase may be a pre-treated or untreated polymer, e.g., polystyrene in the case of microtiter plates (MTP) and nitrocellulose or nylon when test-strips or protein chips are used. In the latter case, glass is also used sometimes. Usually, the binding to the solid support is adsorptive, thereby keeping the functionality of the protein as uncompromised as possible. Different surface modifications of the support material allow for the fact that molecules with a varying degree of hydrophobicity can be adsorbed. In the case of MTPs, hydrophobic (e.g., untreated polystyrene), hydrophilic (e.g., polystyrene with a dense surface layer of $=$ O, $-$ OH, $-$ NH_2, $=$ N) and mixed hydrophobic/hydrophilic (e.g., a reduced load of $=$ O and $-$ OH on polystyrene surface) plate modifications are common. Naturally, strongly hydrophobic material, such as lipids and lipoproteins, preferably adsorbes to hydrophobic surfaces (e.g., Nunc PolySorp® MTP with an IgG binding capacity of 220 ng/cm^2). By analogy, hydrophilic molecules, such as glycolipids or glycoproteins are preferably adsorbed to hydrophilic surfaces (e.g., Nunc MaxiSorp® MTP, with an IgG binding capacity up to 650 ng/cm^2). Other polypeptides and proteins that contain mixed surface functionalities are often best adsorbed to mixed hydrophobic/hydrophilic plates (e.g., Nunc MediSorp® MTP, with an IgG binding capacity <650 ng/cm^2). It should be noted, however, that the binding probability is usually higher for hydrophilic surfaces, so that hydrophilic MTPs are sometimes preferred even in cases of hydrophobic adsorbate.

Nonetheless, the antigen-binding properties of monoclonal antibodies (i.e., highly specific antibodies originating from one cell-line only) can sometimes be affected by adsorbing them directly to a surface. This problem can usually be circumvented by using one out of several indirect binding procedures (Fig. 3).

In one approach, polyclonal antibodies (i.e., those originating from several cell lines, which are therefore more diverse and thus less specific than monoclonal antibodies) directed specifically against the F_c part of the monoclonal antibodies of interest are coated. Since these anti-antibodies will be recognized by any monoclonal antibody from the same animal, this method cannot be used in sandwich-type assays. Instead, it is especially advantageous for competitive assays since there is only one type of antibody involved (*vide infra*).

Alternatively, the monoclonal catcher antibody may be bound via so-called antibody binding proteins, such as protein A, G, or L. Protein A and G specif-

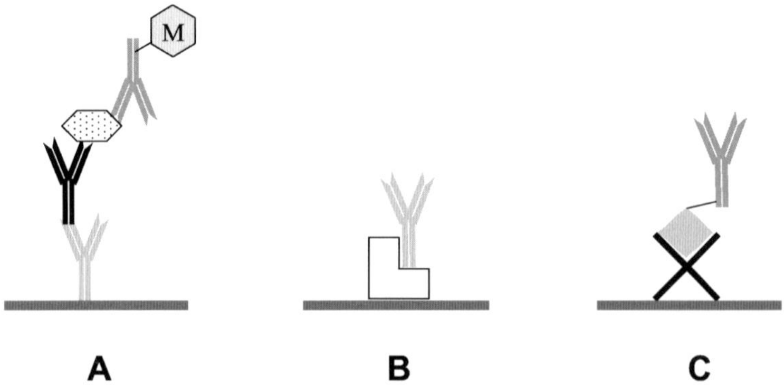

Fig. 3 Different alternatives to immobilize antibodies on a solid surface. The catcher antibody may be bound either via an anti-antibody (**A**), protein A or protein G (**B**) or can be biotinylated to bind to a streptavidin-modified surface (**C**)

ically bind to the F_c part, while protein L binds to the *kappa*-type light chain (Fig. 3B). Binding to one of these proteins is known to maintain the recognition properties of the antibody. Again, for a sandwich assay, this procedure is not very helpful as both catcher and detector antibody will bind to the protein.

A third method, which also works for sandwich assays, is based on the biotin-(strept)avidin system (*vide infra*). Avidin or streptavidin is coated to the solid support. MTP or protein chips with streptavidin surfaces can also be obtained commercially. Prior to use, the monoclonal catcher antibody needs to be conjugated to biotin. Biotinylation of antibodies (using NHS-biotin) is a relatively simple and straightforward process that hardly influences the recognition properties [19]. The immobilization of the catcher antibody then proceeds via binding of the biotin to the streptavidin-coated surfaces (Fig. 3C). The role of this biotin-(strept)avidin binding will be discussed more thoroughly below.

Of special importance to fluorescence immunoassays is the fact that the polystyrene of MTPs shows a small but measurable background emission due to impurities. Given the minute signal intensities from many emission-based assays, this background can seriously hamper the assay performance (*vide infra*). Some improvement can be achieved by using plates with a glass bottom instead, which are offered by some suppliers, or simply by shifting towards longer emission wavelengths.

3.2
Direct and Indirect Assays

All immunoassays can be distinguished between "direct" and "indirect". In a direct assay, the analyte itself is coated and the first specific antibody carries the marker unit (e.g., fluorophore, enzyme, etc.). Indirect assays, by contrast,

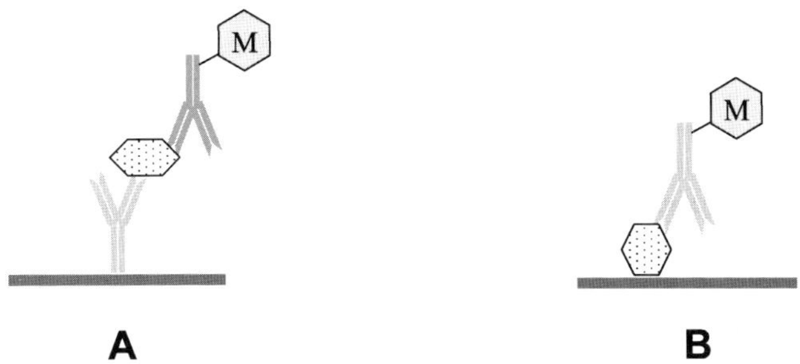

Fig. 4 Indirect (**A**) and direct (**B**) immunoassay

make use of a specific catcher antibody (primary antibody), which is coated. After incubation of the analyte, the labelled detector antibody (secondary antibody), which is also specific but usually for another epitope of the analyte, binds to the analyte (Fig. 4).

This sandwich-type architecture is usually more sensitive, delivers a higher signal intensity, and is therefore often preferred [20]. Naturally, the choice between the two approaches primarily depends on the specific analytical problem to be addressed.

3.3 Competitive and Non-Competitive Assays

Apart from the direct or indirect format, immunoassays can also be classified into "competitive" and "non-competitive". In a competitive assay, the analyte from the sample competes with a labelled antigen that was added to the assay (Fig. 5A,B).

The same concentration of labelled antigen is added to each sample and the surface coated with specific antibodies directed against the analyte. If

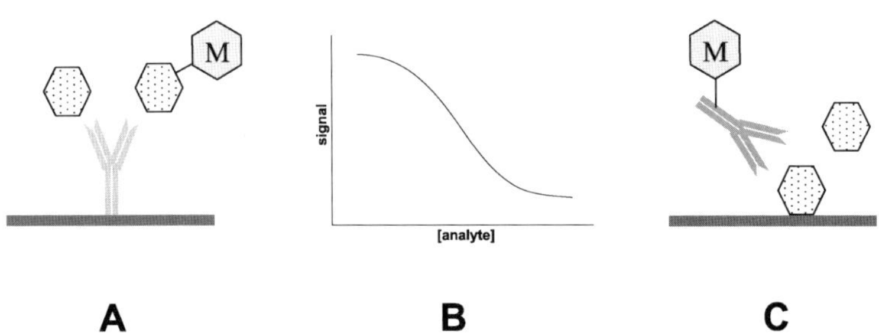

Fig. 5 A competitive immunoassay is based on the competition either between analyte and labelled antigen for the immobilized antibody (**A**) or between immobilized antigen and analyte for a labelled antibody (**C**). In both cases, the signal intensity drops with increasing analyte concentration (**B**)

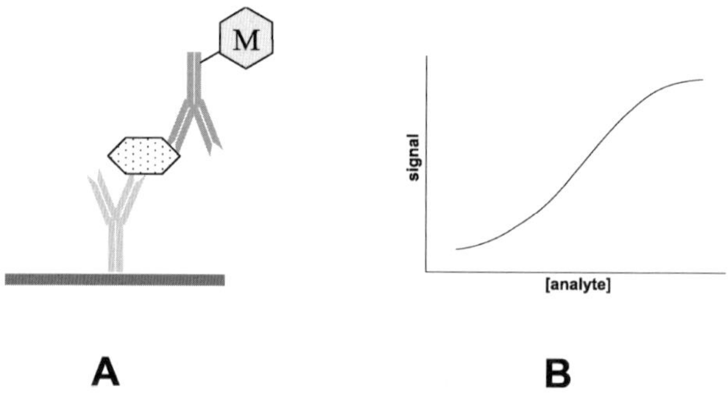

Fig. 6 A non-competitive assay typically has a sandwich format, i.e., the analyte is "sandwiched" between catcher and labelled detector antibody (**A**). The signal obtained from the label increases with the analyte concentration (**B**)

the sample contains no analyte, the labelled antigen can bind to all binding sites, leading to a maximum signal. The more analyte that is contained in the sample, the higher the proportion of analyte bound, i.e., the lower the concentration of bound antigen labels and, thus, signal intensity. For the determination of low-molecular analytes it is preferable to use an alternative setup where the antigen itself is immobilized on the plate/surface and competes with the analyte for the labelled antibody (Fig. 5C). Thus, antibodies that bind to the analyte are not fixed to the surface and will be washed away in a final washing step.

The sandwich type, where two antigen-specific antibodies form a sandwich with respect to the analyte, is the most common non-competitive assay format. In contrast to competitive assays, the assay signal (e.g., fluorescence) increases with increasing analyte concentration (Fig. 6).

Clearly, this sandwich format is only applicable if the analyte has two distinctly accessible epitopes, in other words, if the analyte is not too small. For this reason, haptens and other low-molecular weight molecules are normally not detected by sandwich immunoassays.

3.4
FRET-Based Immunoassays

Förster resonance energy transfer (FRET) is due to dipole–dipole interaction between an optically excited donor molecule and an acceptor in the electronic ground state. FRET is effective at distances comparable with many bound Ab–Ag pairs ($\sim$10–50 Å). To use this effect in immunoassays, which was first suggested by Ullman, Schwarzberg and Rubenstein in 1976, is therefore an obvious possibility [5]. The typical FRET immunoassay is of the competitive type with donor-labelled catcher antibodies (directed against the analyte) being bound to the solid phase, e.g., on an MTP. In a second step, acceptor-dye labelled analyte (or a biochemical analogue of the analyte) is incubated. FRET is enabled if the choice of donor and acceptor is such that the absorption spectrum of the acceptor and the normalized emission spectrum of the donor overlap sufficiently. Upon optical excitation, donor emission is small because the excited donors transfer their energy radiationless to the acceptors. Acceptor fluorescence, in turn, is enhanced. In a last step, the sample is added and the actual assay carried out. As the unlabelled analyte from the sample displaces the labelled one, the intensity of the acceptor emission decreases with increasing analyte content in the sample. Concomitantly, the donor emission increases up to the value of the unquenched emission. The difference in donor emission with and without sample added correlates with the antigen content in the sample. Cy5/Cy5.5 is a popular donor-acceptor pair because, in this case, FRET is effective up to 75 Å and can, therefore, not only be used for IgG antibodies but also for some larger Ab–Ag pairs [21].

A much higher sensitivity can be obtained if a long-lived energy donor is combined with a short-lived acceptor and a time-gated measurement of the delayed fluorescence of the acceptor is recorded. This was successfully demonstrated for donors like europium or terbium chelates and acceptors like tetramethylrhodamine, Alexa 546 and Cy5 [22, 23]. If the time gate is set off against the excitation light pulse, short-lived background emission originating from directly excited acceptor molecules, proteins, or impurities of the MTP can be suppressed efficiently.

3.5 Fluorescence Polarization Immunoassays

Competitive immunoassays of this type make use of fluorescence anisotropy measurements. Fluorescence anisotropy is a measure of deviation from a fast and completely free statistical rotation of all fluorophores. Therefore, any form of immobilizing fluorophore-conjugates, such as binding of a freely rotating dye-labelled antigen to their respective antibody, introduces anisotropy.

Thus, fluorophore-labelled antigen is added to a sample and excited by vertically polarized light and the emission is detected in 90° geometry both in parallel (p) and in vertical (v) orientation with respect to the polarization of the excitation beam. The anisotropy, r, can then be calculated from these emission intensities:

$$r = \frac{I_\mathrm{p} - I_\mathrm{v}}{I_\mathrm{p} + 2I_\mathrm{v}} . \tag{2}$$

Completely polarized (i.e., anisotropic) light results in $I_\mathrm{v} = 0$ and therefore in a maximum anisotropy of $r = 1$. In terms of immunoassays this means that all emitting conjugates are bound. If, by contrast, all fluorophores have a full degree of rotational freedom (unbound case), the time between photon absorption and emission (i.e., the fluorescence lifetime) is sufficient for some molecular rotation to occur. Thereby, the original polarization is reduced, $I_v > 0$ and thus $r < 1$.

The anisotropy of a mixture of free and bound species is given by their respective anisotropies (r_f and r_b) weighted with their relative fluorescence intensities (f_f and f_b) [24]:

$$r = r_f f_f + r_b f_b \, . \tag{3}$$

Anisotropy is particularly small if (a) the fluorescent conjugate is small in size, (b) solvent viscosity is low, and (c) the emission lifetime is long.

In a competitive fluorescence polarization assay, the catcher antibody is immobilized first. The assay mixture contains fluorophore-labelled antigen. The binding reaction between these two components will lead to a maximum of anisotropy. If free antigen from the sample is added, it will displace some of the labelled antigen from the binding sites of the antibodies. Therefore, the labelled antigen is free to rotate and the anisotropy decreases.

Fluorescence polarization immunoassays have been successfully performed for a wide range of analytes [25–27]. In many cases, the preferred labels are small organic molecules with emission lifetimes of a few nanoseconds (e.g., fluorescein). Thus, these assays are limited to the detection of low-molecular-weight analytes: only small analytes rotate fast enough for their emission to be depolarized (i.e., isotropic) in the unbound state. Analytes of high molecular weight (i.e., slow rotation) require much longer decay times of the label, typically on the order of >1 μs.

3.6 The Biotin-(Strept)avidin System

For some assays it is advantageous to include a much stronger, non-immunological binding in the setup. In such cases, one reaction partner is conjugated to the *N*-hydroxysuccinimide ester of biotin (NHS-biotin) or to PEG-ylated or otherwise lengthened biotin derivatives, e.g., succinimidyl-6-(biotinamido)hexanoate (NHS-LC-biotin). A number of proteins are capable of binding specifically to biotin: the strongest binding by far can be achieved by avidin (from chicken albumen) and streptavidin (from *Streptomyces avidinii*). These two proteins consist of four identical sub-units with a molecular weight of 15.6 and 14 kDa, respectively. One sub-unit binds to one biotin with the binding site of streptavidin being located deeper within the protein. Streptavidin-biotin complexes are of exceptional stability ($K_D \sim 10^{-15}$) and can not be separated by [11]:

- short-time heating to >100 °C;
- pH 2.5–13;
- 1% (w/v) of surfactant (e.g., SDS, Tween 20, Triton X-100);
- 8 M Guanidinium hydrochloride (pH ∼7).

Examples for the use of biotin-streptavidin binding in immunoassays are (a) the indirect immobilizing of a catcher antibody or antigen to a solid support, (b) the labelled avidin-biotin system (LAB), where a biotinylated antibody is allowed to incubate and bind with its immobilized target antigen first and then a dye or enzyme-labelled avidin conjugate is allowed to interact with the available biotin sites an the antibody and (c) the bridged avidin-biotin system (BAB), which uses avidin's multiple biotin binding sites to create an assay of higher sensitivity. In the latter case, a biotinylated antibody is again allowed to bind its target, but next an unmodified avidin is introduced to bind with the biotin binding sites. In a third step, a biotinylated enzyme or fluorescent marker is added to provide a means of detection. Due to the multiple biotin binding sites of the avidin, a higher signal intensity is enabled. One of the most sensitive setups is known as (d) the avidin-biotin complex (ABC). In this case, a sensitivity enhancement is achieved by forming a polymer of biotinylated enzyme or fluorophore with avidin first: if the biotinylated conjugate and avidin are mixed in the proper proportion, the multiple binding sites on avidin create a linking matrix that results in a high-molecular-weight complex. After this, an antigen-bound, biotinylated antibody, directed against the analyte, is added to the complex that binds to any remaining biotin-binding sites. This structure is then employed as a "detector-antibody" in an immunoassay and generates a marked signal enhancement over alternative methods [19].

3.7 Aspects of Measuring Fluorescence in Immunoassays

Fluorescence from MTP-based immunoassays is typically excited from the top and can be read either from the bottom (transparent MTP) or from the top (white or black MTP). In most cases, a buffer solution will be present in the plate cavities because several dyes (e.g., FITC) are not good emitters under dry conditions. Since, in most cases, the fluorescent label will be fixed to the Ab–Ag-recognition complex on the wall of the well, "bottom reading" is usually the preferred method. However, an optimum of sufficient signal and low background, the latter heavily depending on the type of MTP but also on the manufacturer, has to be found for every type of label. Two principal types of plate reader systems have to be distinguished: (a) readers based on interference filters, and (b) readers using excitation and emission monochromators, thus recording complete emission/excitation spectra. In practical applications, the much higher throughput and signal intensities of filter-based plate readers have to be weighed against the advantage of obtaining spectra. In commercial applications, e.g., in diagnostic laboratories, the former are clearly preferred. In addition to these systems, a few assay formats (e.g., some FRET-based assays) do not make use of the markers' stationary emission but of their emission decay. In most of these cases, such as assays using lanthanide chelates, time-gated emission is measured. Recording of time-resolved emission spectra and subsequent designation of emission lifetimes to two different analytes was reported for a qualitative antigen recognition reaction [28]. Quantitative assays based on fluorescence lifetimes and amplitudes are more difficult to establish due to changes in these parameters upon binding and weak signal intensities, which lead to extended measurement times.

Another question relates to the kind of microtiter plates used for fluorescence immunoassays. Four types of plates are common: transparent plates are intended for applications where a low background fluorescence is necessary and a "bottom-read" detection system is used. This is the case with most fluorescence immunoassays including those based on time-resolved emission. White plates, by contrast, give the highest possible reflection of the fluorescent signal combined with a background fluorescence that may still be moderate. These plates are suitable for most applications using "top-reading". Black plates are known to give the lowest possible background fluorescence of all polystyrene MTP while at the same time minimizing the backscatter light generated by the excitation beam. Again, these plates are usable for "top-reading" only. Recently, black polystyrene plates with a glass bottom have been commercialized. These are noted for their extremely low levels of background luminescence and are advantageous in "bottom-read" cases with short-wavelength excitation. Immobilizing proteins on the glass surfaces, however, can prove to be less efficient.

In ELISAs and other enzyme-based assays that rely on an absorption measurement of a reaction product between enzyme-label of the antibody and an added substrate, the use of transparent polystyrene plates is most common.

4
Marker Systems and Conjugation

In order to detect and quantify the surface-bound antibody-antigen complexes one part of this complex, most often the last antibody ("detector antibody"), needs to be labelled with a signal-generating marker. These markers or labels may be of different composition and structure, may be bound covalently or adsorptively, may either generate direct signals (e.g., optical absoption, fluorescence, chemiluminescence, delayed emission, radioactive radiation) or indirect signals (e.g., enzymatic generation of dyed or fluorescent products). The ideal marker has to meet the following requirements:

- it must me simple and, above all, enable sensitive detection,
- biological function not or only minimally affected, i.e., sensitivity and specificity of the antibody are not compromised,
- conjugation to the protein stable under assay conditions,

- long-term stability of the conjugate,
- well-targeted binding, e.g., binding to F_{ab} part of the antibody not dominant,
- simple removal of unbound marker possible,
- non-toxic.

Since the 1970s, immunoassays in average diagnostic laboratories have primarily been based on enzymatic reactions between an enzyme label, e.g., horseradish peroxidase (HRP) or alkaline phosphatase (AP), and a substrate, e.g., tetramethyl benzidine (TMB) or *p*-nitrophenyl phosphate, that is added after the affinity reaction has taken place. This reaction leads to the formation of a colored product that is usually detected by a single wavelength absorption measurement. By virtue of the cascading character of the enzymatic reaction, these enzyme-linked immunosorbent assays (ELISA) show very good signal-to-noise (S/N) values that are superior by several orders of magnitude to any non-cascading method using, e.g., static fluorescence (*vide infra*). This advantage also translates into very low limits of detection with respect to the analyte. However, several enzymes have been developed that lead to fluorescent products after reacting with appropriate substrate. Examples for such systems are resorufin phosphate (for alkaline phosphatase) and resorufin-β-D-galactopyranoside (for β-galactose). Systems of this kind are thoroughly discussed in another chapter.

With regard to practical applications in medical or veterinary diagnostics, any fluorescence-based marker system will have to be measured against the corresponding ELISA or EIA and a commercial breakthrough can only be expected if a clear advantage over the latter can be proved.

The first successful labels to be used both in fluorescence immunoassays (FIA) and in immunofluorescence were fluorescein and its derivatives (e.g., fluorescein isothiocyanate isomer I) which have been employed as markers since 1953. However, fluoresceins are far from being optimal markers because of their marked photobleaching and the strong pH dependence of their spectra, resulting from the existence of four differently protonated species. Though they are generally less sensitive to pH and more photostable, rhodamine dyes (e.g., tetramethylrhodamine R isomer), by contrast, are more difficult to use because of their planar hydrophobic structure resulting in limited water solubility, increased non-specific binding of the labelled protein species, and quenching of fluorescence as a consequence of dimerization of bound multiple labels [29].

The disadvantages of the xanthene dyes were overcome by more recent developments, such as the low-molecular-weight cyanine dyes, especially Cy3, Cy5, Cy5.5, and Cy7, which have been available since the early 1990s. Not only are these more photostable but by the introduction of negatively charged sulfonate groups directly to the ring system, a higher water solubility and less dye–dye interactions, which lead to emission quenching, were realized [29].

The above cyanine dyes can also be applied to microarray applications because they remain fluorescent in the dry state. Further developments include the Alexa fluorescent labels from Molecular Probes that emit between 442 nm and 775 nm, Fluka's Atto dyes, and many others (Fig. 7). All of these labels are characterized by a high degree of photostability, a high brightness ($\varepsilon \times \Phi_F$) and can be obtained with different functional groups allowing for different coupling protocols to be used.

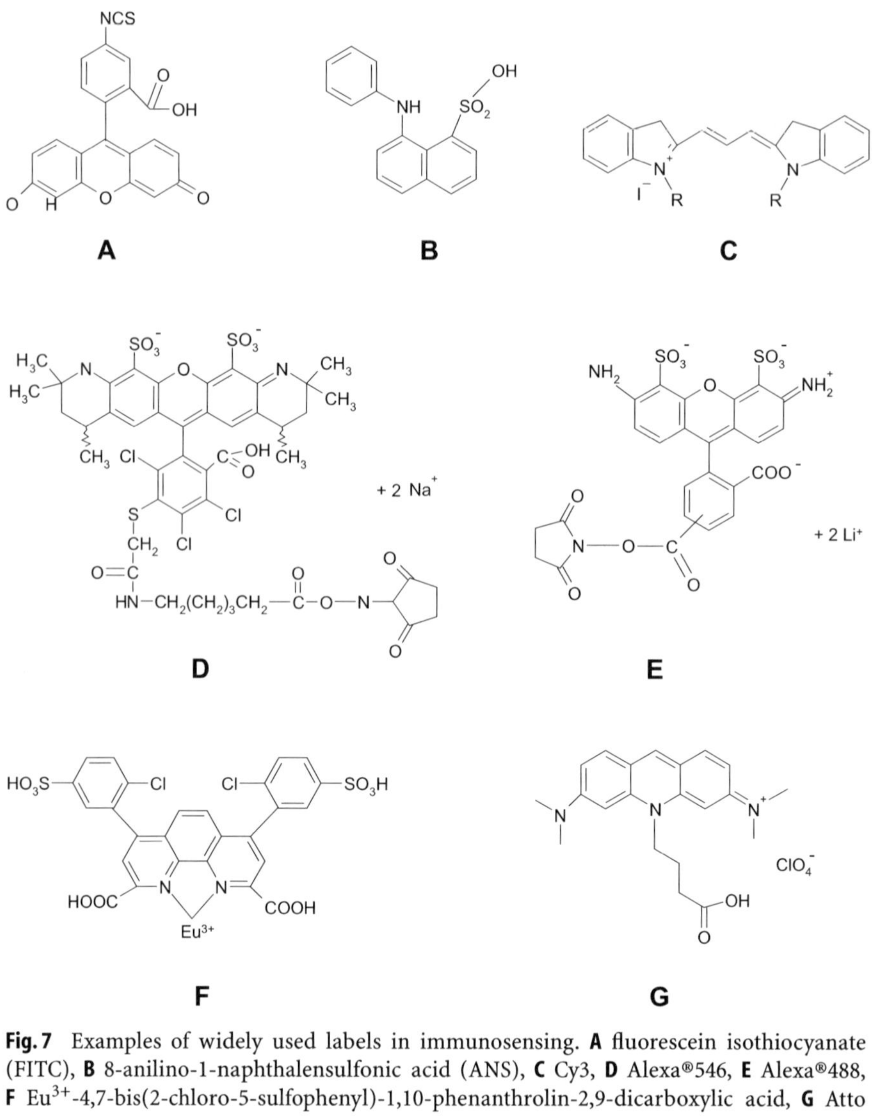

Fig. 7 Examples of widely used labels in immunosensing. **A** fluorescein isothiocyanate (FITC), **B** 8-anilino-1-naphthalensulfonic acid (ANS), **C** Cy3, **D** Alexa®546, **E** Alexa®488, **F** Eu^{3+}-4,7-bis(2-chloro-5-sulfophenyl)-1,10-phenanthrolin-2,9-dicarboxylic acid, **G** Atto 495

Generally, there is a noticeable trend towards the development and use of labelling compounds that emit at longer wavelengths. Although detector systems are usually less sensitive in the NIR than in the Vis region of the optical spectrum, there are two vital advantages in using long-wavelength emitters in immunoassays: (a) background fluorescence from the biomaterials, the buffer and plastic components (e.g., from the microtiter plate) will be considerably reduced and (b) whenever more parameters are to be measured simultaneously by differently emitting dyes, a better spectral separation is enabled with sets of dyes that also include NIR emitters. Unfortunately, infrared fluorescent organic dyes used to have the disadvantages of low Φ_F and reduced chemical and photostability. Recently, progress in the development of more stable low-molecular-weight NIR/IR emitting dyes was made [30–33] and some high-quality dyes and conjugates are now readily available, e.g., the IRDyes from LI-COR. Ongoing work in this field is not limited to molecular organic dyes but also includes particles, such as metal nanoshells [34].

Another trend to be monitored is directed towards enhancing the usually small emission signals in FIA by using extended multichromophoric labels. These can be either of natural origin, such as phycoerythrin, a phycobiliprotein that was first used in 1982 and which is obtained from photosynthetic bacteria but can nowadays also be produced synthetically [35, 36]. Well-known synthetic systems are often based on the incorporation of low-molecular-dye units in extended dextran structures [37, 38]. Recently, the use of a microcrystalline particle of a fluorescein precursor could be shown to produce outstanding results and signal intensities were increased by factors of up to several 10^4 as compared to FITC labelling [39, 40].

All of the aforementioned markers are predominantly used for measuring static fluorescence. A different approach to emission-based immunosensing has been taken by using metal-ligand complexes (MLC) of ruthenium or chelates and oxides of Eu(III) [24, 41, 42]. Here, forbidden transitions, such as $^5D_0 \rightarrow {}^7F_2$ in Eu_2O_3 lead to a slow emissive decay that can be quantified by gated measurements, i.e., within a certain time window a cumulative measurement is carried out [10]. This time window, in turn, is set off by a fixed delay time against the excitation pulse. The advantage of the offset is that both short-lived background emission and scattering light can be eliminated. Together with state-of-the-art detection equipment, this method allows for a very sensitive signal detection. Recent improvements were found to result in limits of detection as low as 40 fg/mL [43].

The original reactive groups on fluoresceins and rhodamines were isothiocyanates that react with free amino groups on the protein. While sulfonyl chlorides are also applicable for labelling amino groups, the labelling process is more difficult to control. Nowadays, the preferred reactive group for labelling amino groups on biological macromolecules are succinimidyl esters, facilitating a reaction that can be easily controlled and results in the formation of a peptide bond [19, 44].

In some diagnostic areas, plate-based laboratory analytics is being replaced by lateral-flow tests ("rapid tests"), which can be used on the spot without any instrumentation. This is particularly advantageous when time is a critical issue, e.g., in acute cardiac diagnostics, or when the end-user is meant to perform the test on his own, as in pregnancy testing. In these lateral-flow tests, a few drops of the fluid sample (whole blood, plasma, urine, saliva, etc.) are put on the conjugate pad that is part of a test strip. It contains colloidal gold conjugates (cgc) between 20–40 nm gold particles and specific antibodies directed against the analyte from the sample. The analyte binds to the conjugates and the resulting affinity complex is washed out of the conjugate pad into the nitrocellulose- or nylon-made test strip along which it then travels by diffusion. After a few centimeters, a line of catcher antibodies is crossed where the complex gets bound and is thereby stopped. The concentrated presence of gold particles leads to the formation of a visible line. Practically all lateral flow tests are based on colloidal gold or dyed polymer beads. Recently, however, fluorescent material was used instead and an improved LOD was reported [45].

Present and future developments in the field of fluorescent markers primarily focus on reliable and user-friendly ways of signal amplification, achievable either by extended multichromophoric systems, such as novel polymeric or dendritic systems, or by completely new concepts, such as the use of dissolvable organic nanoparticles [46]. A further goal will be the realization of multiplexed protein analysis which necessitates markers that show readily distinguishable emission properties. Semiconductor quantum dots have shown narrow emission bands, a large Stokes' shift, broad absorption and a high brightness and have therefore come into focus. So far, the use of quantum dots in immunoassays is rare [47] and their inherent intensity variation (photobrightening) may prove to be a problem in many quantitative assays. Alternative labels, e.g., fluorescent microspheres, for achieving a multiplexed protein detection have also been reported [48].

It should be noted that from the applicative point of view, multiplexed immunoassays will only be competitive with state-of-the-art ELISAs if they are faster, cheaper, and more reliable. Probably, this is the ultimate challenge.

5 Assay Validation

5.1 Validation Parameters and Guidelines

Validating immunoassays is a requirement of the European Directive 98/79 EC on in vitro diagnostic tests passed by the European Parliament and Council. Validation has the superior goal to describe how trustworthy the results from an assay are in terms of the underlying medical, medicinal, or other interpretation. It should also be understood that validation always refers to the intended use. This means, e.g., that IVDs for home use require additional validation steps, such as an evaluation of operability by lay persons.

Although there is no mandatory regulation yet on what parameters to include and what specific validation procedure to adopt, the ICH Harmonized Tripartite Guideline "Validation of Analytical Procedures: Text and Methodology Q2(R1)" (formerly known as Q2A and Q2B) together with the FDA's "Guidance for Industry – Bioanalytical Method Validation" present a readily applicable guide [49, 50]. The following definitions and recommendations for validation parameters and procedures have been agreed on and should be the basis for a validation protocol:

(a) Specificity or Selectivity

"Specificity" is the ability to assess unequivocally the analyte in the presence of components which may be expected to be present. Typically these might include impurities, degradants, matrix, etc. [49].

Similar to the IUPAC, the FDA further distinguishes between specificity and selectivity: "Specificity" is an evaluation of the response to a single analyte, whereas "selectivity" is the evaluation of a response to a group of analytes that may not be distinguished from each other [50, 51].

If typical impurities are known and available, pure analyte should be spiked with appropriate levels of the impurity. It should be shown that the assay results are unaffected by these impurities. If impurities cannot be used directly, the assay should be compared with a second, validated analytical procedure. The two results then need to be compared.

In many immunoassay applications, using highly heterogeneous samples (e.g., plasma, urine, saliva, etc.) these recommendations do not seem very practicable, though. In those cases it has become common practice to check sensitivity against matrix proteins, etc., by using analyte-free samples that can be obtained in some cases. Additionally, if the base level of the analyte to be detected is known (e.g., a certain protein in healthy subjects), sufficiently pooled sample material can also be useful.

(b) Accuracy or Trueness

The "accuracy" of an assay expresses the closeness of agreement between the value which is accepted either as a conventional true value or an accepted reference value and the value found. "Accuracy" is sometimes also termed "trueness". Accuracy should be assessed using a minimum of nine determinations over at least three concentration levels covering the specified range (e.g., three concentrations and three replicates each). Accuracy should be reported as percent recovery by the assay of known added amount of analyte in the sample. Alternatively, the difference between the mean and the accepted true value together with the confidence intervals may be reported [49].

Note: (i) Accuracy should not be confused with "precision". (ii) If the true value is controversial, international reference material should be used.

(c) Linearity

The "linearity" of an analytical procedure is its ability (within a given range) to obtain test results that are directly proportional to the concentration of analyte in a sample.

Linearity should be evaluated visually by a plot of signals (e.g., fluorescence intensity) as a function of at least five analyte concentrations. A regression line should be calculated, e.g., by the method of least squares. Values for the correlation coefficient, y-intercept, slope of the regression, and residual sum of squares should be included in the validation report. Since immunoassays do not always show linearity, their analytical response should then be described by an appropriate non-linear function [49].

(d) Precision

"Precision" expresses the closeness of agreement (degree of scatter) between a series of measurements obtained from multiple sampling of the same homogenous sample under the prescribed conditions. Precision may be considered at three levels: (i) "repeatability" (one laboratory/location, one operator, same day), (ii) "intermediate precision" (one laboratory/location, different days, analysts, equipment, production lots, etc.) and (iii) "reproducibility" (precision between two or more different laboratories/locations, usually applied to standardization of a methodology).

In the context of immunoassays, "repeatability" is identical with "intra-assay precision". It should be assessed using a minimum of nine determinations covering the specified range for the procedure (e.g., three concentrations/three replicates each). Alternatively, a minimum of six determinations at 100% of the test concentration is deemed to be sufficient.

"Intermediate precision" should be investigated to establish the effects of random events within one and the same laboratory/location on the precision of the analytical procedure. These effects need not be studied individually, though. Within the framework of immunoassays the investigation of "inter-assay precision" is much more common but less stringent. At any rate, the conditions should be reported as specific as possible.

"Reproducibility" can only be assessed by inter-laboratory trials or round robin tests.

All precision data should be reported by way of standard deviation, relative standard deviation (coefficient of variation), and confidence interval for each type of precision investigated [49].

The results of a precision analysis should also be reviewed for the presence of outliers. Since outliers cannot be defined arbitrarily, they should be assessed using acceptable methods such as the Henning test or Tukey's rule: For the Henning test, the mean $\langle x \rangle$ and standard deviation σ of a test series are both calculated without the conspicuous value x_1. If it is found that $3\sigma \geq |x_1 - \langle x \rangle|$ then x_1 is understood to be an outlier and may be removed from the analysis. Tukey's rule, by contrast, states that observed values lying at least 1.5 times the interquartile range (= the difference between first and third quartile, i.e., a box containing the middle 50% of data points) beyond one of the quartiles Q_1 and Q_3 can be considered outliers [52–54]. Alternative testing methods, e.g., Grubb's test, are also acceptable. The final decision whether outliers should be removed has to be met and statistically justified by the analyst.

(e) Range

The "range" of an immunoassay is the interval between the upper and lower concentration (including both values) for which it has been demonstrated that the analytical procedure has a suitable level of precision, accuracy, and linearity.

(f) Limit of Detection (LOD)

LOD is defined as the lowest amount of analyte in a sample that can be detected but not necessarily quantitated as an exact value.

Several approaches are possible for estimating LOD. A purely visual evaluation may be used for non-instrumental methods only, e.g., for lateral flow tests. Another elementary approach is based on determining the signal-to-noise (S/N) ratio from samples with known low concentrations of analyte (signal) and blank samples. An S/N of 3 : 1 or 2 : 1 is generally considered acceptable for determining LOD. A more refined method relies on the slope S of the calibration curve (= signal over analyte concentration) and a standard deviation σ of the response:

$$\text{LOD} = \frac{3.3\sigma}{S}\,, \tag{4}$$

σ may be calculated from measuring the signal from an appropriate number of blank samples. Alternatively, if the calibration curve is linear, a specific calibration curve for samples containing analyte in amounts close to the LOD can be used. Either the residual standard deviation of the regression line or the standard deviation of the y-intercepts of regression lines may be used for σ. For immunoassays, the slope method is usually more practical.

If not assessed visually, the estimate for LOD may subsequently be validated by the independent analysis of a suitable number of samples known to be near the LOD.

It should be noted that alternative methods of determining LOD are acceptable if properly described in the validation protocol [49].

(g) Limit of Quantitation (LOQ)

LOQ is the lowest amount of analyte in a sample that can be quantitatively determined with suitable precision and accuracy.

Several approaches are possible for estimating LOQ. A visual evaluation may be acceptable in some cases for non-instrumental methods (e.g., some semi-quantitative lateral flow tests). LOQ can also be determined from the signal-to-noise (S/N) ratio from samples with known low concentrations of analyte (signal) and blank samples. An S/N of 10 : 1 is considered to be acceptable for determining LOQ. Alternatively, the same method as for LOD (using the slope S of the calibration curve and a standard deviation σ of the response) can be used (*vide supra*). LOQ can then be estimated by Eq. 5:

$$\mathrm{LOQ} = \frac{10\sigma}{S}\,. \tag{5}$$

The limit of quantitation should be subsequently validated by the analysis of a suitable number of samples known to be near the quantitation limit.

(h) Robustness

The robustness of an assay is defined as the capacity of the assay to remain unaffected by small (but deliberate) variations in assay parameters and provides an indication of its reliability during normal use.

Estimation of assay robustness should be considered during the development phase. Typical parameters that may undergo small-scale variation in immunoassays are: concentration, temperature, pH, and stability of reference solutions.

It should be noted that for a clinical evaluation of quantitative immunoassays with samples from real patients and the possibility of alternative estimation of the presence/absence of a medical condition, so-called "Receiver-Operator-Characteristics" (ROC) play an important role. They allow an insight in the trade-off between sensitivity and specificity of an assay and are discussed in chapter 40 in more detail [47]. The Clinical and Laboratory Standards Institute (USA) has established guidelines (in document GP10) that provide further guidance on the use and utility of ROC curves [55].

5.2 Standards and References

Well-defined compounds of high quality with established high purity and stability are available for many analytes and are often adequate as standards.

Three types of such reference preparations are commonly used for standardization of immunoassay kits:

In order to calibrate a new method for biological analytes, an *"International Standard"* (IS) or an *"International Reference Preparation"* (IRP) must be used, if available. IS are collected, aliquoted and extensively tested for potency and stability by responsibility of the World Health Organization International Laboratory for Biological Standards. Lists of existing preparations, which are available in limited quantity for a nominal charge for calibration of national or laboratory standards or reference preparations, can be obtained from the WHO [56]. IRPs, by contrast, are preparations that do not meet the demanding criteria for an IS but are, nonetheless, useful for method-to-method standardization [57].

"Reference preparations" are not as extensively tested as IS or IRP. Potency and purity data are usually provided by the producer. Reference preparations are particularly useful for substances that are (a) unable to be completely characterized by chemical or physical means alone, (b) heterogeneous (e.g., glycoproteins), (c) difficult to isolate in pure form (e.g., synthetic proteins), (d) scarce or expensive (e.g., hormone preparations), (e) unstable or easily altered during isolation (e.g., human growth hormone), and (f) for those difficult or expensive to be assayed or characterized (e.g., prolactin). Reference preparations are distributed, e.g., via institutions such as the Division of Biological Standards, National Institute of Medical Research, London [57].

"In-house or 'Working' Reference Preparations" are those that may have been produced by the company or laboratory developing or performing the immunoassay. Often, these materials are acquired without reliable potency estimates but are calibrated by reference to an international standard [57].

"Working Standards" are, in many respects the most widely used form of standard. They are not strictly reference preparations but are used in every routine assay. Although there is no extensive testing and validation required, the producer/laboratory must assume responsibility for maintaining the appropriate quality [20].

Further guidance on reference measurement procedures and reference materials for samples of biological origin can be obtained from European standards EN 12286:1998 (together with EN 12286A1:2000) and EN 12287:1999.

It should be noted, however, that reference preparations may not exist for immunoassays developed for substances that have not been measured by this method before. Under these circumstances, the first publication often serves as a reference point for the assays developed subsequently [20].

If calibrators or control materials are part of an immunoassay (as is common practice), these materials have to be checked against independent reference materials as well. More detailed information on this aspect can be found in the international standard ISO 17511:2003 ("Metrological traceability of values assigned to calibrators and control materials").

5.3 Shelf-Life Evaluation

Shelf life or stability of a product may be understood as the time that essential performance characteristics are maintained under specific handling conditions. The Food and Drug Administration (FDA) as well as many manufacturers require that a product must recover at least 90% of the initial performance value throughout its life [58]. Government regulations are often more specific about the requirement for expiration dates on pharmaceutical materials. For example, § 211.166 of cGMP regulations published by the FDA stipulates a written testing program designed to assess the stability characteristics of all clinical products and lists five criteria. It refers, however, to drug products for administration to humans and animals only [59]. Nonetheless, it should be consulted and may serve as a broad guideline for stability investigations of immunoassays.

Without doubt, *"real-time stability tests"* are the "gold standard" in determining shelf life. For most applications, especially during assay development, they are not practical, though. For validating reference materials, by contrast, real-life stability testing is an essential requirement. In real-life testing, special attention should be paid to possible drifts or changes in the testing method/instrumentation over the testing period [20].

"Accelerated stability testing" is the method most often used in assay development because it is able to provide a fast indication of product shelf life, thereby shortening the product-development schedule. It is based on stressing the product (i.e., the assay or its components) at ≥4 elevated temperatures so that the amount of heat input required to cause product failure is determined. This information is then extrapolated to predict the product's shelf life. Very high temperatures (>55 °C), denaturing proteinaceous components, should be avoided [20, 58]. Product ageing, i.e., the chemical decomposition of assay components, can be described by a temperature-dependent rate constant k and an activation energy E_a. If a complete assay is tested, the rate constants can be understood as signal fall-off, e.g., decrease of emission, per stressing time with reference to an identically performed unstressed assay. For example, a 2% decrease over 100 days would mean $k = 0.02/100\ \text{day}^{-1}$. For two temperatures, T_1 and T_2, the corresponding rate constants, k_1 and k_2, and the activation energy are linked by the Arrhenius' equation:

$$\ln\left(\frac{k_2}{k_1}\right) = -\frac{E_a}{R}\left(\frac{1}{T_2} - \frac{1}{T_1}\right) . \qquad (6)$$

Plotting $\ln(k_2/k_1)$ against $(1/T_2 - 1/T_1)$ results in a straight line with a slope of $-E_a/R$, R being the gas constant. Since E_a is now known, rate constants for any other temperature can now be calculated by means of Eq. 6. Assuming a tolerable decrease to 90% of the original signal (assay) or concentration (assay component) together with a first-order kinetics, Anderson and Scott derived a simple formula for estimating the accepted shelf life from the rate constant k at a given temperature T [58]:

$$t_{90} = \frac{0.105}{k} \, . \tag{7}$$

A similar procedure for assay components is described in the European standard EN 13640:2002 ("Stability testing of in-vitro diagnostic reagents", Annex A) that is applicable to stability testing in Europe.

Several alternative and/or supplemental procedures are also common. If, for example, the shelf-life is known for several temperatures, a semilog plot of shelf life against storage temperature should result in a straight line that can then be extrapolated to the desired (i.e., lower) temperature. If only two rate constants referring to temperatures 10 °C apart are known, the Q rule can be applied, which states that a product degradation rate decreases by a constant factor (Q_{10}) when the storage temperature is lowered by 10 °C [20]. Thus:

$$Q_{10} = \frac{k_{\mathrm{T}+10\,^{\circ}\mathrm{C}}}{k_{\mathrm{T}}} \, . \tag{8}$$

Very common are also bracket tables. This technique assumes that for a given mixture of substances, the activation energies for degradation are between 10 and 20 kcal (42–84 kJ). This assumption is corroborated for proteins and enzymes [58, 60]. Table 1 provides an example for a bracket table for a projection of refrigerated shelf life [58]. Conservative use requires that projections be taken from the 47.5 °C values (and lower) since storage at 60 °C may lead to denaturation effects.

Table 1 Bracket table for the projection of refrigerated shelf life by means of accelerated ageing at elevated storage temperatures. Acceptable performance after the number of days of stress in the 20-kcal column predicts that it is "possible" that the selected claim will be observed while acceptable performance for the duration indicated in the 10-kcal column predicts that it is "probable" that the selected claim will be observed [58]

Storage temperature [°C]	Days of stress to predict stability at 5 °C for the time of: 6 months		1 year		2 years		3 years	
	20 kcal	10 kcal	20 kcal	10 kcal	20 kcal	10 kcal	20 kcal	10 kcal
14.5	55.3	100	111	201	221	402	332	603
25	16.1	54	32	108	64	217	97	326
35.5	5.1	30.6	10	61	20	122	31	183
47.5	1.5	16.6	3	32	6	66	9	100
60	0.5	9.2	0.9	18	1.9	37	2.8	55

From empirical experience, Deshpande [20] quotes the following guidelines for approximating the shelf life at a desired temperature:

1 month at 50 °C is equivalent to 1 year at room temperature,
2 months at 50 °C is equivalent to 2 years at room temperature,
3 month at 37 °C is equivalent to 1 year at room temperature,
6 months at 37 °C is equivalent to 2 years at room temperature,
7 days at 37 °C is equivalent to 1 year at 4 °C,
3 days at 37 °C is equivalent to 3–6 months at 4 °C.

Clearly, these values are not too stringent with the latter two roughly corresponding to those from the 20 kcal activation energy from the bracket table (Table 1).

6 Conclusions and Outlook

The field of fluorescence immunoassays has recently drawn much attention and many new concepts will be discussed in the following chapters. Established methods (e.g., organic marker synthesis, FRET, nanoparticles, etc.) as well as new effects from physical chemistry (e.g., surface-enhanced fluorescence and surface-enhanced Raman scattering) have the potential to be incorporated into new concepts of labelling and detection methods. The primary goal of all new approaches to emission-based assays is obviously to get an edge over traditional ELISA systems in terms of detection limit, practicability, specificity, price and/or speed. Without doubt, this is a considerable challenge, especially as far as signal intensities and the price for detection technology are concerned.

In terms of regulatory and validation aspects, many requirements originally intended for pharmaceutical products have been adopted, at least in the form of guidelines, for the development of immunoassays. It may be anticipated that this trend will continue and, sooner or later, guidelines will be turned into more specific requirements.

Disclaimer. It should be noted that company and brand names mentioned in this chapter are not to be understood as recommendations but as examples only. A superior performance of these products should not automatically be concluded.

Acknowledgements I wish to thank the Bundesministerium für Bildung und Forschung (BMBF) for the financial support for some of our research work (FKZ: 13N8852), the Federal Institute for Materials Research and Testing (BAM) for the fruitful collaboration and scientific exchange and, last but not least, my company for giving me the freedom to publish.

References

1. Niemeyer CM, Adler M, Wacker R (2005) Trends Biotechnol 23:208
2. Tzanavaras PD, Verdoukas A, Themelis DG (2005) Anal Sci 21:1515
3. Templin MF, Stoll D, Schrenk M, Traub PC, Vöhringer CF, Joos TO (2002) Trends Biotechnol 20:160
4. Homola J (2003) Anal Bioanal Chem 377:528
5. Ullman EF, Schwarzberg M, Rubenstein KE (1976) J Biol Chem 251:4172
6. Diamandis EP, Christopoulus TK (2007) Immunoassay. Academic Press, San Diego
7. Lee BK, Richards FM (1971) J Mol Biol 55:379
8. Mach H, Middaugh CR, Lewis R (1992) Anal Biochem 200:74
9. Gill SC, Von Hippel PH (1989) Anal Biochem 182:319
10. Feng J, Shan G, Maquieira A, Koivunen ME, Guo B, Hammock BD, Kennedy EM (2003) Anal Chem 75:5282
11. Holtzhauer M (1996) Immunchemie. In: Holtzhauer M (ed) Methoden in der Proteinanalytik. Springer, Berlin Heidelberg New York, pp 379–425 (in German)
12. Holtzhauer M (2006) Basic Methods for the Biochemical Lab, Chaps. 1.1 and 9.2. 1st English edition. Springer, Berlin Heidelberg New York
13. Janin J, Chothia C (1990) J Biol Chem 265(27):16027
14. Rees DC, Lipscomb WN (1982) J Mol Biol 160:475
15. Janin J, Chothia C (1976) J Mol Biol 100:197
16. Ardelt W, Laskowski M (1985) Biochemistry 24:2274
17. Friguet B, Djavadi-Ohiniance L, Goldbeg ME (1989) Res Immunol 140:355
18. Gruber R (2007) Antikörper. In: Raem AM, Rauch P (eds) Immunoassays, 1st edn. Elsevier Publishers, Munich, pp 19–50 (in German)
19. Hermanson GT (1996) Bioconjugate Techniques. Academic Press, San Diego
20. Deshpande SS (2003) Enzyme Immunoassays: From Concept to Product Development. Springer, Berlin Heidelberg New York
21. Schobel U, Egelhaaf H-J, Brecht A, Oelkrug D, Gauglitz G (1999) Bioconjugate Chem 10:1107
22. Qin Q-P, Peltola O, Pettersson K (2003) Clin Chem 49:1105
23. Blomberg K, Hurskainen P, Hemmilä I (1999) Clin Chem 45:855
24. Lakowicz JR (ed) (2006) Principles of Fluorescence Spectroscopy. 3rd edn. Springer, Berlin Heidelberg New York
25. Winkler M, Schumann G, Petersen D, Oellerich M, Wonigeit K (1992) Clin Chem 38:123
26. De Kanel J, Dunlap L, Hall TD (1989) Clin Chem 35:2110
27. Turek TC, Small EC, Bryant RW, Hill WAG (2001) Anal Biochem 299:45
28. Galla K, Arden-Jacob J, Deltau G, Drexhage KH, Martin M, Sauer M, Wolfrum J, Seeger S (1994) J Fluorescence 4:111
29. Mujumdar RB, Ernst LA, Mujumdar SR, Lewis CJ, Waggoner AS (1993) Bioconjugate Chem 4:105
30. Sowell J, Strekowski L, Patonay G (2002) J Biomedical Optics 7:571
31. Zhu L, Stryjewski W, Lassiter S, Soper SA (2003) Anal Chem 75:2280
32. Oswald B, Gruber M, Böhmer M, Lehmann F, Probst M, Wolfbeis OS (2001) Photochem Photobiol 74:237
33. Zhu L, Stryjewski WJ, Soper SA (2004) Anal Biochem 330:206
34. Hirsch LR, Jackson JB, Lee A, Halas NJ, West JL (2003) Anal Chem 75:2377
35. Oi VT, Glazer AN, Stryer L (1982) J Cell Biol 93:981
36. Haugland RP (2005) The Handbook: A Guide to Fluorescent Probes and Labeling Technologies, 10th edn. Invitrogen Corp, USA

37. Water-soluble, polymer-based reagents and conjugates comprising moieties derived from divinyl sulfone, US Patent 5 543 332, European patents PCT/DK92/00206 and WO 93/01498
38. A method for the preparation of water-soluble cross-linked conjugates, European patents PCT/DK99/00426 and WO 00/07019
39. Brümmel Y, Chan CP-Y, Renneberg R, Thünemann A, Seydack M (2004) Langmuir 20:9371
40. Chan CP-Y, Brümmel Y, Seydack M, Sin K-K, Wong L-W, Merisko-Liversidge E, Trau D, Renneberg R (2004) Anal Chem 76:3638
41. Diamandis EP, Christopoulos TK (1990) Anal Chem 62:1149A
42. Hemmila I, Laitala V (2005) J Fluoresc 15:529
43. Matsuya T, Tashiro S, Hoshino N, Shibata N, Nagasaki Y, Kataoka K (2003) Anal Chem 75:6124
44. Waggoner A (2006) Curr Opin Chem Biol 10:62
45. Bonenberger J, Doumanas M (2006) IVD Technology, 6:41
46. Seydack M (2005) Biosens Bioelectron 20:2454
47. Goldman ER, Mednitz IL, Mattoussi H (2006) Anal Bioanal Chem 384:560
48. Biagini RE, Sammons DL, Smith JP, MacKenzie BA, Striley CAF, Robertson SA, Snawder JE, Quinn CP (2005) Anal Bioanal Chem 382:1027
49. The European Agency for the Evaluation of Medicinal Products. The International Conference on Harmonization of Technical Requirements for Registration of Pharmaceuticals for Human Use (2005) Validation of Analytical Procedures – Text and Methodology Q2(R1). Downloadable from: http://www.ich.org/LOB/media/MEDIA417.pdf. Last visited: 1 November 2007
50. The Food and Drug Administration (FDA), Center for Drug Evaluation and Research (2001) Guidance for Industry – Bioanalytical Method Validation. Downloadable from: http://www.fda.gov/CDER/GUIDANCE/4252fnl.htm. Last visited: 1 November 2007
51. The Clinical and Laboratory Standards Institute (2004) Assessment of the Clincial Accuracy of Laboratory Tests Using Receiver Operating Characteristic (ROC) Plots. NCCLS document GP10. Wayne PA Downloadable from: http://www.nccls.org. Last visited: 1 November 2007
52. Derzko AN (2005) IVD Technology 3:45
53. Polifke T, Rauch P (2007) Auswertung und Validierung. In: Raem AM, Rauch P (eds) Immunoassays. Elsevier Publishers, Munich, (in German, Chap. 8)
54. Tukey JW (1977) Exploratory Data Analysis. Addison-Wesley, Reading, MA
55. The Food and Drug Administration (FDA), Center for Devices and Radiological Health (2003) Statistical Guide on Reporting Results from Studies Evaluating Diagnostics Tests; Draft Guidance for Industry and FDA Reviewers. Rockville, MD. Downloadable from: http://www.fda.gov/cdrh/osb/guidance/1428.html. Last visited: 1 November 2007
56. World Health Organization (2006) WHO International Biological Reference Preparations. Downloadable from: http://www.who.int. Last visited: 1 November 2007
57. Bangham DR (1988) Reference preparations and matrix effects. In: Collins WP (ed) Complementary Immunoassays. Wiley, New York, pp 13–25
58. Anderson G, Scott M (1991) Clin Chem 37:398
59. 21 Code of Federal Regulations Parts 210 and 211, Part 211, Subpart I – Laboratory Controls, § 211.166. Downloadable from: http://www.fda.gov/cder/dmpq/cgmpregs.htm#211.166. Last visited: 1 November 2007
60. Porterfield RI, Capone JJ (1984) Med Device Diagnostic Ind 4:45
61. Raem AM (2007) Einführung in Immunoassays. In: Raem AM, Rauch P (eds) Immunoassys. Elsevier Publishers, Munich (in German, Chap. 1)

Springer Ser Fluoresc (2008) 6: 429–447
DOI 10.1007/4243_2008_038

Published online: 14 March 2008

Time-Resolved Fluorometric Immunoassays; Instrumentation, Applications, Unresolved Issues and Future Trends

Ilkka Hemmilä

PerkinElmer Life and Analytical Sciences, Wallac Oy, PO Box 10, FIN 20101 Turku, Finland
Ilkka.hemmila@perkinelmer.com

Abstract Lanthanide, due to their unique luminescent properties, has provided a new tool to make sensitive and robust assay technologies. The long decay time in millisecond time domain allows very easy and efficient temporal gating to get clean and sensitive readings. The various chelate labels and assay technologies developed are reviewed here. Standardization of immunological assays is a major problem in diagnostics, and is addressed in time-resolved immunoassays with ability to create clean signals and standardized readings. The most recent applications of the technologies in preventive screening are given.

Keywords Fluorescence immunoassays · Lanthanide chelates · Population screening · Standardisation · Time-resolved fluorometry

1 Immunoassays as Analytical Methods in Diagnostics

Immunoassays are analytical methods in which antibodies are employed as specific catching reagents to allow highly specific recognition and detection of biologically relevant analytes in samples. Since the introduction of diagnostic assays based on antibodies in the 1960s, immunoassays have gained a well-established role in diagnostics to measure a wide variety of clinically relevant compounds.

Immunoassays cannot always be regarded as quantitative analytical methods. Sometimes, for example, in serological assays, just the overall immuno-response against an antigen is measured; this is the avidity, or the sum of all antibodies binding under the conditions used. Nevertheless, using suitable controls and standards, immunoassays are today seen as accurate and reliable methods employed in diagnosis, monitoring, and screening.

Traditionally, immunoassay technologies are divided into different categories based on the technology used or assay steps required. Heterogeneous assays require physical separation of bound and unbound antibodies before label detection, whereas in homogeneous (or non-separation) assays, specific interactions can be followed without a physical separation step. Based on the target, immunological technologies can be divided into immunofluorescence (IF) assays (e.g., identification of cell type by IF) and into in vitro quantitative immunoassays. Further on, based on the analyte in question, and availability of specific antibodies, immunoassays can be divided into competitive assays (used for small haptenic antigens having only one available epitopic site) and non-competitive immunometric assays (generally applied to larger antigens with several epitopic sites).

The introduction of monoclonal antibodies in the 1970s by Köhler and Milstein [1] was an important milestone in the development of immunodiagnostics. Monoclonal antibodies provide an unlimited source of binding reagents with a constant quality, thus greatly improving assay quality and consistency.

The label allows an accurate and sensitive detection of the complexes, occupied sites in non-competitive assays or unoccupied sites in competitive assays. Radioisotopes were the labels of choice in original immunoassays—and are still in use despite the development of non-radioisotopic alternatives. In routine clinical diagnostics, the development has resulted in non-isotopic alternatives, such as enzymes measured either using photometric, luminometric or fluorometric substrates, or luminescent compounds, fluorescent compounds, or lanthanide chelates with time-resolved (TR) fluorometric detection [2, 3].

Temporal resolution brings another dimension to the fluorometric analysis, i.e., time, and potentially improves the assays in terms of specificity and sensitivity. When combined with long decay-time probes composed of lanthanide chelates, time resolution provides a simple way of eliminating practically all background-related fluorescence noise by simple gating [2, 3].

2
Time-Resolved Fluorometry

Fluorometry is a very versatile technology that can provide much information on the sample, and can, under favorable conditions, reach the absolute sensitivity of a single-molecule detection. In practical use, however, when analyzing impure biological samples, fluorometry is seriously limited by the problem of background. The background noise variously associated with detection, optics, scattering, and sample autofluorescence greatly reduces the practical sensitivity of fluorometric analysis [2].

Time-resolved fluorometry, which may be achieved by simple detector gating, can be a very efficient tool for eliminating all background-related interferences, provided that the specific signal has a decay time that substantially differs from the average background noise [2]. Figure 1 shows an example of the time-gated detection of a typical europium chelate, demonstrating the complete discrimination between a specific signal and background both in terms of spectral filtering and temporal filtering. In diagnostics applications, time-resolved fluorometric methods became a practical solution only after the development of long-decay time labels based on lanthanide chelates [3].

In the field of diagnostics, time resolution in the millisecond time domain has enabled background-free detection and facilitated achieving the sensitivity range required by demanding applications. In addition, the combination

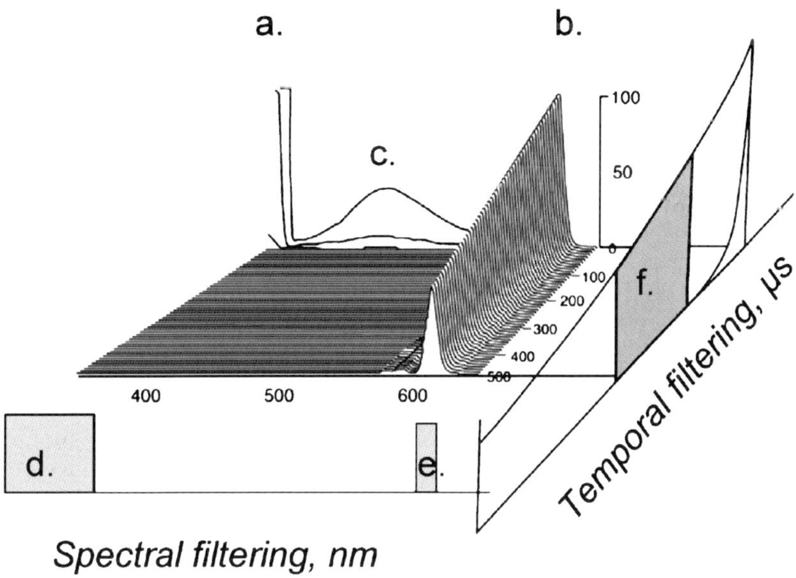

Fig. 1 Emission-decay profile of a europium β-diketone chelate. Pulsed excitation at 340 nm produces scattering (*a*), Eu emission (*b*) and autofluorescence (*c*). Time-resolved fluorometry applies excitation filter (*d*), emission filter (*e*), and gating window (*f*)

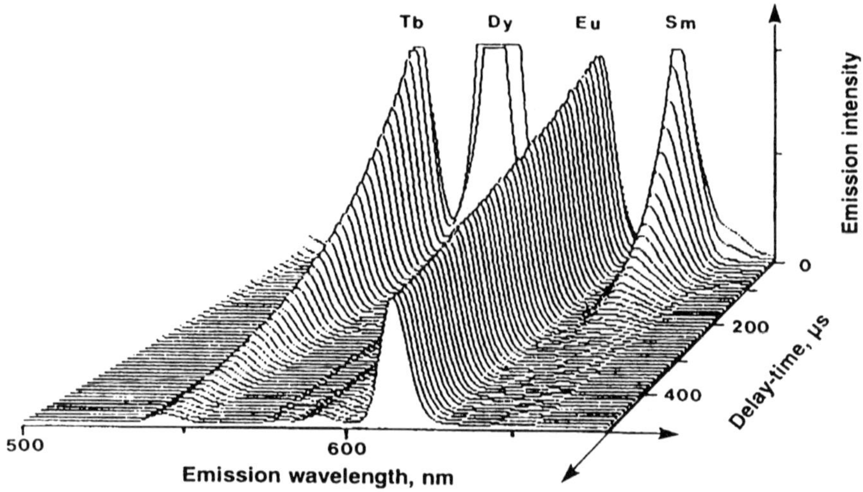

Fig. 2 Emission-decay profile of a mixture of Eu, Sm, Tb, and Dy as pivaloyltrifluoroacetone chelates, excited at 310 nm, showing their typical major emission lines and decays ranging from a few microseconds with Dy to about 700 microseconds with Eu

of spectral and temporal resolution can be used to achieve more specific detection and even multi-label detection. Figure 2 gives an example of a mixture of four different lanthanides, each of which has its own characteristic emission profile and decay time. Sensitive and specific detection of each one is achievable by a combination of spectral filtering and temporal filtering. As technological possibilities for acquiring accurate temporal data become more accessible, it can be postulated that time-gated detection will also provide a tool to allow the use of the nanosecond time domain in assays. It has already been shown in fluorescence life-time imaging (FLIM) that the decay time can be used as a simple parameter not affected by regular interfering factors.

2.1 Time-Resolving Fluorometers

Detection of long-lasting emission, such as phosphorescence, was the area in which simple, millisecond time gating instruments were first used; the time gating was accomplished initially by using rotating choppers [4]. Technologies to acquire temporal data on photoluminescence analysis have subsequently ranged from time-correlated single-photon counting or phase resolution by different frequencies to electronically or mechanically gated detection [5–8]. Millisecond time-domain fluorometry is practically the only gating technology applied in current diagnostic applications. In cellular and histological applications, there is a desire to follow intracellular biochemical reactions in situ, and fluorescence lifetime imaging (FLIM) is the method used in these applications [7]. FLIM is applied in complex cellular assays by measuring energy transfer for example between different genetic variants of green fluorescent proteins (GFP).

Instruments functioning in a gating mode in the millisecond time domain generally employ pulsed excitation sources, and electronically gated detection is accomplished with photomultiplier tubes. A xenon flash lamp is the most commonly used pulsed excitation source [9, 10] and is today applied in various diagnostic readers manufactured by Wallac (PerkinElmer Life and Analytical Sciences, Turku, Finland), such as the 1234 DELFIA® fluorometer, Victor™ multilabel reader and the automated analyzers DELFIA Xpress and AutoDELFIA®. This configuration is also commonly adopted in various microtitration plate readers supplied by other vendors.

A nitrogen laser emitting at 337.1 nm is a powerful source of short-duration excitation, suitable for lanthanide chelate fluorometry [11]. In addition to the Kryptor® fluorometer used for the Trace technology by Brahms Diagnostica GmbH, research plate counters for example from BMG Labtech Ltd are equipped with pulsed nitrogen lasers.

In time-resolved fluorometric imaging instruments, gated detection is generally accomplished either mechanically [12] by a liquid crystal shutter [13] or by electronically gated intensified CCD [14]. These technologies are not, however, in routine diagnostic use.

2.2
Standardization of TR Reading

The standardization of fluorometric analysis has long remained a challenge. Since the introduction of quinine sulfate, traditionally used as a fluorometric standard, various other solid standards, more easily applied, have been made available to test instrument performance in terms of reproducibility and stability. The National Institute of Standards and Technology (NIST) also provides a fluorescein standard solution as a more appropriate standard solution. As opposed to regular fluorometric analysis where the analog signal is highly dependent upon the excitation power and gain setting, and is routinely expressed as relative fluorescence units, time-resolved readers generally function in a photon-counting mode and the response is given as counts, enableing better standardization. In a pulsed mode, samples are generally excited with a preset excitation energy by controlling the exact number of pulses with a feedback control mechanism. With 1-kHz xenon flash lamp excitation, the calculated imprecision is around 0.1% when using 1-s reading without further calibration [10].

Furthermore, in the commonly applied DELFIA technology, the final reading is reduced to the quantitation of the total number of lanthanide ions left in the assay well, and the instrument and assays are calibrated using international lanthanide standard preparations (Eu, Sm, Tb, and Dy acidic solution made for atomic absorption calibration) after appropriate dilution into a fluorescence enhancement solution.

3
Lanthanide Chelates and Assay Technologies

The first (and still widely applied TR fluorometric assay system, DELFIA) is based on dissociative fluorescence enhancement [15]. The system applies stable, non-fluorescent polyaminopolycarboxylate-based labelling reagents, originally a derivative of diethylenetriamine tetraacetic acid (DTTA) [16], as carriers of lanthanide ions in the assay, and an acid dissociation and ligand exchange with β-diketone taking place in the enhancement solution. Currently, alternative labels have been introduced, e.g., a phenyl-pyridine iminodiacetic acid derivative for oligo- and polynucleotide applications [17], a simple nonderivatized DTPA chelate for cell labelling [18], a neutral derivative of DTTA for applications where a negative charge may cause a problem [19], and derivatized DTPA to provide a kinetically and thermodynamically more stable label.

In addition to the chelates employed in DELFIA, an increasing number of fluorescent stable chelates have been developed, published and commercialized for different applications, mainly for homogeneous high-throughput screening (HTS) assays applied in drug discovery. Examples of such structures include the macrocyclic cryptates [20] commercially applied by CisBio International in the HTRF screening system, and Brahms Diagnostica as the Trace method in the Kryptor fluorometer. Fluorescent terbium chelates based on DTPA analogue attached with a chromogenic antenna moiety [21], are used for example in the LanthaScreen™ HTS system by Invitrogen company. Several types of Eu chelates of terpyridine derivatives have been synthesized and some of them also commercialized by Amersham GE and by PerkinElmer [22–24]. The rapidly growing number of structures and publications is outside the scope of the current review. A comprehensive review of the different structure classes [8] and commercial drug discovery applications [25] can be found in recent publications.

In addition to the stable chelates, nano to micrometer-sized latex particles impregnated with fluorescent lanthanide chelates have attracted a lot of attention [26] recently. In polymeric form, lanthanides benefit from their large Stoke's shift, which makes them much less vulnerable to inner-filter quenching. The advantage achieved with the impregnated bead is also related to keeping the chelates stable and avoiding aqueous quenching, hence enabling the use of lanthanides more prone to aqueous quenching, such as ytterbium, neodymium, or erbium.

Nanoparticles impregnated with Eu, Tb, Sm, and Dy and their applications in immunoassays have recently been described [27]. Eu nanobeads are also commercially available from Seradyn and others. Highly sensitive immunofluorometric assays based on Eu nanoparticles have been reported for prostate-specific antigen (PSA) [26] and adenovirus [28]. Cummins et al. [29] used an indium tin oxide (ITO) (quartz) waveguide as a new platform for immunosensors with fluorescent Eu nanobead labels (Seradyn) in a competitive atrazine immunoassay.

3.1 Assay Technologies

The various chelate technologies and respective immunoassay technologies developed for time-resolved fluorometry have recently been extensively reviewed [8]. As described above, with the DELFIA methodology, the lanthanide ions are carried through the binding stages of the assay by non-fluorescent complexes and the signal is generated following an acidic ligand-exchange reaction that takes place at the end of the assay. Additionally, lanthanides and their chelates are applied in assay types that are essentially the same as those employing regular fluorochromes, including homogeneous assays based on energy transfer (TR-FRET), on signal modulation or on external or internal quenching [30].

The advantage of DELFIA relates partly to highly hydrophilic, small, charged labelling reagents allowing even high labelling ratios to be used. On the other hand, the separate enhancement allows the use of the most highly fluorescent chelates in the detection, and the fluorescent chelates are in homogeneous solution thus avoiding surface quality problems. The disadvantage relates to the extra step required making the technology strictly heterogeneous requiring efficient washing steps.

Stable fluorescent chelates allow enhancement-free technologies where the signal is measured directly from the applied solid surface. This technology has the main advantage in simplicity, and it also allows direct localization. The chelates applied so far have lower intensities, and the aromatic structures required for light collection tend to cause somewhat more hydrophobic interactions and non-specific background. The signal intensity has been overcome by using nano-particles creating highly sensitive alternative, bringing along, however, the practical consequences of handling large particles as labels.

Homogeneous assays, in particular, benefit from temporal discrimination because all the assay and sample components are present during the measurement and potentially interfere with the assays. Lanthanides bring two specific advantages into FRET: firstly the energy transfer from the long-decay donor to a short-decay acceptor creates a medium-long decay emission, enabling its complete discrimination from directly excited acceptor emission; secondly, lanthanides have a very characteristic line-type emission profile that provides several low background spectral areas for energy-transfer measurements from the visible to the near-infrared area. The third potential advantage of lanthanides is related to their different energy-transfer processes even enabling energy transfer from the non-emittive lanthanide energy levels [30–32].

TR-FRET in immunoassays was first described by Mathis based on the use of Eu chelates (cryptate) and allophycocyanin (APC) [33]. The system has since been applied in drug-discovery applications by CisBio International, and in diagnostics by Brahms Diagnostica. Other donor-acceptor pairs applied include an Eu chelate and cyanine-5 [34], applied for urinary albumin [35], Eu chelate and Alexa 647 [36], Tb-chelate and Alexa 546 [37], and Tb-chelate with various other acceptors [25].

The advantage of homogeneous assay technologies, such as TR-FRET, relate to the easy and manipulation-free process, rapid solution phase reaction and its suitability to miniaturization. The limitation of FRET relates to the critical energy transfer distance, R_0, which is generally below 100 Å and thus makes assay development cumbersome, particularly for large antigens. As exemplified in Fig. 3, sandwich assay of a large antigen requires careful selection of antibody epitopic sites to create a signal. In addition, the

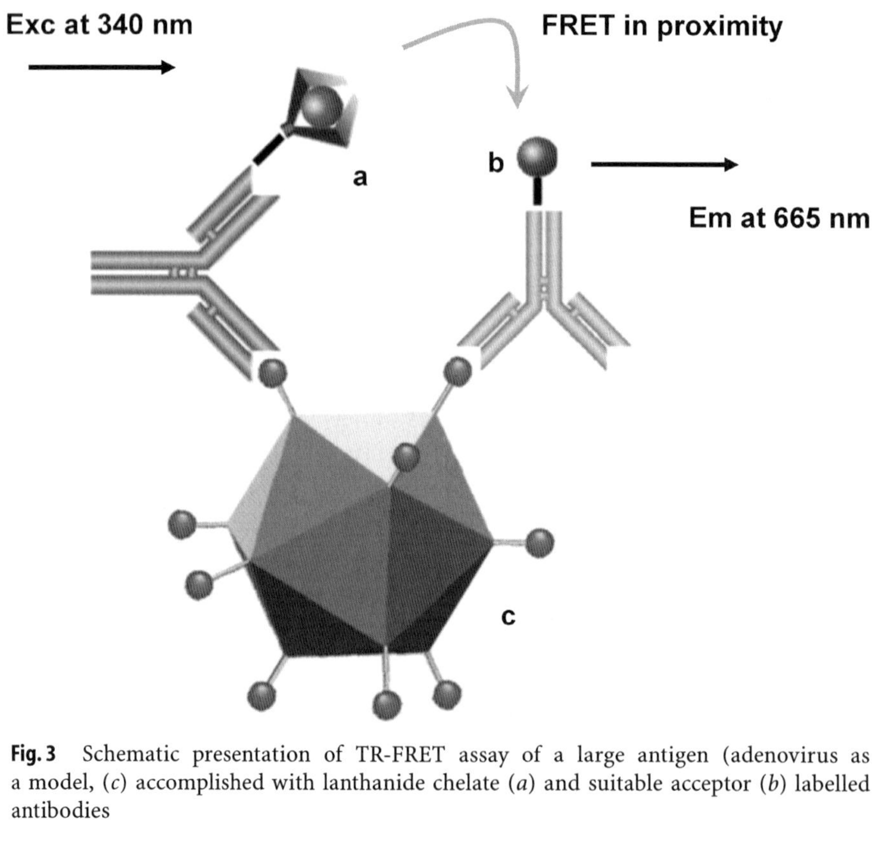

Fig. 3 Schematic presentation of TR-FRET assay of a large antigen (adenovirus as a model, (*c*) accomplished with lanthanide chelate (*a*) and suitable acceptor (*b*) labelled antibodies

antibody-carrying donor chelate must be pure and have high affinity because the chelates create tiny but detectable background interfering energy-transfer measurement.

4 Problems of Standardization in Clinical Immunoassays

The standardization of clinical assays, and immunological assays in particular, has been a major concern of clinicians and authorities for decades. The problems associated with the standardization of diagnostic immunoassays, especially those of protein antigens, are reviewed in several recent articles [38–40]. To a great extent, the standardization of TR-IFMA assays is associated with the same concerns and difficulties as other immunoassays based on the same principle of antibody recognition.

Unlike bioassays, which measure the biological function of the compound in a suitable host system, immunoassays are structural assays that measure the occupancy of the epitopic sites of respective antigens. Accordingly, immunoassays may measure a biologically irrelevant fraction, form, or metabolite of the antigen. The difference between a structural assay and a bioassay is exemplified for example in a recent TR-IFMA of growth hormone (GH) [41].

The standardization of even relatively simple analytes, such as thyroxin (T4), free thyroxin (fT4) and free triiodothyronine (fT3), is a challenge [42]. Free hormone analysis, in particular, creates a difficult analytical problem being an area in which cumbersome dialysis tests are still regarded as the golden standard.

Similarly, in addition to the problems caused by typical cross reactivities with related steroids, the analysis of testosterone involves an analytical problem relating to the actual active form. In particular, this applies to the cross reactivity with dihydrotestosterone, as its biological activity is higher than that of testosterone. Testosterone exists in samples in free form (about 2%), as an albumin-bound fraction (regarded as a bioavailable fraction) and complexed with sex-hormone binding globulin (SHBG). Giton et al. [43] analyzed the bioavailability of testosterone by measuring it and SHBG by AutoDELFIA. According to their conclusion, information of association constants for each patient would be needed to obtain more accurate estimation of hormone bioavailability.

Protein assays have to take into account all the isoforms (post-translational modifications) which may alter within individuals, or depending on age and patient condition. Selva et al. used TR-IFMA to study SHBG isoforms [44]. Like steroid hormones, proteins may also be present in blood circulation complexed to carrier proteins, and the identification of the bioactive, or bioavailable forms may be difficult. In addition, the equilibrium between the forms may be rather complicated, and any environmental factors, such as dilution, assay buffer, ionic strength, detergents, and other manipulations can have a profound effect on the balance.

Prostate-specific antigen, PSA, is an example of a protein present in serum in a free active format, and also complexed to both α_2-macroglobulin and α_1-antichymotrypsin. The measured values generally relate either to the free format, regarded as the most relevant form in the diagnosis of prostate cancer, or the total PSA, which also includes the α_1-antichymotrypsin complex. In addition, the free PSA can consist of pro PSA, bioactive PSA, and non-activated PSA, and the storage time and temperature have been found to have a profound effect on the final results [45]. Sometimes the free and complexed forms of the antigen can even be used as markers for totally different diseases. Pregnancy-associated protein A (PAPP-A) is commonly used as a marker for Down's syndrome (Sect. 5.1) where it is present primarily in a heterotetrameric form with eosinophil major basic protein, whereas in acute coronary syndrome, TR-IFMA of non-complexed PAPP-A can be used as a point-of-care test [46].

Establishing the reference intervals for an immunoassay is generally done separately for each type of immunoassay in each laboratory. Moreover, typical normal values, which may depend on age, nutritional status, physical activity, etc., are associated with each individual person and ethnic group. One example is the TR-IFMA assay of growth hormone to diagnose acromegaly. The analysis was found to require separate reference intervals for men as compared to women [47].

Early detection of a disease, such as cancer, or disclosing the risk factors related to pregnancy, may require repeated measurements to reveal a minor but relevant change in the biomarker concentration. 17-Hydroxyprogesterone is generally used to screen for congenital hyperplasia. The gestational age and birth weight have an effect on its relevant cut-off determination [48]. Placental protein 13 (PP13) used in screening for risk levels of pre-eclampsia, shows typical rise-and-decay patterns during normal and high-risk pregnancy, and its use during the first trimester requires repeated measurements.

In addition, the instability and short half-life of many of the biologically active compounds should be kept in mind. These place high requirements on sample storage. Also, the measured values may correlate with (but do not give absolute values for) the concentrations of the measured substances actually in circulation.

4.1 Interfering Effects

Immunofluorometric assays, like all antibody-based analysis, are prone to a variety of biological interferences. Time resolution in the detection phase greatly alleviates many of the optical interferences, such as background fluorescence. Still, numerous biological interferences remain, making the accurate analysis of the analyte in the sample demanding. Biological interferences include, among others, non-specific interactions, complement recruitment, rheumatoid factors, human anti-mouse antibodies (HAMA) and heterophilic cross-reactive antibodies present in patient serum.

Gibney et al. [49] used polyethylene precipitation to get rid of autoantibody-related interferences in prolactin TR-IFMA. They found that in some patients prolactin is present in serum as the high-molecular-weight macroprolactin. Macroprolactin gives a positive response in the assay regardless of its greatly reduced bio-activity, which, hence, leads to a false diagnosis.

In the TR-IFMA of cardiac troponin I with fluorescent Eu chelates and surface measurement, the design of the antibodies has been observed to be an important factor in avoiding the effect of auto-antibodies [50, 51]. Antibody engineering has been applied to produce antibodies less susceptible to interference. These include Fab fragments with no constant region for complement binding, phage display selected human Fab fragments, and humanized antibodies. Warren et al. used in vivo biotinylated single-chain antibodies to avoid interference derived from heterophilic antibodies in a TR-IFMA of carcinoembryonic antigen, CEA [52].

5 Immunofluorometric Applications

The extent of the applications of immunoassays and TR-IFMA assays is beyond the scope of the present chapter. However, a snapshot of the application field is given based on the most recent publications. This review is organized to follow the novel trend in health care, away from reactive diagnostic practices towards proactive and preventive health care. This includes risk-assessment analysis to find disease predisposition and to screen for diseases primarily before symptoms emerge. During the course of human life, risks can be categorized based on the life cycle from conception to birth, and from the early development of a child to the assessment of age-related risk factors.

5.1 Prenatal Assays and Maternal Health Concerns

In pre-natal screening, the aim is to find and lower health related risks on the child and mother. Maternal health has become an area of intense biomarker search both because of severe conditions like pre-eclampsia, fetal growth retardation (IUGR), pre-term birth as well as congenital malfunctions such as chromosomal aneubloidies (Down's syndrome), and congenital heart disease.

Placental growth hormone (PGH) secreted to maternal blood was tested as a marker for monitoring vaginal delivery and measured by an automated TR-IFMA assay on AutoDELFIA [53]. Placental protein 13 (PP13) is a novel biomarker suggested for pregnancy follow-up. Ish-Shalom et al. studied the function of another pregnancy- related plasma protein, PP14, and its lectin binding and the mechanism of T-cell inhibition using Eu-labelled streptavidin as the label [54]. Other markers used in monitoring ectopic pregnancy are invasive trophoblastic antigen (ITA) and free βhCG, generally combined with ultrasound examination [55].

Signaling via the IGF-1 receptor is crucial for normal prenatal and postnatal growth. Altered IGF-1 signaling is a result of polyphormic forms of the respective gene. Raile et al. monitored the rhGH-mutation of IGF-R [56]. IGF-R polymorphism is reportedly a causative factor to IUGR [57].

The matrix metalloproteinase ADAM-12 has been used as a marker for both Down's syndrome and for pre-eclampsia [58]. Pregnancy-associated plasma protein A is reported to be a marker of both pre-eclampsia and myocardial infarction [59]. Wojdeman et al. used both automated DELFIA and a TR-FRET assay of Brahms (Kryptor) to analyze the Down's syndrome markers PAPP-A and a form of human chorionic gonadotropin (sshCG) together with an ultrasound translucency test [60]. Weinans et al. compared invasive trophoblast antigen (ITA), i.e., hyperglycosylated hCG, to free βhCG and PAPP-A as first-trimester markers for Down's syndrome. PAPP-A was shown to be the best marker, and ITA correlated well with free βhCG [61].

Hypothyroidism screening in pregnancy is also being considered since the maternal thyroid function is of critical importance to the infant, and maternal thyroid dysfunction is even suggested to be a causative factor for pre-eclampsia and IUGR. Buimer et al. (2005) used DELFIA fT4 and TSH assays and luminoimmunoassay (Brahms) for thyroid peroxidase antibody (TPO Ab) screening to study the association of transient hypothyroxinemia with fetal disorders [62]. In their study of TR-IFMAs of TSH, TPO Ab, and TgAb, Brix et al. excluded the connection between low birth weight and subsequent autoimmunity [63].

5.2 Neonatal Screening

The mission in neonatal screening is to identify those newborn babies who are at high risk of developing a severe congenital disease that could be avoided, cured, or alleviated with known medication or dietary treatment if diagnosed in time. Neonatal screening is mainly performed by using dried blood spots taken from the infant's heel or from cord blood. Dried blood spots are a convenient method to collect, store, and deliver samples to centralized screening laboratories.

Disturbances of the fetal thyroid function are generally discovered by measuring neonatal TSH and or T4 levels. If not detected and treated with thyroid replacement therapy at an early stage, hypothyroidism causes severe symptoms and leads to developmental disorders [64, 65].

Thyroid screening is in most countries accomplished by analyzing blood-spot samples either for thyroxin or TSH. TR-IFMAs are generally applied, as recently reported for screening, e.g., in Australia [66] and in Denmark [67]. TSH measurements are frequently made not only to detect hypothyroidism but also to diagnose and screen euthyroid and hyperthyroid patients as well as for some non-thyroidal illnesses and carcinoma. TSH measurements demand high-sensitivity immunoassays, and assay sensitivities have greatly improved over the years [68, 69]. One of the factors that may affect the screening outcome is birth weight, which has been found to constitute a risk of false-negative results in congenital hypothyroidism screening when measuring THS and T4 by TR-FRET [70].

In addition to thyroid function, congenital adrenal hyperplasia (CAH) is commonly screened from neonates by measuring 17-α-hydroxyprogesterone from dried blood spots [48]. CAH relates to type II 3β-hydroxysteroid dehydrogenase deficiency. Its accurate detection is essential for the normal development of the child and thus a false-negative result may be critical [71]. Johanssen measured serum LH and FSH, and, retrospectively, blood spot 17 OHP [72] to follow up delayed detection of CAH.

Lysosomal storage disorders (LSD) are a group of diseases related to inborn errors in metabolism. They are chronic, progressive diseases having a devastating impact on the newborn. In addition to assays of the respective enzymatic metabolic activities, LSDs are screened by mass spectrometry and also by immunoassays with DELFIA design being suitable for several applications [73]. The research group of Hopwood has recently published methods to analyze LSDs on dried blood spots based on the DELFIA technology, including screening for Fabry disease by monitoring α-galactosidase [74], Gaucher disease by measuring β-glucosidase [75], mucopolysaccharidosis MPS-I by measuring α-L-idusonidase [76], MPS-II by iduronate-2-sulfatase [77] and MPS VI by *N*-acetylgalactosamine 4-sulfatase [78] and lysosome-associated membrane protein-1 (LAMP-1) [79].

5.3 Risk Assessment of Adult Population

Finding cancer early, when it is still curable, is a big challenge for diagnostics. There are only few definite and generally accepted analytes to enable early detection of cancer and that can be used in screening mode. Most of the analytes, such as PSA for prostate cancer, and CEA for breast-cancer screening, are relatively unspecific, and their use in large-scale screening is still controversial.

Prostate cancer is a good example of an analyte that is a focus in a continuous struggle towards more sensitive and specific early diagnosis. Free and total PSA are most commonly used markers, and there are numerous TR-IFMA reports on their use, also commercialized as dual-label assays based on Eu and Sm [80–82]. One of the challenges in the use of PSA in prostate-cancer screening is the difficulty to differentiate between benign growth and malignant cancer. The TR-IFMA of urokinase plasminogen activator receptor (uPAR) in urine is reported for cancer studies [83] and, recently, CEA TR-IFMA has been used in the follow-up of recurrence of colorectal cancer [84].

In addition to the screening of congenital risk factors in the early stages of development, there are numerous health risk factors followed, affecting aging populations, such as type 2 diabetes, coronary diseases, altered bone metabolism, a variety of cancers, and Alzheimer's disease. In addition to diagnosing, many of these conditions are also approached in view of preventive medicine and early detection, and in many of them, TR-IFMA has played a role.

SHBG and testosterone TR-FIAs have been used in a study to assess the correlation of metabolic syndrome and smoking with hypogonadism [85], and assays of testosterone and IGFBP-1 in the risk assessment of metabolic syndrome and cardiovascular mortality [86]. Early arthritis and articular cartilage metabolism are also diagnosed by measuring circulating type II collagen by TR-IFMA [87].

Altered bone metabolism affects, in particular, postmenopausal women, and is followed e.g., by urinary osteocalcin measurements based on an SA plate and biotinylated peptide and Eu-labelled antibody [88]. Sipilä et al. assessed muscle strength and risk for bone fracture in elderly women by monitoring estradiol, testosterone, and SHBG levels by TR-IFMA [89]. TSH assay has also been employed to assess bone mineral density, turnover, and risk for fracture in T therapy [90].

In the treatment of Alzheimer's disease, the role of testosterone and LH in the cognition and mood of patients with mild progression of the disease has been reported [91]. Assay of β-amyloid peptide has also been used to study Alzheimer's disease. Cancellotti et al. studied altered glycosylation of PrP proteins that can have different neuronal trafficking in the brain but do not acquire scrapie-like properties [92].

6 Future Trends and Challenges

Health care and diagnostics in the future will face the challenges imposed by economic factors, particularly the need for less-expensive and more-efficient tools to diagnose or to prevent diseases. In terms of assays, the trend is towards more-simple assays, miniaturization, and more-economic assays. There is also a trend to migrate from hospital laboratories to decentralized diagnosis, assays done in doctors' offices or by direct access testing. Multiplexing is a fashionable term that reflects the desire to analyze several factors simultaneously. Multiplexing is also needed to diagnose diseases, or groups of symptoms having different origins, referring to difficulties in diagnosing those based on only a single marker. When going to personalized medicine, and in an attempt to find genetically driven risk factors in the early days of life, the various analyses of DNA sequences will become a part of routine analysis. In addition, for a long time there has been a desire to perform diagnoses using circulating cells, e.g., hunting fetal cells in a mother's blood circulation, or rare cancer cells to be analyzed. The strong development of various forms of in vivo imaging, drug targeting, and localized testing in the whole body is also an area affecting future diagnostics.

Lanthanides and time-resolved fluorometry can provide tools for miniaturization and limited multiplexing. Since the same economic factors are strongly encountered in the field of drug discovery and addressed in high-throughput screening, the price per assay can be reduced by decreasing assay volumes from about 200 μl in a regular 96-well plate and the respective single-use assay cuvettes to a few microliters in 1536-well plates. This development is very much supported by TR-FRET development [25] with its ability to perform in small volumes in a homogeneous format, and even bring the multiplexing feature to the same well [93]. Figure 4 shows the energy-transfer measurement wavelengths used when using europium chelate as the energy donor.

Several groups have recently concentrated on developing highly luminescent beads based on lanthanide chelates, and Eu chelates in particular. Some of these particles are already commercially available, such as beads from Serono company, and are applied in various forms of TR-IFMA. Particles con-

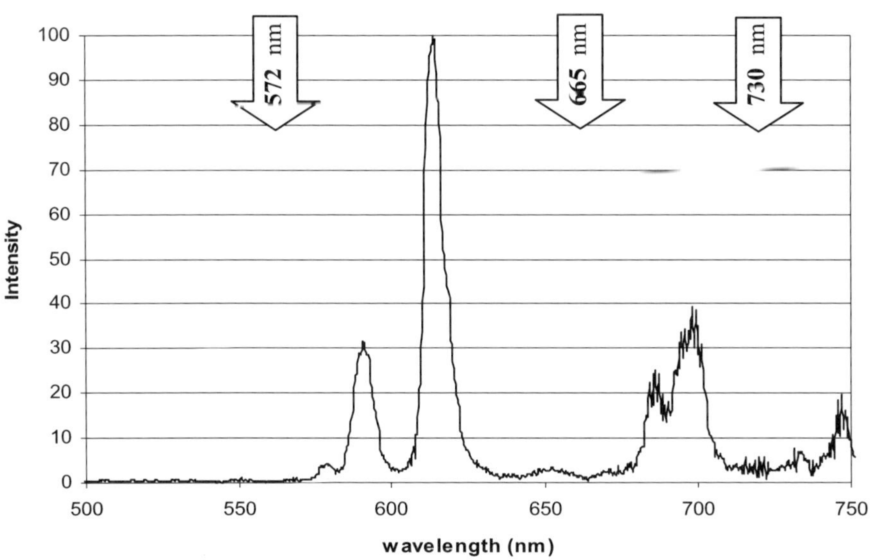

Fig. 4 Suitable acceptor-emission wavelengths for energy transfer (TR-FRET) from a europium chelate to various organic acceptor molecules

taining a large number of chelates can be made, and the chelates are well protected from aqueous environment, thus avoiding both dissociation and quenching caused by O-H overtones. Such beads have been applied both in imaging and in immunoassays.

Development of novel binding reagents, genetically engineered antibodies and their fractions by in vitro immunization and phage display selection, artificial binders such as aptamers, peptide binders, peptide nucleic acids (PNA), molecular imprints and so on, bring new tools to the further development of assays. In vivo biotinylated single-chain antibodies are used as capture reagents in a TR-IFMA of CEA [52]. Single-chain antibodies, as well as antibodies devoid of the CH_2 domain [94], are used to avoid interferences caused by heterophilic antibodies.

One example of the possibilities provided by antibody engineering is the TR-IFMA of small drug molecules and steroids developed based on antibodies recognizing the occupied site on the primary antibody, and accomplished by measuring TR-FRET of the formed complex [95].

Approaching the cell, and whole body imaging with modern modality tools such as luminescence imaging, PET, SPECT and MRI—the latter with lanthanide labels—will become more widely used in scanning whole bodies and diagnosing and localizing diseases. Luminescence imaging entails problems of light excitation and penetration into tissues. The modern nanoparticles, such as quantum dots, upconverting crystals, and lanthanide chelate beads (those with Yb, Nd) may give practical tools for that, too.

References

1. Köhler G, Milstein C (1975) Nature 256:495
2. Soini E, Hemmilä I (1979) Clin Chem 25:353
3. Hemmilä I (1991) Applications of Fluorescence in Immunoassays. Wiley Interscience, New York
4. Weissman SI (1940) J Chem Phys 10:214
5. Yguerabide J (1972) Methods Enzymol 26:498
6. Schneckenburger H (1984) Proc SPIE 491:363
7. Herman P, Maliwal BP, Lin HJ, Lakowicz JR (2001) J Microsc 203:176
8. Hemmilä I, Mukkala V-M (2001) Crit Rev Clin Lab Sci 38:441
9. Stanley PE (1999) Luminescence 14:201
10. Soini E, Kojola H (1983) Clin Chem 29:65
11. Diamandis EP (1988) Clin Biohem 21:139
12. Seveus L, Väisälä M, Syrjänen S, Sandberg M, Kuusisto A, Harju R, Salo J, Hemmilä I, Kojola H, Soini E (1992) Cytometry 13:329
13. Verwoerd NP, Hennink EJ, Bonnet J, van der Geest CR, Tanke HJ (1994) Cytometry 16:113
14. Hennink EJ, De Haas R, Verwoerd NP, Tanke HJ (1996) Cytometry 24:312
15. Hemmilä I, Dakubu S, Mukkala V-M, Siitari H, Lövgren T (1984) Anal Biochem 137:335

16. Mukkala V-M, Mikola H, Hemmilä I (1989) Anal Biochem 176:319
17. Takalo H, Mukkala V-M, Mikola H, Liitti P, Hemmilä I (1994) Bioconjugate Chem 5:278
18. Blomberg K, Granberg C, Hemmilä I, Lövgren T (1986) J Immunol Methods 86:225
19. Peuralahti J, Meriö L, Mukkala V-M, Blomberg K, Hovinen J (2006) Bioorg Med Chem Lett 16:4760
20. Mathis G (1993) Clin Chem 39:1953
21. Li M, Selvin PR (1995) J Am Chem Soc 117:8132
22. Mukkala V-M, Helenius M, Hemmilä I, Kankare J, Takalo H (1993) Helv Chim Acta 76:1361
23. Cooper ME, Sammes PG (2000) J Chem Soc Perkin Trans 2:1695
24. Wang Z, Yuan J, Matsumoto K (2005) Luminescence 20:347
25. Comley J (2006) Drug Discovery World Spring 2006:22
26. Härmä H, Soukka T, Lövgren T (2001) Clin Chem 47:561
27. Huhtinen P, Kivelä M, Kuronen O, Hagren V, Takalo H, Tenhu H, Lövgren T, Härmä H (2005) Anal Chem 77:2643
28. Valanne A, Huopalahti S, Soukka T, Vaininpää R, Lövgren T, Härmä H (2005) J Clin Virol 33:217
29. Cummins CM, Koivunen ME, Stephanian A, Gee SJ, Hammock BD, Kennedy IM (2006) Biosens Bioelectron 21:1077
30. Hemmilä I, Laitala V (2005) J Fluorescence 15:529
31. Laitala V, Hemmilä I (2005) Anal Chem 77:1483
32. Laitala V, Hemmilä I (2005) Anal Chim Acta 551:73
33. Mathis G (1993) Clin Chem 39:1953
34. Kimura H, Matsumoto K, Mukaida M (2005) J Anal Toxicol 29:799
35. Qin Q-P, Peltola O, Pettersson K (2003) Clin Chem 49:1105
36. Zhang WX, Wang R, Wisniewski D, Marcy AI, LoGrasso P, Lisnock J-M, Cummings RT, Thompson JE (2005) Anal Biochem 343:76
37. Leifert WR, Bailey K, Cooper TH, Aloia AL, Glatz RV, McMurchie EJ (2006) Anal Biochem 355:201
38. Panteghini M, Forest JC (2005) Clin Chim Acta 355:1
39. Stenman U-H (2001) Clin Chem 47:815
40. Barker PE, Wagner PD, Stein SE, Bunk DM, Srivastava S, Omenn GS (2006) Clin Chem 52:1669
41. Chalet EA, Travaglino P, Pagani S, Bozzola E, Marino E, Berensztein E, Maceiras M, Tauber M, Rivarola MA, Belgorosky A, Bozzola M (2006) J Endocrinol Invest 29:109
42. Taimela E, Kairisto V, Koskinen P, Leino A, Irjala K (1997) Eur J Clin Chem Clin Biochem 35:889
43. Giton F, Fiet J, Guéchot J, Ibrahim F, Bronsard F, Chopin D, Raynaud J-P (2006) Clin Chem 52:474
44. Selva DM, Bassas L, Munell F, Mata A, Tekpetey F, Lewis JG, Hammond GL (2005) J Clin Endocrinol Metab 90:6275
45. Ulmert D, Becker C, Nilsson J-Å, Piironen T, Björk T, Hugosson J, Berglund G, Lilja H (2006) Clin Chem 52:235
46. Wittfooth S, Qin Q-P, Lund J, Tierala I, Pulkki K, Taralo H, Pettersson K (2006) Clin Chem 52:1794
47. Markkanen H, Pekkarinen T, Välimäki MJ, Alfthan H, Kauppinen-Mäkelin R, Sane T, Stenman U-H (2006) Clin Chem 52:468
48. Van der Kamp HJ, Oudshoorn CGM, Elvers BH, Van Baarle M, Otten BJ, Wit JM, Verkerk PH (2005) J Clin Endocrinol Metab 90:3904

49. Gibney J, Smith TP, McKenna TJ (2005) J Clin Endocrinol Metab 90:3927
50. Eriksson S, Halenius H, Pulkki K, Hellman J, Pettersson K (2005) Clin Chem 51:839
51. Eriksson S, Ilva T, Becker C, Lund J, Porela P, Pulkki K, Voipio-Pulkki L-M, Pettersson K (2005) Clin Chem 51:848
52. Warren DJ, Bjerner J, Paus E, Børmer OP, Nustad K (2005) Clin Chem 51:830
53. Fuglsang J, Sandager P, Møller N, Fisker S, Ørskov H, Ovesen P (2006) Eur J Endocrinol 154:449
54. Ish-Shalom E, Gargir A, André S, Borovsky Z, Ochanuna Z, Gabius H-J, Tykocinski ML, Ramichmilewitz J (2006) Glycobiology 16:173
55. Cabar FR, Pereira PP, Schultz R, Zugaib M (2006) Hum Reprod 21:2426
56. Raile K, Klammt J, Schneider A, Keller A, Laue S, Smith R, Pfäffle R, Kratzsch J, Keller E, Kiess W (2006) J Clin Endocrinol Metab 91:2264
57. Walenkamp MJ, van der Kamp HJ, Pereira AM, Kant SG, Van Duyvenvoorde HA, Kruithof MF, Breuning MH, Romijn JA, Karperien M, Wit JM (2006) J Clin Endocrinol Metab 91:3062
58. Laigaard J, Sørensen T, Placing S, Holck P, Fröhlich C, Wøjdemann KR, Sundberg K, Shalmi A-C, Tabor A, Nørgaard-Pedersen B, Ottesen B, Christiansen M, Wewer UM (2005) Obstet Gynecol 106:144
59. Qin Q-P, Kokkala S, Lund J, Tamm N, Voipio-Pulkki L-M, Pettersson K (2005) Clin Chem 51:75
60. Wojdemann KR, Larsen SO, Rode L, Shami A, Sundberg K, Christiansen M, Tabor A (2006) Scand J Clin Lab Invest 66:101
61. Weinans MJN, Sancken U, Pandian R, Van de Ouweland JMW, De Bruijn HWA, Holm JP, Mantingh A (2005) Clin Chem 51:1276
62. Buimer M, Van Wassenaer AG, Ganzevoort W, Wolf H, Bleker OP, Kok JH (2005) Obstet Gynecol 106:973
63. Brix TH, Hansen PS, Rudbeck AB, Hansen JB, Skytthe A, Kyvik KO, Hegedüs L (2006) J Clin Endocrinol Metab 91:3499
64. Kempers MJE, Van Trotsenburg ASP, Van Tijn DA, Bakker E, Wiedijk BM, Endert E, De Vijlder JM, Vulsma T (2005) J Clin Endocrinol Metab 90:4094
65. Wekking EM, Appelhof BC, Fliers E, Schene AH, Huyser J, Tijssen JGP, Wiersinga WM (2005) Eur J Endocrinol 153:747
66. McElduff A, McElduff P, Wiley V, Wilcken B (2005) J Clin Endocrinol Metab 90:6361
67. Schmidt DR, Hogh B, Andersen O, Fuchs J, Fledelius J, Petersen E (2006) Arch Dis Child 91:661
68. Taimela E, Aalto M, Viikari P, Nuutila J, Irjala K (1995) Scand J Clin Lab Invest 55:537
69. Hebles-Duvison M, Cruz-Ruiz M, Vazquez-Rubio R, Macarro-Sancho I, Silva-Mejias C, Recio-Quijano F (1996) Clin Chem 42:670
70. Tylek-Lománska D, Kumorowicz-Kopiek M, Starzyk J (2006) J Med Screening 12:166
71. Votava F, Török D, Kováks J, Möslinger D, Baumgartner-Parzer SM, Sólyom J, Pribilincová Z, Battelino T, Lebl J, Frisch H, Waldhauser F (2005) Eur J Endocrinol 152:869
72. Johannsen TH, Mallet D, Dige-Petersen H, Müller J, Main KM, Morel Y, Forest MG (2005) J Clin Endocrinol Metab 90:2056
73. Parkinson-Lawrence E, Fuller M, Hopwood JJ, Meikle PJ, Brooks DA (2006) Clin Chem 52:1660
74. Fuller M, Lovejoy M, Brooks DA, Harkin ML, Hopwood JJ, Meikle PJ (2004) Clin Chem 50:1979
75. Fuller M, Lovejoy M, Hopwood JJ, Meikle PJ (2005) Clin Chem 51:2200

76. Fuller M, Brooks DA, Evangelista M, Hein LK, Hopwood JJ, Meikle PJ (2005) Mol Gen Metab 84:18
77. Hein LK, Meikle PJ, Dean CJ, Bockmann MR, Auclair D, Hopwood JJ, Brooks DA (2005) Clin Chim Acta 353:67
78. Dean CJ, Bockmann MR, Hopwood JJ, Brooks DA, Meikle PJ (2006) Clin Chem 52:643
79. Parkinson-Lawrence EJ, Dean CJ, Chang M, Hopwood JJ, Meikle PJ, Brooks DA (2005) Cell Immunol 236:161
80. Piironen T, Häse A, Huland H, Steuber T, Christensen IJ, Brünner N, Danø K, Høyer-Hansen G, Lilja H (2006) Clin Chem 52:838
81. Aus G, Damber J-E, Khatami A, Lilja H, Stranne J, Hugosson J (2005) Arch Int Med 165:1857
82. Safarinejad MR (2006) Ann Onc 17:1166
83. Resnati M, Pallavicini I, Daverio R, Sidenius N, Bonini P, Blasi F (2006) J Immunol Methods 308:192
84. Fernandes LC, Kim SB, Saad SS, Matos D (2006) World J Gastroenerol 12:3891
85. Laaksonen DE, Niskanen L, Punnonen K, Nyyssönen K, Tuomainen T-P, Valkonen V-P, Salonen JT (2005) J Clin Endocrinol Metab 90:712
86. Kalme T, Seppälä M, Qing Q, Koistinen R, Nissinen A, Harrela M, Loukovaara M, Leinonen P, Tuomilehto J (2005) J Clin Endocrinol Metab 90:1550
87. Valverde-Franco G, Binette GS, Li W, Wang H, Chai S, Laflamme F, Tran-Khanh N, Quenneville E, Meijers T, Poole AR, Mort JS, Bushmann MD, Henderson JE (2006) Hum Mol Genet 15:1783
88. Ivaska KK, Käkönen S-M, Gerdhem P, Obrant KJ, Pettersson K, Väänänen HK (2005) Clin Chem 51:618
89. Sipilä S, Heikkinen E, Cheng S, Suominen H, Saari P, Kovanen V, Alén M, Rantanen T (2006) J Gerontol A Biol Sci Med Sci 61:92
90. Heijckmann AC, Huiberts MSP, Geusens P, De Vries J, Menheere PPCA, Wolffenbuttel BHR (2005) Eur J Endocrinol 153:23
91. Lu PH, Masterman DA, Mulnard R, Cotman C, Miller B, Yaffe K, Reback E, Porter V, Swerdloff R, Cummings JL (2005) Arch Neurol 63:177
92. Cancellotti E, Wiseman F, Tuzi NL, Baybutt H, Monaghan P, Aitchison L, Simpson J, Manson JC (2005) J Biol Chem 280:42909
93. Hemmilä I, Laitala V (2005) J Fluorescence 15:529
94. Glaser SM, Hughes IE, Hopp JR, Hathaway K, Perret D, Reff MF (2005) J Biol Chem 280:41494
95. Pulli T, Höyhtyä M, Söderlund H, Takkinen K (2005) Anal Chem 77:2637

Springer Ser Fluoresc (2008) 6: 449–468
DOI 10.1007/4243_2008_041

Published online: 6 March 2008

Particle-Based Assays: Applications and Unresolved Issues

Matthias Seydack

8sens.biognostic GmbH, Robert-Roessle-Str. 10, 13125 Berlin, Germany
seydack@biognostic.de

Abstract For many years, immunoassays have been the most widely used tests for the detection and quantification of diagnostically important proteins, hormones, enzymes, and haptens. In most state-of-the-art applications of immunoassays, the marker is attached to the last binding antibody, which is, therefore, often referred to as the "detector antibody". Depending on the type of marker, immunoassays can be classified as radioimmunoassays (RIAs), enzyme immunoassays (EIAs), enzyme-linked immunosorbent assays (ELISAs), fluorescent immunoassays (FIAs), etc. This article is exclusively concerned with emission-based immunoassays where, with one exception, particles serve as markers and the obtained signal is either fluorescence or luminescence (FIA). In these assays antibodies are normally conjugated covalently to the particle, e.g., by way of carbodiimide coupling, thus forming a layer of affinity molecules on the particle surface.

Keywords Fluorescence · Immunoassays · Luminescence · Particles

Abbreviations

d Diameter
ELISA Enzyme-linked immunosorbent assay
EIA Enzyme immunoassay

ε	Molar decadic extinction coefficient
F/P	Fluorescent molecule-to-protein ratio
FIA	Fluorescence immunoassay
FRET	Förster-type resonance energy transfer
Φ_{em}	Quantum yield of emission (fluorescence or luminescence)
LOD	Limit of detection
MTP	Microtiter plate
n	Number of particles
RIA	Radioimmunoassay
ROC	Receiver-operator characteristic
S/N	Signal-to-noise ratio
SEM	Scanning electron microscopy
TEM	Transmission electron microscopy

1 Introduction

For many years, immunoassays have been the most widely used tests for the detection and quantification of diagnostically important proteins, hormones, enzymes, and haptens. These analyte molecules are usually associated with certain medical conditions, such as troponin T, troponin I, myoglobin and the fatty acid binding protein (FABP) with acute myocardiac infarction, the C-reactive protein (CRP) with inflammatory diseases and human chorionic gonadotropin (hCG) with pregnancy, to name just a few. The sample material to be used may be practically any body fluid or dilution of such, or other fluid sample. The basis of all immunoreactions is the existence of antibodies with antigen-specific binding sites. Therefore, the right antibody can be used to detect its matching antigen and vice versa. Thus, the two common features of immunoassays are: (a) at least one antigen–antibody affinity reaction, and (b) the need of some marker in order to monitor and quantify whether this affinity reaction has taken place.

The first immunoassays were performed in the late 1950s and used radioactive ^{131}I as markers [1]. More recently, ^{125}I and ^{57}Co came to be used. These radio-immunoassays offered very high sensitivities, which have, in some cases, been unmatched to the present day. For obvious reasons, however, the use of radioactive material is nowadays avoided.

In most state-of-the-art applications, the marker is attached to the last binding antibody, which is, therefore, often referred to as the "detector antibody". Depending on the type of marker, immunoassays can be classified as radio-immunoassays (RIAs), enzyme immunoassays (EIAs), enzyme-linked immunosorbent assays (ELISAs), fluorescent immunoassays (FIAs), etc. Moreover, there are numerous assay formats: depending on the type of analyte, its binding and adsorption properties and the desired limits of detection (LODs), it may be useful to employ three or even four different antibodies

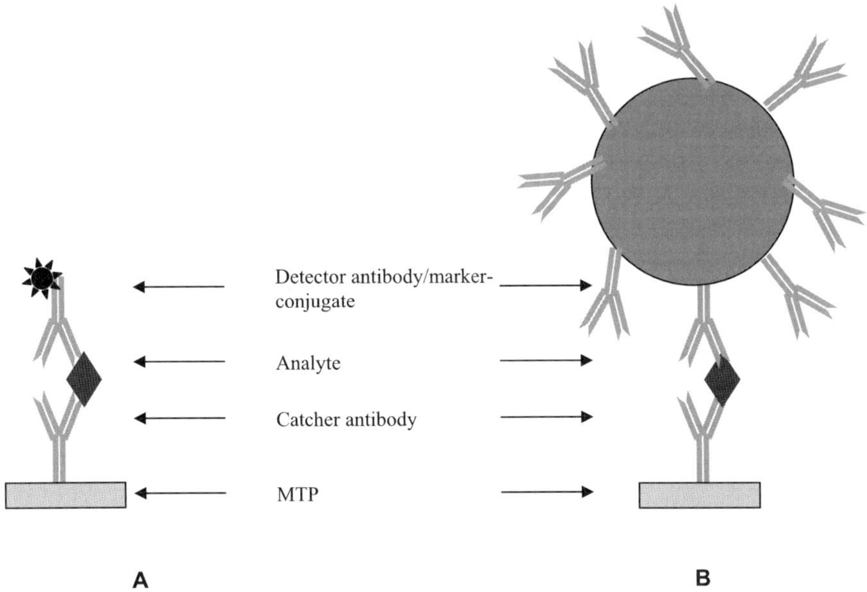

Fig. 1 Scheme of a sandwich-type fluorescence immunoassay (FIA): the catcher antibody is adsorbed on the microtiter plate (*MTP*); after blocking (not shown), incubation with the sample and the detector antibody/marker-conjugate, the presence of the marker can be detected. **A** A characteristic set-up for molecular markers. **B** Larger, particulate markers as they are discussed in this chapter (not to scale). In FIA, markers typically belong to one of the following substance groups: organic dyes, lanthanide chelates, polystyrene particles containing molecular dyes, quantum dots, other organic/inorganic nano- or microparticles

in three or more separate binding events, sometimes including a very tight biotin-streptavidin binding or Förster-type resonance energy transfer (FRET) between two different markers. One of the most common and most sensitive assay formats is the sandwich immunoassay (Fig. 1A).

This article is exclusively concerned with emission-based immunoassays where, with one exception, particles serve as markers and the obtained signal is either fluorescence or luminescence (FIA). Antibodies are normally conjugated covalently to the particle, e.g., by way of carbodiimide coupling, thus forming a layer of affinity molecules on the particle surface (Fig. 1B).

2 Performance Criteria of Immunoassays

In assay development quite a number of different quality criteria are commonly used in order to describe and compare the assay performance. The most important ones are (a) LOD, (b) assay accuracy, (c) assay precision, (d) limit of quantitation (usually defined by the influence of a Hook effect that leads to a signal decrease at very high analyte concentrations), (e) specificity (especially against other proteins from the sample), (f) signal linearity, (g) assay drift, and (h) aging behavior (e.g., accelerated and/or real-time aging).

Once an assay is optimized (as regards to antibody selection, coating procedure, incubation times, concentration of reagents, composition of buffers, etc.) the diagnostic performance needs to be estimated by means of clinical testing against other, reliable references. Even if the test is not quantitative but has only a yes/no-character, e.g., if its goal is to determine whether the concentration of a certain protein in the patient's blood is above a medically meaningful threshold value, four different cases may be distinguished:

1. test positive, reference positive ("true positive")
2. test negative, reference positive ("false negative")
3. test positive, reference negative ("false positive")
4. test negative, reference negative ("true negative")

Obviously, these data strongly depend on the threshold level chosen. This dependence and the overall diagnostic quality of the test are easily depicted and quantified by means of a so-called receiver-operating characteristic (ROC). This method originates from radar detection of airplanes in World War II and has been used in medical diagnostics since the 1970s [2, 3]. Drawing on a sufficiently high number of patient samples and a reliable reference, a clear dependence between specificity and sensitivity can be found for each threshold level (Fig. 2).

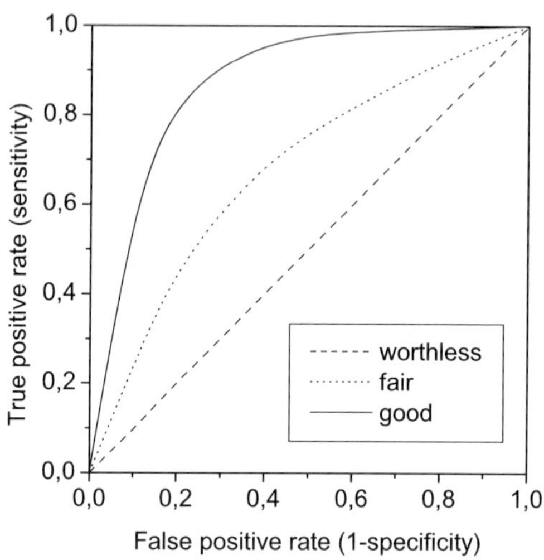

Fig. 2 Scheme of Receiver-operator characteristic (ROC) curves showing the trade-off between sensitivity and specificity (i.e., between the true positive and the false positive rates). The closer the *curve* follows the *left-hand border* and then the *top border* of the ROC space, the more accurate the test

The integral over every ROC curve, ideally 1.0, is a reasonably simple way to express the diagnostic value of the test. ROC integrals above 0.9 are usually considered to denote an excellent performance.

As a first approximation, it is the antibody–antigen reaction that governs the assay performance and the selection of suitable antibodies or pairs of antibodies (sandwich format) is a vital first step. This affinity reaction, however, is influenced by the number of markers bound to one antibody, the character of their binding (covalent or electrostatic), the assay format (solid or fluid phase, sandwich, etc.) and the physical and chemical nature of the immediate environment [e.g., buffers, coating concentrations, microtiter plate (MTP) etc.]. Moreover, quality (e.g., emission bandwidth and maximum) and intensity of the obtained signal, are predominantly determined by the type of marker.

From a regulatory point of view, immunoassays are in-vitro diagnostic devices (IVD), which are thus regulated by European Directive 98/79/EG (Europe) and, correspondingly, by 21 CFR 860 (USA). Therefore, no quality issue (*vide supra*) should be compromised by the inclusion of particulate markers.

3
Why Use Particles?

The most widely used labels today are enzymes (e.g., horseradish peroxidase, alkaline phosphatase). Quantification in those tests is based on an enzymatic reaction instigated by adding a suitable substrate [e.g., tetramethylbenzidin (TMB)]. Thereby, in these ELISAs a change in color (i.e., optical absorption) is measured which can be correlated with the amount of bound complexes. By inclusion of a series of multienzyme cascades the sensitivity of ELISAs can be further increased by a factor of up to 250 [4]. However, there are a few drawbacks: (a) a relatively high sample volume is needed, (b) ELISAs can not be developed into a "multiplexable" assay (i.e., they are limited to one analyte per reaction volume), and (c) it is difficult to further improve the LOD because the enzymatic reaction needs to be stopped at a more or less arbitrary point of time, which introduces some error.

In comparison, the "classical" FIA makes use of organic fluorophores, e.g., fluorescein-isothiocyanate (FITC), Alexa 488, etc., which are covalently bound to the detector antibody. Depending on fluorophore size, binding properties and concentration ratios, 4–8 of these fluorophores bind to one detector antibody (i.e. F/P = 4–8). Unfortunately, using fluorescence-based assays like that, usually affords LODs that are at least one order of magnitude above those of enzymatic assays, even though fluorescence is known to be more sensitive than absorption measurements used in ELISAs or EIAs. There are three fundamental reasons for this. First, it should be borne in mind that the recognition properties of every antibody deteriorates roughly with the number of molecules bound to it. In line with this biochemical argument one might want to decrease the F/P. Second, however, the number of signal-generating molecules in FIAs is already tiny in comparison to enzymatic assays that are based on a cascading reaction to afford colored reaction products. Therefore, further decreasing the F/P in FIA is not a reasonable strategy but would further diminish the signal-to-noise (S/N) ratio from the physical side. The third reason is purely photophysical: in every ensemble of

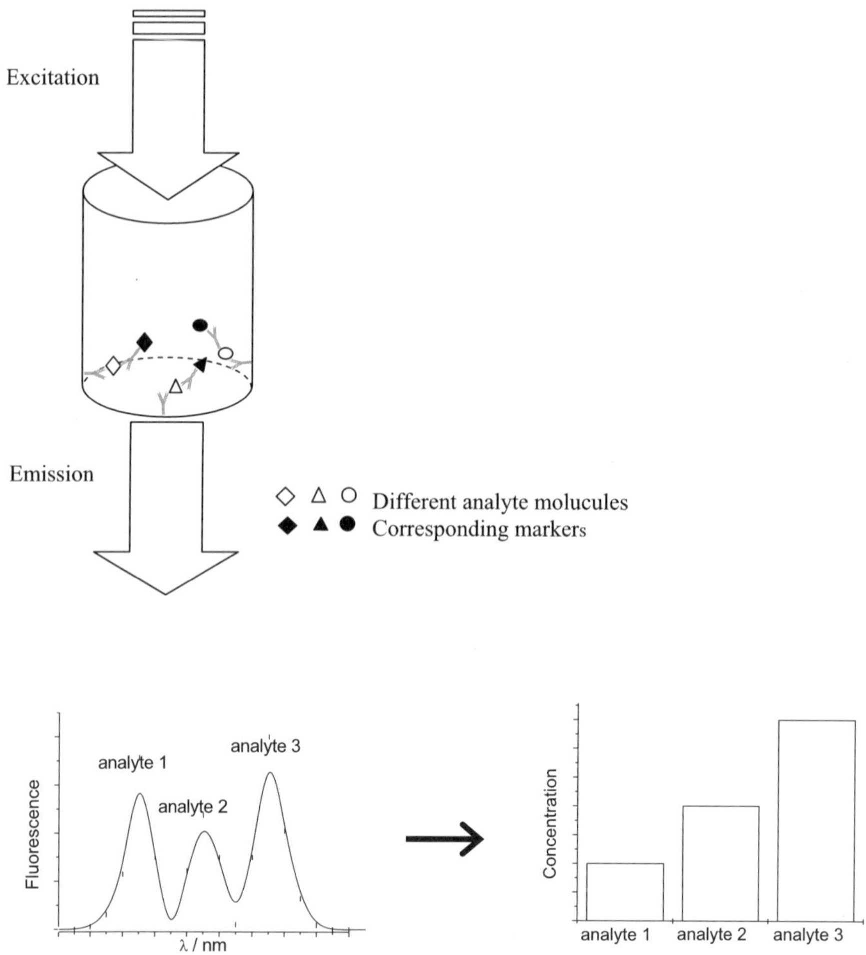

Fig. 3 Scheme of spectral multiplexing for three analyte molecules. The affinity reaction in the MTP well is detected by recording an emission spectrum that contains the emission bands of the three particulate markers, each corresponding to one analyte. Together with previously taken calibration data (which allow for different extinctions and emission quantum yields of the markers as well as for different binding properties and concentrations of the catcher and detector antibodies), the concentrations of each analyte can be calculated from the spectrum

dye molecules at close distance, there will always be some energy migration and emission filter effects. In effect, adopting the opposite strategy and increasing the F/P thereby not only impairs the biochemical properties of the antibodies but also lowers the mean dye–dye distances, which in turn leads to emission quenching within the dye pool attached to every respective antibody. So, again, the S/N will be lowered. In fact, an F/P of 4–8 represents the optimum for molecular, fluorescent markers [5, 6].

A promising strategy to avoid the drawbacks of both ELISAs and FIAs is the use of particulate, emissive marker systems instead of molecular ones. In general, these systems will enable the ideal case of only one binding site between antibody and particle. Owing to the completely different composition and photophysics of these systems in comparison to molecules, energy migration can also be effectively avoided. In addition, some of the particle-based assays also incorporate strategies to multiply the signals. In most particle-based assays the antibodies will be grouped as an outer layer on top of the particle. Alternatively, for very small particles where a protein "layer" can not be realized due to steric effects, e.g., quantum dots (QDs)(*vide infra*) which are often comparable in size to proteins, there may also be a 1 : 1 or 2 : 1 ratio between particle and antibody. Another, long-sought objective of immunosensing has been the simultaneous detection of more than one analyte, especially of those that are usually grouped for diagnostic purposes, e.g., autoantibodies, allergens, thyroid function tests or panels of markers with prognostic value in cardiology and other diagnostic areas. This multiplexing strategy may be accomplished by spectral, lifetime or spatial separation of the markers. In a typical spectral multiplexing experiment, each type of detector antibody is labeled by a particle of a different emission wavelength and every MTP well is then read out by recording an emission spectrum (Fig. 3).

In this case, it is not the high S/N, but narrow emission bands and broad, unstructured absorption features that are required. To date, these features can only be found for inorganic particles, e.g., QDs , which are discussed below in more detail. It should be noted, however, that the fundamental problem of cross-reactivity in a reaction mix of several types of antibodies and analyte molecules will be found in all multiplexed immunoassays. This can not easily be circumvented and will usually lead to increased LOD.

4
General Aspects of Particle-Based Assays

Most particle-based assays that make use of the emissive nature of the marker are either not widely used yet or are still in the research stadium, the only exception being fluorescent latices and lanthanide-based systems.

Independent of the particulate system used, there are a few general quality issues that need to be addressed, which reflect the colloidal nature of the marker system. The first issue is the homogeneity of the colloidal suspension: in theory, strict monodispersity and a narrow size distribution are ideal to avoid size-dependent effects (size bias both in the labeling process and in the affinity reaction itself can not generally be precluded; sedimentation may also be problematic). Real systems, however, will show some degree of polydispersity. This is even true of such stable and widely used colloids as commercially available gold sol. Closely linked with this aspect is the question of how this distribution changes (a) if the particles themselves need to be engineered in a more complex way, such as core-shell structures, and (b) when a layer of antibodies is getting attached to the surface.

4.1 The Size "Problem"

Obviously, the size of the particulate marker is the most straightforward parameter to have an influence on the FIA. Depending on the particles' composition, the character of this influence may be different, though. In the case of polymeric beads with embedded dyes, larger particles may deliver larger emission signals but, for steric reasons, block neighboring binding sites on the MTP and thus make measurements of high analyte concentrations impossible. Particles of higher density, such as microcrystalline organic dyes, will show the same phenomenon but may additionally be prone to precipitation. QDs, finally, will show even more comprehensive spectroscopic changes, e.g., a red-shifted emission, on increasing their core size or when aggregating. In this context, the surface chemistry must not be neglected (cf. Sect. 4.3): extended polymeric layers, often used as coupling interface between particle and biological recognition unit, may considerably increase the overall size of the particle, albeit without altering their emission properties.

Therefore, it is necessary to ask what exactly "particle size" is. There are quite a few answers to this—at first glance obvious—question and several aspects need to be taken into consideration when comparing sizes. First, there will always be a size distribution of some width, which renders the term "size" open to further interpretation. Second, the particles may not be ideal spheroids. Any deviation from the spherical shape results in data that refer to an "equivalent sphere" or "effective sphere". Third, the results from size measurements may depend on the method used. While for very small particles (<100 nm diameter) dynamic light scattering, which always measures hydrodynamic diameters, and scanning electron microscopy (SEM) or transmission electron microscopy (TEM) are the methods of choice, larger particles may also be measured using static light scattering or chromatographic methods. It is only fair to assume that the data generated by different methods are not necessarily identical in terms of polydispersity or mean diameters. This has been shown, e.g., in a comparative study of static and dynamic light scattering [7]. Fourth, even for spheroid particles there is a fundamental difference in what is considered to be a "mean diameter" (which is the parameter usually compared). This becomes obvious if we compare, e.g., three particles of the sizes 1, 2 and 3 units. The number mean is, of course, $\frac{\sum d}{n} = 2.0$ units. It is often expressed in the form $D[1,0]$. This notation reflects the fact that a sum of diameters of the power 1 (i.e., the diameter itself) is divided by a number (or, in other words, a "diameter" of the power 0). This value is a reasonable measure if the emission properties of the particle are only correlated to the number of particles, not to their size or volume. In reality, though, this is often not the case (*vide supra*). Let us take polymeric particles with embedded dyes (organic or inorganic) as an example. In this case, neglecting excited energy migration for a moment, the volume of the particle is correlated to its overall emission intensity. Now, it seems more reasonable to consider the "number-volume mean" of the above three particle sizes, which is $D[3,0] = \sqrt[3]{\frac{\sum d^3}{n}} = 2.29$ units. Similarly, when comparing particles in terms of their antibody binding capacity or if their emission emanates from the outer shell only, the surface plays a central role. Thus, the "number-area mean" $D[2,0] = \sqrt{\frac{\sum d^2}{n}} = 2.16$ units should be discussed. The problem with these simple means of the form $D[X,0]$ is that they all contain the number of particles and particle counting is not normally carried out for large numbers and for very small particles. Hence, the concept of moment means was introduced. The only two moments to have practical relevance are:

$$D[3,2] = \frac{\sum d^3}{\sum d^2} \text{ – surface area moment (also known as Sauter mean diameter)}$$

and

$$D[4,3] = \frac{\sum d^4}{\sum d^3} \text{ – Volume or mass moment mean (De Brouckere mean diameter)} .$$

These means represent—in analogy to the moments of inertia in mechanics—the center of gravity of the surface area and volume (or mass) distribution, respectively. The advantage of using the latter two distributions is twofold: (a) the formulae do not contain the particle numbers anymore and (b) light diffraction measurements produce an initial distribution around volume terms and calculation of $D[4,3]$ is therefore straightforward, while all other means have to be recalculated from the original data and are thus prone to some statistical error. So, what is the right mean? Obviously, there is no final answer to this. If, e.g., particle sizes are measured by an electron microscope, what is measured is $D[1,0]$; if image analysis is used $D[2,0]$ is generated; electrozone sensing measures the volume of all particles which is then divided by the particle number, thus affording $D[3,0]$. In contrast, diffraction-based measurements, as used by most particle sizers, offers $D[4,3]$. While all these values are "correct", special care should be taken when comparing data from different sources and particle standards, such as those offered by the National Institute of Standards and Technology (NIST), need to be used as reference.

4.2 Colloidal Stability and Aggregation

Many micro- and nanoparticles are produced (by milling, precipitation or reduction) in an hydrophobic environment. Since (a) practically all immunoassays are performed in aqueous, i.e., polar, environments, where nonpolar particles immediately aggregate, and (b) in order to enable protein conjugation, polar functional groups are typically incorporated in the surface layer of the particle. This can either be done by adsorption of surfactants and/or polyelectrolytes or, in the case of very small particles, also by adsorption of ions (e.g., citrate). Therefore, prior to antibody conjugation, the suspension is more or less colloidally stable. Depending on the surface nature, this stabilization can be achieved either electrostatically (ions, polyions, ionic surfactants), sterically (polymers, nonionic surfactants) or in a combined way.

Nonetheless, aggregation remains a critical issue whenever this suspension is disturbed by adding salt or polymers or sometimes only by too intense centrifugation. In this respect, the following cases are of special concern: (a) the production of complex particle architectures (e.g., core-shell particles, polyelectrolytic layer-by-layer structures) that have to undergo numerous engineering steps, (b) the process of antibody conjunction by means of linker molecules, and (c) storage effects whenever antibodies are attached, because antibodies consist of differently charged domains, which may lead to charge patch flocculation.

Although many authors do not include aggregation studies in their work, it should be noted that these effects may be drastic. Notably, a layer-by-layer architecture on particle surfaces was found to result in two different types of aggregation: a very tight, irreversible charge patch flocculation and a more flexible bridging process when one polyelectrolyte molecule binds to two particles [8].

4.3 Influence of Surface Chemistry

In most techniques mentioned there is the need to encapsulate the signal-generating particle for reasons of colloidal and, in some cases, thermodynamic stability as well as in order to generate an interface for the antibodies to be attached to. In principle, many different strategies may be or have already been employed, from PEGylation, dendritic and biomimetic architectures to a layerwise assembly of polyelectrolytes, to name just a few [9–11].

As proteins are generally susceptible to any changes in their environment, care has to be taken to avoid adverse effects on the sensitivity and specificity of the antibodies attached to the particle surface. One of the dominant disturbances in particle-based assays can be expected to arise from the surface charge density: while for the binding of the antibodies to the particle, reactive groups (e.g., $-NH_2$ and $-COOH$ in the case of a carbodiimide binding) need to be polarized to form a covalent bond, some of these groups will remain unbound. These unbound groups may then act electrostatically on the nearby antibodies. Therefore, reducing the surface density of reactive groups on the particle is a helpful strategy in optimizing the assay performance.

That said, it might be anticipated that assays based on covalently bound antibodies should produce better results than those with noncovalent, but purely electrostatic attachment. Although there are only few comparative studies, this does not seem to be the case. Brümmel et al. found even better results, i.e., lower LOD, for electrostatic attachment [12]. Thus, it may be assumed that the whole picture is yet more complex. In particular, the coupling procedure itself is known to have, mainly by inducing aggregation, some influence on the overall test as well. The dominant effect here is charge-related flocculation whenever some of the particles bear opposite charges or differently charged domains for some time. Therefore, it is important to ensure a fast coupling procedure.

As far as the surface chemistry is concerned, the most promising results were obtained when antibodies were attached to a "near-biological" particle surface. Such biomimetic surfaces may either consist of cell-membrane mimetic polymers, e.g., 2-methacryloyloxyethyl phosphorylcholine (MPC). MPC-containing block-*co*-polymers were even used to synthesize particles of submicron size [13]. In a more simple approach the use of polymeric detergents, e.g., alkylated polyethyleneimines or grafted polyethyleneglycols, was also shown to have a positive effect on the assay performance [12].

5 Issues Concerning Specific Markers

5.1 Fluorescent Polymeric Beads

The use of polymeric microspheres, typically polystyrene, for immunosensing purposes started in the 1950s with the latex agglutination test: latex micro- or nanoparticles were covered with a layer of antibodies. Upon adding a sample containing the antibody-specific antigen above a certain threshold concentration, the formation of particle aggregates can be observed visually by a sudden onset of turbidity.

Nowadays, polymeric particles doped with dyes are widely used in several emission-based assay formats [14, 15]. The main advantage of using emissive nanobeads is the possibility to "attach" several 10^3 dye molecules to one antibody, thereby increasing the S/N without too much emission quenching. Recent developments have focused on the inclusion of near-infrared dyes, phosphorescent ruthenium(II) bipyridyl complexes and lanthanide chelates (*vide supra*) as well as the engineering of pH-sensitive emission characteristics.

The application of doped latex beads in standard assays (i.e., in MTPs or along a nitrocellulose/nylon membrane in lateral flow tests) is basically faced with two latex-specific challenges. First, assays based on pure latex exhibit a very high degree of nonspecific binding due to hydrophobic interaction between the particle surface and ubiquitous plasma proteins from the sample. This can be reduced, e.g., by coating the particles either with polymers of reduced hydrophobicity or with those mimicking a cell wall. Alternatively, the whole polymer bead can be made of an appropriate polymer, such as a hydrogel. Second, it should be noted that dyed polymer beads below 1 μm (diameter) are more difficult to produce as uniform spheroids. Quantitative applications that need to make use of very small particles (especially in-vivo imaging) are therefore disadvantaged.

5.2 Lanthanide-Based Systems

Lanthanide labeling, the second most common labeling technique in MTP-based assays (behind nonfluorescent enzymatic markers) allows for the best LOD at present. Recent improvements reported by Matsuya et al. afforded the detection of as little as 40 fg/mL α-fetoprotein in a sandwich-type immunoassay [16]. Although lanthanide ions themselves, e.g., Eu^{3+} but also Dy^{3+}, Tb^{3+} and Sb^{3+}, are only weak absorbers in the visual range of the electromagnetic spectrum (due to spectroscopically forbidden transitions), the presence of energetically matched chelators or ligands allows for their applicability as markers: light absorption takes place in the ligand, followed by intersystem crossing to a ligand triplet state and a subsequent excitation energy transfer to an otherwise spectroscopically forbidden lanthanide state. The latter state shows long-lived emission at visible wavelengths and can be detected by time-gated measurements allowing a minimization of interference by background emission from tryptophan units (ubiquitous in any protein) and scattered light [5, 17]. In many recent investigations, these lanthanide ions were embedded in polymeric beads to avoid environmental effects on the emission lifetime. While typical Eu^{3+} labels require excitation wavelengths between 330 and 370 nm, the use of europium-tetracycline complexes (EuTc) shifts this absorption to 405 nm, thereby matching the emission line of blue diode lasers. These EuTc complexes can easily be embedded into different polymeric matrices [18].

By virtue of the spectroscopically forbidden transitions, these systems will not exhibit intra- or interparticle energy migration within the lanthanide moiety. Marked effects on the emission lifetime can thus be excluded. Instead, high lanthanide concentrations will lead to an enhanced signal. Its intensity can be expected to be roughly proportional to the volume of the particles used.

Since lanthanide-based assays are based, in effect, on intensity measurements (within a certain time interval), the homogeneity of particle sizes and of lanthanide concentrations within these particles are crucial factors. In this context, it should be noted that most producers of latices are capable of engineering particle sizes of 1 μm and above at very high reproducibility. Smaller particles, that may be needed in lateral or vertical flow applications, are still a challenge.

With the recent development of the *up-converting phosphor technology* (UPT) the interest in lanthanide-based markers for bioanalytical purposes has gained further momentum [19]. UPT particles possess a core-shell structure, where the core is made up of ceramic material, e.g., yttrium oxysulfide Y_2O_2S, doped with two different types of rare earth ions, e.g., Yb^{3+} and Er^{3+}. The core is embedded in a silica shell which improves both chemical and colloidal stability and reduces unspecific binding of proteins to a minimum. On excitation with an infrared light ($\lambda = 980$ nm) Yb^{3+} gets optically excited and

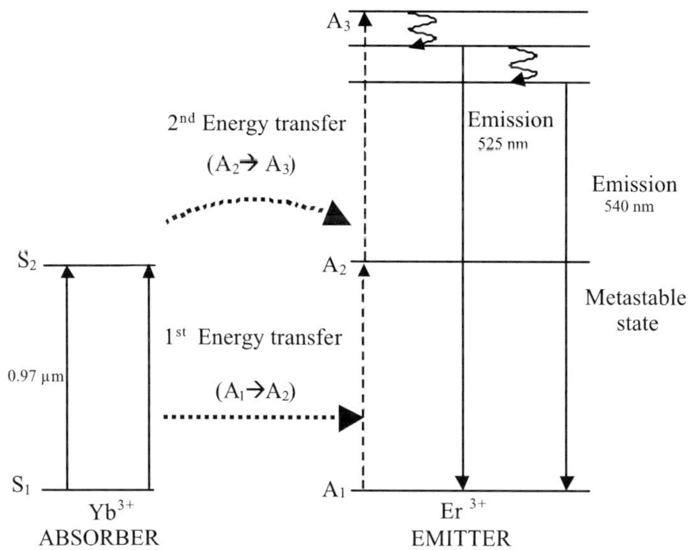

Fig. 4 Scheme of two-photon up-conversion process: up-converting phosphor technology (UPT) particles contain two different lanthanides, the infrared (IR) absorber (Yb^{3+}) and the visible emitter (Er^{3+}). Yb^{3+} sequentially absorbs two IR photons and transfers their energy (via a metastable intermediate state, therefore "phosphor") to the Er^{3+} species. The Er^{3+} emission is in the visible range of the spectrum. (Adapted from: Niedbala [19])

transfers its excitation energy to a metastable (i.e., long-living) energy state of the Er^{3+} ion. During the lifetime of this metastable state a second excitation and transfer step to the Er^{3+} ion takes place. Thereby, Er^{3+} is excited to an energetically higher "up-converted" emissive state (Fig. 4). Accordingly, the luminescence from this state is observed at shorter, visible wavelengths ($\lambda = 550$ nm). Since (a) biological material does not up-convert, and (b) the excitation wavelength is very long, autofluorescence is completely suppressed. The use of different rare earth ions facilitates several different emission wavelengths, thus providing an elegant way to establish signal multiplexing in immunoassays.

So far, UPT particles have successfully been used in immunohistochemistry, microarrays, the antibody-based detection of *Escherichia coli* and drugs of abuse, but also in the detection of specific nucleic acid sequences [20–22]. Sensitivities were several orders of magnitude better than those of conventional tests. Mouse αDIG-labeled UPT particles, for example, were used in a lateral flow format to detect gel-purified amplicons of as little as 0.1 pg or 0.3 amol by means of an immunoreaction [21].

Even more than in FRET-based assay applications, the spectroscopic performance of UPT particles strongly depends on the homogeneity of the distances between energy donator (Yb^{3+}) and acceptor (Er^{3+}), the photon flux density of the exciting light source (pulsed laser) and the sensitivity of the detector.

5.3 Dissolvable Nanoparticles

Organic markers, nonemissive as a particle but dissolvable into a fluorescent derivative, have only recently been incorporated into immunoassays [23, 24]. In these experiments, hydrophobic fluorescein diacetate (FDA) was first milled in a solution of amphiphilic polymeric surfactant, e.g., alkylated polyethyleneimines or grafted polyethylene glycols. This resulted in a colloidally stable suspension containing particles of $d{\sim}107$ nm in diameter (volume mean) that showed only negligible leakage of FDA when stored in aqueous buffer. At the same time, the polymer provided an interface for the conjugation of antibodies. This labeling system— particle, polymeric layer and antibody layer—was now used as a "detector antibody unit" in an MTP-based sandwich immunoassay. After the affinity reaction and washing steps, the nonemissive particles were dissolved and converted into highly fluorescent, dianionic fluorescein by means of adding a mixture of dimethylsulfoxide (DMSO) and 1 N NaOH (1 : 1) (Fig. 5). For the above particle size, an effective F/P of 2.6×10^6 was calculated. After some optimization, this model assay afforded an LOD of as little as 0.06 ng/mL or 37 amol mouse IgG—a value superior to many ELISAs—even though polyclonal, i.e., not highly specific, antibodies were used as the surface layer on the particles. In com-

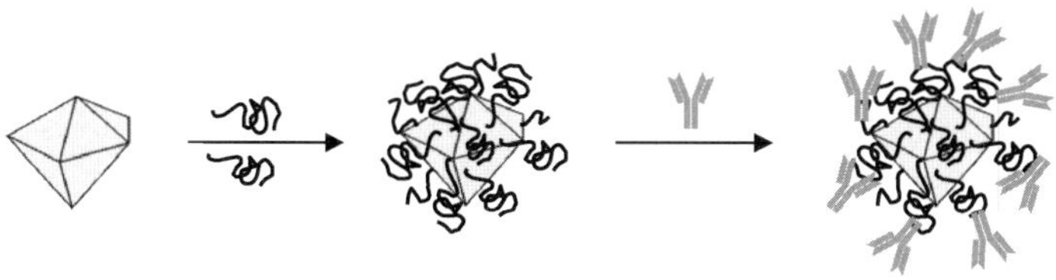

A

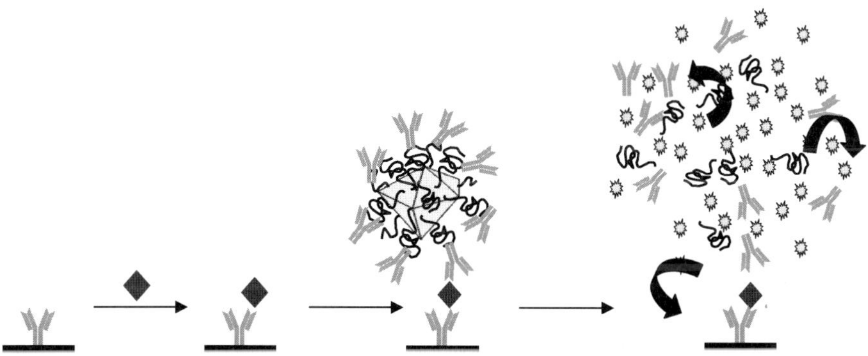

B

Fig. 5 **A** Amphiphilic polymeric detergent is adsorbed on non-fluorescent, hydrophobic fluorescein diacetate (FDA) nanoparticles, followed by the conjugation to antibodies. **B** In a sandwich-type immunoassay these functionalized particles form the "detector antibody unit". After the affinity reaction the FDA core is dissolved and converted into dianionic fluorescein. (Adapted from Brümmel [24])

parison to more traditional FITC-labeling of the same detector antibody, the fluorescence signal was amplified by a factor of up to 2.7×10^3. Naturally, this factor as well as the effective F/P depends on the particle size. Accordingly, in experiments with $d{\sim}500$ nm, values of up to 1.8×10^5 were reported. It should be noted that over several orders of magnitude no concentration quenching was observed when the emission from the MTP was excited and monitored in front-face mode. Interestingly, the amplification compared to FITC-labeling also depended on the analyte concentration. This effect could be elucidated by studying the binding kinetics using surface plasmon resonance, which suggested that above a certain threshold concentration of analyte, the particles show multivalent binding behavior, i.e., one particle binds to two to three catcher antibody/analyte complexes [8, 25, 26].

Dissolvable particles of this size are typically produced by milling, rather than by seeding and precipitation. So, the exact reproducibility of the milling conditions with respect to energy input, batch volume, viscosity and concentration of the detergent is vital in order to guarantee acceptable batch-to-batch variations. Nonetheless, bi- or even multimodal size distributions will be the norm. Therefore, a size separation by one or more centrifugation steps is usually unavoidable. Further, it should be understood that the micromechanical properties of the organic material are decisive for the result of the milling process, thereby limiting the choice of dyes to be used.

5.4 Quantum Dots and Related Systems

Semiconducting luminescent material in the size range of $\sim$1–10 nm offers unique spectroscopic properties, which is due to the fact that this size range is equal to or smaller than the spatial spread of the excitonic wavefunction of the excited particle. This leads to a quantum confinement effect, similar to the "particle in the box" problem, resulting in a splitting of electronic levels, narrow emission bands and a very high brightness ($\varepsilon \times \Phi_{em}$). In contrast to bulk semiconductors of the same material, there is no band structure in a QD anymore. Rather, the particle behaves like an intermediate between an atom and an extended semiconducting material. QDs are strictly size-governed systems, e.g., the smaller the core, the more blue-shifted the emission maximum. As a result, the spectroscopic properties of QDs can be engineered over a wide range [27]. Since "naked" dots are prone to oxidative shrinking, they need to be capped by a shell of an insulating or wide-bandgap material that is crystallographically matched to the luminescent core. The synthesis commonly proceeds in nonpolar organic solvents, e.g., toluene. In order to transfer the dots to an aqueous environment, wherein biomolecules can be attached, an organic layer bearing functional groups needs to be adsorbed onto this shell. The most commonly used core materials so far are either III–V or II–VI semiconductors, such as CdSe, which is then capped by an insulator, e.g., ZnS. Common examples for materials used as the surface layer are tri-*n*-octylphosphine oxide (TOPO) and hexadecylamine [28]. The ability of such core-shell structures to maintain the functional properties of the biomolecule have been known since 1998 when colloidal QDs were first used as biolabels and in an in-vitro agglutination test [29, 30]. The methodology of QD syntheses has made considerable progress since then and elongated core-shell nanoparticles, so-called "quantum-well–quantum dots" (also referred to as "nano-onions") and related topologies have been reported [31, 32].

Interest in QDs for bioassays is mainly—in comparison to organic dyes—due to their narrow emission bands and their broad but energetically well-separated, rather unstructured absorption spectra. Together with only minimal photobleaching, this mix of properties makes QDs interesting candidates for multiplexing approaches in bioassays.

For quantitative applications like immunoassays, however, there are a number of critical issues that need to be taken into account. The main problem is their photophysical instability: QDs are sensitive to pre-illumination. In particular, exposing them to UV light increases their apparent quantum yield of emission. This photobrightening process is a consequence of a competition between band-edge luminescence and emission due to recombination of trapped states due to surface impurities, which are ubiquitous in any crystal. As a rule, QD cores are grown kinetically (i.e., relatively fast), and so the amount of surface impurities will always be significant. Thus, emission intensities in QD-based FIAs can only be compared if (a) the QDs have on average the same concentration of surface impurities and (b) all wells in the MTP are pre-illuminated in exactly the same way (i.e., same duration, photon number, wavelength and time interval between illumination and measurement). Whereas (a) can be achieved by means of an internal standard, (b) is more difficult to accomplish. A second issue is the overall emission intensity. Owing to their small sizes, usually only one to two QDs can be expected to bind to one antibody. Despite their outstanding emission properties, this very small ratio may easily result in weak emission signals. Finally, a third issue is a result of the fact that QDs are among the most complex structures feasible as markers. This complexity arises from the necessity to control not only the nucleation process of the particle core, but also the uniform growth of the shell material, passivating the shell, the adsorption of the outer organic layer and finally the binding of recognition molecules to the latter. To do this in a reproducible way for several production batches is an enormous challenge.

5.5 Particle-Enhanced Fluorescence

An alternative approach to using nanostructures in immunoassays is based on the observation that organic dye molecules change their spectroscopic behavior in the vicinity of electron-rich *noble metal particles* and surfaces. Of special interest is the increase in the quantum yield of fluorescence that has been reported when poorly or nonemissive molecules approach gold or silver nanoparticles (other effects observed in the vicinity of noble metal particles are surface-enhanced Raman scattering (SERS) increased rates of Förster resonance energy transfer and multiphoton excitation). This phenomenon can be observed at molecule-particle distances of less than 20–30 nm and was first theoretically predicted in the 1980s by Gersten and Nitzan [33, 34]. It is ascribed to a local enhancement of the electromagnetic field which leads

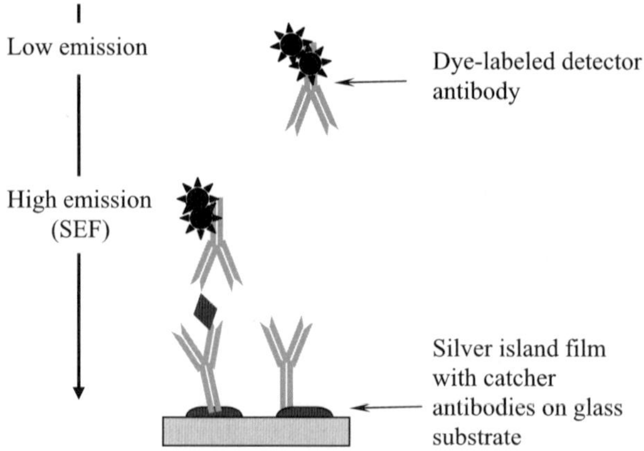

Fig. 6 Scheme of a sandwich immunoassay using surface-enhanced fluorescence (*SEF*). On a glass substrate spotted with silver islands ("particles"), a layer of catcher antibodies is adsorbed. Next, analyte-containing sample and the dye-labeled detector antibodies are incubated. Only in the case of a complete sandwich architecture are the dye labels in close proximity to the silver islands, thereby showing an emission increase by a factor of 10–15 due to SEF. (Adapted from Aslan et al. [36])

to an additional radiative decay rate. Recently, this surface-enhanced fluorescence (SEF) effect was utilized in protein assays where well-characterized silver islands, instead of particles, were placed on a glass substrate prior to immobilizing catcher antibodies and conducting the assay. The detector antibodies were marked with a poorly fluorescent organic dye. Upon binding, the dye molecule enters the SEF-active zone and, when optically excited, starts to fluoresce (Fig. 6) [35–37].

SEF was demonstrated to work in many different set-ups, e.g., in a solution-based sensing platform using SEF instigated by an affinity reaction between silver spheres coated with biotinylated BSA and silica-coated silver spheres with Cy3-labeled streptavidin [38]. In this case, the emission enhancement was only moderate (three- to fivefold). However, this is not a really quantitative assay format, because the cascade can not be stopped easily.

6 Conclusion and Outlook

The incorporation of particles as signal-generating elements in FIAs is a promising strategy to improve the performance of such assays, even beyond that of ELISAs and EIAs. Some of the particulate markers have already come very close to this goal. In addition to this, inorganic, emissive particles are, at present, the only systems that have the potential to realize a spectral multiplexing strategy, whereby several analyte molecules in one sample can be quantified with one measurement, i.e., by measuring a luminescence spectrum (Fig. 3).

The most important issues that will decide whether this class of markers will receive more attention from assay producers and end-users are (a) a uniform and reproducible production of these nanoparticles, (b) the colloidal stability of the dispersions, both in terms of storage time and with regard to the conjugation process, (c) the S/N that can be obtained by these markers, (d) the applicability of standard MTP reader systems or availability of suitable new instruments and, of course, (e) the simplicity of use.

In a more general context, the overall aim of all immunosensing reactions is to detect very small quantities of analyte in a very specific way, i.e., against a background of many more or less similar (proteinaceous) molecules. Obviously, a "natural barrier" is defined by the affinity properties of the antibodies themselves. Coupling markers to them generally deteriorates their performance and leads to more unspecific binding. So, the ultimate challenge would be to establish a procedure whereby to eliminate at least part of these unspecific reactions and, at the same time, generate a spectrally narrow and highly amplified emission signal. Using emissive nanoparticles can fulfil only some of these requests. The additional incorporation of photochemical, polymeric or enzymatic reactions as well as of dendritic structures may hold promise for further improvement.

Acknowledgements The author wishes to thank the German Federal Ministry of Education and Research (BMBF) for the generous funding of our research work, the Hong Kong University of Science and Technology (HKUST) for awarding me several research stays and finally 8sens.biognostic GmbH (Berlin) for giving me the chance to publish some of our research results. Further, I am indebted to Prof Reinhard Renneberg (HKUST) and Dr Ute Resch-Genger (Federal Institute for Materials Research and Testing, Germany) for many inspiring and fruitful discussions in the fields of bioanalytics, nanoparticles and optical spectroscopy.

References

1. Berson SA, Yalow RS (1959) J Clin Invest 38:1996
2. McFall RM, Treat TA (1999) Annu Rev Psychol 50215
3. Zweig MH, Campbell G (1993) Clin Chem 39:561
4. Brooks JL, Mirhabibollahi B, Kroll RG (1991) J Immunol Methods 140:79
5. Seydack M (2005) Biosens Bioelectron 20:2454
6. Davies C (1994) Immunoassy design. In: Wild D (ed) The immunoassay handbook. Stockton, New York, p 15
7. Schuhmann R, Müller RH (1995) Pharm Ind 57:1 (in German)
8. Brümmel Y, Chan CP-Y, Renneberg R, Thünemann A, Seydack M (2004) Langmuir 20:9371
9. Chávez JL, Wong JL, Jovanovic AV, Sinner EK, Duran RS (2005) IEE Proc Nanobiotechnol 152:73
10. Nagasaki Y, Kataoka K (2005) IEE Proc Nanobiotechnol 152:89

11. Park J, Korosawa S, Watanabe J, Ishihara K (2004) Anal Chem 76:2649
12. Brümmel Y, Chan CP-Y, Rennberg R, Thuenemann A (2004) Langmuir 20:9371
13. Park J, Kurosawa S, Watanabe J, Ishihara K (2004) Anal Chem 76:2649
14. Medina MB (2006) J Agric Food Chem 54:4937
15. Hall M, Kazakova I, Yao Y-M (1999) Anal Biochem 22:165
16. Matsuya T, Tashiro S, Hoshino N, Shibata N, Nagasaki Y, Kataoka K (2003) Anal Chem 75:6124
17. Dickson EFG, Pollak A, Diamandis EP (1995) J Photochem Photobiol B 27:3
18. Wolfbeis OS, Böhmer M, Dürlop A, Enderlein J, Gruber M, Klimant I, Krause C, Kürner J, Liebsch G, Lin Z, Oswald B, Wu M (2002) Advanced luminescent labels, probes and beads and their application to luminescence bioassay and imaging. In: Kraayenhof A, Visser AJWG, Gerritsen HC (eds) Fluorescence spectroscopy, imaging and probes. Springer, Berlin, Heidelberg New York, pp 3
19. Suyver JF, Aebischer A, Biner D, Gerner P, Grimm J, Heer S, Krämer KW, Reinhard C, Güdel HU (2005) Opt Mater 27:1111
20. Niedbala RS, Feindt H, Kardos K, Vail T, Burton J, Bielska B (2001) Anal Biochem 293:22
21. Corstjens PLAM, Li S, Zuiderwijk M, Kardos K, Abrams WR, Niedbala RS, Tanke HJ (2005) IEE Proc Nanobiotechnol 152:64
22. Corstjens PLAM, Zuiderwijk M, Brink A, Li S, Feindt H, Niedbala RS, Tanke HJ (2001) Clin Chem 47:1885
23. Yang W, Trau D, Renneberg R, Yu NT, Caruso F (2001) J Colloid Interface Sci 234:356
24. Trau D, Yang W, Seydack M, Caruso F, Yu N-T, Renneberg R (2002) Anal Chem 74:5480
25. Chan CP-Y, Brümmel Y, Seydack M, Sin K-K, Wong L-W, Merisko-Liversidge E, Trau D, Renneberg R (2004) Anal Chem 13:3638
26. Brümmel Y (2005) PhD thesis, Potsdam University (in German)
27. Willard DM (2003) Anal Bioanal Chem 376:284
28. Riegler J, Nann TH (2004) Anal Bioanal Chem 379:913
29. Bruchez M Jr, Moronne M, Gin P, Weiss S, Alivisatos AP (1998) Science 281:2013
30. Chan WCW, Nie S (1998) Science 281:2016
31. Riegler J, Nick P, Kielmann U, Nann TH (2003) J Nanosci Nanotechnol 3:380
32. Weller H (1998) Curr Opin Colloid Interface Sci 3:194
33. Gersten J, Nitzan A (1981) J Chem Phys 75:1139
34. Weitz DA, Garoff S, Gersten J, Nitzan A (1981) J Chem Phys 78:5324
35. Lakowicz JR (2001) Anal Biochem 298:1
36. Lakowicz JR, Shen Y, D'Auria S, Malicka J, Fang J, Gryczynski Z, Gryczynski I (2002) Anal Biochem 301:261
37. Aslan K, Gryczynski I, Malicka J, Matveeva E, Lakowicz JR, Geddes CD (2005) Curr Opin Biotechnol 16:55
38. Aslan K, Lakowicz JR, Szmacinski H, Geddes CD (2004) J Fluoresc 14:677

Springer Ser Fluoresc (2008) 6: 469–485
DOI 10.1007/4243_2008_058

Published online: 20 May 2008

Advances in Fluorescence Enzyme Detection Methods

James N. Miller

Department of Chemistry, Loughborough University, Loughborough LE11 3TU, UK
J.N.Miller@lboro.ac.uk

Abstract This chapter summarises recent progress in the use of fluorescence to study enzyme-catalyzed reactions. Most such studies now use non- or weakly-fluorigenic substrates converted by an enzyme to a strongly-fluorescent product. Some of these substrates are based on established fluorophores, such as fluorescein, rhodamine, coumarins, or resorufin, while others use energy transfer principles. The latter include low molecular weight substrates containing a fluorescent donor and a quenching acceptor: the enzyme separates the two moieties and the donor fluorescence is enhanced. In other cases, proteins or polysaccharides over-labelled with one fluorophore are internally quenched, but the enzyme fragments them to yield enhanced fluorescence. Fluorescence polarisation changes are also useful: kinases as well as hydrolytic enzymes can be studied using this method. Substrates giving insoluble reaction products have found much use in combination with separation and blotting methods. An extremely important area is the study of intra-cellular reactions: the cell permeability and compatibility of the reagents is crucial. Lastly, some examples of multiplexed enzyme assays are summarised. These are facilitated by modern data handling methods which allow even strongly overlapping spectra to be resolved into several components.

Keywords Chemometric methods · Fluorigenic substrates · Fluorescence polarisation · Fluorescence resonance energy transfer · Intracellular enzyme assays

1 Introduction

The detection of many enzymes is one of the numerous areas where fluorescence spectroscopy has made a great and growing contribution in recent years. Most of the applications are based on the established advantages of fluorescence – extreme sensitivity, good spectroscopic selectivity, and adaptability to a wide variety of liquid and solid sample types and sampling devices. These spectroscopic benefits complement those of all enzyme assays whose specificity and signal amplifying effects are also well known. Combined fluorescence enzyme assays are thus expected to be more sensitive than methods using fluorescence labels alone. At the same time, many of the recent developments in fluorescence studies, such as the use of long-wavelength measurements, the application of some elegant energy transfer or fluorescence polarization principles, the use of confocal fluorescence microscopy, the detection of high-resolution separations, the use of micro-fluidic systems and flow cytometry, and the application of accessible chemometric methods to resolve overlapping spectra, have been brought into use in enzyme chemistry. As a result, excellent fluorescence enzyme assays are now available for countless applications in food chemistry, clinical and veterinary medicine, environmental analysis, high throughput screening (an area of huge current importance and interest) and basic biochemical research including, very importantly, intra-cellular measurements. Moreover, the same techniques can often be used when enzymes are used as labels in immunoassays and nucleic acid based methods, which are covered separately in this volume: such applications probably account for a significant proportion of modern enzyme analyses. In these and other applications where large numbers of samples are processed, the favoured instrumental format is the use of microtiter plates: 96-, 384-, and 1536-well plates are now routinely measurable in dedicated instruments. Moreover, these detectors combine adaptability to the different plate formats with excellent sensitivity and a range of light sources and accessories, including polarisation measurements and chemi/bioluminescence detection. Some plate readers with fluorescence lifetime capabilities are now also available and lifetime-based assays have been shown to offer good alternatives to steady state measurements. Microtiter plates designed specifically for fluorescence studies are also widely available. In applications using conventional cuvettes, a great range of fluorescence instruments is available, again with the required sensitivity, polarisation accessories, data handling facilities, and so on.

Nowadays, the great majority of fluorescence enzyme assays are based on the use of *fluorigenic substrates* [1]. These are molecules which, on reaction with an enzyme, show greatly enhanced and/or altered fluorescence properties. In many cases, the substrates themselves are effectively non-fluorescent in the conditions of the assay, so that all the measured fluorescence is due to the product of the reaction. The principles of such assays have been well known for many years, but new and better substrates for an increasing range of enzyme specificities are continually being developed and, in many cases, are available in kit form.

Before the use of fluorigenic substrates is reviewed in more detail, two other approaches to fluorescence enzyme assays should be recalled. Many enzyme-catalysed reactions involve the production or consumption of the nicotinamide adenine co-factors NADH and NADPH, and can be monitored by measuring the absorbance of these molecules at 340 nm, a wavelength at which their oxidised forms, NAD^+ and $NADP^+$, respectively, show negligible absorption. NADH and NADPH are also moderately fluorescent, with an emission maximum of ca. 460 nm in neutral aqueous solution. This fluorescence can be used to follow the kinetics of a great number of oxidoreductase enzymes, and of other enzymes whose reactions couple with the latter [2]. Many other natural products have absorption and fluorescence wavelengths similar to 340 and 460 nm respectively, however, so the use of this fluorescence in the analysis of complex bio-samples is restricted, unless a good deal of preliminary sample extraction and purification is acceptable to remove such interferences. Moreover, the NAD(P)H excitation and emission maxima may change when these cofactors bind to proteins, or in the presence of some metal ions: again, this is not ideal for routine analytical methods. Flavin co-enzymes, such as flavin mononucleotide, FMN, exhibit intense fluorescence usually at about 530 nm, but again protein-binding effects interfere with potential applications.

A different approach can be used to follow enzymatic reactions in which amino-acids or small peptides are generated from proteins, as the fluorescamine [3] or *o*-phthalaldehyde (OPA) [4] reactions, developed over 30 years ago, can be used to detect the low molecular weight reaction products. In each case, amino-groups of the products react with non-fluorescent reagents to give fluorescent products, while the original proteins either do not react or react much more slowly. Neither of these systems is ideal for the continuous monitoring of enzyme reactions. The fluorescamine reaction is effectively instantaneous, but the products are not very stable or very strongly fluorescent. The OPA (and other dialdehyde) reactions may require several minutes for completion, depending on the nature of the peptides under study, and require the use of unpleasant nucleophiles, such as thiols or cyanide. In each case, the fluorescence intensity also depends on the peptides being labelled. Fluorescamine and OPA react with the products of any reaction in which proteins are degraded to small peptides, and thus do not provide for the detection of specific proteases in complex samples. In modern practice, proteolytic enzymes are studied using fluorigenic substrates, at least some of which provide excellent specificity: this topic is treated in detail below.

It should be noted that very sensitive enzyme determinations can also be achieved with the aid of chemiluminescence (CL) as opposed to fluorescence detection methods. The CL approach often utilises one of a number of luciferase enzymes from a variety of species, and is valuable in a range of gene assays, immunoassays, etc. Chemiluminogenic substrates are available for enzymes, such as alkaline phosphatase, and peroxidase is often determined with the aid of luminol reactions. These techniques are not covered in this chapter, but it is worth noting that, while CL detection is frequently very sensitive and utilises simple instrumentation, it does not match fluorescence in terms of the different photochemical approaches available. Moreover, the number and wavelength range of CL systems are limited, and multiplexed assays are problematical as CL spectral scans are normally unavailable.

2 Fluorigenic Substrates – Introduction

As previously noted, most modern fluorescence enzyme assays utilise fluorigenic substrates. The broadest definition of such molecules is that each is designed as a substrate for a specific enzyme or group of enzymes, and that on reaction with such enzymes they undergo a significant change in their fluorescence properties. In practice, the outcome of the enzyme reaction is almost always a large increase in the intensity of the measured fluorescence, but in some systems a considerable shift in emission wavelength is just as important, and, in a small number of cases, the substrate and reaction products have similar fluorescence properties and thus have to be separated before the products are measured.

The requirements for a good fluorigenic substrate merit further discussion. Substrate stability is an important factor. Some of the substrates used a few decades ago tended to undergo spontaneous degradation, thus giving a high background fluorescence, but modern reagents are much better in that respect. Substrates providing low values of the Michaelis constant, K_m, are obviously desirable. In most cases, the reaction product should be freely water-soluble, although systems producing insoluble products have also been used (see below). The fluorescence generated should be measurable at the pH of the enzyme reaction, to allow continuous monitoring without the addition of acid or alkali for pH adjustment. The product fluorophore should normally be stable and have a high molar absorptivity (ε) and fluorescence quantum yield (φ_f), but care must be taken to make intensity measurements only over the linear response range of the measuring system: strongly absorbing fluorophores are most likely to give a non-linear response because of inner filter artefacts. Systems providing fluorescence at longer wavelengths, where autofluorescence from the sample matrix is likely to be less, are desirable, except when polarisation measurements are used (see below).

Many fluorigenic substrates have been based on fluorescein and rhodamine derivatives. A simple example is provided by fluorescein diphosphate, which has been used for over 40 years, but is still regarded as one of the most sensitive substrates for alkaline phosphatase [5]. This colourless substrate incorporates two phosphate groups which prevent the ionisation of the fluorescein molecule and thus abolish its fluorescence. The enzyme converts it in alkaline solution to fluorescein itself, which is brightly fluorescent in the conditions of the assay ($\varepsilon \sim 90\,000$, $\varphi_f \sim 0.9$). In practice, fluorescein monophosphate will be formed as an intermediate (or may be present as an impurity in the original substrate): this compound is also fluorescent, but it is usually possible to use reaction conditions that ensure its complete conversion to fluorescein. Among many applications of this reaction, its use in micro-fluidic systems, with argon-ion laser excitation of the fluorescein, is noteworthy [6]. Other di-substituted fluorescein derivatives are available for the detection of the enzymes β-glucosidase, β-galactosidase and β-glucuronidase. They have the disadvantage that a pH change may be needed after the enzyme reaction to yield the full fluorescein emission intensity. Some of these reagents have been further modified to facilitate intra-cellular studies (see below).

Several di-substituted rhodamine derivatives have been synthesised as substrates for peptidases. The amino groups of these rhodamine dye form peptide bonds with a range of model peptides, and these bonds are cleaved by matching serine and cysteine proteases to yield the parent dye rhodamine 110 [7]. (Again, the intermediate mono-peptide derivative shows some fluorescence). This fluorophore has similar fluorescence properties to fluorescein and can be studied using the same instrumentation, but with the advantage that a pH range of ca. 3–9 can be used. Many proteolytic enzymes and their inhibitors have been studied using this approach. For example, polyfluorobenzoyl-rhodamine 110 substrates can be used to assay caspase-3, a marker for cell apoptosis [8]. Naphthofluorescein di-phosphate has also been synthesised as a long-wavelength substrate for alkaline phosphatase [9]. The naphthofluorescein product fluorophore has some interesting properties: like many other naphthalene derivatives, its excitation and emission wavelengths are environment dependent. In alkaline aqueous solution, they are ca. 595 and 660 nm respectively, but the addition of 5% w/v of the detergent CHAPS shifts them to 630 and 680 nm, and also produces a signal enhancement of ca. 50%. Naphthofluorescein could thus be detected at sub-nanomolar levels using a very simple instrument with a small diode laser light source, and a simple assay for the alkaline phosphatase inhibitor theophylline, which reduces the amount of naphthofluorescein produced from the diphosphate, was demonstrated.

A second and very widely used group of fluorigenic substrates is based on the coumarin nucleus. These systems differ from those based on xanthene dyes in a number of ways. Perhaps the most important difference is that in many cases the substrates themselves are fluorescent [10]. Fortunately, their excitation and emission wavelengths are very different from those of the enzyme reaction product. For example, a typical substrate conjugate of 4-methylumbelliferone (see below) in alkaline solution has an excitation maximum at ca. 320 nm and fluoresces at about 380 nm. In the same conditions, the reaction product, 4-methylumbelliferone itself, has excitation and emission maxima of ca. 360 and 450 nm, respectively. The absorbance of the conjugate at 360 nm is negligible, and its fluorescence at 450 nm is very weak indeed. Thus, there is no serious interference from unreacted substrate in most practical enzyme assays. These wavelengths naturally vary among coumarins with different substituent groups, but in each case the reaction product has substantially longer excitation and emission wavelengths than the substrate. However, compared with fluorescein- and rhodamine-based substrates, the coumarin substrates are less intensely fluorescent, having ε values of only 15–20 000 and often lower quantum yields also. This may limit their applicability if extreme sensitivity is required, for example, in capillary electrophoresis or micro-fluidics detection systems. Different coumarin derivatives have widely different pK_a values, a factor that may markedly influence the choice of substrate for a particular assay.

The best known coumarin derivatives are those based on 7-hydroxy-4-methyl coumarin, widely known as 4-methylumbelliferone (MU). Ionisation of the phenolic group is required for fluorescence of this compound: since its pK_a value is 7.8, measurements must be made at pHs > 9. Just as in the case of fluorescein di-phosphate (see above), this high pH presents no problem in assays of alkaline phosphatase using 4-methylumbelliferyl phosphate (MUP) as the substrate. However, better MU derivatives with lower pK_a values are available as the result of work in the laboratories of Molecular Probes Inc. For coumarins, as in other groups of fluorophores, the incorporation of fluorine substituents in aromatic systems has also been shown to enhance photostability and fluorescence intensity with little or no effect on the excitation and emission spectra. For example, 6,8-difluoro-4-methylumbelliferone (DiFMU) has a quantum yield higher by 40% than that of MU and a pK_a value of 4.9 [11]. Thus, it can be detected at pHs > 5.5: this allows acid as well as alkaline phosphatases to be assayed rapidly and continuously, using 6,8-difluoro4-methylumbelliferone phosphate (DiFMUP) as the substrate [12].

Many coumarin-based substrates are available. As well as alkaline and acid phosphatases, β-glucosidase, β-galactosidase and β-glucuronidase can be assayed using MU derivatives. Several 7-amino-4-methyl coumarin (AMC) derivatives are used to study peptidase enzymes [13]. As with the rhodamine 110 substrates discussed above, the AMC substrates use individual peptides attached via the amino-group to provide specificity for a range of proteases: the fluorescent product has properties similar to those of MU. Substrates based on 7-amino-4-trifluoromethyl coumarin yield on proteolysis longer excitation and emission wavelengths (ca. 400 and 500 nm respectively). 7-Amino-4-chloromethyl coumarin (CMAC) provides substrates suitable for intracellular studies (see below).

A third group of substrates of importance consists of those based on the orange-red fluorophore resorufin. This molecule has excitation and emission wavelengths of ca. 565 and 585 nm respectively, with $\varepsilon \sim 73\,000$ and $\varphi_f \sim 0.75$ [14]. Although its Stokes shift is relatively small, like that of fluorescein and other xanthene dyes, its longer emission wavelength reduces the problems of autofluorecence background from biological samples, and its pK_a value of ca. 6.0 makes it easier to use at physiological pH values. Over many years, this molecule has been used to assay dehydrogenase reactions involving NAD(P)H [2]: these co-factors produce resorufin by reduction from its *N*-oxide, also known as resazurin. (Although resorufin is not exceptionally stable – it can be further reduced by thiols, etc. to hydroresorufin, and it is also susceptible to re-oxidation to the *N*-oxide – this is not a serious problem in most practical enzyme assays). More recently, fluorigenic resorufin derivatives have been synthesised: as with other substrates, their basis is the blocking of the ionisable phenolic group of the phenoxazine nucleus, which is essential for fluorescence. Enzymatic removal of the blocking group restores the fluorescence. Thus, resorufin galactoside has been used to detect β-galactosidase [15], an enzyme commonly used as a label in homogeneous and heterogeneous immunoassays, and range of resorufin ethers has been used to study cytochrome P-450 isoenzymes [16].

A striking recent development in the use of resorufin derivatives has been the introduction by Molecular Probes Inc. of 10-acetyl-3,7-dihydroxyphenoxazine (Amplex Red®) [17]. This molecule is oxidised to resorufin by the action of hydrogen peroxide in the presence of a peroxidase enzyme, but shows great stability when the enzyme is absent. This substrate represents a major advance over earlier fluorigenic substrates for peroxidases, such as 3-methoxy-4-hydroxyphenylacetic acid (homovanillic acid), in terms of stability, sensitivity, and its absence from biological samples, and it finds many applications in the determination of both peroxidase enzymes and hydrogen peroxide. Peroxidases are widely used as labels in heterogeneous enzyme immunoassays, while H_2O_2 is generated by a number of oxidase reactions of great importance. In addition, coupled reactions can be used for other important assays based on the same detection system. For example, acetylcholinesterase can be measured by taking advantage of the fact that its reaction product, choline, can be oxidised by choline oxidase to yield betaine and the all-important hydrogen peroxide [18].

It should be emphasised that, while the relatively small number of fluorophore families discussed above provide the majority of practical modern enzyme assays, a substantial number of other fluorigenic substrates have been synthesised: this is a very dynamic area of enzyme assay research.

3 Fluorigenic Substrates Utilising Energy Transfer

The phenomenon of fluorescence resonance energy transfer (FRET) has been widely studied and much used in bioanalytical measurements. In simple terms, FRET involves the absorption of a light quantum by a donor moiety (D), and the non-radiative transfer of the excitation energy through space to an acceptor moiety (A). A requirement is that the emission spectrum of D must overlap the absorption spectrum of A, so that there is sufficient energy from the former to excite the latter. The result is that the fluorescence of the donor is quenched, while that of the acceptor may be enhanced. In practice, non-fluorescent species are sometimes used as "dark" acceptors, and the FRET phenomenon is usually monitored by the study of D only. The crucial feature of FRET is that its efficiency is strongly distance dependent: for singlet–singlet energy transfer, the efficiency is proportional to the inverse 6th power of the D–A separation. In practice this means that the energy transfer is only observed when the D–A distance is very small, usually ca. 4–10 nm, and that energy transfer can only be detected between fluorophores which are part of the same molecule or molecular or particle-based complex [19].

In the field of enzyme assays, two distinct and elegant applications of FRET have been adopted. In the first, fairly low molecular weight substrates containing both the D and A moieties in close proximity are synthesised: the D and A groups have different chromophores. When a substrate of this type is excited at a wavelength corresponding to a D absorption band, the D fluorescence is minimised by energy transfer to A. Since D and A are chemically distinct, this type of energy transfer is known as a heterotransfer. When an appropriate enzyme specifically cleaves the substrate into two fragments, the D and A moieties are separated and the energy transfer no longer occurs, so a great enhancement of D fluorescence is observed (Fig. 1). An example of this approach is provided by the commercially available substrate for HIV protease. This has the amino-acid sequence Arg-Glu-Ser-Gln-Asn-Tyr-Pro-Ile-Val-Gln-Lys-Arg, with the fluorescent donor group EDANS ([(2-aminoethylamino]naphthalene-sulfonic acid) attached to the Glu residue and the dark acceptor group DABCYL (4-(4-dimethylaminophenylazoo)benzoic acid) attached to the lysine residue. The protease cleaves this virtually non-fluorescent peptide between the tyrosine and proline residues to expose EDANS fluorescence with excitation and emission wavelengths of ca. 335 and 500 nm, respectively [20]. A similar system is available for the determination of β-secretase (also known as BACE1) and its inhibitors. This proteolytic enzyme contributes to the formation of amyloid beta peptide and, hence, amyloid plaques in the brains of Alzheimer's disease patients. Thus, the identification of inhibitors of possible therapeutic value is of great importance [21].

This principle can be extended to other enzyme classes, and their inhibitors. Phospholipase A2 can be determined with the aid of a phospholipid

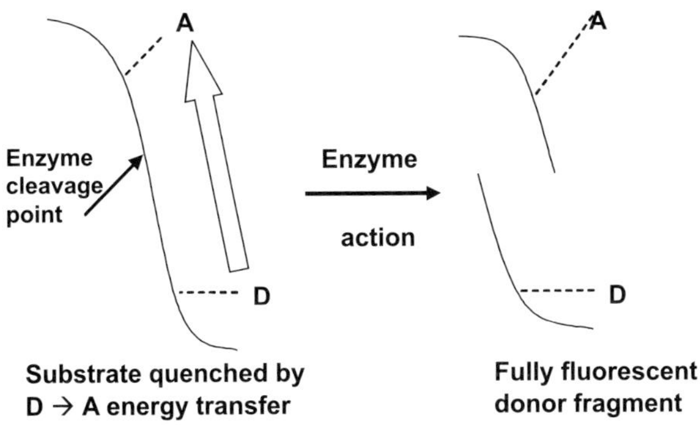

Fig. 1 Determination of a proteolytic enzyme using a model peptide substrate labelled with a donor fluorophore and an acceptor group

substrate containing the dye Bodipy-FL attached to the lipid end of the molecule and a quenching 2,4-dinitrophenyl group attached to the other end [22]. The action of the enzyme separates the fluorophore and the quencher, and thus produces increased fluorescence.

The second quite distinct application of FRET utilises substrates (often based on widely available proteins such as α-casein) that are heavily labelled with a single fluorophore. Again, the fluorescence of such molecules is quenched by energy transfer from one label group to another, the D and A groups in this case being chemically identical: clearly this will only occur if the excitation and emission spectra of the fluorophore used overlap quite strongly. Such energy transfers have been called homotransfers [19]. (Different D and A fluorophores can be used in a similar way to label proteins, etc., but this is not normally advantageous and the utility of the doubly-labelled conjugate critically depends on the ratio with which the two fluorophores bind to the protein). If the protein is degraded by a protease, each of the peptides produced probably only contains one fluorophore group at most, so the FRET effect disappears, and a large increase in fluorescence intensity can be detected (Fig. 2). Such methods are invaluable for the study of a variety of proteases and their inhibitors, an area of great importance in the high throughput screening of candidate drug molecules. Again, the assay requires careful design, depending on the enzymes to be studied, as the following examples show.

Over-labelled substrates have been shown to provide a suitable detection method for the monitoring of capillary electrophoresis separations of proteases with laser excitation. A Bodipy-FL-casein substrate with 4–10 fluorophore groups per casein molecule was used to demonstrate this approach [23]. (Note that dye-binding ratios of this type are only average values, and that their estimation involves a number of possibly unwarranted assumptions about the spectroscopic properties of both the protein and the

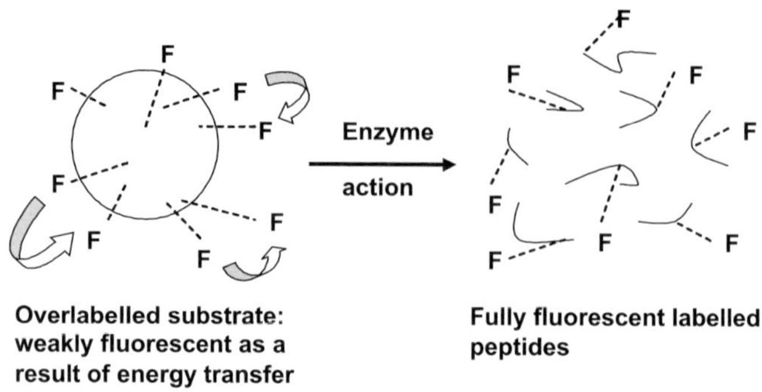

Fig. 2 Determination of a proteolytic enzyme using an overlabelled fluorigenic substrate

fluorophore). The effects of time and enzyme concentration on the digestion of the substrate were studied, using trypsin as an example enzyme. A more specialised and specific application of the same detection principle uses a heavily over-labelled gelatin-fluorescein conjugate to detect the metallo-enzymes collagenase and gelatinase [24]. One possible restriction on this approach to protease assays is that the rate of proteolysis of the substrate may be reduced if the amino-acid residues used in the fluorescence labelling reaction are the same as those normally targeted by the enzyme. However, the range of labelling chemistries available should allow this problem to be circumvented.

As with the heterotransfer systems described above, the use of these homotransfer enzyme assays is not confined to the study of proteases. A conjugate of *Micrococcus lysodeikticus* cell walls heavily labelled with fluorescein provides a very sensitive substrate for lysozyme, an enzyme of clinical significance in its own right, and one also used as a label in immunoassays. Starch labelled with BODIPY FL is similarly used to detect amylase [25], and the same dye attached to a phospholipid provides a quenched substrate for phospholipases. In this case the substrate can be inserted into cell membranes, providing direct information on cell functions and activities [26].

4 Fluorescence Polarisation Enzyme Assays

Another elegant and widely-used measurement principle in fluorescence spectroscopy is that of fluorescence polarisation. In essence, it is a method in which the rotational relaxation time of a fluorophore is compared with its excited state lifetime. A population of small molecules excited with polarised light, and hence initially oriented in relation to the electric vector of the excitation beam, will normally lose this orientation during a typical fluorescence lifetime of a few nanoseconds. The emitted fluorescence will thus show little or no polarisation. By contrast, large fluorophores, or small ones bound to larger molecules or particles, will retain some of their original orientation during the excited state lifetime, so their emitted fluorescence will be partially polarised. Such measurements thus allow any process in which a large fluorophore is converted to a small one, or vice-versa, to be monitored. The polarisation is measured by the inclusion of polarising elements in the normal optical pathway of a fluorescence spectrometer, and dedicated instruments, often using electro-optic polarisers, are also available. Samples are excited with vertically polarised light, and two emission intensities are measured with the polariser element in the emission beam being vertically and horizontally oriented in turn. The measured polarisation, p, is the difference between these two signals divided by their sum: this ratio can in theory vary between -0.33 and $+0.5$. The polarisation values and changes observed are usually more modest than theory suggests, with common values between 0 and 0.3[19]. To achieve the necessary measurement precision (ca. ± 0.002 units), good temperature control is necessary. However, changes in polarisation are in principle independent of the absolute fluorescence intensity of the sample, and polarisation methods also tolerate quite high sample absorbance values. They are well suited to microtiter plate measurements and, as with energy transfer assays, they are finding many uses in high throughput screening systems [27].

The fluorescence polarisation measurement principle can be successfully applied to a variety of enzyme assays. An attractive example is a series of methods that utilises synthetic peptide substrates labelled with fluorescein to study kinases with phosphorylating activity. The labelling of the substrate occurs at an amino-acid residue sufficiently distant from the phosphorylation site to prevent subsequent interference with the kinase reaction. Peptides phosphorylated by the latter reaction bind strongly to nanoparticles which have been surface modified by the addition of trivalent metal ion complexes. (The latter are similar to those used for immobilised metal affinity chromatography). The fluorescence polarisation of the bound phosphorylated peptides is significantly larger than that of the unbound unreacted ones, and polarisation measurements provide a direct measure of kinase activity. Serine, threonine and tyrosine phosphorylation can all be detected with equal ease (this is not the case for the fluorescence polarisation immunoassay methods for kinase enzymes), so that numerous kinases and their inhibitors can be detected [28].

Proteolytic enzymes can also be detected using fluorescence polarisation. The principle is that a fluorophore attached to a large molecule should show an appreciable polarisation of fluorescence, while after enzyme degradation it will be attached to a much smaller peptide with little or no polarisation effect. Fluorescent protein conjugates for this application must be designed and prepared with care. They must clearly exhibit a substantial fluorescence at the optimal pH of the protease to be studied. The degree of labelling also requires careful control. If the protein is over-labelled, energy transfer causes a reduced signal intensity (see above) and also causes some depolarisation of the fluorescence *before* enzyme fragmentation. A weakly labelled protein conjugate may not provide a sufficient intensity of fluorescence (polarisation measurements sacrifice some intensity in any case because of the effects of the polarising filters and the selective excitation of oriented fluorophores) or sufficient changes in polarisation. If the fluorophore groups covalently bound to the protein have significant freedom of rotation, they will give low p values and the changes in polarisation produced by proteolytic digestion will be small: labels with short links between the fluorophore itself and the group that reacts with protein surface amino acids are thus recommended. Longer-wavelength fluorophores tend to have shorter fluorescence lifetimes, sometimes < 1 ns. Such short periods minimise the differences between the polarisation of fluorescence of small and large molecules: this is one of the relatively few areas where longer wavelength fluorophores are less useful than the more conventional labels emitting in the yellow–green spectral regions (see Goulko et al., 2008, in this volume).

Several applications of this protease assay methodology have utilised the boron-containing label Bodipy-FL®. This compound has been designed to have an emission spectrum closely similar to that of fluorescein and has the high molar absorptivity (ca. 8×10^4 cm^{-1} M^{-1}) and quantum yield (~0.9) necessary for sensitive measurements. It has the advantage that it is fluorescent across the pH range 2–11, so that it can be used in assays of acidic, neutral and alkaline proteases, and it is little affected by organic solvents, which is a useful attribute in high throughput screening. Its lifetime (6 ns) is greater than that of fluorescein and it is rather more photostable. Attachment of the dye at an optimum dye-to-protein ratio to α-casein using a succinimidyl ester reaction gives a conjugate that has been used to provide rapid and sensitive assays for pepsin, papain, proteinase K and *Streptomyces Griseus* alkaline protease at pHs of 2.0, 6.0, 7.4 and 11.0, respectively [29].

A different approach to protease detection using fluorescence polarisation was described by Levine et al. [30], who labelled a peptide substrate specific for a protease from human cytomegalovirus. The peptide was biotinylated at the amino-terminus and labelled with fluorescein at the carboxy-terminus. After incubation with the enzyme, avidin was added. The intact doubly-labelled peptide bound to the avidin with a large increase in polarisation, but the fluorescent fragment of the cleaved peptide could not bind to the protein, so no change in polarisation was observed. This double-labelling approach should be applicable to many protease assays where the enzyme specificity is known and a model substrate can be synthesised and labelled at two well-separated positions.

Fluorescence polarisation methods of enzyme assay have been combined with confocal microscopy to image enzyme activity [31]. In a model system bovine serum albumin labelled with Bodipy-FL was attached to agarose beads and used as a substrate for trypsin and proteinase K. Each enzyme was shown to produce small labelled protein fragments with reduced fluorescence polarisation, and with high microscope apertures images were obtained at depths of over 200 μm.

5 Fluorigenic Substrates Yielding Insoluble Products

The reactions described thus far in this chapter involve soluble fluorigenic substrates that yield soluble fluorescent products. However, in a number of important areas there is value in using reactions yielding insoluble products. These areas include the localisation of enzymes after separation by electrophoresis, isoelectric focusing or thin layer chromatography; the detection of proteins and nucleotides in Western blotting and related methods; and use as detection systems in immuno-histochemistry and in situ hybridisation experiments. In all these cases, the enzyme activity must be detected at specific sites on a TLC plate, electrophoresis gel, tissue section, etc. Therefore, just as in non-biospecific staining methods using coloured dyes, the detection reaction must generate a product that is insoluble as well as fluorescent. Here as elsewhere it is expected that fluorescence detection will be more sensitive than colorimetric methods, whether the latter use non-specific or enzyme-based staining methods. In many applications the precipitation process is designed to detect alkaline phosphatase, a very common labelling reagent for antibodies, nucleotides, etc. An extremely elegant substrate for alkaline phosphatase, ELF 97 (the acronym standing for Enzyme Labelled Fluorescence), has been developed by Molecular Probes Inc. [32]. This molecule, 2-(5-chloro-2-phosphoryl-oxyphenyl)-6-chloro-4(3*H*)-quinazolinone, is converted by the enzyme from a weakly blue fluorescent phosphate to a strongly fluorescent green alcohol, which is insoluble and stabilised by intra-molecular hydrogen bonding. The product fluorophore has a large Stokes shift (excitation and emission wavelengths of ca. 350 and 530 nm, respectively), which facilitates multi-colour applications, and can be used in a variety of phosphatase-based assays, including some secondary detection systems for in situ hybridisation studies, immuno-histochemistry and cell labelling. Its photostability is so good that repeated photography of the completed reaction is simple, and autofluorescence from other sample components can be removed by deliberately photo-bleaching them with minimal effect on the phosphatase product. Analogous substrates are available for other enzymes, including esterases, lipases, and β-glucuronidase.

The use of soluble enzymes for the detection of molecules already attached to a solid surface, for example, in enzyme immunoassays, is very well established and is not treated here.

6 Intracellular Enzyme Assays

One area of fluorescence enzyme assays where major advances have recently been made has been the ability to make intracellular measurements. In many cases, the enzyme activities studied and the fluorescent moieties used are similar to those described in the previous sections, although intracellular studies place extra demands on the properties of the substrates. It is necessary for the substrate to be inserted into the cell, preferably from the culture medium, with minimal disruption of the cell function; the substrate and its fluorescent product must not leak out of the cell to a significant extent, as this would give misleading results in the enzyme assays. Several methods for tackling these problems have been developed. One approach, which may appear relatively obvious, is to use lipophilic fluorescent substrates, and C_8 and C_{12} derivatives of β-galactosidase substrates, such as fluorescein and resorufin β-galactosides[33]. The same enzyme can be assayed in live cells using a C_{17} derivative of a difluoro-umbelliferyl galactoside. Pentafluoro-benzyl derivatives of fluorescein-based substrates also show improved cell insertion and product retention properties, because of the relatively non-polar pentafluorobenzyl group [34]. In other cases, a more subtle approach to the problems of cell penetration and retention has been used. Substrates containing chloromethyl groups are found to give good cell retention properties, even surviving a sequence of cell divisions, possibly as a result of an intracellular enzyme-mediated reaction with glutathione. This principle has been used in the development of chloromethyl derivatives of fluorescein- and AMC-based substrates [35].

7 Multiplexed Enzyme Assays

Since enzymes, and often their inhibitors, are very specific in their activities, the selectivity of fluorescence spectroscopy raises the possibility of screening for two or more enzymes simultaneously in the same solution, provided the fluorescence signals from the substrates in use can be resolved, and the participants in the different reactions do not interfere with each other. (Other methods of multiplex screening, such as the use of two or more sets of reagents on a single microtiter plate, or the use of flow-based methods to separate the detected fluorophores on a time basis, will not be discussed here). Ideally, we could imagine a system in which two or more distinct fluorescent products are produced by the separate enzyme reactions, the fluorophores having a common excitation wavelength and well separated emission spectra. Very rapid spectral scans, the use of array detectors or even simple optical filter systems, would then allow the enzymes or inhibition processes to be monitored more or less simultaneously. In practice, despite the large number of fluorigenic substrates now available, this simple situation is not likely to arise. In the visible and near-IR regions of the spectrum, most fluorophores have only modest Stokes shifts, and two or more fluorophores excited at a single wavelength are likely to have moderately or even heavily overlapping emission spectra. (Tandem labelling might overcome this problem, but the technique has not been much applied to conventional enzyme assays). However, modern data handling methods may be able to overcome this problem: in particular partial least squares (PLS) statistical methods [36] have been shown to be capable of resolving the contributions from overlapping spectra with an accuracy that is more than sufficient for many screening procedures.

Figure 3 shows the strongly overlapping spectra of α-casein conjugates of the fluorophores fluorescein, BODIPY FL and BODIPY 493, each conjugate being a possible substrate for proteolytic enzymes. The two BODIPY dyes have almost identical maximum emission wavelengths (though different emission profiles), and that of fluorescein is only about 10 nm greater. Nonetheless, the application of the so-called PLS-1 method allowed the satisfactory resolution of mixtures of these three dyes. The calibration set consisted of 25 mixtures of the fluorophores, including some where one component was present at much lower levels than the other two. After the removal of two possible outliers from this set, the calibration lines for each of the fluorophores were calculated, and used to predict the solute concentrations in 20 further mixtures. The results for this prediction set, which again included

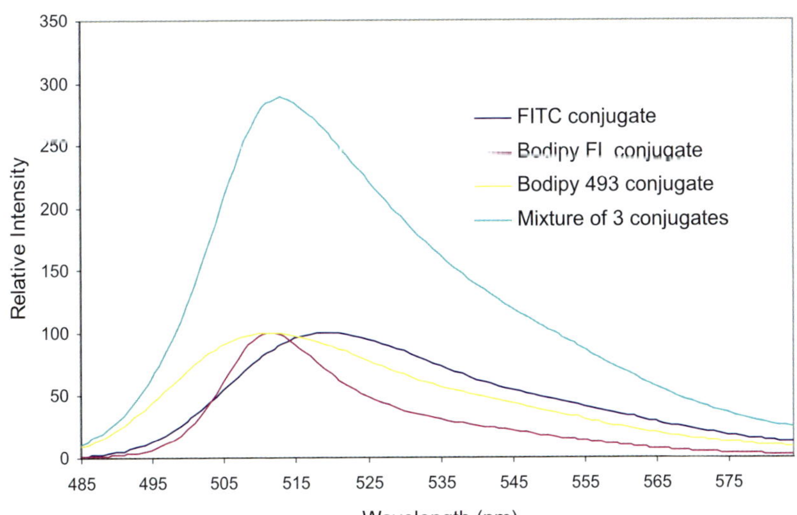

Fig. 3 Strongly overlapping emission spectra of three fluorigenic α-casein lablled enzyme substrates

mixtures with highly diverse concentrations, were excellent, each of the components in each mixture being determined with recoveries between 91 and 109.5%: in the great majority of cases the error in the analysis was less than 5%. Results of this quality are sufficient for screening purposes. The same principles were successfully applied to the simultaneous determination of two inhibitors. Alkaline phosphatase and its inhibitor sodium vanadate and an alkaline protease and its inhibitor 3-nitrophenyl boronic acid were determined using fluorescein diphosphate and BODIPY FL substrates, respectively. These results confirm the applicability of purely chemometric methods to multi-analyte enzyme assays for screening purposes [37].

8 Conclusions

This brief review has only surveyed a fraction of the fluorescence enzyme assay methods available, and an even smaller fraction of the many application areas of such techniques. It is evident that here, as in many fields of biochemistry and molecular biology, the sensitivity and selectivity of fluorescence methods, allied to their ease of use, their adaptability to a range of sampling systems, and the availability of a variety of suitable instrumentation, provide a powerful tool. The continuing development of new methodologies, fluorophores and instruments ensures that this area of fluorescence spectroscopy will continue to grow and flourish.

References

1. Leaback DH (1969) An Introduction to the Fluorimetric Assay of Enzyme Activities. Koch-Light Laboratories, Colnbrook, UK
2. Guilbault GG (1976) Handbook of Enzymatic Methods of Analysis. Marcel Dekker, New York
3. Froelich PM, Murphy LD (1977) Anal Chem 49:1606
4. Roth M (1971) Anal Chem 43:880
5. Rotman B, Zderic J, Edelstein M (1963) Proc Nat Acad Sci USA 50:1
6. Muakami Y, Morita T, Kanekiyo T, Tamiya E (2001) Biosens Bioelectron 16:1009
7. Leytus SP, Melhado LL, Mangel WF (1983) Biochem J 209:299
8. Zhang NZ, Kasibhatla S, Guastella J, Tseng B, Drewe J, Cai SX (2003) Bioconjugate Chem 14:458
9. Sarpara GH, Hu SJ, Palmer DA, French MT, Evans M, Miller JN (1999) Anal Commun 36:19
10. Leaback DH, Walker PG (1961) Biochem J 78:151
11. Sun WC, Gee KR, Haugland RP (1998) Bioorg Med Chem Lett 8:3107
12. Gee KR, Sun WC, Bhalgat MK, Upson RH, Klaubert DH, Latham KA, Haugland RP (1999) Anal Biochem 273:41
13. Garcia-Echeverria C, Rich DH (1992) FEBS Lett 297:100

14. Guilbault GG, Kramer DN (1964) Anal Chem 36:2497
15. Eggerston MJ, Craig DB (1999) Biomed Chromatogr 13:516
16. Wolf CR, Seilman S, Oesch F, Mayer RT, Burke MD (1986) Biochem J 240:27
17. Zhou M, Diwu Z, Panchuk-Voloshina N, Haugland RP (1997) Anal Biochem 253:162
18. Zhou M, Zhang C, Haugland RP (2000) Proc SPIE Int Soc Opt Eng 3926:166
19. Lakowicz JR (1999) Principles of Fluorescence Spectroscopy, 2nd edn. Kluwer Academic/Plenum Press, New York
20. Matayoshi ED, Wang GT, Krafft GA, Erickson J (1990) Science 247:954
21. Grüninger-Leitch F, Schlatter D, Küng E, Nelböck P, Döbeli H (2002) J Biol Chem 277:4687
22. Hendrickson HS, Hendrickson EK, Johnson ID, Farber SA (1999) Anal Biochem 276:27
23. Welder F, Moody E, Colyer CL (2002) Electrophoresis 23:1585
24. Oh LY, Larsen PH, Krekoski CA, Edwards DR, Donovan F, Werb Z, Yong VW (1999) J Neurosci 19:8464
25. Haugland RP (2002) Handbook of Fluorescent Probes and Research Products, 9th edn. Molecular Probes, Eugene
26. Hendrickson HS (1994) Anal Biochem 219:1
27. Zaman GJR, Garritsen A, De Boer T, van Boeckel CAA (2003) Comb Chem & High Throughput Screening 6:313
28. Loomans EEMG, van Doommaien AM, Wat JWY, Zaman GJR (2003) Assay & Drug Development Technologies 1:445
29. Scade SZ, Jolley ME, Sarauer BJ, Simonsen LG (1996) Anal Biochem 243:1
30. Levine LM, Michener ML, Toth MV, Holwerda BC (1997) Anal Biochem 247:83
31. Bigelow CE, Vishrasrao HD, Frelinger JG, Foster TF (2004) J Microsc 215:24
32. Paragas VB, Zhang Y-Z, Haugland RP, Singer VL (1997) J Histochem Cytochem 45:345
33. Haugland RP (1995) Biotech Histochem 70:243
34. Arttamangkul S, Bhalgat MK, Haugland RP, Diwu Z, Liu J, Klaubert DH (1999) Anal Biochem 269:410
35. Rosser BG, Powers SP, Gores GH (1993) J Biol Chem 268:23593
36. Miller JN, Miller JC (2005) Statistics and Chemometrics for Analytical Chemistry, 5th edn. Pearson Education, Harlow, UK
37. Kownarumit S (2006) PhD Dissertation. Loughborough University, Loughborough, UK

Part VI
Quantitative PCR

Springer Ser Fluoresc (2008) 6: 489–508
DOI 10.1007/4243_2008_046

Published online: 13 March 2008

Quantitative Real-Time PCR: Fluorescent Probe Options and Issues

Marcia J. Holden[1] (✉) · Lili Wang[2]

[1]National Institute of Standards and Technology, Biochemical Science Division, 100 Bureau Drive, Mail Stop 8311, Gaithersburg, MD 20899, USA
marcia.holden@nist.gov

[2]National Institute of Standards and Technology, Biochemical Science Division, 100 Bureau Drive, Mail Stop 8312, Gaithersburg, MD 20899, USA

Abstract Fluorescence has played a vital role in the development of polymerase chain reaction (PCR)-based DNA amplification. In qualitative PCR, an end point reaction, the amplified DNA, is visualized using DNA intercalating fluorescent dyes. Creative uses of nucleotide probes with fluorescent tags have been developed for real-time quantitative PCR. These probes take advantage of the behavior and properties of fluorophores. There are advantages and disadvantages to various probe types as well as design considerations. Attention to these issues will help in the development of robust and accurate DNA quantification using real-time PCR.

Keywords FRET · Hybridization probe · Molecular beacon · Real-time quantitative PCR · Scorpion primer · SYBR Green I · TaqMan probe

Abbreviations

Ct Cycle threshold
CTAB Cetyltrimethylammonium bromide
DNA Deoxyribonucleic acid
FRET Fluorescence resonance energy transfer
PCR Polymerase chain reaction
T_m Melting temperature

1 PCR

The polymerase chain reaction (PCR) has become one of the most important tools in molecular biology in the last 30 years, with rapidly expanding uses in areas such as gene expression, disease detection and monitoring, species identification, forensics, single nucleotide polymorphisms, mutation detection, and evolutionary studies. The core of the technology, first developed in 1985 [1], is copying of specific sequences of DNA using the enzyme DNA polymerase in conjunction with primers (single-stranded DNA consisting of 20 to 25 bases) that match the DNA sequence at each end of the region to be amplified. The reaction mixture consists of the DNA polymerase, two primers (one for each strand of DNA), individual nucleotide triphosphates (building blocks for the copied DNA), buffer (with salts and additives), and template DNA. This reaction mixture goes through a temperature cycle wherein the double-stranded DNA is denatured at 95 °C into single strands by breaking of the hydrogen bonds between the bases. The temperature is lowered so that the primers, in high concentrations, can bind to the matching sequence on the template DNA. When the primers have bound then DNA polymerase utilizes that primer position as a place to recreate the double strand. The enzyme incorporates nucleotides to the primer with the order dictated by the sequence of the template DNA. This constitutes a single cycle of amplification. In the subsequent cycles the primers can bind to template DNA or to a previously amplified DNA fragment called an amplicon. In every cycle, in theory, it is possible to copy every previously amplified fragment of DNA, and reactions frequently involve 30 to 50 cycles. Billions to trillions of copies are produced which are easily visualized, for example, by electrophoresis in an agarose gel in the presence of ethidium bromide, which fluoresces upon intercalation in the DNA, or by capillary electrophoresis using a dye-labeled primer.

Two technological developments were implemented to increase the utility of the PCR. The first was the adoption of the thermostable DNA polymerase [2]. DNA polymerase would normally be degraded by heating to 95 °C and so the reaction had to be stopped during each cycle to add more polymerase. Thermostable DNA polymerases with a variety of properties have been isolated and cloned from thermal vent microbes whose enzymes must survive the high temperatures in which the organisms live. The second development was the creation of superior heat blocks. Precise temperature control and rapid temperature cycling were important for consistent results. At this stage of development PCR provided qualitative information, the accuracy of which was dependent on good design of the PCR protocol and the quality of the DNA template. Among the most important components are the specificity of the primers and the choice of the thermocycling temperatures. Primers with sequence similarity to other regions of DNA will result in multiple products. Temperatures for the annealing phases, if too low, can also lead to nonspecific amplification.

2 Real-Time Quantitative PCR (rt-Q-PCR)

Fluorescence in the form of ethidium bromide intercalated into DNA amplicons was important to visualize products in qualitative PCR, but fluorescence serves a vital role in quantitative PCR detection strategies as well. Instrumentation utilizing CCD cameras to measure fluorescence signals in every reaction and in each cycle of PCR opened the door to quantitative real-time PCR. Equipment manufacturers have developed a variety of platforms with various optical and heating/cooling options combined with plastic tubes, 96/384-well plates, or glass capillaries for holding the reactions. The speed of a completed protocol and the ability to monitor it in real time has driven different approaches. Machines vary in their ability to support the use of the proliferating fluorescence detection options and the ability to multiplex (more than one PCR reaction in each tube). Determining the needs of a laboratory and investigating the capabilities of each platform before purchase is therefore important.

The Q-PCR reaction mixture has the same components as a qualitative reaction with the addition of a detection probe, either DNA intercalating fluorescent dyes or specific single-stranded DNA with covalently linked fluorophore with or without an acceptor. The optimization of the reaction (concentrations of reaction components, timing, temperature) is as important for quantitative as qualitative PCR. It is out of the scope of this chapter to discuss in detail the many aspects of the PCR that must be considered during development and validation of a specific protocol, and instead we will focus on the fluorescent probe options and issues. Recent comprehensive sources of information include two volumes dedicated to quantitative PCR [3, 4].

Q-PCR real-time platforms monitor the fluorescence signal at some point during each PCR cycle, the timing being dependent on the nature of the fluorescent probe type used in the reaction (Fig. 1). Initially the signal is below a preselected baseline or threshold that is set above background noise, as there are few amplicons contributing to the signal (lag phase). But as the number of completed cycles increases the number of amplicons, the fluorescence signal will cross this threshold during the beginning of the log phase of the amplification. In later cycles a plateau is reached where the signal is relatively constant. No additional signal is generated for several possible reasons including exhaustion of the reagents. The amplification cycle at which the fluorescence in a particular reaction crosses the selected threshold is known as the cycle threshold, or Ct, and is in the early log phase of amplification. The initial copy number of target DNA sequences added to the reaction deter-

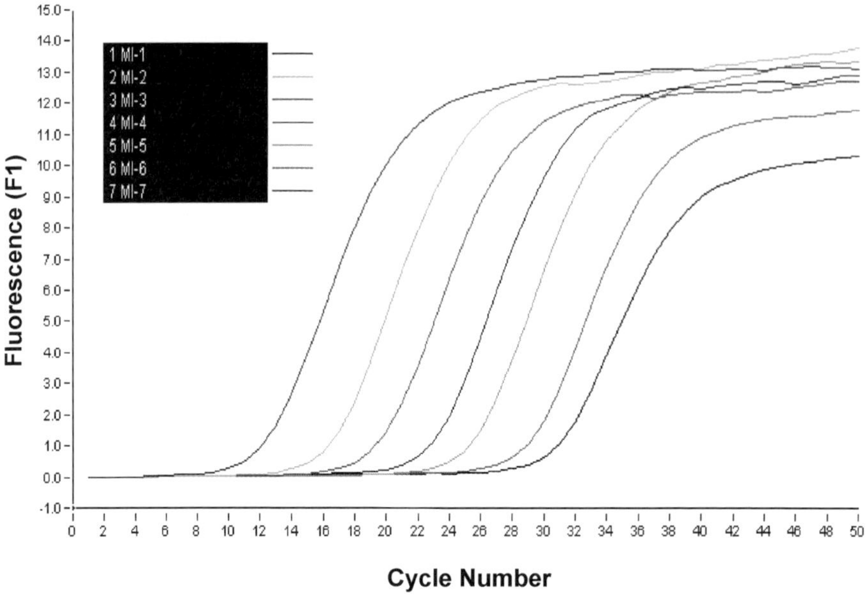

Fig. 1 Quantitative real-time amplification plot of a representative standard curve (tenfold dilutions of a plasmid). The insert graph is a plot of the Ct values vs the log of the concentration where the slope of the curve is –3.386

mines when that threshold is reached, all other things being equal. The Ct value is inversely proportional to the concentration of DNA targets. This is the basis of quantitation. Absolute quantitation involves the use of DNA or RNA standards, validated for the specific detection. An amplification efficiency of 100% equals a doubling of the number of amplicons in every cycle. Plots of the Ct values versus the log of the concentration of the target will yield a curve with a slope of [–3.32] when efficiency is equal to 100%. Two DNA samples or standards that differ in concentration by tenfold will have Ct values that differ by 3.3 cycles. Figure 1 shows the change in fluorescence for a series of tenfold dilutions of a plasmid. A plot of this data gives a slope of –3.386, for an efficiency of 97.4%.

3
Fluorescence Detection in Quantitative PCR: Nonspecific Detection

Fluorescence detection methods are utilized in real-time PCR due to the need for high sensitivity and a large dynamic range [5, 6]. The first attempts utilized ethidium bromide again as the source of fluorescence [7, 8]. SYBR Green I, which binds to the minor groove of the DNA double helix, has replaced ethidium bromide as an intercalating fluorescent dye [9, 10]. SYBR

Green I and related molecules represent a detection strategy that is nonspecific, as any double-stranded DNA will bind SYBR Green. Therefore, the specificity of the PCR becomes an important issue. Amplification artifacts, such as primer dimers, will add to the signal [11, 12]. Running a melting protocol at the end of the amplification cycle protocol should help in validating the specificity of the reaction.

While it is a simple strategy for the detection of PCR amplicons, in practice the use of SYBR Green I is not necessarily straightforward. Diluted SYBR Green I can be stable for up to 3 weeks, but its breakdown products can be inhibitory to PCR and are increased by alkaline conditions [13]. The dye itself can be inhibitory depending on the concentration [6] and can affect the optimum magnesium concentration [13]. Additives may be necessary to optimize the reaction. SYBR Green I can also affect the melting temperature (T_m) of the products or even whether melting curve analysis can detect the product at all [14]. While SYBR Green has been used for multiplexing in cases where the T_m values of the products are sufficiently different to distinguish the amplicons, Giglio et al. [15] caution that there can be preferential dye binding to specific fragments. Further, some new versions of SYBR Green I introduced very recently address some of the problems noted above. Also a recent investigation into the use of the dye SYTO9 suggests that it might be a more stable and predictable alternative to SYBR Green I [14].

Nonspecific DNA intercalating/binding dyes offer a cost saving over specific fluorescent probes, discussed below, when running quantitative PCR assays. Alternatives to dyes, such as labeled primers, may not justify the cost (over the dyes), but they offer the possibility of multiplexing, that is, more than one PCR per tube.

4
Fluorescence Detection in Quantitative PCR: Specific Detection

All of the rest of the fluorescence detection strategies involve the use of oligonucleotide probes, complementary to a portion of the amplified target, which offer specificity. These oligonucleotide probes contain both a fluorophore donor and an acceptor (emissive or nonemissive) that interact through a fluorescence resonance energy transfer (FRET) mechanism [16–20]. When compared to the DNA binding dyes, fluorophores linked to oligonucleotides offer higher sequence specificity and are less susceptible to contamination, such as primer–dimer formation in the case of SYBR Green I [11, 12], and are somewhat easier for the detection of single nucleotide polymorphisms [18, 21, 22].

FRET is the underlying mechanism for various real-time PCR methods employing a variety of probe design tactics (Fig. 2) including TaqMan probes [19, 23], molecular beacons [16, 24, 25], hybridization probes [26, 27],

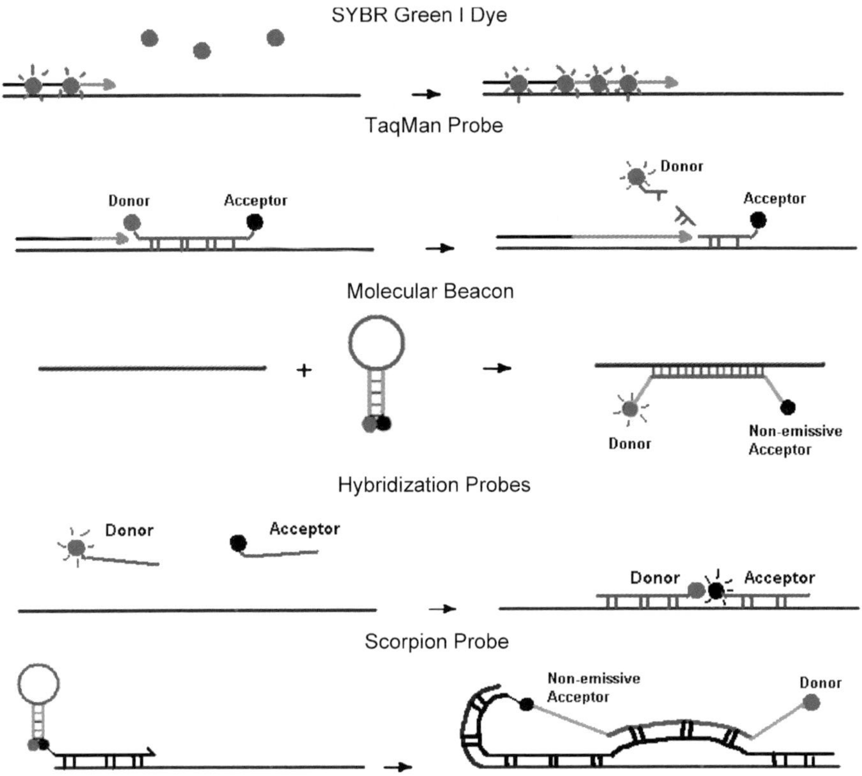

Fig. 2 Depictions of the more commonly used types of fluorescent detection probes used in quantitative real-time PCR

Scorpion primers [28], and strategies where fluorescence is quenched by neighboring nucleotides [18, 29]. FRET occurs between an energy donor and a suitable energy acceptor as a result of long-range dipole–dipole interactions between the two fluorophores. This type of energy transfer can take place over separation distances of 20 to 90 Å [30]. Contact quenching comes about when the donor and acceptor/quencher are in very close contact and quenching is more efficient than that of FRET with the same donor–acceptor pair [31]. The efficiency of FRET depends on the overlap of the emission spectra of the donor and the absorption spectra of the acceptor, whereas contact quenching does not.

4.1 Molecular Beacons

Because of the probe design strategy, molecular beacons generally give low initial fluorescence background. These probes are composed of a stem–loop structure. The loop portion contains the sequence for hybridization to the amplified target while the stem is a double-stranded sequence complementary only to itself and not the target. The stem structure of the molecular beacons warrants efficient fluorescence contact quenching by the proximal acceptor. Upon hybridization to the amplified templates during PCR, donor fluorescence enhancement occurs due to the large separation distance between the donor and the acceptor. The molecular beacon is displaced from the target strand by DNA polymerase during the elongation phase of amplification. The donor–acceptor pair can have a significant impact on the T_m of the probe and should be considered in addition to the contribution of the nucleotides themselves [31]. This is particularly important because there is a competition between the formation of the stem and the binding of the loop sequence to the target DNA.

4.2 TaqMan Probe

The first of the current generation of real-time quantitative PCR assays combined a dual-labeled fluorogenic probe, TaqMan, with the use of the $5' \rightarrow 3'$ exonuclease activity of Taq DNA polymerase [23] and this approach remains the most popular option. By comparison to probes such as molecular beacons, TaqMan probes give much higher background signals in that the intramolecular separation distance between a donor and an acceptor in the unhybridized state results in only partial donor signal reduction. The random coil behavior brings the acceptor into the range of the donor for FRET. The amount of resonance energy transfer is considerably lower than in the case of the molecular beacon, because of the lack of the stem to bring the donor and acceptor into close proximity when not bound to the template DNA/amplicons. During PCR, TaqMan probes are hybridized to the amplicons and subsequently cleaved for removal from the templates by $5' \rightarrow 3'$ exonuclease activity of the Taq polymerase [32]. As a result, donor fluorescence enhancement takes place because of the cleavage of the donor fluorophore from the oligonucleotide probe, and therefore separation from the acceptor fluorophore. An assumption in these assays is that cleavage of the TaqMan probe is efficient.

Some modifications have been developed to address the background issue with TaqMan probes. For instance, a 3′ minor groove binder (MGB, 1,2-dihydro-(3*H*)-pyrrolo[3,2-*e*]indole-7-carboxylate) is added to the conventional TaqMan probe design to enhance the binding affinity of short TaqMan MGB probes (8–16 mers) to their templates [33, 34]. In comparison to the conventional TaqMan probes, the short TaqMan MGB probes give much lower background signals due to the relatively short intramolecular separation distance between a donor and an acceptor in the unhybridized state. The short probe lengths warrant their use in mutation detections [35]. Their sensitivity in that respect can cause problems unless possible mismatches are well understood [36]. Other strategies for shortening probe length include the use of locked nucleic acids (LNAs) incorporated into oligonucleotides [37]. LNAs are nucleic acid analogs with an O2 to C4 methylene linkage that has the effect of reducing conformational flexibility and raising the T_m of the oligonucleotide.

4.3 Hybridization FRET Probes

Hybridization FRET probes are composed of two separate oligonucleotides with one fluorophore attached to the 3′ end of one nucleotide and another fluorophore linked to the 5′ end of the second oligonucleotide. In the unhybridized state the donor–acceptor pair also gives low fluorescence background. When there are sufficient amplified templates present, the two oligonucleotide probes are hybridized to the template adjacently to each other (head-to-tail), which brings the donor and acceptor into close proximity to enhance FRET efficiency. The donor fluorophore is excited and the fluorescence signal from the acceptor fluorophore is detected.

4.4 Scorpion Primers and Shared Stem Molecular Beacons

More recently, a variety of new probe designs have been explored for further improvement of the sequence-specific probes described above on the detection sensitivity of quantitative real-time PCR. Scorpion probes are self-probing combinations of primer and probe [28]. The probe and primer are separated by a PCR blocker to prevent construction of a double strand in the probe region. The primer segment binds in the first annealing phase followed by extension. During a second denaturing and annealing the probe portion can then bind to the newly synthesized target on the same strand. Detection of a target sequence is converted into an intramolecular event with improved kinetics and thermodynamics. The original variation featured a stem–loop type of probe like a molecular beacon with the donor fluorophore and acceptor in close proximity. When the probe is bound to the target the acceptor is somewhat removed from the fluorophore, but still in the vicinity. So a more sensitive variation was developed wherein the acceptor was on a second oligonucleotide complementary to the probe sequence [38]. This is known as a duplex scorpion and upon denaturation the acceptor is released. The link between the primer and probe automatically brings the probe into the proximity of the target and is always available for binding. This links amplification with detection in a 1:1 fashion which is not guaranteed with other probe systems, especially with hybridization FRET probes that require two separate bindings to adjacent segments of the same target. The proximity of target to probe was believed to play a role in the performance of a Scorpion probe as compared to TaqMan or beacon probes when running very fast cycling times [39]. Scorpions have shown utility for the detection of splice variants and mutation detection [39, 40].

Additionally, Kong and coworkers reported shared-stem molecular beacon probes that combined properties of a TaqMan probe (cleavage) and a conventional molecular beacon (stem) [41]. The authors have shown that the signal-to-background ratios are superior to that of conventional molecular beacons using this probe design strategy. Understandably, fluorescence from a fluorophore cleaved from a nucleotide probe ought to be greater than a fluorophore separated from an acceptor by ~20 nucleic bases like in the case of a conventional molecular beacon.

4.5 Nucleotide-Based Quenching

It is well known that guanosine and guanine cause fluorescence quenching of many commonly used fluorescent dyes, such as fluorescein, coumarin, BODIPY FL, TAMRA, JOE, HEX, TET, ROX, and some of the Alexa dyes [42, 43]. The quenching mechanisms were suggested to be due to photoinduced electron transfer from guanine to the singlet excited state of dye molecules [42, 44]. The degree of quenching depended on various factors, including guanine position relative to the fluorophore attachment site, number of guanines, and the attachment site of the fluorophore in the oligonucleotide (including attachment chemistry). The change in fluorescence intensity attributed to guanine quenching has been exploited to develop novel detection assays for DNA and RNA molecules [18, 29, 45, 46]. One tactic executed by LUX™ fluorogenic primer technology from Invitrogen is to design a molecular beacon primer with the labeling site of a fluorophore in the hairpin region and opposite to several guanosine residues [18]. This design results in highly quenched fluorescence prior to PCR. When the molecular beacon primer is open for the extension during PCR, the fluorescence from the fluorophore is restored.

4.6 Probe Design Issues and Validation

Commonly used, sequence-specific oligonucleotide probes are described above for quantitative real-time PCR. These oligonucleotide probes are generally required for multiplex real-time PCR applications. In order to achieve optimal detection sensitivity, knowledge of the advantages and limitations of each probe design is also required. Below we will use our experimental results to show likely problems associated with some probe design strategies.

In the following study, we investigated three probe design strategies commonly used in quantitative PCR for sensitivity in detection of the PCR amplicon [47]. A plasmid with a 120 base pair insert served as the DNA template. The probes included TaqMan, conventional molecular beacon (MB),

Table 1 Oligonucleotide sequences

Name	Fluorophore	Sequence
Amplicon	None	5′ AGGACGTGGACCAGAGATCGAATGACCATCGTG TGCTGACTCCAGAGGTTGCAGTCAGCGAGTGCA TCAGGTGTTGTAGCCTGATCCCTGTTCCGAAGT ACCTATCGTCGAGCGGTCTGT 3′
Forward primer	None	5′ ACAGACCGCTCGACGATAGG 3′
Reverse primer	None	5′ AGGACGTGGACCAGAGATCG 3′
TaqMan	5′-Fluorescein 3′-Rhodamine	5′ ACTTCGGAACAGGGATCAGGCTACA 3′
ATssMB	5′-Fluorescein 3′-Dabsyl	5′ ACTTCGGAACAGGGATCAGGCTACAccgaagt 3′
GCssMB	5′-Fluorescein 3′-Dabsyl	5′ CGGAACAGGGATCAGGCTACAACAgttccg 3′
MB	5′-Fluorescein 3′-Dabsyl	5′ ccgccCTCTGGAGTCAGCACACGATGGTCAggcgg 3′
GC-TaqMan	5′-Fluorescein 3′-Rhodamine	5′ CGGAACAGGGATCAGGCTACAACAC 3′
TaqssMB	None	5′ AGGTGTTGTAGCCTGATCCCTGTTCCGAAGTACC TAT 3′
CompMB	None	5′ ATCGAATGACCATCGTGTGCTGACTCCAGAGGT TGCAGTCAGCGAGTGCATCAGGTGTTGTAG 3′

and shared-stem molecular beacon (ATssMB and GCssMB). The shared-stem beacon probe briefly described above [41] combines the properties of a TaqMan probe and a conventional molecular beacon. The sequences of the primers, the real-time PCR amplicon, and various PCR probes are given in Table 1 together with two control oligonucleotides, TaqssMB and CompMB. The lengths and locations of the primers and probes with respect to the amplicon are shown in Fig. 3. With the use of both TaqMan and shared-stem molecular beacon probes, signal amplification relies on hybridization with the amplicon and hydrolysis of the probe by the 5′-exonuclease activity of

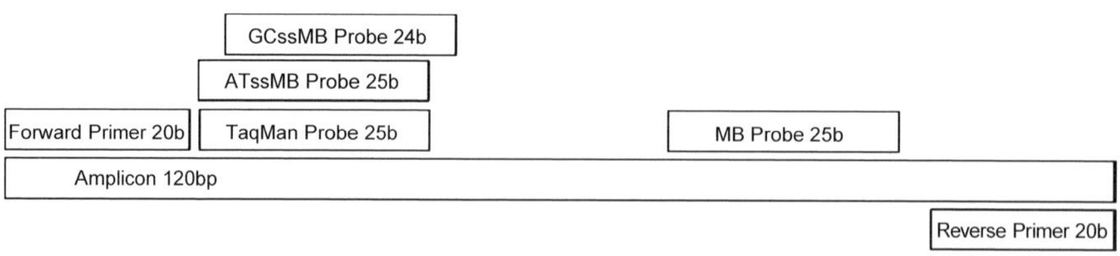

Fig. 3 The positions and lengths of various real-time PCR probes and forward and reverse primers with respect to the amplified region of the model DNA plasmid. The length includes only the regions involved in hybridization

Taq DNA polymerase (55 °C); therefore, these probes were placed close to the forward primer to enhance hydrolysis efficiency. For the molecular beacon probe, the final signal depends on the amplicon concentration and hybridization efficiency with the amplicon. In this case the molecular beacon probe is displaced from the target by Taq DNA polymerase (72 °C). The location of the probe on the target was therefore positioned some distance from the primer. The thermal cycling profiles using these probes were different to reflect the differences in the nature of the probes [47].

Figure 4A shows fluorescence intensities after completion of PCR (gray columns) for the four PCR probes. The black columns in the same figure show the initial fluorescence background averaged over the first five PCR cycles. To account for possible (1) unequal concentrations of probes and/or (2) differences in fluorescence quantum yields of fluorescein in each case due to the microenvironment induced by the nearby nucleotide sequence, the signal-to-background ratio (S/N) was calculated and is displayed in Fig. 4B for comparison of the four probes. The ratio trails in the following order: MB > GCssMB > ATssMB > TaqMan. The conventional molecular beacon probe, MB, gives a low fluorescence background when compared to a TaqMan probe in that fluorescence from the fluorophore is highly quenched by the adjacent quencher as seen in Fig. 4A (black column). Additionally, the guanine bases in the stem portion of the beacon quench the fluorescence of fluorescein. The amount of fluorescence quenching is, in general, proportional to the number of nearby guanine bases (in the region of five to six bases) [43, 46, 48]. Based on these principles, we expected the fluorescence background of the three molecular beacon probes to be in the order of MB < GCssMB < ATssMB. The fluorescence background from GCssMB (Fig. 4A) is surprisingly the lowest among the three beacon probes used in the study. On the other hand, with the use of TaqMan, ATssMB, and GCssMB probes the signal enhancement is critically dependent on the hydrolysis activity of the polymerase during PCR. The final signals after PCR should be comparable for the three probes assuming 100% cleavage efficiency (a single nucleic base conjugated with a fluorescein). Yet, they are very different (Fig. 4A). The signal from ATssMB is the highest and that from GCssMB is the lowest. The fluorescence signals of the post-PCR filtrates obtained by using Microcon YM-3 centrifugal filter devices (molecular weight cutoff of 3000 Da, ten single-stranded nucleotides) are very close to those after PCR, inferring that the three probes are lysed during PCR to be equal to or smaller than ten single-stranded nucleotides in size.

By design, the GCssMB probe should give a better signal-to-background ratio than the ATssMB probe if it is hydrolyzed completely during the PCR and obeys the quenching rule by the number of guanine bases in the stem portion. We measured fluorescence quantum yields of the post-PCR filtrates which are given in Table 2 for TaqMan, ATssMB, and GCssMB probes. The yields for TaqMan and ATssMB probes are relatively close. Although lower

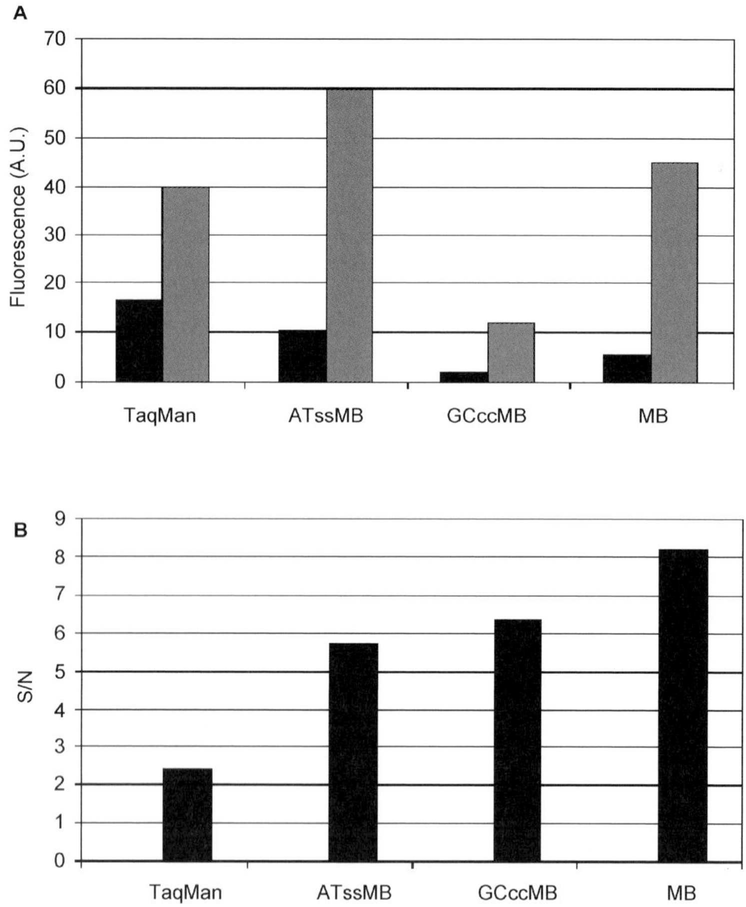

Fig. 4 **A** Fluorescence signals after completion of PCR (*gray columns*) for various real-time PCR probes. The black columns show the initial fluorescence background averaged over the first five PCR cycles. AU, arbitrary units. **B** Signal-to-background ratios (S/N) obtained using the data given in **A** for the four real-time PCR probes

than expected, it is evident that hydrolysis cleavage takes place during PCR. The yield for GCssMB is, nonetheless, much lower than for the other two. When a GCssMB probe is hybridized to the amplicon, its location is shifted from the primer by four nucleic acid bases more than ATssMB and TaqMan probes. We synthesized a control probe, GC-TaqMan (see Table 1), which has the identical sequence at the 5′ end as the GCssMB probe and no 3′ end stem sequence, to verify if the hydrolysis reaction does indeed take place during real-time PCR. The quantum yield determined for the post-PCR filtrate is 0.59 using GC-TaqMan as the probe. This suggests that the polymerase can effectively cleave the probe and result in an increase of the fluorescence signal because of physical separation of the fluorophore from the quencher.

Table 2 Relative fluorescence quantum yields determined for control samples and filtrates after PCR

	Sample	Quantum yield[b]
Control	dC-fluorescein	0.81
	dA-fluorescein	0.81
	dCG-fluorescein	0.20
	dCGG-fluorescein	0.23
	dCGGA-fluorescein	0.42
Probe and duplex with complementary strand[a]	TaqMan	0.19
	TaqMan/TaqssMB	0.65
	ATssMB	0.081
	ATssMB/TaqssMB	0.78
	GCssMB	0.028
	GCssMB/TaqssMB	0.43
	MB	0.050
	MB/CompMB	0.64
	GC-TaqMan	0.11
	GC-TaqMan/TaqssMB	0.49
Samples after enzymatic digestion	GCssMB + SVP	0.75
	GCssMB + BSPD	0.77
	GCssMB + SVP + BSPD	0.71
Post-PCR filtrate	TaqMan	0.38
	ATssMB	0.47
	GCssMB	0.12
	GC-TaqMan	0.59

[a] The samples in 20 mM Tris-HCl, pH 8.4, 50 mM KCl, 2 mM $MgCl_2$ went through the following protocol: 25 °C for 30 s, 95 °C for 2 min, then decreasing the temperature to 25 °C at the rate of 0.2 °C/s, incubation for 8 min at 25 °C

[b] The quantum yields were averaged over several experimental repeats with a standard deviation of less than 5% for the control, probe alone, and duplexes with the complementary strand and samples after enzymatic digestion, and with standard deviations of $\leq$20% for post-PCR filtrates

Melting curve measurements were then performed to ascertain that GCssMB behaved similarly to ATssMB and was able to hybridize to its template. The obtained T_m values for both GCssMB (60.5 °C) and ATssMB (58.8 °C) are similar and 8–10 °C lower than those of their duplexes with the complementary target (TaqssMB): 68.4 °C and 69.0 °C for GCssMB duplex and ATssMB duplex, respectively. These temperatures are generally consistent with the probe design strategies. The results imply that GCssMB and ATssMB should behave similarly in real-time PCR and be able to hybridize to the complementary strand.

We further asked the question whether C-linked fluorescein resulting from the complete hydrolysis of the GCssMB probe would be quenched. Both dC-fluorescein and dA-fluorescein were synthesized to serve as controls. Their fluorescence quantum yields measured against the reference standard are the same (0.81) and given in Table 2. We measured the quantum yields of GCssMB alone in Tris buffer, pH 8.4, and GCssMB probe digested by two different enzymes either separately or jointly (Table 2). Enzymatic digestion by snake venom phosphodiesterase (SVP) starts from the 5′ end of the nucleotides and digestion catalyzed by bovine spleen phosphodiesterase (BSPD) originates from the 3′ end of the nucleotides. The yields for digested GCssMB are close to each other and above 0.70, inferring that fully cleaved GCssMB should fluoresce strongly. Three additional control samples, dCG-fluorescein, dCGG-fluorescein, and dCGGA-fluorescein, were made to show the likely outcomes of partial hydrolyses of GCssMB serving as the probe in PCR. The fluorescence quantum yields of these controls are given in Table 2. The yields for dCG-fluorescein (0.20) and dCGG-fluorescein (0.23) are about four times lower than for dC-fluorescein (0.81), and the yield of dCGGA-fluorescein (0.42) is about half of that of dC-fluorescein. These results point out that a low fluorescence signal after PCR using GCssMB as the probe is most likely due to partial hydrolysis of the probe. Although the quantum yield determined for GCssMB after PCR (0.12) is much lower than anticipated, it is more than four times higher than that of GCssMB in the stem–loop state (0.028). When a GCssMB probe was hybridized to its complementary oligonucleotide, TaqssMB, the quantum yield of the formed duplex (GCssMB/TaqssMB) was measured to be 0.43. This yield is much higher than that after PCR (0.12). The measured yield (0.43) in the duplex form is comparable to that of the duplex GC-TaqMan/TaqssMB (0.49) (Table 2) due to the same microenvironment fluorophores experienced in both cases. Interestingly, the quantum yield of the four PCR probes increases in the order of GCssMB < MB < ATssMB < TaqMan. The trend is the same as that shown in Fig. 4A (black columns).

It is worth noting that the control probe (GC-TaqMan) would be a better probe than TaqMan when comparing the quantum yields of probes alone and post-PCR filtrates (Table 2). The fluorescence signal increases fivefold after PCR for GC-TaqMan and about twofold for TaqMan. The results strongly suggest that one could utilize surrounding nucleotide sequences to improve assay sensitivities when designing PCR probes.

We also prepared a tenfold dilution series of the preamplified DNA construct ($1 \times 10^{13}/\mu L$) to compare the sensitivity of various PCR probes. With the same amount of starting material, the signal difference between the second and the first PCR amplification cycles is likely to show the sensitivity of the probe in cases where there is an observable signal difference between one PCR cycle and the next cycle. Figure 5 displays differences in measured fluorescence signals as a function of the concentration (logarithmic copy numbers) of the starting construct for TaqMan (a), ATssMB (b), GCssMB (c), and MB (d). With a starting copy number of $1 \times 10^{12}/\mu L$, the signal difference between the second and first amplification cycles is distinguishable from that of less starting material, for instance, $1 \times 10^{11}/\mu L$ using MB, ATssMB, and

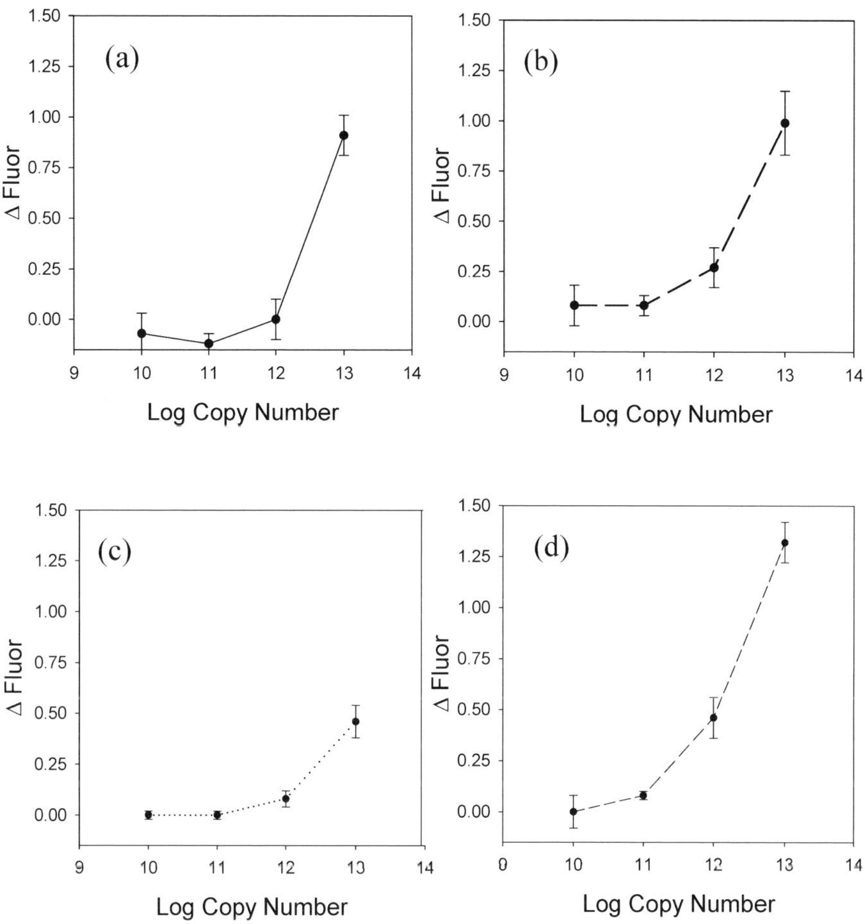

Fig. 5 Differences in fluorescence signals between the second and the first PCR amplification cycles as a function of the concentration (copy number) of the starting amplicons for TaqMan (**a**), ATssMB (**b**), GCssMB (**c**), and MB (**d**). The plots include the standard deviations from nine replicates for each concentration

GCssMB probes. Due to an intrinsic high fluorescence background associated with the TaqMan probe, the signal difference was not significant enough to differentiate the starting material from 1×10^{12} to 1×10^{11} copy/μL. The overall sensitivities for the four probe types are in the order of MB > ATssMB > GCssMB > TaqMan. The sensitivity result is, in general, consistent with that of the signal-to-background ratio shown in Fig. 4B.

Two key points can be learned from this study. One is that the complete hydrolysis generally assumed for the TaqMan probe strategy is not likely to be true. Second, in order to increase the detection sensitivity and signal-to-background ratio of real-time PCR, it is critical to decrease the fluorescence background of probes through careful placement of reporting fluorophores in the oligonucleotide microenvironment.

The second example is dedicated to the issue of guanine-induced quenching, which can be especially problematic in detection methods employing FRET as decreases in the donor fluorescence could be due to both resonance energy transfer and quenching by the microenvironment. Since fluorescein (FAM) and Alexa-488 are commonly used donor fluorophores [26, 49], for instance, with the use of LightCycler technology, we investigated the influence of the overhang region of the complementary strand on the resulting fluorescence from a hybridizing probe [48].

A series of target oligonucleotides, each with a unique 3′ overhang (four bases long), were hybridized to either 5′ fluorescein or Alexa-488 labeled probes, and the changes in fluorescence intensity and anisotropy were monitored. The four-base overhang serves as a good model for target molecules analyzed using real-time PCR in that significant quenching was observed in the presence of guanine bases in the overhang region, close to the fluorophore labeling nucleotide [29, 46]. The probe sequence was derived from the genome of the bacterium *Bacillus globigii*, and is detailed in Table 3 to-

Table 3 Nomenclature and base sequences of the oligonucleotides used in this study

Annotation	Sequence
Probes	
Alexa probe	5′-/**Alexa-488**/TGC GCC CAT TTT TCA AGC TGC G-3′
Fl probe	5′-/**Fluorescein**/TGC GCC CAT TTT TCA AGC TGC G-3′
Target [a]	
No overhang	5′ –CGC AGC TTG AAA AAT GGG CGC A-3′
TTGT	5′ –CGC AGC TTG AAA AAT GGG CGC ATG TT-3′
TGTT	5′ –CGC AGC TTG AAA AAT GGG CGC ATT GT-3′
GTTT	5′ –CGC AGC TTG AAA AAT GGG CGC ATT TG-3′
TTTG	5′ –CGC AGC TTG AAA AAT GGG CGC AGT TT-3′
GGTT	5′ –CGC AGC TTG AAA AAT GGG CGC ATT GG-3′
TGTG	5′ –CGC AGC TTG AAA AAT GGG CGC AGT GT-3′
GTGT	5′ –CGC AGC TTG AAA AAT GGG CGC ATG TG-3′
TTGG	5′ –CGC AGC TTG AAA AAT GGG CGC AGG TT-3′
GGGT	5′ –CGC AGC TTG AAA AAT GGG CGC ATG GG-3′
TGGG	5′ –CGC AGC TTG AAA AAT GGG CGC AGG GT-3′
AAAA	5′ –CGC AGC TTG AAA AAT GGG CGC AAA AA-3′
CCCC	5′ –CGC AGC TTG AAA AAT GGG CGC ACC CC-3′
TTTT	5′ –CGC AGC TTG AAA AAT GGG CGC ATT TT-3′
GGGG	5′ –CGC AGC TTG AAA AAT GGG CGC AGG GG-3′

[a] The "target" annotation is given as the sequence of the 3′ overhang region (four bases) read from the 3′ to 5′ end. Bold font in the sequences designates a 5′ covalently bound fluorophore attached through an aminohexylphosphate linker

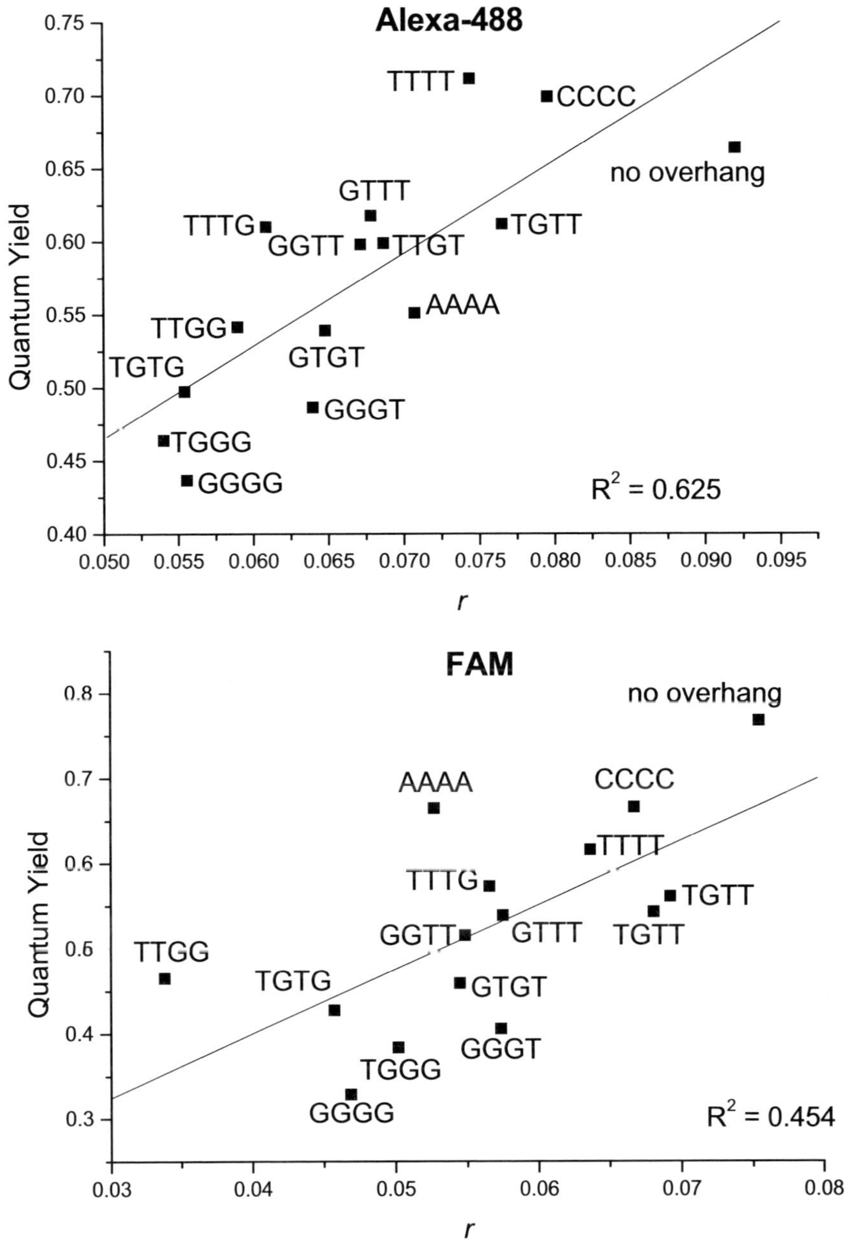

Fig. 6 The relationship between fluorescence quantum yield and anisotropy for the hybridization reactions employing Alexa-488- and FAM-labeled probes. The anisotropy data for each hybridized oligonucleotide shown is mostly the average from two independent reactions (four repeats for the TTTT, TTTG, and GGGG incorporated duplexes), where each spectrum was recorded in duplicate, generating S.E.M. values typically <10% of the total *r* value

gether with the target sequences. All sequences were analyzed using MFOLD Web Server (Version 3.1) [50] to ensure that potential secondary structures did not complicate the results. We found that the number of guanine bases in the overhang region of the target oligonucleotides is proportional to the amount of fluorescence quenching observed for both the FAM and Alexa-488 dyes (Fig. 6). FAM appeared to be more sensitive to guanine-induced quenching with three and four guanine bases resulting in a greater than twofold decrease in the quantum yield of the fluorophore compared to the no-overhang target. In addition, we found that adenine bases caused fluorescence quenching of the Alexa-488-labeled probe, whereas the FAM-labeled probe appeared insensitive. The quenching data, generated with the steady-state fluorescence measurements, also displayed a linear correlation with those obtained using a fluorescent thermal cycler, suggesting the applicability to real-time PCR measurements (data not shown). Anisotropy data from the series of duplexes correlated with the fluorescence quantum yield (Fig. 6), suggesting that quenching was accompanied by increased dye mobility.

Nazarenko and colleagues [43] found that fluorescence quenching of fluorescein that was attached next to the 3′ end of the probe oligonucleotide by a 5′ one to two guanine base overhang on the complementary target strand was less than that by a terminal G-C pair, though the magnitude of quenching was still significant. Comparison with most reported data is problematic due to the differences in oligonucleotide sequence and furthermore because the data are often not quantitative, i.e., are expressed relative to the single-stranded probe. Our study and other reported work point out that, while designing nucleotide probes for real-time PCR, one needs to pay special attention to about four to five nucleotide bases near the fluorophore labeling nucleotide either at the 3′ end or 5′ end of an oligonucleotide. Even in the case of designing hybridization probes, where a one to five base separation between donor and acceptor is recommended by the manufacturer with the use of LightCycler technology, one should still be able to move donor and acceptor nucleotides around to avoid the quenching effect and enhance energy transfer efficiency. A larger separation distance between the donor–acceptor pair may still allow for adequate transfer efficiency [49] in quantitative real-time PCR.

5 Conclusions

There is a variety of real-time quantitative PCR equipment platforms and amplification fluorescence detection strategies combined with a basic amplification protocol. The diversity of fluorescent tools that are available offers options for many different types of detections. Validation of the fluorescent probe for a specific detection is an important part of the validation of a PCR protocol.

For trace detections the sensitivity of the detection is critical. An example is the quantification of the adventitious presence of genetically modified organisms (GMOs) in grain or food products. Validation of a PCR method also involves the determination of the limit of detection (LOD) and the limit of quantification (LOQ). The sensitivity of the fluorescent probe contributes to the LOD and LOQ. How specific probe types work is fairly straightforward, but as we have tried to convey in this chapter, the performance of a specific probe in the detection of an amplicon should not be assumed. Is there sufficient hydrolysis of a TaqMan-style probe using the 5′-nuclease assay? Is the background fluorescence of a probe significantly lowering the signal-to-noise ratio? Will the probe–quencher combination of a molecular beacon significantly affect the T_m of the probe? Is the design of the assay sufficiently robust to perform if there are slight variations in the performance of the thermocycler platform or variations in the reagent mix? Does the melting curve analysis with SYBR Green detection show all the products? Analysis of the performance of the fluorescent probe as part of the validation process in method development will pay dividends in accuracy and precision of the quantitative PCR assay.

References

1. Saiki RK, Scharf S, Faloona FA, Mullis KB (1985) Science 230:1350
2. Saiki RF, Gelfand DH, Stoffel S, Scharf S, Higuchi R, Horn GT, Mullis KB, Erlich HA (1988) Science 239:487
3. Bustin SA (ed) (2004) A–Z of quantitative PCR. International University Line, La Jolla
4. Edwards K, Logan J, Saunders N (eds) (2004) Real-time PCR: an essential guide. Horizon Bioscience, Norfolk
5. Meuer S, Wittwer CT, Nakagawara K-I (eds) (1999) Rapid cycle real-time PCR: methods and applications. Springer, Heidelberg
6. Wittwer CT, Herrmann MG, Moss AA, Rasmussen RP (1997) Biotechniques 22:130
7. Higuchi R, Dollinger G, Walsh PS, Griffith R (1992) Biotechnology 10:413
8. Higuchi R, Fockler C, Dollinger G, Watson R (1993) Biotechnology (NY) 11:1026
9. Nakao M, Janssen JWG, Flohr T, Bartram CR (2000) Cancer Res 60:328
10. Dhar AK, Roux MM, Klimpel KR (2001) J Clin Microbiol 39:2835
11. Vandesompele J, De Paepe A, Speleman F (2002) Anal Biochem 303:95–98
12. Mouillesseaux KP, Klimpel KR, Dhar AK (2003) J Virol Methods 111:121
13. Karsai A, Muller S, Platz S, Hauser M-T (2002) Biotechniques 32:790
14. Monis PT, Giglio S, Saint CP (2005) Anal Biochem 340:24
15. Giglio S, Monis PT, Saint CP (2003) Nucleic Acids Res 31:e136
16. Tyagi S, Kramer FR (1996) Nat Biotechnol 14:303
17. Szuhai K, Sandhaus E, Kolkman-Uljee SM, Lemaitre M, Truffert JC, Dirks RW, Tanke HJ, Fleuren GJ, Schuuring E, Raap AK (2001) Am J Pathol 159:16
18. Nazarenko I, Lowe B, Darfler M, Ikonomi P, Schuster D, Rashtchian A (2002) Nucleic Acids Res 30:e37
19. Lie YS, Petropoulos CJ (1998) Curr Opin Biotechnol 9:43
20. Barragan E, Bolufer P, Moreno I, Martin G, Nomdedeu J, Brunet S, Fernandez P, Rivas C, Sanz MA (2001) Leuk Lymphoma 42:747

21. Marras SAE, Kramer FR, Tyagi S (1999) Genet Anal Biomol E 14:151
22. Breen G, Harold D, Ralston S, Shaw D, St Clair D (2000) Biotechniques 28:464
23. Heid CA, Stevens J, Livak KJ, Williams PM (1996) Genome Res 6:986
24. Lewin SR, Vesanen M, Kostrikis L, Hurley A, Duran M, Zhang L, Ho DD, Markowitz M (1999) J Virol 73:6099
25. Szuhai K, Ouweland J, Dirks R, Lemaitre M, Truffert J, Janssen G, Tanke H, Holme E, Maassen J, Raap A (2001) Nucleic Acids Res 29:e13
26. Eckert C, Landt O, Taube T, Seeger K, Beyermann B, Proba J, Henze G (2000) Leukemia 14:316
27. Cardullo RA, Agrawal S, Flores C, Zamecnik PC, Wolf DE (1988) Proc Nat Acad Sci USA 85:8790
28. Whitcombe D, Theaker J, Guy SP, Brown T, Little S (1999) Nat Biotechnol 17:804
29. Kurata S, Kanagawa T, Yamada K, Torimura M, Yokomaku T, Kamagata Y, Kurane R (2001) Nucleic Acids Res 29:e34
30. Lakowicz JR (1999) Principles of fluorescence spectroscopy. Plenum Press, New York
31. Marras SAE, Kramer FR, Tyagi S (2002) Nucleic Acids Res 30:e122
32. Holland PM, Abramson RD, Watson R, Gelfand DH (1991) Proc Nat Acad Sci USA 88:7276
33. Kutyavin IV, Afonina IA, Mills A, Gorn VV, Lukhtanov EA, Belousov ES, Singer MJ, Walburger DK, Lokhov SG, Gall AA, Dempcy R, Reed MW, Meyer RB, Hedgpeth J (2000) Nucleic Acids Res 28:655
34. Kulesh DA, Baker RO, Loveless BM, Norwood D, Zwiers SH, Mucker E, Hartmann C, Herrera R, Miller D, Christensen D, Wasieloski LP Jr, Huggins J, Jahrling PB (2004) J Clin Microbiol 42:601
35. Luderer R, Verheul A, Kortlandt W (2004) Clin Chem 50:787
36. Whiley DM, Sloots TP (2006) J Clin Virol 35:81
37. Kennedy B, Arar K, Reja V, Henry RJ (2006) Anal Chem 348:294
38. Solinas A, Brown LJ, McKeen C, Mellow JM, Nicol JTG, Thelwell N, Brown T (2001) Nucleic Acids Res 29:e96
39. Thelwell N, Millington S, Solinas A, Booth J, Brown T (2000) Nucleic Acids Res 28:3752
40. Taveau M, Stockholm D, Spencer M, Richard I (2002) Anal Biochem 305:227
41. Kong D-M, Gu L, Shen H-S, Mi H-F (2002) Chem Commun 8:854
42. Torimura M, Kurata S, Yamada K, Yokomaku T, Kamagata Y, Kanagawa T, Kurane R (2001) Anal Sci 17:155
43. Nazarenko I, Pires R, Lowe RB, Obaidy M, Rashtchian A (2002) Nucleic Acids Res 30:2089
44. Seidel CAM, Schulz S, Sauer MHM (1996) J Phys Chem 100:5541
45. Knemeyer JP, Marme N, Sauer M (2000) Anal Chem 72:3717
46. Crockett AO, Wittwer CT (2001) Anal Biochem 290:89
47. Wang L, Blasic JR Jr, Holden MJ, Pires R (2005) Anal Biochem 344:257
48. Noble JE, Wang L, Cole KD, Gaigalas AK (2005) Biophys Chem 113:255
49. Wang L, Gaigalas AK, Blasic JR Jr, Holden MJ, Gallagher DT, Pires R (2003) Biopolym (Biospectrosc) 72:401
50. Zuker M (2003) Nucleic Acids Res 31:3406

Part VII
Fluorescence In Situ Hybridization and Immunohistochemistry

Springer Ser Fluoresc (2008) 6: 511–530
DOI 10.1007/4243_2008_043

Published online: 4 April 2008

Cellular Bioimaging in Fluorescent Cancer Biomarker Evaluation: Validation, Technologies and Standards Development

Yan Xiao[1] · Peter E. Barker[2] (✉)

[1]SAIC, 4301 N. Fairfax Drive, Suite 200, Arlington, VA 22203, USA

[2]Biochemical Sciences Division, National Institute of Standards and Technology,
100 Bureau Drive, Gaithersburg, MD 20899-8311, USA
peter.barker@nist.gov

Abstract Here we summarize the approach to analytical validation of cellular cancer biomarkers visualized and quantified by fluorescence in-situ hybridization (FISH) and immunohistochemical fluorescence methods. The value of cellular cancer biomarker analytes and the role of analysis of cellular heterogeneity is emphasized for FISH and fluorescence immunohistochemistry (IHC). Finally, the timing of cancer cell biomarker standards in the development of screening assays is discussed in light of our experience developing DNA probes and IgY avian antibodies for analysis of prostate cancer gene fusions and quantitative screening of the enzyme telomerase.

Keywords Biomarker · Cancer · Fluorescence in-situ hybridization · HER2 · Immunohistochemistry · Telomerase · Validation

Abbreviations

PPV	Positive predictive value
CdSe/ZnS	Cadmium–selenium, zinc sulfide core/shell quantum dot
FISH	Fluorescence in-situ hybridization
NIST	US National Institute of Standards and Technology
NCI	US National Cancer Institute
ISCN	International System for Chromosome Nomenclature
cRNA	Complementary ribonucleic acid
UTP	Uridine triphosphate
dUTP	Deoxyuridine triphosphate
BCR	Human breakpoint cluster region gene
ABL	Abelson-related human oncogene
TMPRSS2	Transmembrane protease serine 2
ETV1	E26-Specific (ETS)-virus like oncogene 1
ERG	ETS-related gene
RT-PCR	Reverse transcriptase polymerase chain reaction
BAC	Bacterial artificial chromosome
C_0t1	Index of repeat number of a DNA by hybridization
CML	Chronic myelogenous leukemia
FDA	US Food and Drug Agency
PSA	Prostate specific antigen, a serum biomarker for prostate cancer
CA-125	A serum biomarker for ovarian cancer
IgY	An immunoglobulin molecule of avian origin
SKY	Spectral karyotyping
M-FISH	Multicolor fluorescence in-situ hybridization
CGH	Comparative genomic hybridization
SPR	Surface plasmon resonance
ELISA	Enzyme-linked immunosorbent assay
HUPO	Human Proteome Organization
IHC	Immunohistochemistry
FACS	Fluorescence-activated cell sorting
POC	Point-of-care

1 Introduction: Validation, Heterogeneity and Cellular Cancer Biomarkers

Research defining, quantifying and standardizing early cancer biomarker measurements has led to fresh perspectives in the emerging field of translational medicine recently. Validation represents the crucial bridge between discovery and the clinical utility of novel cancer biomarkers. While biomarker initiatives have generated many new discoveries, validation is often the stage at which promising new cancer biomarkers and their measurement technologies falter [1].

Validation also distinguishes cancer biomarkers in research and discovery mode from those of rigorously defined clinical merit. As third-party reimbursement for medical diagnostic tests may be contingent upon whether cancer biomarkers are validated, this area of work has commercial impact as well. Analytical validation is thus fundamental to improving the clinical utility of cellular biomarkers for early cancer detection.

Cancer biomarker validation consists of an analytical (technical) validation phase, and a subsequent clinical (biological) validation phase [2, 3]. Performance metrics of measurement systems are verified during analytical validation, with associated metrology including standard materials. Clinical validation addresses clinical performance of a cancer biomarker, and whether the biomarker measurement effectively distinguishes patients from healthy control subjects under defined measurement conditions. The specificity, sensitivity and positive predictive value of a cancer biomarker are the quantitative measures of clinical utility.

An often overlooked issue is that validation implies both a defined assay and its coupling with a specific intended clinical use. Not only must the biomarker be validated under specified measurement conditions and methods, but its use must be defined if performance metrics are to apply.

As novel genomic and proteomic cancer biomarkers are discovered and evaluated, the importance of quantitative measurement of analytical and clinical performance will increase. Quantitative measurements are the means by which new cancer biomarkers can be objectively compared among each other, and with cancer biomarkers previously approved by the US Food and Drug Agency (FDA), and in standard-of-care medical practice.

Both ensemble and single molecule detection technologies are used in cancer biomarker evaluation. Here, results in which the cancer biomarker signal is observed in the context of three-dimensional tissue and cellular architecture, will be our focus.

Cancer is a disease among cells characterized by heterogeneous biomarker signals. Thus, the context of fluorescence measurements is important. This inherent heterogeneity in biological systems presents a challenge for classical metrology.

Heterogeneity itself may be a distinguishing feature of cancer cells. For example, in a single tumor, chromosome instability may generate a multitude of different karyotypes, gene copy numbers or gene expression values among cells [4]. Diagnostic biopsies often include infiltrating healthy cells admixed with tumor cells.

Crucial information on analyte heterogeneity among cells in a population may be lost when tissue specimens are homogenized, for example, for gene expression [complementary DNA (cDNA)] arrays or enzymatic analysis. In both instances, results represent averages among cells of a population, and are only as good as the purity of cell sampling allows. In tumors, the few cells with significantly different analyte concentrations (extra copies of a gene, overexpression of an enzyme, or loss of heterozygosity) may be far more critical diagnostically than an ensemble average value for the entire cell population in a biospecimen. Thus, for tumors, cell populational heterogeneity in specimens is a consideration, and should be quantified in the measurement system for many biomarker applications.

Ideally, the analytical features required for these types of cancer biomarker detection are: (1) quantitative signal detection, sometimes at the single molecule level, with favorable signal-to-noise ratios, (2) definition of cell types that give rise to biomarker signal in mixed or heterogeneous cell populations, and (3) multiplex capability; that is, quantification of multiple optical signals simultaneously, or as more often the case, consecutively. Each feature has associated technical challenges.

Although many detection modalities exist, such detection in our laboratory is generally implemented by fluorescence measurements using specific nucleic acid or protein affinity probes (antibodies) in fixed cells with detection by photostable fluorophores such as semiconductor nanocrystals [5–7].

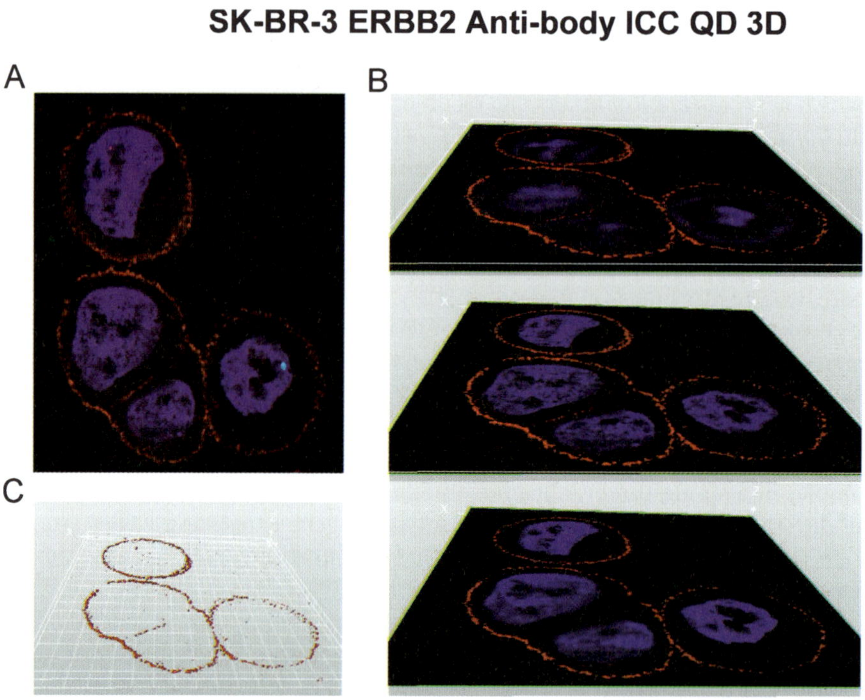

Fig. 1 Example of deconvolution imaging of HER2 protein receptor (*red*) on the surface of human breast cancer cell line SK-BR-3. *Blue* 4′,6-diamidino-2-phenylindole-stained interphase nuclei

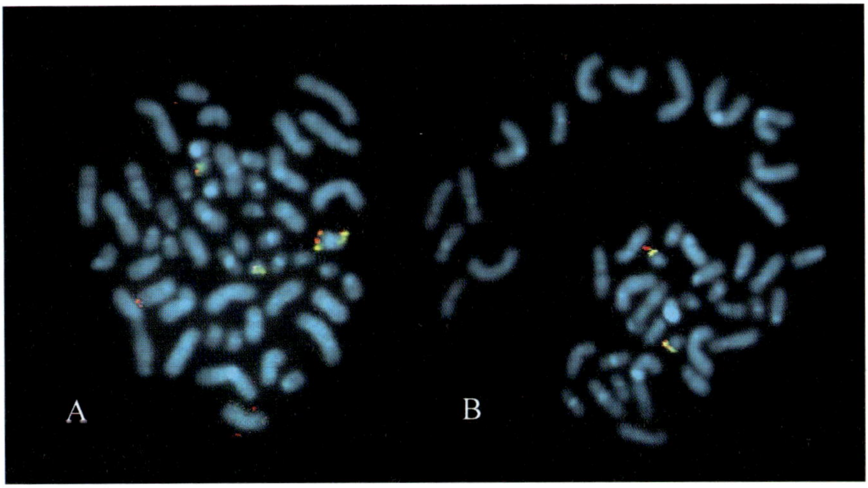

Fig. 2 **A** Fluorescence in-situ hybridization (FISH) analysis of DNA in the transmembrane protease serine 2 (TMPRSS2)-E26-Specific-related gene (ERG) region in VCaP human prostate cancer cells. **B** Normal human metaphase. *Red* and *green/yellow* FISH probes, *blue* metaphase chromosomes

The US National Cancer Institute (NCI) has defined the cancer biomarker [8]. For technology evaluation discussed here, a *biomarker* is defined as "a biological analyte the metrology of which generates actionable data of established accuracy, precision and reproducibility". We focus discussion on cancer detection biomarkers and related technology evaluation. Our experience primarily involves development and evaluation of methods and reagents for immunofluorescence detection of cellular proteins with an immunoglobulin molecule of avian origin (IgY) (chick antibodies) (Fig. 1) and fluorescence in-situ hybridization with DNA probes (FISH) (Fig. 2), and related applications involving quantitative detection with CdSe/ZnS quantum dot–streptavidin conjugates [5]. The considerations represented here reflect these biases, and experiences in our work on cellular cancer biomarkers.

2
Continuous and Discrete Values for Cancer Biomarkers, Cellular Heterogeneity and Measurement Issues

Ease of screening for early cancer biomarkers may dictate which tissues are sampled. Sampling body fluids such as blood, urine and saliva are generally considered minimally invasive. Dilution of disease-specific signals, local concentration of analytes and complexity of the matrix will all impact the ability to measure an early cancer biomarker.

Although clinical proteomics, especially mass spectrometry of patient sera and plasma, has drawn much recent attention as the source of novel early cancer biomarkers [9, 10], significant measurement challenges face validation and implementation of this class of cancer biomarkers [11, 12]. These include development of measurement technologies that are capable of spanning twelve or more orders of magnitude from common, high-concentration serum proteins such as albumin, to low-concentration-range analytes such as the interleukins [13–15]. Thus, the dynamic range needed for proteomics measurement and discovery work may exceed that available with currently used platforms.

In some instances, a biomarker indicating cancer may involve an abnormal concentration of structurally normal proteins. For analyte concentration to differentiate health from disease, a discriminant threshold value is required. Therefore, a threshold concentration value with defined uncertainty limits must be established and validated. For analytes of molecular structure indistinguishable between healthy and cancer specimens, the defining feature of the cancer biomarker may be concentration alone. There may be complications with such an approach. Recent discussions of prostate specific antigen (PSA) [16] and the ovarian cancer biomarker CA-125 [17] as early cancer biomarkers are cases in point.

The diagnostic scenario is confounded if analyte concentrations in normal individuals vary for reasons other than the presence of disease. Concentration stability in healthy individuals is often the exception. Normal variability may arise from, for example, physical exercise, nutrition, and menstrual cycle.

Unlike the variation of protein concentrations in body fluids, cells may be associated with a more limited range of analyte concentrations. It is attractive to consider the rare cancer cell as a collection platform for cancer-specific analytes where possible. The technologies and approaches by which cells could be collected have been considered elsewhere [18].

Unlike cancer biomarkers dependent on defined thresholds for clinical utility, cancer-specific structural changes at the level of the cell offer a more definitive opportunity for diagnostic precision. The classic example is the chromosomal change in chronic myelocytic leukemia (CML), the disease-specific gene fusion between the Abelson-related human oncogene (ABL) and human breakpoint cluster region gene (BCR) genes manifest as a microscopically detected chromosome translocation [19]. This diagnostic genetic rearrangement can also be detected by polymerase chain reaction (PCR) methods at the DNA or RNA levels [20], and at the protein level, with affinity reagents. In cancer-specific gene fusion events, the chromosomal break sites may be imprecise from patient to patient, necessitating multiple PCR reactions to cover the diagnostic spectrum. When gene fusions involve a broad range of breakpoints, a lower resolution technology such as FISH may accommodate clinical variability while providing adequate diagnostic data for clinical decisions.

Whereas a number of such gene fusions have been documented in the lymphomas, leukemias and in Ewing's sarcoma [21], only recently has gene expression and high-dimensional data analyses led to an unexpectedly common chromosome rearrangement in up to 50% of prostate cancers [22]. This, as is the CML gene fusion, is an example where analyte (in prostate cancer, a gene fusion specific to the cancer cells, often involving the transmembrane protease serine (2TMPRSS2) and E26-Specific-related genes (ERGs) on chromosome 21, Fig. 2, [22]), is structurally distinct from related DNA molecules from chromosome 21 from healthy individuals. Because the rearrangement is tumor-specific and diagnostic, issues defining a threshold and limits separating healthy from disease levels of biomarker are obviated, and the test results are binary instead of continuously variable and therefore relative to an established threshold and its measurement error limits.

The problem of dilution in serum proteomics may be mitigated if the diagnostic analytes are concentrated from specimen fluids in the laboratory, as is done by various affinity columns to remove immunoglobulins and albumin. In recent work [23], immunoaffinity columns have been used to select serum subfractions of greatest diagnostic interest in discovery proteomics work.

Alternatively, by focusing diagnostics on analytes preconcentrated within the individual cancer cell, analyte concentration issues as described above may also be avoided. In effect, capture of rare cancer cells that bear diagnostic signatures of early cancer is an alternative means of achieving biological concentration of diagnostic cancer biomarker molecules. The selection and analysis of cancer biomarkers in rare cancer cells is itself fraught with technological and biological challenges. However, if the rare cell serves to concentrate and localize the analyte for detection, this approach may offer significant advantages over analysis of high-dimensional data likely to result from, for example, serum proteomics via mass spectrometry. For this reason, we have focused our attention of cellular cancer biomarkers detected by fluorescence with DNA hybridization or affinity probes.

3
DNA and Antibody Probes: Fluorescence Reagents for Cancer Biomarker Quantification in Cells

Molecular diagnostics has long relied on fluorescent read-out as a quantitative, low-hazard and environmentally friendly alternative to radioisotope-based detection. Whereas this is sufficient for applications with millisecond read times, the photobleaching of conventional organic fluorophores through the microscope allows limited quantification, except where exposure times are minimal, as in flow- and laser-scanning cytometry [24].

For cytogenetics applications, where quantification has not been emphasized, organic fluorophores have been adequate in the past. As applications become more quantitative, measurement technologies with adequate resolution will also be required. Two applications of molecular cytogenetics biomarkers that require quantification are chromosome-specific telomere measurement and gene copy number determination for amplified cancer genes.

3.1 Standardization in Clinical Cytogenetics and FISH: a Paradigm

The implementation of human cytogenetics from research to clinical settings offers a paradigm for cancer biomarker development. Like emerging cancer biomarker work, technology and standards proceeded hand in hand to bring about clinical utility [25–28]. Before it was possible to comprehensively define disease-specific chromosome changes, the range of normal variation had to be established. To do this, technical breakthroughs were necessary. Technologies included hypotonic treatment to promote metaphase chromosome spreading, consensus on the structure of the standard human karyotype, the discovery of normal human chromosome banding landmarks and normal variants and cytogenetics nomenclature for describing normal and disease-related changes from the standard karyotype.

In-situ hybridization [29] was the first molecular cytogenetics technology, contrasting with morphological cytology methods and chromosome banding, to associate a specific nucleic acid species via hybridization with its three-dimensional locus in a cell or chromosome. Since then [28], a multitude of optical and molecular genetic developments have provided the research and clinical cytogeneticist, and histopathologists, with biomarker detection and quantification tools of increasing molecular sophistication and resolution. This family of technologies includes spectral karyotyping (SKY) [30], multicolor FISH (M-FISH) [31] and comparative genomic hybridization (CGH) [32]. The cost of labor has been reduced by moving some of these technologies to chip platforms that are read automatically and quantitatively. Although genomic DNA standards have been informally discussed at research and clinical meetings for almost a decade, there is at present no universal "standard" human genomic DNA preparation. However, the wide use of gene expression cDNA microarrays has pressed the issue and a number of gene expression microarray standards are under development [33, 34].

Like the current effort to define validated early cancer biomarkers, it was necessary in clinical cytogenetics to develop methods and characterize normal variability in order to interpret disease-specific changes. A significant drawback to some recently touted cancer biomarker discovery efforts is the lack of background detail for normal variability that occurs in healthy human populations. Interpretation of mtDNA changes in tumors is a case in point where putative tumor-specific DNA sequence changes in tumors are superimposed on DNA sequence changes in healthy individuals that occur with age, often in the absence of disease [35]. As above with serum proteomics, the lack of data on the normal variability of the analyte (in this case acquired mtDNA sequence changes with age) compromises the utility of the biomarker as a cancer-specific analyte.

Thus, research programs to define normal analyte variability may prove as important to the success of cellular biomarker development as studies comparing cancer cases with healthy controls. It is perhaps appropriate to bear this in mind when considering claims of the utility of cancer biomarkers such as serum proteomic analytes and acquired mtDNA sequence changes.

3.2
Affinity Reagents for Cancer Biomarker Measurement in Cells

In light of developments in human proteomics, there have been several surprises. First, normal human proteomic variability is poorly defined for serum and plasma. This holds despite the long history of clinical investigation of blood specimens for protein changes associated with disease [14]. Second, the evaluation of affinity reagents (antibodies, affybodies, aptamers, ankyrin-repeat library probes) used to detect human proteins is almost exclusively empirical and more often than not, lacks rigorous quantification or universal standards [36].

The cost and time expended in developing clinically useful antibodies is significant due to the fact that most, for example monoclonal and polyclonal antibodies, require extensive screening before a useful clone is found. Although such methods as surface plasmon resonance (SPR) and enzyme-linked immunosorbent assay (ELISA) can be configured in a high-throughput format to quantify binding properties of antibodies for their target epitopes, in many cases the epitopes are not well defined. It is also known that each antibody must be evaluated for its intended purpose independently. The fact that an affinity reagent can be used successfully in ELISA cannot be simply extrapolated to other methods such as tissue immunochemistry or western blotting. Thus, affinity reagent measurement for cancer biomarker detection would be a productive area for improvement as human proteomics methods evolve and background data accumulate.

For basic clinical proteomic standards, there are a number of national and international efforts to address the lack of certificated standard reference materials for proteomics in international organizations such as the Human Proteome Organization (HUPO) [37] and for US national proteomics needs [15].

Developments in production of affinity reagents and various fluorescence detection methods that preserve tissue architecture include antigen retrieval [38], various dye–antibody or secondary antibody combinations, and slide processing systems for high-throughput preparation of tissue sections. Although dating from three decades ago [39], coupled with high-throughput laser scanning microscopy [19], immunohistochemistry (IHC) and related technologies offer the potential for unparalleled throughput, precision and reproducibility for medical testing based on IHC when coupled with quantitative fluorescence detection.

3.3
Developments in Optical Imaging and Probes: Resolution and Quantification for Clinical Applications

In parallel with genetic and proteomic bioinformatics, physical resources and technologies, optical microscopy technology has also evolved recently in significant ways. Three-dimensional capture of fluorescence information is effected by confocal microscopy [40], and by deconvolution of *z*-plane stacks of images [41] (Fig. 2).

The resolution of optical imaging of fluorescent images is limited, as defined by the Abbe equation, and is roughly one-half the wavelength of excitation illumination [42]. Even the once-insurmountable Abbe limit has receded to advances in optical microscopy by a factor of 4 or more by 4Pi microscopy recently [42–44]. In some systems utilizing specialized fluorescent protein fluorophores and timed laser illumination regimens, optical resolution may approach the nanoscale [45]. While these are emerging research applications, they clearly demonstrate the prospect for and technical feasibility of high-resolution fluorescence imaging for clinical assays resulting from these and future breakthroughs in optical imaging for biological and biomedical applications.

The practicing clinical pathology laboratory or small point-of-care (POC) application currently relies on deposition of a color in histochemical and cytochemical assays rather than quantitative fluorescence microscopy. These imaging modalities are more qualitative than quantitative, and in some applications this is adequate data for medical histopathology. Recent efforts have gone toward improving the interobserver variability inherent in these relatively subjective assessments in conventional histopathology. The advent of human molecular genetics, the human genome databases and biotechnology have generated many measurement modalities, the bioinformatic infrastructure and supporting high-dimensional databases to expedite significantly more quantitative analyses in the medical pathology laboratory. It remains to be seen how much value the application of high specificity, quantitative molecular detection reagents (DNA probes and antibodies, primarily) will bring to clinical pathology, and if the extra cost will be justified by the improved quality, specificity and clinical relevance of such data.

3.4
Sources of Physical Standards: Cell Lines, Faux Tissues and Constructs

Classically, reference standards for mutation detection with optical methods such as bright field and fluorescence microscopy have depended on well-characterized mutant cell lines [46], and more recently isolated DNA from cell lines or patients bearing the mutations of interest. As cloning and the capacity to synthesize in vitro virtually any nucleic acid sequence variant have penetrated the medical standards and diagnostic community, sequence mutation standards of various types have appeared [47, 48].

In clinical oncology, there has been a continuing discussion for at least a decade of the best practice to institute standards for tumor tissue analysis with respect to specific, defined molecular analytes [49]. As HER2 was in the vanguard of molecular diagnostics and therapeutics for breast cancer, it has been among the first analytes for which fixed-cell reference standards have been available [49]. Despite this, much of the analysis of HER2 in FDA-approved tests remains qualitative. In addition to a good fixed-cell standard based on cell lines developed in Europe [50], alternative standard materials have been explored [51–53]. A summary of HER2 testing standards by the College of American Pathologists [49] has appeared recently. A National Institute of Standards and Technology (NIST) Workshop on HER2 testing [54] has resulted in much of the imaging technology development work at NIST with the intent to incorporate these data into a national HER2 testing standard.

Tumors in three dimensions are often not well modeled by cell cultures growing flat in a culture flask, due to differences in penetration of neutral-buffered formalin fixative between thin layers of cells and solid blocks of tissue. Some laboratories rely on bona fide tumor sections as internal physical standards. The fact that such materials are nonrenewable compromises their large-scale use. There have been some interesting technology developments in which multiple cell lines (normal and tumor) are cocultivated in a matrix that encourages the formation of "histoids" or reproducible aggregations of tumor and normal infiltrating cells [51]. This offers the prospect for regeneration of well-defined artificial tumors or standard "faux tissues", that mimic and parallel conditions in which clinical tumors are fixed and sectioned, which might be a good candidate for fixed-cell tumor standards if the conditions for histoid generation technology can be defined and replicated.

The development of high-throughput formats for analyzing tumor tissues has contributed in the research arena. Tissue microarrays are a means for massive parallel analyses, although this is primarily a research method, not a clinical testing modality at present. Whether tissue and expression arrays become established as a standard of care for diagnostic testing, remains to be seen.

3.5 Semiconductor Nanocrystals as Fluorophores: Development, Applications and Areas of Work

A significant component of highly specific optical detection at the level of the cell is the quality, reliability and features of molecular biological probes. For nucleic acid analysis, cloning methods and federally supported generation of bacterial artificial chromosome (BAC) libraries and mapped large-insert clones [55] have been a resource without which many new discoveries and resulting validated clinical tests based on nucleic acid hybridization would not have been possible. In addition, the decreasing cost and availability of custom oligonucleotide synthesis have made probe design easier. Since the analytical validation of a nucleic acid sequence is relatively straightforward and precise, such BAC and related synthetic oligonucleotide probes allow definition of assay performance.

Unlike nucleic-acid-based probes, affinity probes (affybodies, antibodies, ankyrin-repeat library probes, aptamers) are more problematic. A case in point involves a widely studied cancer biomarker, the reverse transcriptase termed telomerase. In normal cells of limited life span, each cell division is accompanied by erosion of chromosome ends [56]. In malignant cells, telomerase activity may often be reactivated to restore eroded chromosome ends, and contributes to the uncontrolled cell proliferation common in cancer.

This potentially important cancer biomarker has been widely studied as a promising early cancer biomarker in bodily fluids as well as in various tumor types with mixed results. Its clinical significance remains moot. Many clinical studies have utilized a commercially available murine antibody for human telomerase. Very recently, this antibody was shown to detect the protein nucleolin [57] rather than telomerase. Nucleolin shares some of its spatial and temporal expression features in cells. As such, the conclusions of prior clinical studies that relied on this antibody may warrant reevaluation.

This experience underscores the need for universal standards for characterization of the properties (specificity, definition of epitopes) of affinity reagents. Recent NCI [36] and NIST [58] workshops address these issues.

Although fluorescence has become an important issue in quantitative histopathology when antibodies [59] or DNA probes are used, in general the clinical community does not depend on rigorously standardized light sources for fluorescence analysis of cytogenetics preparations. The flow cytometry community was in the vanguard in promoting standards for fluorescence in fluorescence-activated cell sorting (FACS) systems [60].

With the advent of the biological application of CdSe/ZnS quantum dots [61, 62], many laboratories have attempted to implement quantum-dot probes for clinical FISH and antibody analysis. An early success with oligonucleotides [63] showed that Y chromosome repeats could be detected in sperm cells with a 23-mer oligonucleotide conjugated directly to a CdSe/ZnS quantum-dot fluorophore. More recently, successful FISH detection of single copy with BACs [5] and multicopy targets with oligonucleotides [64] has been reported.

Although metaphase cells have been a cytogenetics target for in-situ work, with the availability of BAC [55] clones that could be tiled in arrays, karyotyping can be done in an arrayed chip format. Discussion of standards for

CGH-like methods is evolving in the diagnostic community along with the development of nucleic acid standards for high-complexity nucleic acid populations for chip hybridizations such as gene expression microarrays.

Although initially suggested as a robust replacement for organic fluorophores in biooptical imaging, semiconductor nanocrystals development and applications have faced a number of obstacles to widespread application. However, progress continues in adapting nanocrystals to biological and diagnostic, and even therapeutic applications [65]. The problems commercializing this discovery include the difficulty of synthetic chemistry methods. In this realm, the core shell particles were synthesized initially under hazardous conditions [66]. It took some effort to develop the methods to the point at which reasonably monodisperse preparations could be created, and a shell–core structures proved to have better quantum efficiency than cores alone [67]. Once quantum dots could be made reproducibly, the problem became one of functionalizing these inorganic core–shell structures with biomolecules. The layers needed added size to already large particles.

Most recently, quantum dots have become commercially available as streptavidin conjugates or as kits functionalized with carboxyl (–COOH) or amino (–NH_2) surface moieties for linking points to proteins or nucleic acids.

Although the variety of quantum-dot colors and surface chemistries has rendered these reagents more accessible to research and clinical laboratories, the size problems remain an issue, as will the fact that the interaction between quantum dots and living systems is for the most part not well understood, as is the case for many nanomaterials being introduced into commercial products. For in-vivo applications, the known toxicities of heavy metals such as Cd and Pb, and the stability parameters of quantum dots containing these elements, will require more work before such agents can be widely accepted in applications requiring introduction of such materials into patients.

Another issue confounding the application of nucleic acid probes with current generation quantum dots is the poorly defined valency of both streptavidin coatings on the fluorophores and the multiple biotin haptens introduced by some nucleic acid reactions. Some recent successes have utilized biotin or its water-soluble analog biocytin [68] to titrate out the commercially available multivalent reagents, making stoichiometry more manageable.

One potential way around the problem of the large size of quantum dots due to many layers of coatings, is the recent demonstration that certain nucleic acid sequences may catalyze quantum-dot formation directly, allowing the potential for an integrated biomolecule designed into the fluorophore itself, with the prospect of obviating the extra layers that increase size [69, 70].

Thus, there remain some issues to address before quantum-dot fluorophores meet the potential many have predicted. That said, this active area of research and development has owed some of its vigor to the wide interest in exploration of this novel class of materials for quantitative fluorescence measurement.

4 Application to Quantitative Medical Histopathology

These studies on biomarker detection in cells by optical fluorescence methods and nanomaterials fluorophores may have applications beyond the immediate cancer biomarker systems and analytes. Classically, much of medical histopathology has been taught and applied as a means for a clinical observer to reach a medical diagnosis based on his training, experience and the visual appearance of the cells and tissues fixed and stained on glass slides. In many cases, it has not been possible to objectively identify precisely which features of the histopathology images contribute to the resulting diagnosis. Much has been written about the interobserver variability in diagnostic histopathology and recently, some serious attempts to render image classification more rigorously objective have been made [71].

Beyond light microscopy with histological staining as a "classical" cancer biomarker imaging technology, in similar laboratory pathology test settings, clinical assays have also utilized specific antibodies to increase confidence in histopathological diagnoses. Although many of the methods used are either semi- or nonquantitative, this clinical practice is clearly moving in a direction where it may benefit from the wealth of new genomic and proteomic information, and quantitative imaging and related technology. It is our bias that the quantitative and rigorously validated analysis of one or multiple cancer biomarkers in cells, and the thoughtful design and implementation of measurement standards, has great potential to improve the precision, accuracy and reproducibility of classical histopathology.

5 Significant Challenges to Implementation of Standards for Cancer Biomarkers

Assuming the perfect cancer biomarker or, as seems far more likely, panel of multiple biomarkers are discovered and validated both analytically and clinically, there remain a number of challenges that should be borne in mind that may not be obvious to the research scientist.

For example, the difficulty of translating research cancer biomarker protocols to the clinical environment will remain. Here, perhaps the most significant challenge is the development of methods and protocols that address significant variability in preanalytical processing of clinical specimens. This means standardizing, or at very least defining with some rigor, the conditions clinical specimens are exposed to after sampling (ex vivo). Variability in types and treatment of blood for proteomics, for example, has stymied progress in evaluation of what are already complex data sets. Tissue samples can also be subject to a wide range of ex-vivo environmental conditions that are associated with changes in levels of cancer biomarkers from their in-vivo state. No matter how reproducible, quantitative and precise the analytical laboratory methods, if variability in preanalytical processing remains beyond definition and standardization, the output data will fail to meet its potential for supporting improved clinical decisions by managing physicians.

Related factors that the research scientist may not consider are cost and value. If an assay is expensive or goes without third-party reimbursement, the clinical community may not implement or embrace it, no matter how accurate it may be. On the other hand, if the information gained is accurate and highly significant, and costly therapeutics are under consideration, a relatively expensive test may be supported in the clinical community [72]. In any case, assay formats to emphasize high throughput and cost effectiveness should be considered.

In the past, analytical and clinical validation have been performed by the same laboratory that initiates a discovery. This may be scientifically and technologically justified. However, in cancer biomarker validation, it is also important to demonstrate that an assay is transportable to nonspecialist laboratories under conditions likely to apply in cancer biomarker screening centers. If a single laboratory nationwide or worldwide alone is able to run the test, it may not find wide acceptance in the clinical community. Thus, independent validation at both the analytical and clinical levels in multicenter trials has been shown to be necessary to fully evaluate cancer biomarker analysis systems.

There is not adequate space here to address issues in cancer biomarker study design; recent publications and meeting summaries do cover progress in this area [73]. Suffice it to say that appropriate study design in evaluating novel cancer biomarkers is essential to avoid some of the missteps made in cancer biomarker evaluation in the recent past. Such recent issues and debates [11] have led to a robust dialogue on biomarker study design among the academic biostatisticians, cancer biomarker and metrology communities, and in companies developing such cancer biomarkers commercially.

6
Physical Standards for Fluorescent Measurement of Cancer Biomarkers: Now or Later?

Although in retrospect, standards and control materials seem an obvious adjunct that logically should parallel development of cancer biomarker assays, the interrelationship between assay development and standards may be more subtle than it appears. Standards are appropriate when a test, technology and biomarker analyte are well defined (quantitative) and stable, there is established need in the clinical community, and there is consensus on the format of the physical standard.

In design of generic physical standards for biomedical testing, the stakeholder community should be universally encouraged to take the opportunity to express needs. Stakeholders may include private, government and academic scientists and clinicians. It is also important to consider manufacturers, patient advocates and even third-party reimbursement agencies, all of whom have a vested interest in the accuracy of testing based on cancer biomarkers. It is important to understand how standards development and production will be supported by identifying who will benefit from more accurate testing. In some cases, validation and standards will make the difference between tests considered research and therefore nonreimbursable, and tests considered adequately validated for FDA approval and third-party reimbursement.

Considering that 2–3 years are generally required for background and production work for developing, qualifying and properly certifying a physical standard, timeliness is a major factor in planning. If the technology and needs are rapidly evolving, the initiating need may change or may not exist by the time a standard reference material is completed. Whether the need is generic or specific should also be considered when time and resource constraints drive prioritization of standards production.

Although there is no crystal ball to reveal the course of future cancer biomarker work and associated standards needs, there are a number of constraints that assist planners in resource allocation. Because of time and cost constraints in development of physical standards for molecular medical testing, a careful, comprehensive and quantitative needs analysis is most often appropriate, and may be an essential prerequisite to successful, cost-effective development of a universal physical standard.

Finally, for the sake of thoroughness, any cancer biomarker standard should include in its development specific plans for evaluating the impact of the physical standard on the target community, to evaluate whether the effort was successful, and whether the standards are being fully utilized.

7 Summary

We have considered issues related to development and implementation of physical standards in validation of cancer biomarkers, especially those relating to analysis of cellular- and tissue-level heterogeneity detected by fluorescence methods. Here, we consider FISH and immunofluorescence in quantitative histopathology. Whereas much current effort focuses on the measurement of subtle concentration changes in body fluids (clinical proteomics of serum and plasma, for example), the heterogeneity inherent in cancer may offer more opportunity for robust measurement methods when cellular integrity is maintained in the analysis.

Standards in biomedical molecular diagnostics in such areas as early detection of cancer are driven by needs in the commercial, clinical and academic sectors. They are also a function of the state of measurement science at the time, with an emphasis on how truly quantitative the measurement technologies are, how stable the biological analytes are, and whether the availability of a properly certificated standard can be expected to expedite a measureable change for the good in the desired community at the end of the project.

A further requirement is the existence of adequate quantitative data on what defines the limit of healthy values for biomedical analytes.

The development of biotechnological measurement tools for the quantitative measurement of DNA, RNA and protein, has offered an opportunity for the research community to define the limits of measurements associated with healthy individuals. Without these, establishing disease-specific measurement science is difficult, if not impossible. The development of a universally accessible human genome database and technologies for its inexpensive analysis will be key in realizing the promise of molecular diagnostics in early cancer detection through quantitative measures of biomarkers.

Acknowledgements We acknowledge the support of NCI (Early Detection Network, EDRN) and NIST (RM development funds for HER2 testing) in these studies. Certain commercial entities, equipment, or materials may be identified in this paper in order to describe an experimental procedure or concept adequately. Such identification is not intended to imply recommendation or endorsement by the National Institute of Standards and Technology, nor is it intended to imply that the entities, materials, or equipment are necessarily the best available for the purpose.

References

1. Editorial (2006) Nat Biotechnol 24:869
2. Pepe M, Pepe MS, Etzioni R, Feng Z, Potter JD, Thompson ML, Thornquist M, Winget M, Yasui Y (2001) J Natl Cancer Inst 93:1054
3. Barker PE (2003) Ann NY Acad Sci 983:142
4. Jefford CE, Irminger-Finger I (2006) Crit Rev Oncol Hematol 59:1
5. Xiao Y, Barker PE (2004) Nucleic Acids Res 32:1
6. Xiao Y, Barker PE (2004) Minerva Biotecnol 16:1
7. Xiao Y, Telford WG, Ball JC, Locascio LE, Barker PE (2005) Nat Methods 2:723
8. http://www.cancer.gov/Templates/db_alpha.aspx?CdrID=45618. Accessed 14 Mar 2008
9. Petricoin EF, Liotta LA (2006) Scientist 20:33
10. Hu S, Loo JA, Wong DT (2006) Proteomics 6:6326
11. Ransohoff DF (2005) J Natl Cancer Inst 97:315
12. Diamandis E (2003) Clin Chem 49:1272
13. Vitzthum F, Behrens F, Anderson NL, Shaw JH (2005) J Proteome Res 4:1086
14. Anderson NL, Polanski M, Pieper R, Gatlin T, Tirumalai RS, Conrads TP, Veenstra TD, Adkins JN, Pounds JG, Fagan R, Lobley A (2004) Mol Cell Proteomics 3:311
15. Barker PE, Wagner PD, Stein SE, Bunk DM, Srivastava S, Omenn GS (2006) Clin Chem 52:1669

16. Thompson KE, Hernandez J, Canby-Hagino ED, Troyer D, Thompson IM (2005) J Urol 174:553
17. Bast RC Jr, Badgwell D, Lu Z, Marquez R, Rosen D, Liu J, Baggerly KA, Atkinson EN, Skates S, Zhang Z, Lokshin A, Menon U, Jacobs I, Lu K (2005) Int J Gynecol Cancer 15:274
18. Goon PK, Boos CJ, Stonelake PS, Blann AD, Lip GY (2006) Thromb Haemost 96:45
19. Rowley JD (1973) Nature 243:290
20. Miglino M, Varaldo R, Colombo N, Grasso R, Clavio M, Garuti A, Aquino S, Albarello A, Sessarego M, Gobbi M (2006) J Exp Clin Cancer Res 25:321
21. Rowley JD (2001) Nat Rev Cancer 1:245
22. Tomlins SA, Rhodes DR, Perner S, Dhanasekaran SM, Mehra R, Sun XW, Varambally S, Cao X, Tchinda J, Kuefer R, Lee C, Montie JE, Shah RB, Pienta KJ, Rubin MA, Chinnaiyan AM (2005) Science 310:644
23. Gong Y, Li X, Yang B, Ying W, Li D, Zhang Y, Dai S, Cai Y, Wang J, He F, Qian X (2006) J Proteome Res 5:1379
24. Luther E, Kamentsky L, Henriksen M, Holden E (2004) Methods Cell Biol 75:185
25. Hsu TC (1979) Human and mammalian cytogenetics: an historical perspective. Springer Verlag, Basel
26. Levsky JM, Singer RH (2003) J Cell Sci 116:2833
27. Shaffer LG, Tommerup N (2005) ISCN, An international system for human cytogenetic nomenclature (2005). International Standing Committee on Human Cytogenetic Nomenclature. Karger, New York
28. Clinical Laboratory Standards Institute (2004) Fluorescence in situ hybridization (FISH) methods for medical genetics; proposed guideline. Document MM7-A, Wayne, PA
29. Pardue ML, Gall JG (1969) Proc Natl Acad Sci USA 64:600
30. Schrock E, du Manoir S, Veldman T, Schoell B, Wienberg J, Ferguson-Smith MA, Ning Y, Ledbetter DH, Bar-Am I, Soenksen D, Garini Y, Ried T (1996) Science 273:494
31. Walter J, Joffe B, Bolzer A, Albiez H, Benedetti PA, Muller S, Speicher MR, Cremer T, Cremer M, Solovei I (2006) Cytogenet Genome Res 114:367
32. Pinkel D, Albertson DG (2005) Annu Rev Genomics Hum Genet 6:331
33. Cronin M, Ghosh K, Sistare F, Quackenbush J, Vilker V, O'Connell C (2004) Clin Chem 50:1464
34. Salit M (2006) Methods Enzymol 411:63
35. Bua E, Johnson J, Herbst A, Delong B, McKenzie D, Salamat S, Aiken JM (2006) Am J Hum Genet 79:469
36. Haab BB, Paulovich AG, Anderson NL, Clark AM, Downing GJ, Hermjakob H, Labaer J, Uhlen M (2006) Mol Cell Proteomics 5:1996
37. Omenn GS, States DJ, Adamski M, Blackwell TW, Menon R, Hermjakob H, Apweiler R, Haab BB, Simpson RJ, Eddes JS, Kapp EA, Moritz RL, Chan DW, Rai AJ, Admon A, Aebersold R, Eng J, Hancock WS, Hefta SA, Meyer H, Paik YK, Yoo JS, Ping P, Pounds J, Adkins J, Qian X, Wang R, Wasinger V, Wu CY, Zhao X, Zeng R, Archakov A, Tsugita A, Beer I, Pandey A, Pisano M, Andrews P, Tammen H, Speicher DW, Hanash SM (2005) Proteomics 5:3226
38. Rahimi F, Shepherd CE, Halliday GM, Geczy CL, Raftery MJ (2006) Brain Tissue Anal Chem 78:7216
39. Faulk WP, Hijmans W (1972) Prog Allergy 6:9
40. Lee S, Howell BJ (2006) Methods Enzymol 414:468
41. Evans JG, Matsudaira P (2007) Methods Mol Biol 356:33
42. Heintzmann R, Ficz G (2006) Brief Funct Genomic Proteomic 5:289

43. Egner A, Hell SW (2005) Trends Cell Biol 15:207
44. Bewersdorf J, Schmidt R, Hell SW (2006) J Microsc 222:105
45. Betzig E, Patterson GH, Sougrat R, Lindwasser OW, Olenych S, Bonifacino JS, Davidson MW, Lippincott-Schwartz J, Hess HF (2006) Science 313:1642
46. Bernacki SH, Beck JC, Stankovic AK, Williams LO, Amos J, Snow-Bailey K, Farkas DH, Friez MJ, Hantash FM, Matteson KJ, Monaghan KG, Muralidharan K, Pratt VM, Prior TW, Richie KL, Levin BC, Rohlfs EM, Schaefer FV, Shrimpton AE, Spector EB, Stolle CA, Strom CM, Thibodeau SN, Cole EC, Goodman BK, Stenzel TT (2005) Clin Chem 51:2013
47. O'Connell CD, Tully LA, Devaney JM, Marino MA, Jakupciak JP, Atha DH (2004) Mol Diagn 7:85
48. Sunar-Reeder B, Atha DH, Aydemir S, Reeder DJ, Tully L, Khan AR, O'Connell CD (2004) Mol Diagn 8:123
49. Wolff AC, Hammond MEH, Schwartz JN, Hagerty KL, Alred DC, Cote RJ, Dowsett M, Fitzgibbons PL, Hanna WM, Langer A, McShane LM, Paik S, Pegram MD, Perez EA, Press MF, Rhodes A, Sturgeon C, Taube SE, Tubbs R, Vance GH, van de Vijver TM, Hayes DF (2006) Arch Pathol Lab Med 131:18
50. Rhodes A, Borthwick D, Sykes R, Al-Sam S, Paradiso A (2004) Am J Clin Pathol 122:51
51. Ingram M (2006) Replication of biological tissue. US Patent 6 998 264, 14 Feb 2006
52. Sompuram SR, Kodela V, Zhang K, Ramanathan H, Radcliffe G, Falb P, Bogen SA (2002) J Histochem Cytochem 50:1425
53. Braunschweig T, Chung JY, Hewitt SM (2005) Expert Rev Proteomics 2:325
54. Hammond E, Barker PE, Taube S, Gutman S (2003) Appl Immunohistochem Mol Morphol 11:103
55. Cheung VG, Nowak N, Jang W, Kirsch IR, Zhao S, Chen XN, Furey TS, Kim UJ, Kuo WL, Olivier M, Conroy J, Kasprzyk A, Massa H, Yonescu R, Sait S, Thoreen C, Snijders A, Lemyre E, Bailey JA, Bruzel A, Burrill WD, Clegg SM, Collins S, Dhami P, Friedman C, Han CS, Herrick S, Lee J, Ligon AH, Lowry S, Morley M, Narasimhan S, Osoegawa K, Peng Z, Plajzer-Frick I, Quade BJ, Scott D, Sirotkin K, Thorpe AA, Gray JW, Hudson J, Pinkel D, Ried T, Rowen L, Shen-Ong GL, Strausberg RL, Birney E, Callen DF, Cheng JF, Cox DR, Doggett NA, Carter NP, Eichler EE, Haussler D, Korenberg JR, Morton CC, Albertson D, Schuler G, de Jong PJ, Trask BJ (2001) Nature 409:953
56. Blackburn EH, Greider CW, Szostak JW (2006) Nat Med 12:1133
57. Wu YL, Dudognon C, Nguyen E, Hillion J, Pendino F, Tarkanyi I, Aradi J, Lanotte M, Tong JH, Chen GQ, Segal-Bendirdjian E (2006) J Cell Sci 119:2797
58. U.S. Measurement System Workshop Series (2006) http://usms.nist.gov/workshops/AutoantibodyMeasurement.htm. Accessed NIST (Gaithersburg, MD, USA) Jan 24 2008
59. Rimm DL (2006) Nat Biotechnol 24:914
60. U.S. Measurement System Workshop Series (2006) http://usms.nist.gov/workshops/flow-cytometry.htm. Accessed NIST (Gaithersburg, MD, USA) 19 Feb 2008
61. Chan WC, Nie S (1998) Science 281:2016
62. Bruchez M Jr, Moronne M, Gin P, Weiss S, Alivisatos AP (1998) Science 281:2013
63. Pathak S, Choi SK, Arnheim N, Thompson ME (2001) J Am Chem Soc 123:4103
64. Bentolila LA, Weiss S (2005) Cell Biochem Biophys 45:59
65. Michalet X, Pinaud FF, Bentolila LA, Tsay JM, Doose S, Li JJ, Sundaresan G, Wu AM, Gambhir SS, Weiss S (2005) Science 307:538
66. Murray C, Norris D, Bawendi M (1993) J Am Chem Soc 115:8706
67. Hines MA, Guyot-Sionnest P (1996) J Phys Chem 100:468

68. Hinds S, Taft B, Levina L, Sukhovatkin V, Dooley CJ, Roy MD, MacNeil DD, Sargent EH, Kelley SO (2006) J Am Chem Soc 128:64
69. Ma N, Dooley CJ, Kelley SO (2006) J Am Chem Soc 128:12598
70. Chan P, Yuen T, Ruf F, Gonzalez-Maeso J, Sealfon SC (2005) Nucleic Acids Res 33:E161
71. Malpica A, Matisic JP, Niekirk DV, Crum CP, Staerkel GA, Yamal JM, Guillaud MH, Cox DD, Atkinson EN, Adler-Storthz K, Poulin NM, Macaulay CA, Follen M (2005) Gynecol Oncol 99 [Suppl 1]:S38
72. Paik S, Shak S, Tang G, Kim C, Baker J, Cronin M, Baehner FL, Walker MG, Watson D, Park T, Hiller W, Fisher ER, Wickerham DL, Bryant J, Wolmark N (2004) N Engl J Med 351:2817
73. Maruvada P, Srivastava S (2006) Cancer Epidemiol Biomarkers Prev 15:1078

Part VIII
Fluorescence Technologies in Biomedical Diagnostics

Springer Ser Fluoresc (2008) 6: 533–548
DOI 10.1007/4243_2008_045

Published online: 13 March 2008

Fluorescence Techniques in Biomedical Diagnostics: Instrumentation, Analysis and Unresolved Issues

Herbert Schneckenburger

Hochschule Aalen, Institut für Angewandte Forschung, Anton-Huber-Str. 21,
73430 Aalen, Germany
herbert.schneckenburger@htw-aalen.de

Abstract A short overview on principles and applications of fluorescence measurements is given, and three methods are described in detail. These methods include total internal reflection fluorescence (TIRF) microscopy and screening, fluorescence anisotropy as well as Förster resonance energy transfer (FRET) and total internal reflection energy transfer (TIRET) microscopy. All these methods have a high potential in biomedical research and in vitro diagnostics, but also need further optimization and standardization. The present chapter shows some approaches towards those goals and describes possible applications.

Keywords Fluorescence anisotropy · Fluorescence lifetime · Fluorescence microscopy · Fluorescence reader · Förster resonance energy transfer · Total internal reflection

Abbreviations

EBFP	Enhanced blue-fluorescent protein
ECFP	Enhanced cyan-fluorescent protein
EGF	Epidermal growth factor
FLIM	Fluorescence lifetime imaging (microscopy)
FRET	Förster resonance energy transfer
EGFP	Enhanced green-fluorescent protein
FAK	Focal adhesion kinase
HCS	High-content screening
HTS	High-throughput screening
NADH	Reduced nicotinamide adenine dinucleotide
Pax	Paxillin
R123	Rhodamine 123
TIR	Total internal reflection
TIRET	Total internal reflection energy transfer
TIRF	Total internal reflection fluorescence
EYFP	Enhanced yellow fluorescent protein

1
Overview

Fluorescence techniques are very sensitive for analyzing specific molecules, cells and tissues, and therefore are used increasingly in biomedical diagnostics. Fluorescence arises upon absorption of light and is related to an electronic transition from the first excited singlet state (S_1) to the ground state (S_0) of an (organic) molecule. Generally fluence rates or photon fluxes ("fluorescence intensities") on one hand and lifetimes of the excited state ("fluorescence lifetimes") on the other hand are measured as functions of spatial coordinates, wavelength or time. Whereas fluorescence intensity is a measure of the number of excited molecules, fluorescence lifetime corresponds to the reciprocal rate of all (radiative and nonradiative) transitions originating from the excited state S_1, thus characterizing the interaction of a molecular species with its environment.

Fluorescence may result from intrinsic fluorophores, e.g., cellular proteins [1, 2], coenzymes (like reduced nicotinamide adenine dinucleotide, NADH [3–5]) or extracellular fibers [6, 7], as well as from numerous dyes that are used for staining of specific cell sites or organelles [8–10]. The most exciting recent development of fluorescent probes for biological studies has been the introduction of fluorescent proteins. A green fluorescent protein (GFP) is naturally produced by the jellyfish *Aequorea victoria* [11]. After cloning of the GFP gene, GFP variants with different excitation and emission properties have been produced. By fusion of genes coding for a specific cellular protein and a GFP variant, fluorescent protein chimera are created, permitting site-specific tracking in living cells or even whole organisms [12, 13].

Fluorescence techniques are used for in vitro as well as for in vivo measurements. In the latter case, fluorescence of various organs is commonly excited via glass fibers and detected endoscopically [14–17], whereas in the former case fluorescence microscopy or screening techniques are applied. Several advanced methods, e.g., laser scanning or structured illumination techniques [18, 19], evanescent field [20, 21] or fluorescence lifetime imaging (FLIM) [22, 23] measurements, have been used in connection with fluorescence microscopes. For this reason microscopic techniques are in the focus of the present chapter.

2
Methods, Instrumentation and Applications

2.1
Total Internal Reflection Fluorescence (TIRF)

Axial resolution in optical microscopy has been a challenge for more than 20 years. In addition to confocal or multiphoton [24, 25] microscopy, wide field techniques using e.g., structured illumination or evanescent waves have become more and more popular. In TIRF microscopy a light beam propagating through a medium of refractive index n_1 (e.g., glass) meets an interface with a second medium of refractive index $n_2 < n_1$ (e.g., cytoplasm) and is totally reflected at all angles of incidence Θ which are greater than the critical angle $\Theta_c = \arcsin(n_2/n_1)$. Despite being totally reflected, the incident beam establishes an evanescent electromagnetic field that penetrates into the second medium and decays exponentially with the distance z from the interface. According to the relation

$$d = (\lambda/4\pi)\left(n_1^2 \sin^2 \Theta - n_2^2\right)^{-1/2} \tag{1}$$

(with λ corresponding to the wavelength of light), penetration depths d between about 70 nm and more than 200 nm are attained and used to detect fluorophores within or close to the plasma membrane. So far, TIRF microscopy has been used for measuring the topology of cell–substrate contacts [20, 26], protein dynamics [27], membrane-proximal ion fluxes [28, 29], endocytosis or exocytosis [30–32], and membrane-associated photosensitizers [33].

Most TIRF systems presently used (and commercially available) are based on extreme dark-field illumination by a microscope objective lens of high numerical aperture A_N [34]. A laser beam is focused close to the edge of the aperture and falls onto the sample under an angle Θ that exceeds the critical angle Θ_C (typical values: $\Theta = 66$–$72°$ for $\Theta_C \approx 65°$). Although this technique works fairly well in many cases, variation of Θ and therefore of the penetration depth d of the evanescent wave (Eq. 1) is usually not possible, such that cell–substrate topologies cannot be measured in detail. On the other hand, variable-angle TIRF microscopy requires rather complex optical equipments, so far. This problem has recently been overcome [35] by using the illumination device depicted in Fig. 1, which replaces a conventional microscope condenser. In this device, parallel light from a single-mode glass fiber is deflected on an adjustable mirror and focused by a concave mirror onto the sample, which is placed in the centre of a hemispherical glass prism. By imaging a light spot from the adjustable mirror onto the sample, identical parts of the sample are illuminated under variable angles of incidence when the mirror is rotated. Therefore, a series of images can be recorded at variable penetration depth of the evanescent field, and the distance between a cell and its substrate (e.g., microscope object slide) can be calculated from these images with nanometer precision [35]. An example is given in Fig. 2, where the distribution of the photosensitizer protoporphyrin IX (induced by 5-aminolevulinic acid [36]) in close vicinity to the plasma membrane of human glioblastoma cells is depicted (Fig. 2a), and where cell–substrate distances have been calculated from eight images taken at $66° \leq \Theta \leq 73°$ (Fig. 2b). As previously shown [35], fluorescence intensity in TIRF images is proportional to the penetration depth d of the evanescent field which is about

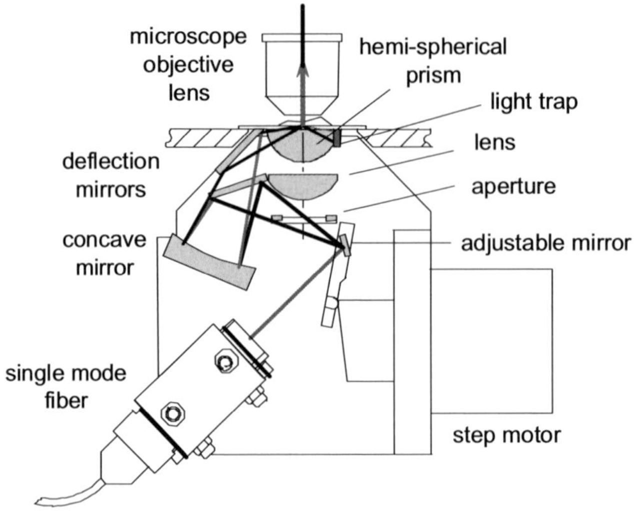

Fig. 1 Illumination device for variable angle total internal reflection fluorescence (TIRFM) microscopy, including an additional light path for transillumination and phase contrast microscopy

100 times smaller than the sample thickness. This also implies that total internal reflection(TIR)-induced fluorescence intensities are about 100 times lower than those of conventional fluorescence images. Conventional charge-coupled device (CCD) cameras are often not sensitive enough to detect those weak fluorescence signals. However, a novel generation of electron-multiplying CCD (EMCCD) cameras with an internal amplification factor around 1000 is well suitable to detect TIRF images with little noise [37] such that these images can be used together with mathematical algorithms for calculation of cell surface topology.

In addition to variable-angle TIRF microscopy, the illumination device depicted in Fig. 1 can also be used for transillumination or phase contrast microscopy (needed e.g., for adjustment of the samples). For this purpose an additional lens has been placed in front of the hemispherical prism, such that both lenses together are used for Köhler's illumination, as well as for imaging a ring-shaped aperture into the microscope aperture (as needed in phase contrast microscopy). Moreover, this kind of TIR illumination can be combined with any microscope objective lens and is not limited to high-aperture (and therefore highly magnifying) lenses.

Meanwhile, TIRF microscopy has been optimized for detection of very small sample volumes [38] and even single molecules [39]. However, only in a few cases have multiple samples (located e.g., on the surface of waveguides) been excited simultaneously by TIR [40, 41]. Therefore, for parallel TIRF screening of a larger number of samples, a microtiter plate reader system has been developed, as depicted in Fig. 3 [42]. This system is based on

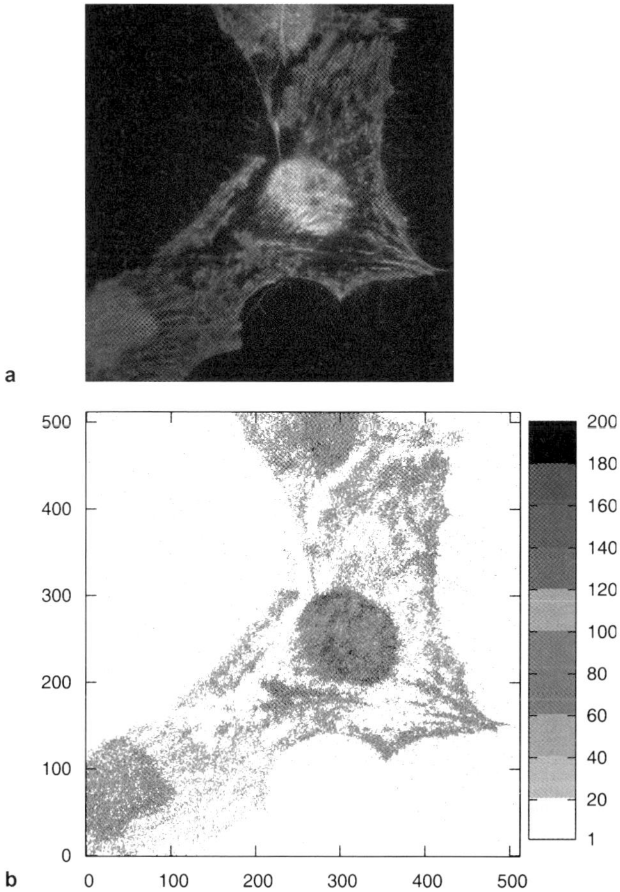

Fig. 2 Fluorescence of 5-aminolevulinic-acid-induced protoporphyrin IX in U373 MG glioblastoma cells upon total internal reflection (TIR) excitation. **a** Fluorescence intensity excited at $\lambda = 391$ nm and $\Theta = 66°$. **b** Cell-substrate distances (in nanometers) calculated from variable-angle TIRFM ($66° \leq \Theta \leq 73°$). Emission measured at $\lambda \geq 590$ nm; image size: 105×105 μm each

the splitting of laser excitation into eight individual beams and multiple TIRs of each beam within the bottom of a microtiter plate (cell substrate), such that up to 96 individual samples are illuminated simultaneously by an evanescent electromagnetic wave. A specific glass bottom of high transmission and appropriate thickness (2 mm in order to excite each sample under TIR conditions) was selected and attached to the 96-well cell culture plate by a noncytotoxic adhesive. Glass rods of rectangular cross section were optically coupled to this bottom for TIR illumination. When using this setup, fluorescence arising from the plasma membrane of living cells was detected simultaneously from all samples using an integrating CCD camera [high-throughput screen-

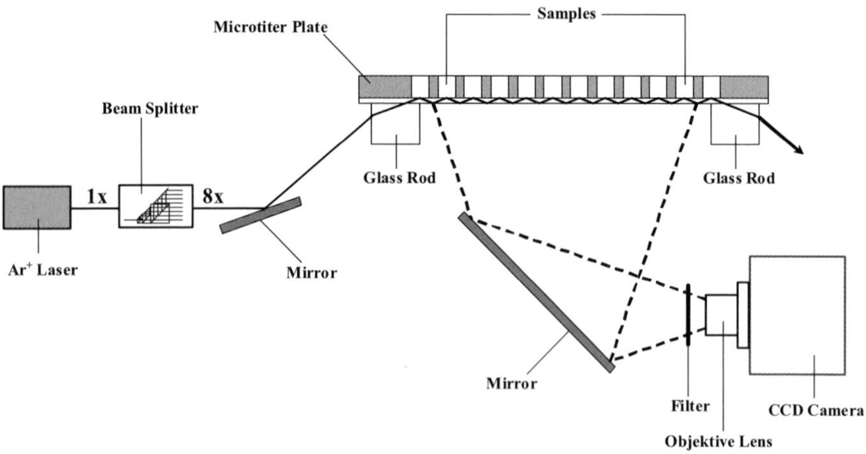

Fig. 3 TIR fluorescence reader using 8× beam splitting and 12-fold TIR of each laser beam to illuminate 96 samples of a microtiter plate (reproduced from [42] with modifications)

ing (HTS)]. So far, membrane-associated fluorescent proteins and fluorescent markers have been quantified, and intracellular translocation of fluorescently tagged protein kinase c has been measured upon activation [42]. If further information about individual samples is required (e.g., fluorescence lifetime or polarization data), HTS can be combined with high-content screening (HCS) if a specific detector is inserted and scanned over the interesting samples of the microtiter plate. This combination has recently been used to examine the membrane dynamics of living cells [43].

2.2 Fluorescence Anisotropy

Excitation of samples with polarized light is used to select molecules whose optical transition dipole moments are parallel to the electrical field vector. Depolarization of fluorescence is often due to rotational diffusion of these molecules during the lifetime of their excited state, dependent on the size and shape of the molecules as well on the viscosity of their environment. Therefore, fluorescence polarization measurements can be used to examine viscosities of cells or cell membranes [1].

Commonly, fluorescence intensities are measured parallel ($I_{||}$) and perpendicular ($I_{\perp}$) to the plane of incidence, and the anisotropy function is determined according to

$$r(t) = \left[I_{||}(t) - I_{\perp}(t)\right] / \left[I_{||}(t) + 2I_{\perp}(t)\right] . \qquad (2)$$

If a sample is excited by a short (picosecond) polarized light pulse, $r(t)$ decreases exponentially (from the initial value r_0) with the rotational diffusion time constant t_r according to

$$r(t) = r_0 \, e^{-t/t_r} , \tag{3}$$

where t_r is expected to increase with cell viscosity. In some cases, however (e.g., if long-shaped molecules are embedded parallel to the membrane lipids), molecular motion is hindered in one direction, and even at longer times after optical excitation some fluorescence anisotropy (r_∞) is maintained. In this case, Eq. 3 can be modified according to

$$r(t) = (r_0 - r_\infty) \, e^{-t/t_r} + r_\infty . \tag{4}$$

Some experimental problems may arise from

1. polarization dependent sensitivities of fluorescence detection
2. photochemical modification (e.g., photobleaching) of the sample during the measurement

The first problem can be resolved, if two polarizations of incident light are used, and if for each polarization, fluorescence intensity is measured. If in a first step for "horizontal" polarization of incident light, fluorescence intensities I_{HV} (vertical) and I_{HH} (horizontal) are detected, a response factor of the detection system $G = I_{HV}/I_{HH}$ can be determined. When using "vertical" polarization of incident light and measuring the fluorescence intensities I_{VV} (vertical) and I_{VH} (horizontal) in a second step, the anisotropy function can be determined as

$$r(t) = [I_{VV}(t) - G \, I_{VH}(t)] \, / \, [I_{VV}(t) + 2G \, I_{VH}(t)] \; . \tag{5}$$

Generally G is wavelength dependent, and deviations from $G = 1$ are most pronounced when monochromators are used.

The second problem is most pronounced, if the two polarizations are measured sequentially, such that photochemical modifications mainly affect the second measurement. This error can be minimized by simultaneous detection of the two polarizations in a two-channel system and by reducing the power of incident light to a minimum (using a highly sensitive detection system). In fluorescence imaging, a polarization-sensitive beam splitter and a dual imaging system may be used such that fluorescence is detected on two sections of a camera system. An experimental setup for polarization imaging in a fluorescence microscope is depicted in Fig. 4. Samples are excited either by conventional epi-illumination or by TIR illumination (for selective measurements of fluorophores located within or close to the plasma membrane). It should be mentioned that for TIR measurements, one polarization of incident light (generally perpendicular to the plane of incidence) should be maintained, such that the G factor has to be determined independently from epi-illumination measurements. Two perpendicularly polarized images are recorded on an EMCCD camera (DV887DC-BV, ANDOR, Belfast, UK) using a dual-view optical system (Optical Insights, Tucson, Arizona, USA), as shown in Fig. 5 for U373 MG human glioblastoma cells incubated with

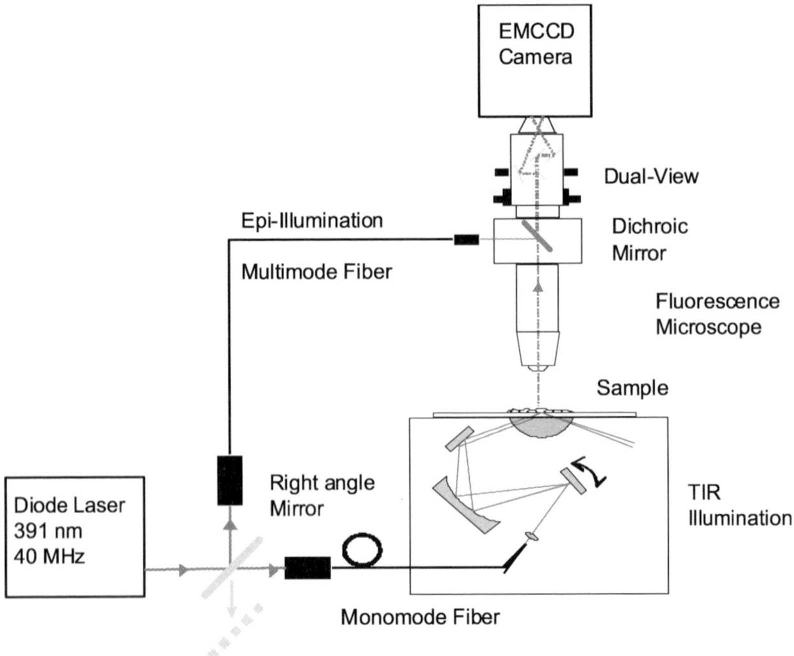

Fig. 4 Schematic set up for polarization imaging with a fluorescence microscope using TIR or epi-illumination and dual-view (simultaneous) detection.*EMCCD* Electron-multiplying charge-coupled device. (Reproduced from [46] with modifications)

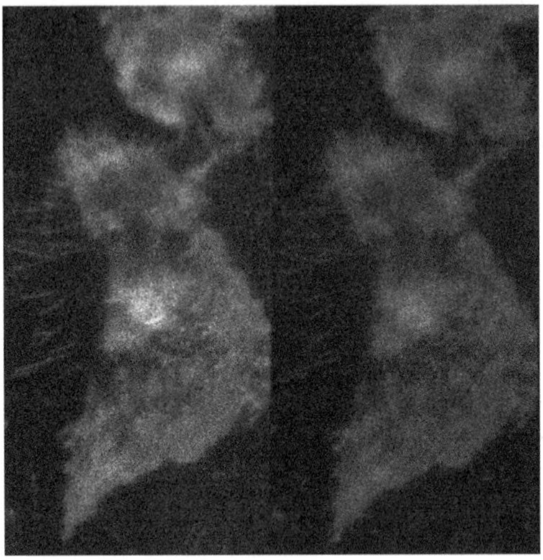

Fig. 5 Fluorescence images of U373 MG human glioblastoma cells incubated with the membrane marker laurdan (8 μM, 60 min) upon TIR illumination; polarizations parallel (*left*) and perpendicular (*right*) to excitation light (391 nm); emission measured at $\lambda \geq 420$ nm; image size: 65 μm × 130 μm each

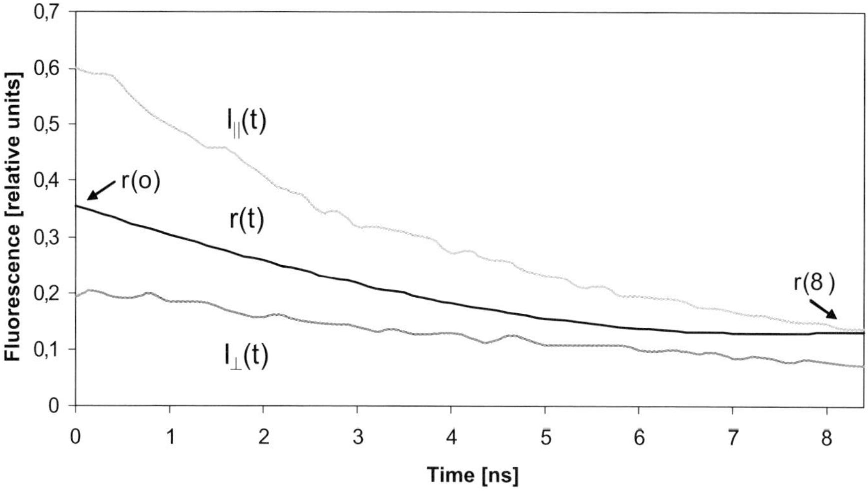

Fig. 6 Time course of fluorescence intensities of U373MG cells incubated with laurdan parallel and perpendicular to the exciting electrical field vector upon epi-illumination at $T = 16\,°C$; anisotropy function $r(t)$. (Reproduced from [46] with modifications)

the membrane marker 6-dodecanoyl-2-dimethylamino-naphthalene (laurdan [44, 45]; 8 μM, 60 min) upon TIR illumination. For time-resolved measurements of fluorescence anisotropy the EMCCD camera has to be replaced by a high-speed detection system, e.g., an image-intensifying camera system with subnanosecond time resolution (Picostar HR 12; LaVision, Göttingen, Germany; time resolution: 200 ps). This camera has also been used for recording the fluorescence decrease of polarized light and the anisotropy function $r(t)$ in the nanosecond range, as depicted in Fig. 6. The rotational diffusion time constant t_r (evaluated according to Eq. 4) depended on temperature as well as on the concentration of cholesterol [46]. In particular, a decrease of t_r from 3.2 ns to 2.3 ns in the temperature range between 24 °C and 32 °C indicated an increase in membrane fluidity (decrease in viscosity) and may be correlated with a phase transition from the rigid gel phase to the more fluid liquid crystalline phase of membrane lipids [44].

2.3 FRET and TIRET Microscopy

One of the most interesting mechanisms used in fluorescence analysis is non-radiative energy transfer between adjacent molecules in their lowest excited electronic state. This energy transfer is often due to an interaction of optical transition dipoles of a donor and an acceptor molecule, which is proportional to r^{-6} (with r being the intermolecular distance) and which requires an overlap of the emission spectrum of the donor and the absorption spectrum of the acceptor according to

$$k_{ET} \sim r^{-6} \int \varepsilon_A(\nu) I_D(\nu) \nu^{-4} \, d\nu \tag{6}$$

with k_{ET} corresponding to the rate of energy transfer, $\varepsilon_A(\nu)$ to the molar extinction coefficient of the acceptor, $I_D(\nu)$ to the fluence of emitted photons of the donor and ν to the frequency of radiation. Owing to the spectral overlap of donor and acceptor frequencies and with reference to Förster who first described this mechanism [47], the term Förster resonance energy transfer (FRET) is commonly used. Although nonradiative intermolecular energy transfer is well documented for photosynthetic antenna systems [48, 49], it has only recently been used as an analytical tool for probing intermolecular distances below 5–10 nm.

Energy-transfer rates k_{ET} can be determined from stationary as well as from time-resolved fluorescence measurements. In the first case, the photon fluxes of donor (I_D) and acceptor (I_A) fluorescence are measured and the ratio

$$I_A/I_D = \tau_0 \left(\eta_A/\eta_D\right) k_{ET} \tag{7}$$

is calculated, if the quantum yields of acceptor and donor fluorescence (η_A, η_D) and the fluorescence lifetime of the donor (τ_0, without acceptor) are either known from the literature or can be determined from additional experiments. Another possibility of calculating k_{ET} is time-resolved fluorescence spectroscopy. If the energy-transfer rate k_{ET} is summed up by the rates of radiative (k_F) and nonradiative (k_{nr}) transitions originating from the excited state S_1 of the donor molecule, the reciprocal fluorescence lifetime of the donor results in $1/\tau = k_F + k_{nr} + k_{ET}$ in the presence and $\tau_0 = k_F + k_{nr}$ in the absence of an acceptor. Therefore the energy-transfer rate can be calculated according to

$$1/\tau - 1/\tau_0 = k_{ET} \, . \tag{8}$$

Energy-transfer measurements using fluorescent dyes in biological systems date back to the 1970s [50]. A few years later energy-transfer studies were performed to measure intermolecular distances in cell membranes [51, 52], structures of actin filaments [53] or enzymatic binding sites [54]. In addition, measurements of nonradiative energy transfer from the coenzyme NADH to the mitochondrial marker rhodamine 123 (R123) proved to be useful for selective detection of the mitochondrial fraction of the coenzyme NADH [5, 55]. Changes of mitochondrial NADH upon inhibition of the respiratory chain appeared to be more pronounced than changes of total cellular NADH (measured by autofluorescence). Therefore energy-transfer spectroscopy with mitochondrial NADH as a donor and R123 as an acceptor may be a sensitive tool for measuring mitochondrial malfunction in various kinds of diseases [56–59].

More recently, mutants of GFP fused with specific cellular proteins were found to be rather ideal species for numerous FRET applications. A direct interaction between the two proteins Bcl-2 and Bax, which are supposed to regulate apoptosis, was proven within individual mitochondria using GFP–Bax and blue fluorescent protein (BFP)–Bcl-2 fusion proteins coexpressed within the same cell [60]. In addition, it was shown that specific amino acid sequences located between BFP and GFP were cleaved by caspase enzymes upon induction of apoptosis: nonradiative energy transfer BFP → GFP disappeared, thus allowing activation of specific caspases to be monitored in vitro and in vivo [61]. Since different GFP mutants can be localized on various sites of a protein, conformational changes of proteins (e.g., of calmodulin upon binding of calcium ions) can be measured selectively [62]. This makes it possible to visualize calcium uptake and distribution in single cells. Further applications of FRET were dedicated to the detection and visualization of GFP-tagged receptors in cells which were focally stimulated by the epidermal growth factor (EGF). Following focal stimulation, energy transfer from GFP to a fluorescent acceptor was measured and visualized, thus proving a rapid and extensive propagation of receptor phosphorylation over the plasma membrane [63]. Similarly, receptor dephosphorylation on the surface of the endoplasmic reticulum was described [64].

Recent FRET technologies include photoquenching FRET [65] and homo-FRET [66]. In the first case, fluorescence increase of a donor upon photoquenching of an acceptor is evaluated, whereas in the second case—using the same molecular species as a donor and as an acceptor—fluorescence polarization may be sensitive to nonradiative energy transfer. In particular, rotational diffusion time changed upon dimerization of GFP-tagged proteins [66]. For probing molecular interactions in close proximity to the plasma membrane, TIRF microscopy and FRET have recently been combined. Total internal reflection energy transfer (TIRET) microscopy proved to be an appropriate method to study focal adhesions and their associated proteins focal adhesion kinase (FAK) and Paxillin (Pax), which play major roles with respect to cell migration, growth, and survival [67]. As depicted in Fig. 7, TIRF microscopy is necessary to select protein fluorescence in focal adhesions (a) and to exclude overlapping fluorescence from inner parts of the cells (b). For evaluation of energy-transfer rates from enhanced cyan fluorescent protein (ECFP)–FAK to Pax-enhanced yellow fluorescent (EYFP) fusion proteins, either fluorescence spectra (Fig. 8) or fluorescence decay kinetics (Fig. 9) can be used. In the first case, the overlapping spectra with maxima around 480 nm (ECFP) and 530 nm (EYFP) have to be deconvoluted, and ECFP as well as EYFP fluorescence intensities (photon fluxes) determined, whereas in the second case, shortening of fluorescence lifetime of the donor (ECFP) must be evaluated. Further applications of TIRET microscopy include studies of the structural arrangement of potassium channels in the plasma membrane with very little superposition by background fluorescence from inner parts

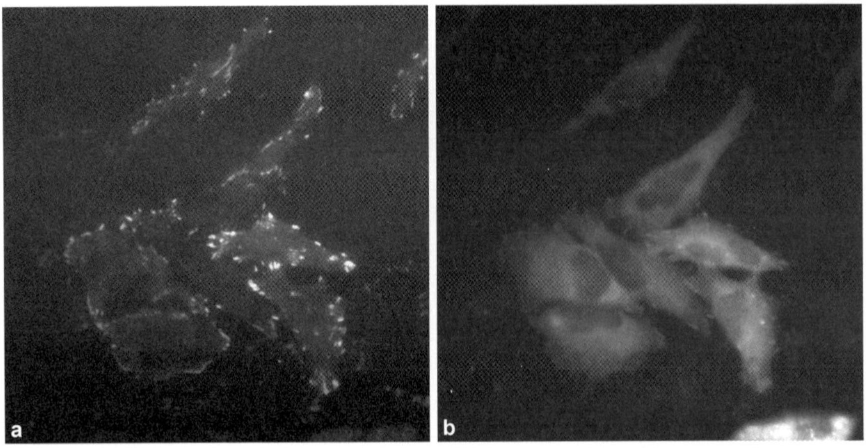

Fig. 7 Fluorescence of HeLa cells transfected with a vector encoding for a paxillin (Pax)–enhanced yellow fluorescent protein (EYFP) fusion protein upon TIR illumination (**a**) and epi-illumination (**b**). Excitation wavelength, 470 nm; detection range, $\lambda \geq 530$ m; image size, 210×210 μm each

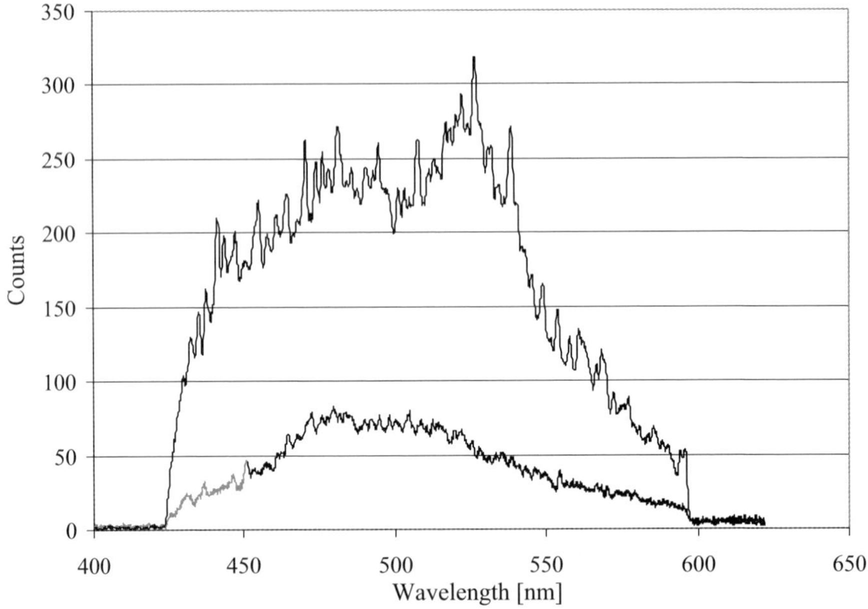

Fig. 8 Fluorescence spectra of a single HeLa cell after transfection with a vector encoding for the enhanced cyan-fluorescent protein (ECFP)–focal adhesion kinase (FAK) fusion protein (*lower curve*) or after cotransfection with vectors encoding for ECFP–FAK and Pax–EYFP fusion proteins (*upper curve*) using TIR excitation. Excitation wavelength, 391 nm; detection range, 420–590 nm. (Reproduced from [68] with modifications)

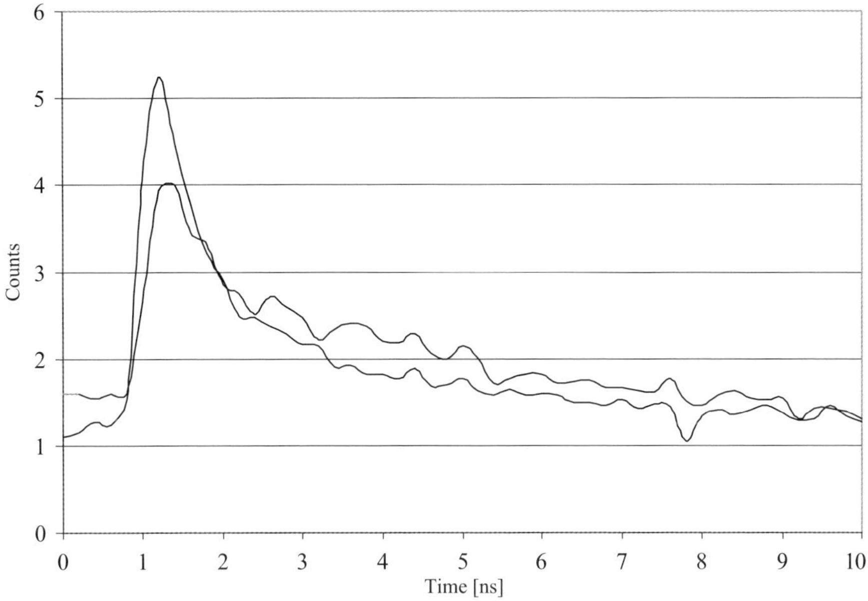

Fig. 9 Fluorescence decay kinetics of single HeLa cells upon transfection with a vector encoding for ECFP–FAK (*upper curve*) or after cotransfection with vectors encoding for ECFP–FAK and Pax–EYFP fusion proteins (*lower curve*) using TIR excitation with picosecond laser pulses. Excitation wavelength, 391 nm; detection range, 476±20 nm. (Reproduced from [68] with modifications)

of the cell [68, 69]. Single-pair measurements between donor and acceptor molecules prove the high sensitivity of the TIRET method [70].

3 Conclusions

The potential of three methods of fluorescence diagnostics has been demonstrated, and numerous applications have been summarized. Suggestions to improve TIRF and polarization microscopy have been given. In particular, technical solutions for variable-angle TIRF microscopy and TIRF screening were described. In polarization microscopy, the problems of polarization-dependent sensitivities and photochemical modifications of the sample during measurements were discussed, and technical solutions were shown. It should be mentioned that in addition to polarization measurements further methods, e.g., measurement of spectral shifts [45, 71] or fluorescence recovery after photobleaching (FRAP) [72] can be used to evaluate membrane dynamics. FRET and TIRET microscopy proved to be valuable methods to probe intracellular distances in cellular diagnostics. Two methods were described to calculate energy-transfer rates. However, for stationary fluorescence measurements, additional data, i.e., fluorescence quantum yields as well as the fluorescence lifetime of the donor, are needed, and overlapping emission spectra of donor and acceptor must be deconvoluted. In fluorescence lifetime measurements, shortening of the donor lifetime is a measure of energy transfer. However, different donor–acceptor distances may result in different lifetimes, and donor–acceptor interactions may be determined from a lifetime distribution rather than from an individual fluorescence lifetime. In addition, evaluation is getting complicated, if a donor species has more than one inherent lifetime, which is the case, e.g., for ECFP [67]. However, these present restrictions will not reduce the high potential of energy-transfer measurements in cellular diagnostics.

Acknowledgements This chapter includes results of current research projects funded by the Bundesministerium für Bildung und Forschung (BMBF; grant no. 17 01B 04), the Ministerium für Wissenschaft, Forschung und Kunst Baden-Württemberg and Landesstiftung Baden-Württemberg GmbH. The author thanks M. Wagner and H.-P. Lassalle for their cooperation in preparing this manuscript.

References

1. Lakowicz JR (1999) Principles of fluorescence spectroscopy. Plenum Press, New York
2. Beechem JM, Brand L (1985) Rev Biochem 54:43
3. Aubin JE (1979) J Histochem Cytochem 27:36
4. Obi-Tabot ET, Hanrahan LM, Cachecho R, Berr ER, Hopkins SR, Chan JCK, Shapiro JM, LaMorte WW (1993) J Surg Res 55:575
5. Gschwend MH, Rüdel R, Strauss WSL, Sailer R, Brinkmeier H, Schneckenburger H (2001) Cell Mol Biol 47:OL95
6. Andersson-Engels S, Johansson J, Svanberg K, Svanberg S (1991) Photochem Photobiol 53:807
7. Fujimori E (1989) Biochim Biophys Acta 998:105
8. Johnson I (1988) Histochem J 30:123
9. Mullins JM (1999) Methods Mol Biol 115:97
10. Zorov DB, Kobrinsky E, Juhaszova M, Scollott SJ (2004) Circ Res 95:239
11. Cody CW, Prasher DC, Westler WM, Prendergast FG, Ward WW (1993) Biochem 32:1212
12. Rizzuto R, Brini M, Pizzo P, Murgia M, Pozzan T (1995) Curr Biol 5:635
13. Ikawa M, Yamada S, Nakanishi T, Okabe M (1999) Curr Top Dev Biol 44:1
14. Stepp H, Sroka R, Baumgartner R (1998) Endoscopy 30:379
15. Bird D, Gu M (2003) Opt Lett 28:1552
16. Zeng H, Petek M, Zorman MT, McWilliams A, Palcic B, Lam S (2004) Opt Lett 29:587
17. D'Hallewin MA, El Khatib S, Leroux A, Bezdetnaya L, Guillemin F (2005) J Urol 174:736
18. Neil MAA, Juskaitsi R, Wilson T (1997) Opt Lett 22:1905
19. Gustavsson MG (2005) Proc Nat Acad Sci USA 102:13081
20. Axelrod D, Thompson NL, Burghardt TP (1983) J Microsc 129:1
21. Schneckenburger H (2005) Curr Opin Biotechnol 16:13

22. Bastiaens PI, Squire A (1999) Trends Cell Biol 9:48
23. Elson D, Requejo-Isidro J, Munro I, Reavell F, Siegel J, Suhling K, Tadrous P, Benninger R, Lanigan P, McGinty J, Talbot C, Bebhinn T, Webb S, Sandison A, Wallace A, Davis D, Lever J, Neil M, Philipps D, Stamp G, French PM (2004) Photochem Photobiol 3:795
24. Denk W, Strickler JH, Webb WW (1990) Science 248:73
25. König K (2000) J Microsc 200:83
26. Truskey GA, Burmeister JS, Grapa E, Reichert WM (1992) J Cell Sci 193:491
27. Sund SE, Axelrod D (2000) Biophys J 79:1655
28. Omann GM, Axelrod D (1996) Biophys J 71:2885
29. Demuro A, Parker I (2004) Biophys J 86:3250
30. Betz WJ, Mao F, Smith CB (1996) Curr Opin Neurobiol 6:356
31. Oheim M, Loerke D, Stühmer W, Chow RH (1998) Eur J Biophys 27:83
32. Beaumont V (2003) Biochem Soc Trans 31:819
33. Sailer R, Strauss WSL, Emmert H, Stock K, Steiner R, Schneckenburger H (2000) Photochem Photobiol 71:460
34. Axelrod D (2001) J Biomed Opt 6:6
35. Stock K, Sailer R, Strauss WSL, Lyttek M, Steiner R, Schneckenburger H (2003) J Microsc 211:19
36. Malik Z, Lugaci H (1987) Br J Cancer 56:589
37. Coates CG, Denvir DJ, McHale NG, Thornbury KG, Hollywood MA (2003) Proc SPIE 5139:56
38. Ruckstuhl T, Seeger S (2004) Opt Lett 29:569
39. Sako Y, Uyemura Z (2002) Cell Struct Funct 27:357
40. Obremski RJ, Silzel JW (2000) US Patent 6 110 749
41. Brandenburg A (2001) Tech Messen 68:513
42. Bruns T, Strauss WSL, Sailer R, Wagner M, Schneckenburger H (2006) J Biomed Opt 11:34011
43. Bruns T, Strauss WSL, Schneckenburger H (2006) Proc SPIE 6191:61910V
44. Parasassi T, de Stasio G, d'Ubaldo A, Gratton E (1990) Biophys J 57:1179
45. Parasassi T, Krasnowska EK, Bagatolli L, Gratton E (1998) J Fluoresc 4:365
46. Wagner M, Weber P, Schneckenburger H (2006) Proc SPIE 6191:619109
47. Förster T (1960) Z Elektrochem 64:157
48. Holzwarth AR, Wendler J, Haehnel W (1985) Biochim Biophys Acta 807:155
49. Berens SJ, Scheele J, Butler WL, Magde D (1985) Photochem Photobiol 42:51
50. Stryer L (1978) Ann Rev Biochem 47:819
51. Uster PS, Pagano RE (1986) J Cell Biol 103:1221
52. Szöllösi J, Damjanovich S, Mulhern SA, Tron L (1987) Prog Biophys Mol Biol 49:65
53. Taylor DL, Reidler J, Spudich A, Stryer L (1981) J Cell Biol 89:65
54. Squier TC, Bigelow DJ, deAncos JG, Inesi G (1987) J Biol Chem 89:362
55. Schneckenburger H, Gschwend MH, Strauss WSL, Sailer R, Kron M, Steeb U, Steiner R (1997) Photochem Photobiol 66:33
56. DiMauro S, Bonilla E, Zeviani M, Nakagawa M, DeVivo DC (1985) 17:521
57. Wallace DC (1992) Annu Rev Biochem 61:1175
58. Luft R (1994) Proc Nat Acad Sci USA 91:8731
59. Schapira AHV (1994) Mov Disord 9:125
60. Mahajan NP, Linder K, Berry G, Gordon GW, Heim R, Herman B (1998) Nat Biotechnol 16:547
61. Mahajan NP, Harrison-Shostak DC, Michaux J, Herman B (1999) Chem Biol 6:401
62. Brasselet S, Peterman EJG, Miyawaki A, Moerner WE (2000) J Phys Chem B 104:3676

63. Verveer PJ, Wouters FS, Reynolds AR Bastiaens PI (2000) Science 290:1567
64. Haj FG, Verveer PJ, Squire A, Neel BG, Bastiaens PI (2002) Science 295:1708
65. Demarco IA, Periasamy A, Booker CF, Day RN (2006) Nat Methods 3:519
66. Gautier I, Tramier M, Durieux C, Coppey J, Pansu RB, Nicolas J-C, Kemnitz K, Coppey-Moisan M (2001) Biophys J 80:3000
67. Angres B, Steuer H, Wagner M, Weber P, Schneckenburger H (2006) Proc SPIE 6191:61910A
68. Riven I, Kalmanzon E, Segev L, Reuveny E (2003) Neuron 38:225
69. Sonnleitner A, Mannuzzu LM, Terakawa S, Isacoff EY (2002) Proc Nat Acad Sci USA 99:12759
70. Suzuki Y, Tani T, Sutoh K, Kamimura S (2002) FEBS Lett 512:235
71. Schneckenburger H, Wagner M, Kretzschmar M, Strauss WSL, Sailer R (2004) Photochem Photobiol Sci 3:817
72. Meyvis TK, De Smedt SC, Van Oostveldt P, Demeester J (1999) Pharm Res 16:1153

Springer Ser Fluoresc (2008) 6: 549–560
DOI 10.1007/4243_2008_052

Published online: 14 May 2008

In-vivo Fluorescence Imaging: Applications, Future Trends & Approaches to Standardization

Vasilis Ntziachristos (✉) · Damon Hyde

Center for Molecular Imaging Research,
Massachusetts General Hospital and Harvard Medical School, Building 149,
13th Street 5406, Charlestown, MA 02129-2060, USA
vasilis@helix.mgh.harvard.edu

Abstract Standards are important for calibration procedures in fluorescence imaging and overall for enabling accurate quantification. However, due to the strong nonlinear dependence of the fluorescence signal on tissue scattering, tissue absorption and activity depth, the construction of standards becomes challenging. So far, most fluorescent standards for diffusive imaging have been based on laboratory solutions that mix scattering, absorbing and fluorescence materials to construct substances of known and stable optical properties. Herein we review the most common characteristics of diffusive imaging and outline strategies to produce materials that can serve as standards in whole body imaging applications.

Keywords Diffusion imaging · Fluorescence · Fluorescence imaging · Scattering · Tissue optical properties

1
Main Text

Fluorescence imaging has emerged as an important modality for in-vivo observations of molecular function from whole animals and entire tissues. This can be primarily achieved by using appropriately engineered fluorescent molecules with the ability to target different cellular markers in-vivo. Because of the low absorption of light by tissue in the near-infrared, light can penetrate for several millimeters to centimeters in tissues and, if of appropriate wavelength, excite various fluorochromes present. These fluorochromes could be intrinsically expressed fluorescent proteins or exogenously administered fluorescent probes that mark different cellular and subcellular processes. The increasing knowledge of molecular pathways associated with normal function and disease, and the improving ability to develop adept molecules with optimal biodistribution and specificity direct by consequence the need for fluorescence imaging to capture and quantify fluorescence from tissues. The major difficulties associated with in-vivo fluorescence imaging are the high tissue scattering and variation of optical properties. As a result, there is strong nonlinear dependence of photon intensity on both the depth of the activity recorded and the optical properties of tissue.

Approximations to the radiative transfer equation have been used for predicting photon propagation in tissues. One of these approximations describes photon propagation as a diffusive process, modeled by the diffusion equation [1]. This modeling approach gives an accurate description of general propagation characteristics when the observations are performed some millimeters away from the photon source. Fluorescence imaging therefore becomes quantitative when such a model is used in the formation of images. This is especially true for macroscopic fluorescence observations where signals may propagate deeper than a few millimeters. The most common approach therefore combines solutions to the diffusion equation with measurements obtained at different projections through tissue to obtain tomographic images through tissues. These approaches typically offer three-dimensional imaging capability, limited sensitivity to the variation of optical properties and improved resolution compared to photographic, single projection imaging approaches.

Since image formation, in its more accurate form, involves a model-based inversion technique, the imaging accuracy depends on several experimental and theoretical parameters. Therefore, the problem of standardization in fluorescence imaging of tissues goes beyond the determination of stable fluorescent substances, described elsewhere in this book. Instead, standards with determined and stable absorption and scattering characteristics, in addition to fluorescent characteristics become important for comparing imaging methods or for calibrating systems and measurements. Furthermore, the availability of phantoms with different optical properties is crucial in diffusion imaging as it is necessary to examine the relative performance of imaging methods that account for the effects of spatially changing absorption and scattering on the fluorescence intensity signal and retrieve true fluorescence concentrations or fluorochrome amounts. Therefore, although fluorescence standards in this case naturally require highly environmentally stable fluorochromes, they further necessitate the use of fluorescent diffusive material of varying optical properties and sizes. This is in contrast to more traditional biomedical instrumentation that assumes a linear imaging response as for example in regard to the measurement of fluorochromes distributed in a clear fluid as in photo-spectrometers, common plate readers or other types of in-vitro or monolayer cellular assays.

Diffusion fluorescence imaging is a new field of imaging sciences and it offers many different optical system and inversion algorithm implementations. Widely accepted standards would be important in this developing field in order to offer reference schemes by which to compare different optical methods and systems; however no such commonly accepted standards are established yet. Herein we explain in more detail the specifics of forming optical images from diffusive media, in order to outline the key challenges that come with identifying standards for diffusive imaging of tissues. This work also summarizes the more common approaches currently used for testing and comparing the performance of various systems and methods.

2 Diffuse Fluorescence Tomography

The techniques involved in imaging fluorescence at depths beyond those reached by microscopy make use of photons that have scattered multiple times and exhibit a diffusive behavior. Photon propagation in tissues is commonly modeled by the diffusion equation which can be expressed as [1]:

$$\begin{aligned} &\nabla \cdot [D_x(r)\nabla \Phi_x(r,\omega)] - \left[\mu_a(r) + \frac{i\omega}{c}\right] \Phi_x(r,\omega) =- S(r,\omega) \\ &D_x(r) = \frac{1}{3\mu_s + (1-\rho)\mu_a}\,. \end{aligned} \tag{1}$$

Here, μ_a and μ_s are the spatially varying absorption and scattering parameters, $\Phi_x(r,\omega)$ is the photon density at a point r, $S(r,\omega)$ is the source term with an intensity modulated at a frequency ω. For constant intensity light $\omega = 0$. The constants ρ and c are, respectively, the mean cosine of the scattering angle of the various cellular structures, and the speed of light within the medium. A simple solution to this equation, in the absence of boundaries, can be derived in the form of a Green's function:

$$g(r, r_s) \sim \frac{\exp(-\,ik\,|r - r_s|)}{|r - r_s|}\,, \tag{2}$$

which represents the attenuation of the light detected (photon density) at a point r due to a point source at a location r_s. k is a factor (wave number) that depends on the tissue optical properties and photon wave frequency and can be computed as $k = \sqrt{\frac{-\mu_a}{D} + \frac{i\omega}{cD}}$. It is readily apparent from this equation that the photon density is nonlinearly related to the distance from the source as well as the optical properties of the tissue. Equation 2 is written for a homogenous unbounded diffusive medium, but even in this general form, it describes well the generic dependence of the signals measured to the underlying medium. For tissue imaging, solutions are reached that implement boundaries. By use of the Kirchoff approximation for example, the Green's function in the presence of an arbitrary surface tissue–air boundary S can be written as [2, 3]:

$$G(r, r_\mathrm{s}) = g(r, r_\mathrm{s}) - \frac{1}{4\pi} \int_S \left[\frac{\partial g(r_\mathrm{b}, r)}{\partial n'} + \frac{1}{C_\mathrm{nd} D} g(r_\mathrm{b}, r) \right] \Phi(r_\mathrm{b}, r_\mathrm{s}) \, \mathrm{d}r_\mathrm{b} \tag{3}$$

with the integral being taken over the surface S, and n' being the normal at each point on that surface. This method is related to the tangent plane method from physical optics, and accounts for the effect of the boundaries by introducing a correction factor to the previous equation for the infinite space solution. Note that the right hand side of this equation utilizes the photon density at the surface of the medium $\Phi(r_\mathrm{b}, r_\mathrm{s})$, due to the source at r_s. In practice, these values are either directly measured or theoretically calculated.

Equations 2 and 3 describe the propagation of light between two points, typically one acting as a source (for example incident light or a fluorescent point) and the other as a detector. For tomography, we are interested in the total signal received due to a fluorochrome distribution [4]

$$O^\mathrm{f}(r', \omega) \sim \frac{\gamma \varepsilon \left[F(r') \right]}{1 - \mathrm{i}\omega\tau} \,, \tag{4}$$

where $F(r')$ is the density of fluorochrome molecules at the location r' in the volume of interest, ω is the modulation frequency of the source, τ is the fluorochrome lifetime, and γ and ε are the quantum efficiency and absorption cross section of the fluorochrome, respectively. In this case, the total fluorescence signal received from a detector at r_d due to a light source incident at r_s is [4]

$$U(r_\mathrm{s}, r_\mathrm{d}) = \int_V G(r', r_\mathrm{d}) O^\mathrm{f}(r', \omega) \Phi(r_\mathrm{s}, r') \, \mathrm{d}r' \,. \tag{5}$$

This equation states that the fluorescence light intensity $U(r_\mathrm{s}, r_\mathrm{d})$ measured at a point r_d due to a point light source located at r_s, is an integral function of the photon density at the excitation wavelength, $\Phi(r_\mathrm{s}, r')$, and the Green's function $G(r', r_\mathrm{d})$ from each fluorochrome at position r' to the detector at r_d. Then, the goal of diffuse fluorescence tomography is to obtain the unknown fluorochrome distribution $O^\mathrm{f}(r', \omega)$ given several boundary measurements $U(r_\mathrm{s}, r_\mathrm{d})$. While direct inversion of the integral Eq. 5 is mathematically challenging, a typical solution approximates $O^\mathrm{f}(r', \omega)$ as piecewise constant, and replaces the integral above with the summation:

$$U(r_\mathrm{s}, r_\mathrm{d}) = \sum_{i=1}^{N} G(r_i, r_\mathrm{d}) O^\mathrm{f}(r_i, \omega) \Phi(r_\mathrm{s}, r_i) \, \mathrm{d}V_i \,,$$

where r_i is the center of the ith voxel, and $\mathrm{d}V_i$ is the associated volume. This equation can be rewritten into the matrix form:

$$\mathbf{W}\mathbf{x} = \mathbf{u}\,,$$

which relates the collected data **u** as the product of a vector **x** containing the unknown parameters, i.e. the image, via the 'weight' matrix **W**. Each of the elements of a row in this matrix represents the sensitivity of a single source-detector to a certain volume element (voxel) in the image. To retrieve the unknown image this system of equations needs to be inverted. In its simplest form this operation can be written as:

$$\hat{\mathbf{x}} = \mathbf{W}^{-1}\mathbf{u}\,.$$

In general however, the matrix **W** is not square; thus a formal inverse does not exist. Additionally, the structure of the matrix **W** is such that the inverse problem is highly ill-posed. This means that direct inversion of the matrix W is likely to result in a solution which is very sensitive to noise and cannot be accurately calculated. Therefore, the goal of numerical inversion methods is to create an approximation to $\mathbf{W}^{-1}$ which results in a solution that is sufficiently robust to noise in the collected data. Such inversions are generally obtained through iterative solutions on the forward problem based on data fitting methods.

3 Relation of Inversion and Standardization

Photons propagating in tissue experience more scattering events the deeper they propagate. The sensitivity of measurements to deeper seated voxels is consequently lower and the resolving power is also decreased with depth. Two objects of the same size and intensity, one located at the center of a diffusive volume and one located at the boundary can be reconstructed to differing sizes due to the effects of depth-dependent resolution. This is demonstrated in Fig. 1. The figure shows the reconstructions for two identically sized tubes, located at different depths (2 mm vs. 12 mm) within a cylindrical diffusive volume. Simulated data was generated using a solution of the diffusion equation, which is the same model used for data inversion as well. Even in this ideal situation, it can clearly be seen that the more superficial tube is resolved with better resolution than the deeper-seated tube. Correspondingly, the maximum reconstructed value drops since the same "fluorochrome amount" is reconstructed over a larger volume for the deeper-seated tube resulting in a reconstruction "dilution".

To improve on depth-dependent loss of resolution or quantification, as seen in Fig. 1, inversion approaches regularly utilize *regularization*. Achieved through a wide variety of methods, regularization serves to make the problem

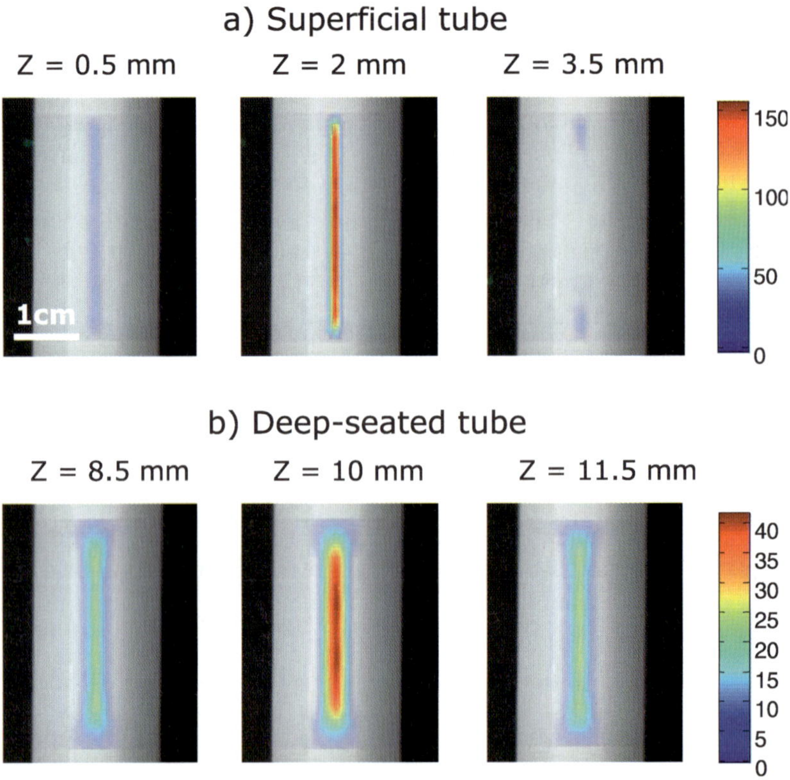

Fig. 1 Reconstructions of a 1.3 mm diameter fluorescent tube embedded in a 2 cm diameter diffusive cylinder. The optical properties of the cylinder simulate the optical properties of the mouse torso ($\mu_a = 0.3\ \text{cm}^{-1}$ $\mu_s' = 10\ \text{cm}^{-1}$). A photograph of the cylinder is shown in the background of all images. The fluorescent tube was implanted **a** 2 mm under the surface and **b** 10 mm under the surface at the center of the cylinder. The reconstructions are based on simulated measurements using an analytical solution to the diffusion equation

less ill-posed by introducing constraints on the solution based on prior knowledge of the inversion problem. Depth dependent regularization for example, makes use of the known fact that there is a depth dependent loss of resolution and has shown improvements in resolution and quantification. For details on the subject see [5, 6]. Many other regularization algorithms exist, ranging from general purpose to highly application-specific implementations, depending on the a-priori knowledge available for inversion.

Another challenge of in-vivo fluorescence imaging is that the optical absorption and scattering parameters vary across tissue. Therefore, the fluorescence intensity recorded varies not only as a function of depth but also as a function of the optical properties of the tissues that the light propagated through. An additional goal of tomographic fluorescence methods therefore is to utilize strategies that account for optical property variation. So far, a common method used for in-vivo imaging uses measurements at both the excitation and emission (fluorescence) wavelengths and reconstructs the ratio of fluorescence to excitation light collected at each detector. Because both signals follow a similar mean path through the tissue, especially in transillumination mode, this method, referred to as the *normalized Born* [7] (or the *Born ratio*), helps to eliminate variations in the fluorescence signal owing to perturbations in the optical coefficients of the media. Additionally, using the normalized Born method ratio makes the system less sensitive to theoretical vs. experimental mismatches. Figure 2 shows a plot of the expected fluorescence signal from a cylindrical fluorescing object placed in the middle of a 2 cm diffusing cylinder. The reduced scattering coefficient of the diffusive medium was 10 cm^{-1} and the absorption coefficient of the medium is varied from 0.1 cm^{-1} to 1.2 cm^{-1}. The fluorescent cylinder is 1.3 mm in diameter and 2.5 cm in length and was located in the center of the 2 cm diameter diffusing cylinder. The measurement was performed with a source and a detector placed on the opposite side on the boundary of the cylinder. In this case, the traditional Born model shows nearly a 3500-fold change from its

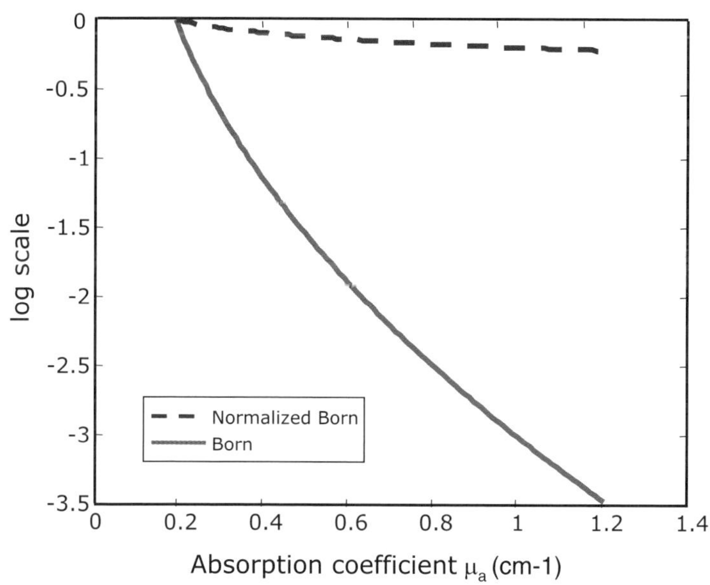

Fig. 2 Expected signal intensity for a fluorescence cylinder of 1.3 mm in diameter, placed in the middle of a 2 cm diameter diffusing cylinder. The *graph* shows expected signal intensity for both Born and normalized Born models, plotted against the absorption coefficient of the diffusing cylinder. Increased absorption leads to a decrease in the expected signal, but the normalized Born model shows a smaller relative decrease, indicating reduced sensitivity to parameter mismatch

minimum to maximum value, as a function of the optical property used, whereas the normalized Born exhibits a mere 1.6-fold change over the same range of absorption coefficient variation. While tissue parameters do not typically deviate by such absorption coefficient extremes, the graph of Fig. 2 indicates that a mismatch in the assumption of optical parameters used in the forward model will have significantly lower effect upon the reconstructions when using the normalized Born model. This generic insensitivity of the normalized Born method is similarly seen for spatially varying changes in optical properties.

4
Current Diffusive Fluorescent Standards

The previous paragraphs demonstrated the main features of model-based diffusive tomography and the typical challenges in regard to depth and optical properties. Spatially dependent regularization schemes and normalization methods have shown benefits in improving imaging performance. Iterative, nonlinear algorithms offer an alternative in improving imaging performance at the expense of significantly higher computational needs [8–10]. Conversely, imaging in the absence of model-based inversions, such as planar (photographic) imaging that has been widely used for fluorescence observations, offers significantly worse quantification and resolution compared to the methods discussed above.

In order to examine the performance of, or experimentally calibrate the various methods examined, it is common to construct phantoms simulating the mouse optical properties, which further contain known fluorochromes at predetermined positions. Ideally, diffusive fluorescence standards would be important for repeating comparisons, however such materials are not generally available for this purpose. Fluorochromes mixed with polyester resin, TiO_2 spheres and absorbing dye to simulate fluorescence, scattering and absorption properties common to tissues have been used for this purpose but the fluorescence yield achieved with this method is significantly lower than from free-fluorochromes and difficult to characterize [11]. The long-term stability of such materials is not well studied either. More commonly, fresh solutions of intralipid, mixed with India ink and freshly prepared fluorochromes at measured concentrations (via a fluorometer or photo-spectrometer) is used for fluorescent phantoms and calibration measurements. The photostability of the fluorochromes employed in the construction of these phantoms is an important parameter in their selection. It should be noted, however, that photobleaching effects, more common in microscopy, are less important in diffusive measurements due to the significantly lower light intensities reaching the fluorochromes when embedded in the diffusive medium.

Characterization of fluorescence phantoms needs to take into account the coupled dependence of fluorescence intensity emitted from these phantoms on the phantom's optical properties. This dependence was described by Eq. 5. For this reason a diffusive fluorescent phantom or standard for diffusive measurements can be best described therefore by the amount of fluorochrome present, the corresponding volume it occupies, the fluorochrome's quantum yield and the phantoms optical properties.

An example of a solid fluorescent phantom and corresponding reconstructions is shown in Fig. 3. The phantom is made of polyester resin mixed with TiO_2 spheres and India Ink particles prior to curing and yields an absorption coefficient of $\mu_a = 0.3\ cm^{-1}$ and reduced scattering coefficient $\mu_s' = 10\ cm^{-1}$. These values in this case represent average optical properties of the mouse torso. Characterization of optical properties is confirmed using time-resolved methods [12, 13]. Four open tubes can contain in this case known amounts of fluorochrome diluted in a scattering fluid, such as a mix of intralipid and India ink at a concentration that mimics the optical properties of tumors. The ability of the tomographic method to resolve true fluorescence concentrations for varying optical properties can be examined by adding solutions of different optical properties and fluorochromes. Similar approaches include phantoms of different sizes and phantoms where the optical properties of the background are also modified.

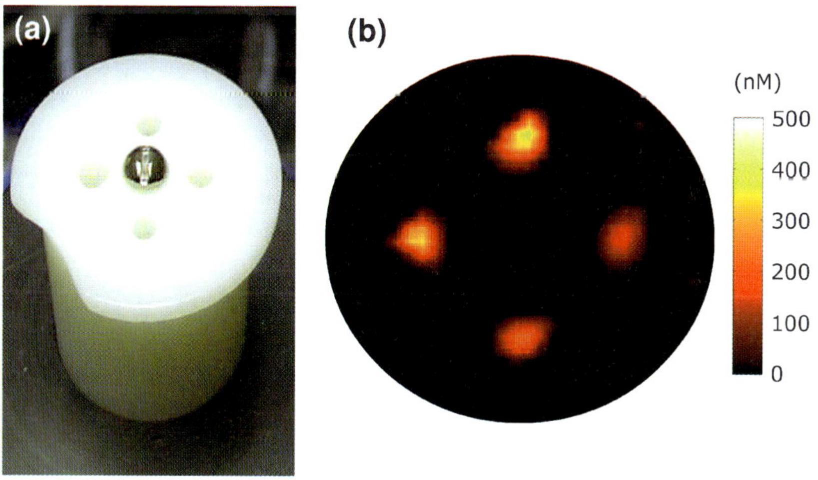

Fig. 3 Example of a fluorescent phantom used as standard. **a** Photograph of a resin tube of characterized optical properties that contained four openings that can be filled with varying amounts of fluorochromes mixed in diffusive fluids to better simulate in-vivo experiments; **b** a corresponding reconstruction of the phantom using experimental measurements when top and left tubes contain 500 nM of Cy5.5 dye and the bottom and right tubes contain 250 nM of Cy5.5. More details are reported in [14]

While fluorescence materials based on resin have not shown good long-term stability, silicone-based materials containing silicone pigments and TiO_2/India ink, have demonstrated better performance as fluorescence standards. When used with hydrophobic fluorochromes suspended in dichloromethane these materials demonstrated good signal intensity and long-term stability [15]. Conversely, the use of hydrophilic fluorochromes did not yield long-term stability and linear characteristics between fluorescence concentration and signal strength. The stability of fluorescence intensity as a function of time for resin and silicone base materials can be seen in Fig. 4, taken from [15]. The stability data from the Cy5.5 fluorochrome (Amersham Biosciences, Piscataway, NJ; excitation peak 675 nm, emission peak 694 nm) in resin (Fig. 4a) showed that there was a 70–80% reduction in intensity in the first 24 h, and then a less rapid but fairly steady degradation after curing with about a 1–2% decrease in intensity each day. The resin material using the AF 750 dye (Invitrogen, Carlsbad, California, excitation peak 749, emission peak 775) demonstrated a significantly faster decay, shown in Fig. 4b, resulting in complete quenching of fluorescence activity during the curing process. At the

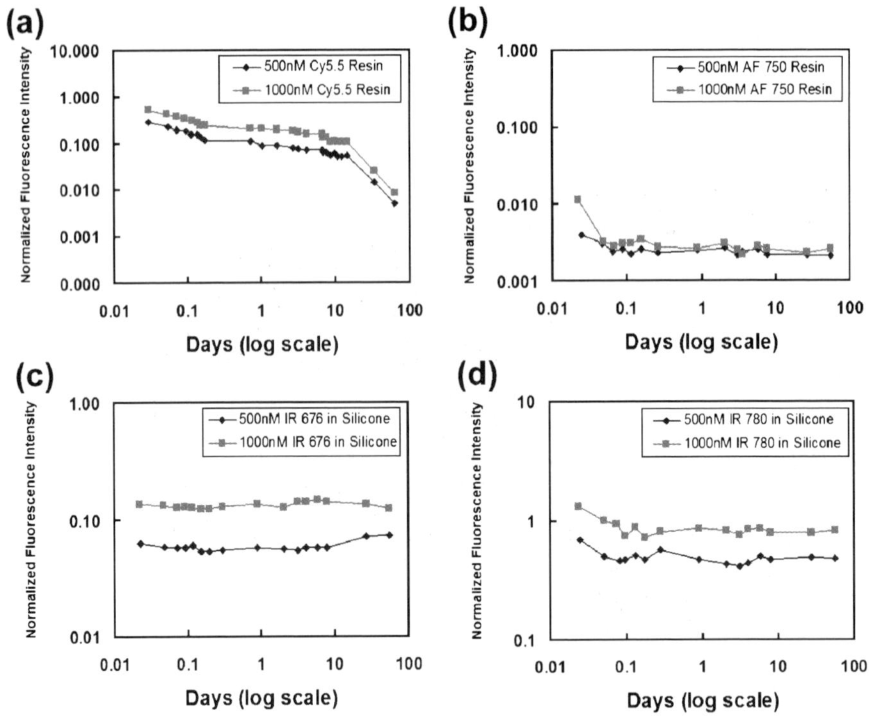

Fig. 4 The fluorescence strength measured from resin- and silicone-based materials as a function of time at dye concentrations of 500 and 1000 nM. The samples were imaged at short intervals (~2 h) during the curing process, and then periodically thereafter over a period of two months. More details are reported in [15]

end of curing only the material's auto-fluorescence remained. Conversely, the results from hydrophobic dyes placed within silicone are shown in Fig. 4c,d. Figure 4c depicts the long-term stability of IR 676 Iodide (Sigma-Aldrich Chemical Company, Inc, Milwaukee, WI; excitation peak 676 nm, emission peak 700 nm) within silicone, showing no intensity degradation during or after the curing process. A similar fluorochrome, the IR 780 (Fig. 4d); (Sigma-Aldrich Chemical Company, Inc, Milwaukee, WI; excitation peak 780 nm, emission peak 799 nm) demonstrated 40–45% strength decay during the curing process but after curing also produced a material with constant long-term fluorescence intensity. The fluorescence measurements shown in Fig. 4, used constant wave (cw) laser-diode illumination for fluorochrome excitation, and CCD camera-based detection, operating in transillumination mode. Laser intensity variations were minimal throughout the measurements but all measurements were corrected for laser intensity variation.

5 Conclusion

While fluorescence imaging has received significant attention for studying molecular function in vivo, the development of accurate, quantitative imaging methods has only recently yielded practical solutions for in-vivo imaging. The development of standards is expected to follow the dissemination of the technology and the wider propagation of FMT systems and methods in the biomedical laboratory. FMT presents the double challenge of calibrating both system (hardware) and algorithmic and software parameters and requires standards that not only offer long term and environmental stability but can be further manufactured at different sizes, shapes and absorption and scattering optical properties. Overall, the development of such materials can facilitate the accurate comparison of methods and systems and can be widely used for the calibration of production units and methods. As such, appropriate fluorescence standards for diffusive imaging need to be developed for accelerating the development of robust optical imaging methods.

References

1. Arridge SR (1999) Optical tomography in medical imaging. Inverse Problems 15:R41–R93
2. Schulz RB, Ripoll J, Ntziachristos V (2003) Noncontact optical tomography of turbid media. Opt Lett 28:1701–1703
3. Ripoll J, Ntziachristos V (2003) Iterative boundary method for diffuse optical tomography. J Opt Soc Am A 20:1103–1110
4. Graves EE, Ripoll J, Weissleder R, Ntziachristos V (2003) A submillimeter resolution fluorescence molecular imaging system for small animal imaging. Med Phys 30:901–911

5. Boverman G, Miller EL, Ang L, Zhang Q, Chaves T, Brooks DH, Boas DA (2005) Quantitative spectroscopic diffuse optical tomography of the breast guided by imperfect a ptiori structural information. Phys Med Biol 50:3941–3956
6. Pogue BW, Paulsen KD (1998) High-resolution near-infrared tomographic imaging simulations of the rat cranium by use of apriori magnetic resonance imaging structural information. Opt Lett 23:1716–1718
7. Ntziachristos V, Weissleder R (2001) Experimental three-dimensional fluorescence reconstruction of diffuse media by use of a normalized Born approximation. Opt Lett 26:893–895
8. Bouman CA, Sauer K (1996) A unified approach to statistical tomography using coordinate descent optimization. IEEE Transact Image Process 5:480–492
9. Klose AD, Hielscher AH (2003) Quasi-Newton methods in optical tomograpic image reconstruction. Inverse Problems 19:387–409
10. Milstein AB, Stott JJ, Oh S, Boas DA, Millane RP, Bouman CA, Webb KJ (2004) Fluorescence optical diffusion tomography using multiple-frequency data. J Opt Soc Am A 21:1035–1049
11. Niedre M (2006) Personal Communication
12. Niedre MJ, Turner GM, Ntziachristos V (2006) Time-resolved imaging of optical coefficients through murine chest cavities. J Biomed Opt 11(6):064017-1-7
13. Patterson MS, Chance B, Wilson BC (1989) Time Resolved Reflectance and Transmittance for the Noninvasive Measurement of Tissue Optical-Properties. Appl Opt 28:2331–2336
14. Ntziachristos V, Tung C, Bremer C, Weissleder R (2002) Fluorescence-mediated tomography resolves protease activity in vivo. Nat Med 8:757–760
15. Baeten J, Niedre M, Dunham J, Ntziachristos V (2007) Development of fluorescent materials for Diffuse Fluorescence Tomography standards and phantoms. Opt Express 15(14):8681–8694

Subject Index

Printing: Krips bv, Meppel, The Netherlands
Binding: Stürtz, Würzburg, Germany